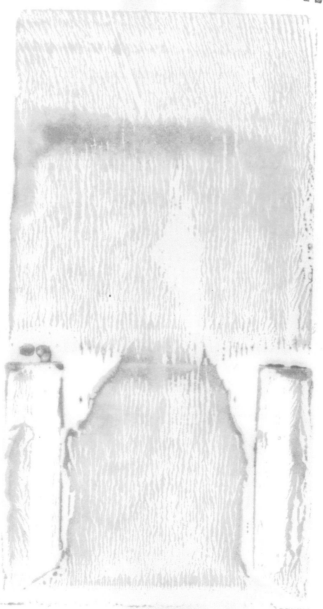

To **George Frank Smith**
  who awakened my interest in natural science

Nuclear physics | an introduction

Nuclear physics | an introduction

Second edition

W. E. Burcham F.R.S.

 Longman

LONGMAN GROUP LIMITED
London
Associated companies, branches and representatives
throughout the world

© W. E. Burcham 1963
This Edition © Longman Group Limited 1973

First published 1963

Second Impression (with problems) 1965

Second Edition 1973

ISBN 0 582 44110 2

Set in Monophoto Plantin
and printed in Great Britain
by William Clowes & Sons, Limited
London, Beccles and Colchester

ERRATA

page 120 line 17   for 'm' read 'mm'

page 184 line 21   for '(a, p)' read '(d, p)'

page 193 line 38   for 'automatic' read 'atomic'

page 200 line 1   insert 'to' after 'particles'

page 261 line 4   for 'N_0' read 'N_A'

page 271 line 11   for 'N_0' read 'N_A'

page 273 line 18   for ' <92' read ' >92'

page 282 equation 11.7   for '$e\ i\mathbf{q}.\mathbf{r}/\hbar$' read '$e^{i\mathbf{q}.\mathbf{r}/\hbar}$'

page 507 lines 24 and 27   for 'MeV c^2' read 'MeV/c^2'

page 507 line 32   for ' $+0{\cdot}6$' read ' $<0{\cdot}6$'

page 515 lines 16 and 17   exchange '⚲' and '⚵'

page 516 line 11   for '(a) π^+' read '(a) π^-'; for '(b) π^-' read '(b) π^+'

page 538 line 33   for '(Ref. 20.3)' read '(Ref. 20.8)'

page 544 line 1   for 'elastic scattering' read 'total interaction'

page 566 line 20   for 'now' read 'no'

Preface to the first edition

This book is intended primarily for the undergraduate who is approaching the end of a first-degree course in physics. It may also be found useful by graduates who are undertaking courses in which a knowledge of nuclear physics is required and by research students who wish to renew their acquaintance with the elementary parts of the subject. An experimental standpoint is adopted, since most nuclear physicists are concerned to some degree with experimental techniques. Mathematical symbols are freely used in a descriptive manner but only straightforward mathematical techniques, such as normally appear in a University Course in physics, are required. It is assumed that the reader has completed a general introductory course in modern physics, including the elements of the special theory of relativity and of atomic spectroscopy.

Nuclear physics is a diffuse and complex subject, which at elementary level is apt to tax the memory rather than the understanding of the student. It is possible to present the results of more than sixty years of experimental and theoretical work in a strictly logical form, but such an approach often conceals or obscures the interest and importance of the historical development. On the other hand a detailed historical treatment is not practicable in a book which is to be of a convenient size and which is to relate to a course which must be given in perhaps one or two University terms at the most. In the present book, therefore, a compromise has been adopted. Part A contains much historical material, particularly in the discussion of radioactivity (ch. 2), and attempts (ch. 3) to pick out the aspects of atomic physics of most significance in nuclear studies. The content of both these chapters may well have been covered by the student in other courses before he enters upon nuclear physics proper, but they have been included for completeness as essential background material. The remainder of the book then attempts to present the subject in something approaching a logical sequence. Chapter 4, completing Part A, gives what today appears to be justifiable prominence to the interaction between atomic and nuclear moments, since this field of work has contributed extensively to our knowledge of nuclear properties. Part B treats the experimental side of the subject in what is considered to be adequate, but not excessive detail, in view of the important advances that have stemmed directly from improvements in technique. In Part C the results of measurements of the static properties of nuclei are reassembled and are set out as far as possible within the coordinating framework now provided by the single-particle shell model. Nuclear reactions are similarly discussed in Part D. The basic structure of nuclei has been left until Part E which is at present a single chapter on nuclear forces. In the original plan of the book it had been hoped to supplement this by a discussion of the Wigner super-multiplet theory and by a chapter on cosmic radiation and mesons, but this plan was abandoned in view of the necessity for keeping the book to a reasonable size. The scope of the work is, therefore, such as to exclude most of high energy physics and although much of what is presented is relevant to work in this subject, the contributions of the high energy physicist to our understanding of nuclear structure regrettably receive scant attention. It is also a matter of regret that room has not been found for an account of the many important applications of nuclear physics in modern technology; a brief account of the best-known of these applications, the nuclear reactor, has been included among the appendices.

In selecting illustrative material for the various sections of the book emphasis has been placed on relevance, significance and ease of assimilation, and published data have been freely used. No attempt is made to give an exhaustive treatment of all types of nuclear process and the number of references quoted has been kept to a minimum. The references given at the ends of the chapters are often to articles of the review type, which are suitable for further reading, and which themselves usually contain extensive reference lists.

I am indebted to a large number of friends and colleagues for assistance at all stages in the preparation of this book; many errors have been avoided by their vigilance, but those that remain are my own responsibility. Chapters have been read by H. Burkhardt, L. Castillejo, K. F. Chackett, G. V. Chester, N. Feather, M. A. Grace, G. W. Greenlees, J. V. Jelley, J. S. C. McKee, P. B. Moon, D. A. O'Connor, R. E. Peierls, W. B. Powell, L. Riddiford, K. F. Smith and J. Walker. I should like to thank G. Pyle for help with the proof-reading and Doreen Kellett for tracing some of the diagrams. I must acknowledge with gratitude the patience and skill of my typists, Susan Wiggin, Susan Tennant, Susan Dorté and my wife Mary, who have had to deal with a frequently amended manuscript. Finally, the kindness, courtesy and efficiency of the publishers, Messrs. Longman, have helped materially throughout the preparation of the work.

Birmingham, March 1962   W.E.B.

Preface to second edition

In this edition an introduction to elementary particle physics has been added. To make way for this without adding unreasonably to the size of the book, severe reductions have been made in the content of Chapter 3 (atomic physics) and in Chapters 6 and 7, in which no pretence to a treatment of modern electronic techniques is made. Nuclear structure physics has advanced considerably in the past eight years and some recent developments (e.g. channelling, germanium counters, (p, 2p) reactions, isobaric analogue levels) have been included. Unfortunately it has been necessary to exclude some of the advances in nuclear structure theory such as the SU(3) classification of energy levels and the application of the theory of superconductivity to the nuclear level spectrum.

A change to SI units has been made, with the help of J. K. Hulbert, to whom I express my gratitude. The change is not complete; some quantities remain in other units for historical reasons or because they are best visualized in terms of g or cm, or because the units form part of what is essentially a proper name, e.g. 21 cm line, 184″ cyclotron. The magnetic force vector **B**, measured in teslas ($1T = 10^4$ gauss), is frequently referred to as a field strength although it is properly a flux density.

Problems originally set in c.g.s. units, or in which atomic masses based on the ^{16}O scale appeared, have generally been converted. For reasons of expense dimensionless graphical scales have not been introduced.

I am grateful to H. Burkhardt, D. C. Colley, J. D. Dowell, M. Jobes and I. R. Kenyon for reading the new chapters, and to Jill Nixon and Susan Hanson for typing the manuscript.

Birmingham, February 1971   W.E.B.

Contents

Contents

Contents

Contents

Contents

Plates

Acknowledgements

For permission to reproduce photographs or to redraw diagrams we are indebted to the following: Academic Press Inc., New York: Ajzenberg-Selove, *Nuclear Spectroscopy* and Coles, *Advances in Electronics*; Akademische Verlagsgesellschaft, Frankfurt: Kopferman, *Kernmomente*; Akademische Verlagsgesellschaft, Leipzig: *Zeit. für Physik. Chemie*; the authors concerned and the American Institute of Physics: *Rev. Sci. Inst.*; the Editor, *Ann. Rev. Nucl. Sci.*; Brookhaven National Laboratory: *Report B.N.L.*; Cambridge University Press: Wilson, *Principles of Cloud Chamber Technique* and Wilkinson, *Ionisation Chambers and Counters*; Clarendon Press, Oxford: Bacon, *Neutron Diffraction* and Heitler, *Quantum Theory of Radiation*; McGraw-Hill Book Co. Inc.: Evans, *The Atomic Nucleus* and White, *Introduction to Atomic Spectra*; The Macmillan Company, New York: Shankland, *Atomic & Nuclear Physics*; Laboratory for Nuclear Science, Massachusetts Institute of Technology: *Progress Report*; National Research Council, Ottawa: *Canadian Journal of Research*; North Holland Publishing Company: *Nucl. Instrum. and Methods*, Siegbahn, *Beta and Gamma Spectroscopy* and Wapstra *et al.*, *Nuclear Spectroscopy Tables*; Pergamon Press, Oxford: *Progress in Nuclear Physics*; Pergamon Press Inc., New York: Jelley, *Cherenkov Radiation* and Littler & Raffle, *An Introduction to Reactor Physics*; Sir Isaac Pitman & Sons Ltd.: Chadwick, *Radioactivity*; Philips Research Laboratories, Eindhoven: *Philips Technical Review*; Professor Powell; the Editors, *Rev. Mod. Phys.* and *Phys. Rev. Letters*; the authors concerned, the Institute of Physics and The Physical Society: *Proc. Phys. Soc.*; Stichting Physica: *Rep. Prog. Phys.*; Royal Danish Academy: *Kgl. Danske Videnskab Selskab*; Royal Swedish Academy of Science: *Arkiv. fur Fysik*; Royal Society: *Proc. Roy. Soc.* A; Editors, *Science Progress*; United Kingdom Atomic Energy Authority: *A.E.R.E. Harwell Reports*; Springer-Verlag, Heidelberg: *Zeits. für Physik* and *Naturwissenschaften*; Taylor & Francis Ltd.: *Phil. Mag.*; United States Atomic Energy Commission: Rossi & Staub, *Ionization Chambers & Counters*; D. Van Nostrand Company Inc.: Korff, *Electron & Nuclear Counters*; John Wiley & Sons, Inc.: Livingston, *High Energy Accelerators*, Hughes, *Neutron Optics*, Blatt & Weisskopf, *Theoretical Nuclear Physics* and Segrè, *Experimental Nuclear Physics*.

We are indebted to the Universities of Birmingham, Cambridge, Glasgow, Hull, Keele and Liverpool for permission to reproduce questions from various past examination papers, and McGraw-Hill Book Company for material from *The Atomic Nucleus* by R. D. Evans, copyright © 1955.

Acknowledgements for new material in second edition

For permission to reproduce photographs or to redraw diagrams for this second edition we are indebted to the following publishers and to the authors concerned in each case: Academic Press Inc., London: Kabir, *The CP Puzzle*; Academic Press Inc., New York: Burhop, *High Energy Physics*; American Association for the Advancement of Science: *Science*; American Institute of Physics: *Phys. Rev.* and *Phys. Rev. Lett.*; W. A. Benjamin Inc.: Segré, *Nuclei and Particles*; Cambridge University Press: Eden, *High Energy Collisions of Elementary Particles*; W. H. Freeman and Company: Wigner, *Violations of Symmetry in Physics*, and Chew et al., *Strongly Interacting Particles*; Institute of Physics and The Physical Society: *Rep. prog. Phys.*; Macmillan Journals Ltd.: *Nature*; National Bureau of Standards: *Rev. Mod. Phys.*; National Research Council of Canada: *Canadian Journal of Physics*; Nordita: *Nuclear Physics*; North-Holland Publishing Company: Bertolini and Coche, *Semiconductor Detectors*, Orphan and Rasmussen, *Nucl. Instrum.* and Bailey *et al*, *Phys. Lett.*; Messrs. Taylor and Francis: *Contemporary Physics* and *Phil. Mag.*

For permission to reproduce extracts from Physics Examination Papers we are indebted to the following: University of Birmingham for 1956, Physics Course IV, Paper III, question 2; 1957, Physics Course IV, Paper III, questions 1 & 2; 1959, Physics Course IV, Paper III, question 2; 1960, Physics Course IV, Paper VII, question 2; 1962, Physics Course III, Paper VII, questions 1 and 2; 1963, Physics Course III, Paper VII, questions 2 & 3; 1969, Physics, 3/0 IV, questions 6 & 7; University of Cambridge for 1967, Part II, Natural Sciences (Theoretical Physics Paper 3, question 8); University of Durham for 1969, Physics Paper III, question 9 and University of East Anglia for 1969, Physics Paper V, question 6.

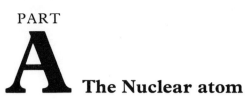

PART

A The Nuclear atom

1 Introduction

1.1 General survey

Nuclear physics was born, in a rather obscure and unrecognized way, in 1896 with Becquerel's discovery of naturally occurring radioactivity. The significance of this phenomenon was perhaps rather overshadowed by Röntgen's discovery of X-rays in the year preceding and by Thomson's demonstration of the existence of the electron in the year following. It was not to be foreseen at that time that the curious radiations from uranium were the first observed manifestations of a specific atomic nucleus, whose existence was not only to coordinate within two decades such diverse phenomena as the X-radiation and the electronic structure of atoms, but also within half a century to place in the hands of man the richest source of energy that seems likely to come his way. The vast and complex subject of nuclear physics that has developed over these years is based on twin structures of experiment and theory which have risen together, sometimes the one seeming the more advanced, sometimes the other. The experimental structure includes the extensive empirical data of atomic spectroscopy and the considerable body of information on natural radioactivity and more lately on nuclear transformations. The theoretical structure is essentially the quantum theory of Planck which grew from its origin in 1900 to maturity in the quantum mechanics of Bohr, Schrödinger, Heisenberg, Dirac, Born and Jordan. This is in turn linked to the observable properties of material particles by the concept of the de Broglie wavelength and by Bohr's principle of complementarity, sometimes referred to as the principle of wave-particle duality.

There is a mathematical theorem which states that no system of electric charges at rest can be stable if the forces between the charges are of 'ordinary' type, i.e. governed as far as distance is concerned by the inverse square law. The alternative possibility of an atomic structure in which extraordinary forces operated was of course ruled out by the detailed verification of Rutherford's hypothesis of the nucleus in 1911. The nuclear atom was conceived in order to explain the results of experiments on the scattering of the α-particles by thin metallic foils. The unexpectedly prolific scattering of these heavy particles through very large angles was inconsistent with a uniform diffuse atomic structure and led both qualitatively and quantitatively to the suggestion of a central nucleus in which should reside the whole of the positive charge and most of the mass of the atom. The exact fit of the experimental results to the scattering law first given by Rutherford and now associated with his name showed that the Coulomb law of force indeed held good down to distances of about 10^{-14} m, many times smaller than known atomic dimensions. No extraordinary forces therefore seemed to operate over the greater part of the atomic volume and stability of the nuclear atom could only be achieved by allowing the electrons to move about the nucleus in mechanically stable orbits. This simple model and the quantum laws given by Planck were used by Bohr in 1913 to explain, with dramatic success, the numerical relationships between the frequencies of the lines in the Balmer series of atomic hydrogen. In the next few years, the Rutherford–Bohr atom, with its central nucleus of mass A units and charge $+Ze$, was used to marshal and interpret much of the evidence of optical spectroscopy. In X-ray spectroscopy the work of Moseley confirmed the role suggested for the charge number Z as the atomic number of the element concerned, and this number was seen to be more fundamental for prediction of chemical properties than the chemical atomic weight of elements which later often proved to be mixtures of isotopes. In radioactive phenomena, the successive transformations of uranium, thorium and actinium could be ascribed to the central nucleus and these elements and their decay products could be accommodated with certainty in the periodic system by application of the displacement law of Russell and Soddy. The suggestion of isotopic constitution arising from this work was directly confirmed by the positive ray parabolas seen by Thomson and accurately established by the mass spectra obtained for the majority of elements by Aston.

Despite this formidable total of positive evidence for the Rutherford–Bohr atom it was evident from the beginning that the model was unstable according to classical electrodynamics. An electron with charge $-e$ moving in a closed orbit is continuously accelerated towards the centre of force and must radiate, according to electromagnetic theory, at the rate

$$\frac{dW}{dt} = \frac{\mu_0 e^2 f^2}{6\pi c} \text{ J s}^{-1} \tag{1.1}$$

where f is the instantaneous value of the acceleration and c is the velocity of light. The orbit should consequently contract indefinitely. The denial of this consequence was Bohr's first postulate, namely that the state of motion of an electron in an atom should be *stationary* (non-radiating). The second postulate, that the quantization of angular momentum should determine the actual orbits from the infinity of possibilities, and the third postulate, that the frequency of the light emitted in an atomic transition between stationary states of energies E_i and E_f should be given by the relation

$$hv = E_i - E_f \tag{1.2}$$

where h is Planck's constant, provided the quantitative basis for the theory. Although Bohr always insisted on the necessity for a correspondence between classical and quantum theories in the limit of large quantum numbers the juxtaposition of classical and non-classical concepts in the Bohr theory of the atom became less attractive as the development of theory proceeded. There was in particular the difficulty that despite the many successes of the concept of photons with quantized energy, yet light appeared to be propagated in all respects as a wave motion. The way to a new approach to atomic phenomena was opened by de Broglie, who in 1924 suggested that electrons might behave as waves. The experimental verification of de Broglie's hypothesis of material waves with wavelength

$$\lambda = \frac{h}{p} \tag{1.3}$$

for a particle of momentum p is well known; the theoretical techniques of analysis according to de Broglie's ideas were developed under the name of *wave mechanics* by Schrödinger in 1926. At about the same time an alternative and more fundamental approach to the same problem was made by Heisenberg who realized that the wave-like properties of matter implied a revision of traditional methods of thought, at least in so far as atomic phenomena were concerned. Heisenberg proposed that detailed pictures and models which did not correspond to experimentally observable quantities should not appear in the theory. Such ideas as orbits of definite radius for electrons in an atom or protons in a nucleus are then excluded and the theory treats only *observable* quantities such as energy and momentum. Orbital frequencies no longer have a meaning and are replaced by the experimentally observed radiative transition probabilities. It was soon realized that the techniques appropriate for describing observables of this kind were those of matrix algebra and these have been adopted by modern *quantum mechanics*; it has been shown that the wave mechanics of Schrödinger is a precisely equivalent method. It is crucial to both systems of calculation that electron distributions shall be interpreted statistically, so that it is no longer possible to say with precision that an electron is at a given point in space, but only that the probability of it being there is known. Both systems also lead to Heisenberg's famous *uncertainty principle* according to which the precision of measurement of connected mechanical quantities is limited by the value of Planck's constant, i.e.

$$\Delta E \cdot \Delta t > \hbar$$
$$\Delta p_x \cdot \Delta x > \hbar \tag{1.4}$$

where ΔE and Δt are the uncertainty in energy and time of observation in a given state of a particle, Δp_x and Δx are the uncertainty in linear momentum and position and $\hbar = h/2\pi$. The uncertainty relations are intimately connected with the wave description of matter and rest in fact on the supposition of wave-particle duality; the 'complementarity' of the measurements implied in eq. (1.4) has been emphasized by Bohr.

Much of nuclear physics is concerned with particles moving slowly relative to light, and for problems of this sort, among which are found many collision phenomena, the original non-relativistic methods are adequate. Einstein's celebrated energy–mass relation of special relativity

$$E = mc^2 \qquad (1.5)$$

was involved in de Broglie's proposition of material waves, but it remained for Dirac to set up the relativistic form of quantum mechanics and to show that it predicted the existence of an anti-electron or positron. This particle was discovered in the cosmic radiation in 1932. Pauli had shown by considering the statistical distribution of electrons among possible states of motion that a new degree of freedom, beyond those given by spatial coordinates, was necessary. When Goudsmit and Uhlenbeck suggested that this was in fact an electron spin, it was found that this new property could be satisfactorily fitted into the Dirac theory. The *Pauli exclusion principle* as now formulated, states that for particles with half-integral spin, only two (with spins opposed) are permitted in any one state of motion. This principle, on which is based the current interpretation of the periodic system of the elements, ranks with Heisenberg's uncertainty principle as one of the fundamental laws of nature; together they underlie the theoretical structure within which an understanding of the phenomena of nuclear physics is to be achieved.

The development of nuclear physics itself was not remarkable over the period of the great theoretical advances just discussed. In 1919 Rutherford achieved the disintegration of the nitrogen nucleus by α-particle bombardment in an apparatus of rugged simplicity, and although this great experiment showed how the structure, as distinct from the existence, of the nucleus might be studied, the subsequent decade saw only a gradual increase in experimental activity. Natural α-particles remained the main nuclear projectiles until well into the 1930s; they were used by Chadwick in 1932 in the experiments which led to the discovery of the neutron, by Curie and Joliot (1934) in the first demonstration of 'artificial' radioactivity and (for the production of neutrons) by Hahn and Strassmann in the experiments which first showed the phenomena of fission (1938–9). By this time, however, the development of nuclear accelerators had been rewarded and once the artificial disintegration of lithium by protons had been announced by Cockcroft and Walton (1932) progress became extremely rapid. The greater part of our present-day detailed knowledge of nuclear properties is derived from accelerator experiments. That this knowledge can be fitted with some success into a reasonable theory of nuclear structure is primarily due to the early emergence of the neutron–proton nuclear model, and to the existence of well-tried theoretical methods for its development. The assumption of this model had no serious consequence for the theory of β-decay for it was evident for many reasons that nuclei could not reasonably contain electrons. The suggestion by Pauli of a light particle, the neutrino, which would be created together with an electron in the transformation of a neutron into a proton in beta decay enabled Fermi (1934) to provide the basic framework of the necessary theory of this process. The neutrino has proved one of the most elusive of fundamental particles but positive evidence of its existence has now been obtained and its role in nuclear processes has assumed major significance.

The discovery of the fission of uranium and similar bodies by thermal neutrons has had much less influence on the development of an understanding of the nucleus than might have been expected from the overwhelming economic impact of the process. The importance of Anderson's discovery of the positron (1932), of Neddermeyer and Anderson's discovery of the μ-meson (1936) and of Powell's discovery of the parent π-meson (1947), all in the cosmic radiation, is much greater. The π-meson is particularly interesting since it may play the part of an exchange particle in theories of nuclear forces of the type first suggested by Yukawa. Despite the low intensity of the cosmic radiation, it has proved the source of many new unstable particles in addition to the π- and μ-mesons and although many of these particles are now produced copiously by accelerators, the natural cosmic radiation is still the only means of studying nuclear phenomena at energies of about 300 GeV and above. The need for such studies, and for the great variety of high energy experiments now being conducted with accelerating machines, is founded on a desire to understand the relation between the new particles and the fields which describe them in as much detail as the relation between electrons and the electromagnetic field is understood. One result of such an understanding might be explicit knowledge of the force between nucleons, which is responsible for the stability of complex nuclei. This force is known at the moment only semi-empirically, but the consequence for nuclear structure of many of its general properties can be worked out and the behaviour of stable nuclei can be used as a check on the validity of deductions from high energy nucleon–nucleon or meson–nucleon experiments. It is unwise therefore to draw any sharp lines of distinction between low energy nuclear physics, high energy physics and cosmic ray physics, although experimentally they may appear rather sharply differentiated. In content of information these fields of study overlap and mutually fertilize one another and the future is likely to see them merge even more closely in the pursuit of a deeper understanding of the structure of matter. In the future, as in the past, it is to be hoped and expected that the apparent complexity of many nuclear phenomena will be resolved by simplifying developments in both theory and experiment.

The main events in the history of nuclear physics are summarized in Table 1.1, which is taken mainly from the similar table given by Evans (Ref. 1.4).

1.2 The fundamental particles, their interactions and the conservation laws

1.2.1 The particles and their properties

The fundamental particles of physics may be described, rather naively, as those which have no obvious substructure. A complex nucleus is not a fundamental particle, because it can be thought of as an assembly of neutrons and

TABLE 1.1 Chronology of main advances in the growth of nuclear physics

Advance	Date	Physicist
Periodic system of the elements	1868	Mendeléev
Discovery of X-rays	1895	Röntgen
Discovery of radioactivity	1896	Becquerel
Discovery of electron	1897	J. J. Thomson[a]
The quantum hypothesis	1900	Planck
Mass–energy relation	1905	Einstein
The expansion chamber	1911	Wilson
Isotopes suggested	1911	Soddy
Nuclear hypothesis	1911	Rutherford
Nuclear atom model	1913	Bohr
Atomic numbers from X-ray spectra	1913	Moseley
Positive ray parabolas for neon isotopes	1913	J. J. Thomson
Transmutation of nitrogen by α-particles	1919	Rutherford
Mass spectograph	1919	Aston
Wavelength of material particles	1924	de Broglie
The wave equation	1926	Schrödinger
Diffraction of electrons	1927	Davisson and Germer; G. P. Thomson
Uncertainty principle	1927	Heisenberg
Wave mechanical barrier penetration	1928	Gamow, Condon, Gurney
The cyclotron	1930	Lawrence
The electrostatic generator	1931	Van de Graaff
Discovery of deuterium	1932	Urey
Discovery of the neutron	1932	Chadwick
Transmutation of lithium by artificially accelerated protons	1932	Cockcroft and Walton
Discovery of the positron	1932	Anderson
Hypothesis of the neutrino	1933	Pauli
Neutrino theory of beta decay	1934	Fermi
Discovery of artificial radioactivity	1934	Curie and Joliot
Neutron-induced activity	1934	Fermi
Hypothesis of heavy quanta (mesons)	1935	Yukawa
Discovery of the μ-meson	1936	Anderson and Neddermeyer
Magnetic resonance principle	1938	Rabi
Discovery of fission	1939	Hahn and Strassmann
The principle of phase-stable accelerators	1945	McMillan, Veksler
Discovery of π-meson	1947	Powell
Discovery of strange particles	1947	Rochester and Butler
Use of space–time diagrams	1949	Feynman
Production of π°-mesons	1950	Bjorklund et al.
Hypothesis of associated production	1952	Pais
Discovery of hyperfragments	1953	Danysz and Pniewski
Strangeness	1953	Gell–Mann; Nakano and Nishijima
Hypothesis of K°	1955	Gell–Mann and Pais
Discovery of antiproton	1955	Chamberlain et al.
Non-conservation of parity	1956	Lee and Yang; Wu; Garwin
Observation of (anti) neutrino	1956	Reines and Cowan
Prediction of heavy meson	1957	Nambu
Helicity of neutrino	1958	Goldhaber, Grodzins and Sunyar
Hypothesis of conserved vector current	1958	Feynman and Gell–Mann
Discovery of ω-meson	1961	Maglic et al.
Unitary symmetry	1961	Gell–Mann; Ne'eman
Muon neutrino	1962	Danby et al.
Discovery of Ω^-	1964	Barnes et al.
Non-conservation of CP	1964	Cronin, Fitch and Turlay

[a] See *Physics Today*, **19**, July 1966, p. 12

protons of constant composition. A neutron or proton, however, is at present considered fundamental because its resolution into more elementary entities cannot be demonstrated and because it is subject to conservation laws. Such particles are not necessarily stable, but they decay into other particles of the same general type, also in accordance with conservation laws.

The more familiar elementary particles are the following:

(a) the *electron* and *proton*, which are charged particles familiar from atomic physics,

(b) the *neutron*, an uncharged particle produced copiously in nuclear reactions, and an important constituent of complex nuclei,

(c) the *neutrino*, an uncharged particle of zero rest mass invoked in the theory of beta decay,

(d) the *photon* or quantum of radiation, familiar in atomic physics and of similar importance in nuclear physics,

(e) the π- and μ-*mesons*, particles of mass between that of the electron and that of the proton, which are of importance respectively for the theory of nuclear forces and for the understanding of cosmic ray phenomena.

Each of these particles, although conveniently known by that name, participates in the general wave-particle duality which is a characteristic of natural phenomena and is clearly apparent in atomic and sub-atomic events. For the purposes of classification it is most convenient generally to refer to the particle aspect.†

Most of this book is concerned only with the non-strange particles (a)–(e) above and their main properties are listed in Table 1.2. This table is extended in Appendix 8 to include strange particles and some resonances. The properties of *mass*, *charge* and *angular momentum (spin)* are familiar classically and may be defined and measured for macroscopic objects. For such objects the properties assume a continuous range of values, but for the fundamental particles the masses have certain discrete values and charge and spin are multiples of fundamental units, namely the electronic charge and the Planck quantum $\hbar$. The *lifetimes* listed refer to the particles in a free state, in which case they decay exponentially and do not disappear by any type of interaction with other bodies. The decay law is that familiar from natural radioactivity (Sect. 2.3.4), namely

$$N_t = N_0 e^{-t/\tau} \tag{1.6}$$

where N_t is the number of particles existing at (proper) time t, N_0 is the number present at time $t = 0$ and τ is the mean lifetime.

Column 3 of Table 1.2 is headed 'antiparticle'. The presumption that each

† It is no accident that light enters our ordinary experience mainly as a wave motion while electrons are most readily understood as particles. The difference is fundamental and is connected with the fact that the statistical properties of light quanta make it possible to define a *phase*, whereas this quantity is unmeasurable for the de Broglie waves associated with electrons. A discussion of this general question may be found in the articles by R. E. Peierls and others, 'A survey of field theory', *Rep. progr. Phys.*, **18**, 423, 1955.

TABLE 1.2 Properties of some fundamental particles

Name	Particle	Anti-particle	Rest mass in units of electron mass	Charge in units of electron charge	Spin in units of $\hbar$	Mean lifetime (s)	Decay mode
Electron	e^-	e^+	1	± 1	$\frac{1}{2}$	Stable	—
Proton	p	$\bar{p}$	1836	± 1	$\frac{1}{2}$	Stable	—
Neutron	n	$\bar{n}$	1839	0	$\frac{1}{2}$	932	$n \to p + e^- + \bar{\nu}$
Neutrino	ν	$\bar{\nu}$	0	0	$\frac{1}{2}$	Stable	—
Photon	γ	Self	0	0	1	Stable	—
Charged π-meson	π^+	π^-	73	± 1	0	$2 \cdot 6 \times 10^{-8}$	$\pi^+ \to \mu^+ + \nu$
Neutral π-meson	π°	Self	264	0	0	10^{-16}	$\pi^\circ \to 2\gamma$
μ-meson (muon)	μ^-	μ^+	207	± 1	$\frac{1}{2}$	$2 \cdot 2 \times 10^{-6}$	$\mu^+ \to e^+ + \nu + \bar{\nu}$

particle has a counterpart of opposite charge but equal rest mass derives from the theory of the positron given by Dirac and extended to other particles by Pauli and Weisskopf, and from the clear experimental evidence for nearly all the antibodies listed in Table 1.2 and Appendix 8. In the case of the neutron, and similar neutral heavy particles, the distinction from the antiparticle rests on the relative direction of the vectors representing spin and magnetic moment. For the neutrino, the evidence is more subtle and is discussed in Chapter 16; for the photon and neutral meson there is no distinction, and in a formal sense each of these particles is self-conjugate.

The general classification of particles according to mass is as follows:

(*a*) *leptons* (light particles) are the electrons, the neutrinos and (because of their behaviour) the particles historically known as μ-mesons. To emphasize their essentially leptonic nature these latter particles are now called *muons*.

(*b*) *mesons* (particles of intermediate mass) have masses between that of the electron (m_e) and that of the proton (1836 m_e). However, some mesonic resonant states (Chapter 20) have a mass greater than this.

(*c*) *baryons* (heavy particles) which are subdivided into
   (i) *nucleons*, the neutron and proton
   (ii) *hyperons*, the particles of mass greater than that of the neutron (1839 m_e) but which behave as if they contain just one nucleon.

1.2.2 Interactions between particles; the conservation laws

The effect of one particle on another, or their mutual interaction, now appears to be one or more of four types:

(*a*) *Gravitational interactions*, which are familiar in classical physics because of the long range of the inverse square law force.

(*b*) *Electromagnetic interactions*, also familiar in classical physics, and also of long range. They include the forces between charges at rest and in motion, and the effects of the electric and magnetic fields of radiation on a charge distribution.

(*c*) *Weak interactions*, which describe the production and behaviour of leptons (e.g., in nuclear β-decay) and the decay processes of most of the strange particles.

(*d*) *Strong interactions*, among which appear the specifically nuclear forces between nucleons responsible for binding these particles together into a complex nucleus.

The weak and strong interactions are of short range, comparable with nuclear dimensions, and for this reason do not show the macroscopic effects associated with gravity and electrical forces. They were therefore only identified as a result of experiments with nuclei. The relative strengths of the interactions, as seen in such experiments, are discussed in Chapter 21; for the present we note only that gravitational interactions between particles, although of over-whelming importance in astronomy because of their cumulative nature, are entirely negligible at atomic and nuclear distances in comparison with the other interactions and will be disregarded in this book. The identification of the remaining interactions, electromagnetic, weak and strong, permits a further classification of the particles according to their participation in these interactions; this will be discussed in Chapter 21.

We assume that all such interactions are constrained by general principles. Chief among these are the *conservation laws* for the mechanical quantities *total energy*, *linear momentum* and *angular momentum*. These are classically familiar quantities which are known to assume a continuous range of values in macroscopic problems. When quantum mechanics must be applied, i.e. when the de Broglie wavelength of a particle is no longer negligible with respect to the dimensions of the system or enclosure in which it appears, the conservation of energy and of momentum are still assumed to apply, although the limits set by the uncertainty principle become important in particular situations. In many problems, too, such as those of the simple harmonic oscillator or of the hydrogen atom, the energy and angular momentum take *discrete* values and are said to be *quantized*, in a way which is not encountered in classical physics. These conservation laws may be related to certain specific properties of the space–time framework within which we conduct our experiments.

There are also conservation laws of an additive type, based on experience, which relate to numbers of particles and to charge. It seems likely that the total number of nucleons in the universe (with antiparticles counted negatively) is a constant, and that if a new proton is created, it is always accompanied by an antiproton, e.g.

$$p + p(+energy) \longrightarrow p + p + p + \bar{p} \tag{1.7}$$

Decay of a neutron does not alter the number of nucleons:

$$n \longrightarrow p + e^- + \bar{\nu} \tag{1.8}$$

The *conservation of nucleons* with certain necessary extensions to include hyperons, is implicit in our present description of all nuclear processes. There is also excellent evidence (ch. 5) for the joint production (or annihilation) of positron–electron pairs from (or into) radiation under suitable circumstances according to the process

$$\gamma \leftrightarrow e^+ + e^- \tag{1.9}$$

Conservation of electrons is not valid because of decay processes of type (1.8), but if neutrinos and muons are suitably included it is possible reasonably to postulate a *conservation of leptons*.

An even more compelling reason for asserting the conservation of nucleons is that if processes of the type

$$p \rightarrow e^+ + \gamma \tag{1.10}$$

were possible with a lifetime short on the cosmological scale, then the universe as we know it would be evidently decaying away. The conservation of particles also follows directly from the conservation of charge since particles and antiparticles have equal and opposite charges to a high degree of accuracy. Conservation of mass number (A) and of charge number (Z) will be assumed throughout our discussion of nuclear processes; the place of these conservation laws in formal theory will be indicated in Chapter 21.

1.3 Terminology

Although international recommendations on symbols, units and nomenclature exist (International Union of Pure and Applied Physics, S.U.N. Commission Document U.I.P. 11, 1965; see also Signs, Symbols and Abbreviations, Royal Society of London, 1969), these are still not comprehensive and are unfortunately not yet adopted uniformly in the major journals. In the present book the main concern has been to avoid ambiguity in terminology and the resulting symbols are a compromise between the S.U.N. recommendations and the practice of the American Institute of Physics.

1.3.1 Specific definitions

nucleus   a general term for the finite structure of neutrons and protons constituting the centre of force in an atom.

nuclide   a specific nucleus, with given proton number Z and neutron number N.

isotope   one of a group of nuclides each having the same proton number Z.

isotone   one of a group of nuclides each having the same neutron number N.

isobar   one of a group of nuclides each having the same mass number $A(=Z+N)$.

isomer   a nuclide excited to a long-lived state from which beta or gamma decay ensues (ch. 13).

1.3.2 Nomenclature and symbols

atomic number	Z
mass number	A
proton number	Z
neutron number	$N = A - Z$
charge of electron	$-e$
electron mass	m (kg)
electron rest mass	m_e (kg)
mass of a particle x	m_x (kg)
	M_x (atomic mass units)
mass of a neutral atom	$M(A, Z)$ (atomic mass units)
mass of a heavy particle	M (kg)

magnetic moment of a particle x μ_x

Bohr magneton μ_B

nuclear magneton μ_N

principal quantum number n, n_i

orbital quantum number L, l_i

spin quantum number S, s_i

total angular momentum quantum number $\mathcal{J}, j_i$

magnetic quantum number M, m_i

nuclear spin quantum number I or $\mathcal{J}$

isobaric spin quantum number T or I

hyperfine quantum number F

rotational quantum number $\mathcal{J}, K$

quadrupole moment Q

Rydberg constant R_∞

Mechanical quantities

velocity v

momentum p

kinetic energy T

potential energy V

total energy E or W

Nuclear radii are conveniently measured in femtometers ($1 \text{ fm} = 10^{-15}$ m) and nuclear areas in *barns* ($1 \text{b} = 10^{-28}$ m^2).

Symbols for *chemical elements* are written:

mass number   state of ionization

$$^{14}_{7}\text{N}^{6+}_{2}$$

atomic number   atoms per molecule

A two-body *nuclear reaction*, in which a particle a bombards a nucleus X with the result that a residual nucleus Y and a particle b are produced, i.e.

$$X + a \rightarrow Y + b$$

will be abbreviated $X(a, b)Y$.

One *curie* (Ci) of a radioactive nuclide is that quantity in which the number of disintegrations per second is $3 \cdot 700 \times 10^{10}$.

References

The literature of nuclear physics is very extensive. The following books, among many others, provide a general background.

1. PEIERLS, R. E. *The Laws of Nature*, Allen and Unwin, 1955.
2. FRENCH, A. P. *Principles of Modern Physics*, Wiley, 1958.

A general coverage of nuclear physics, mainly from the experimental point of view, is given by

3. SHANKLAND, R. S. *Atomic and Nuclear Physics*, Macmillan, 1960.

4. EVANS, R. D. *The Atomic Nucleus*, McGraw Hill, 1955.
5. MEYERHOF, W. E. *Elements of Nuclear Physics*, McGraw Hill, 1967.
6. SEGRÈ, E. *Nuclei and Particles*, Benjamin, 1964.
7. ENGE, H. A. *Introduction to Nuclear Physics*, Addison-Wesley, 1966.

Among theoretical works are:

8. BLATT, J. M. and WEISSKOPF, V. F. *Theoretical Nuclear Physics*, Wiley 1952.
9. ELTON, L. R. B. *Introductory Nuclear Theory* (2nd edn.), Pitman, 1965.
10. PRESTON, M. A. *Physics of the Nucleus*, Addison-Wesley, 1962.
11. DE BENEDETTI, S. *Nuclear Interactions*, Wiley, 1964.

General and historical articles which may be consulted include:

12. BEYER, R. T. *Foundations of Nuclear Physics*, Dover Publications Inc., 1949.
13. BOHR, N. 'The Rutherford Memorial Lecture 1958' *Proc. phys. Soc.*, **78**, 1083, 1961.
14. ANDERSON, C. D. 'Early work on the positron and the muon', *Amer. jnl. Phys.*, **29**, 825, 1961.
15. ZUCKER, A. and BROMLEY, D. A. 'Nuclear Physics—a Status Report', *Science*, **149**, 1197, 1965.
16. WEISSKOPF, V. F. 'Three Steps in the Structure of Matter', *Physics Today*, **23**, No. 8, Aug. 1970.

Compilations of nuclear techniques and data:

17. SIEGBAHN, K. (ed.) *Alpha, Beta and Gamma Ray Spectroscopy*, North Holland Publishing Co., 1965.
18. LEDERER, M., HOLLANDER, J. M. and PERLMAN, I. *Table of Isotopes* (6th edn.), Wiley, 1967.
19. *Nuclear Data* (journal), Academic Press. Light nuclei are treated in detail by F. Ajzenberg-Selove and T. Lauritsen ($A = 5$–20) and P. M. Endt and C. van der Leun ($A = 21$–40) in articles in the journal *Nuclear Physics*.
20. 'Review of Particle Properties', Particle Data Group, *Phys. Lett.*, **33B**, 1, 1970.

2 Natural radioactivity and the discovery of the nucleus

The radioactivity found in naturally occurring elements is not a basically different phenomenon from the radioactivity now known to be a property of several hundred artificially produced nuclei. It might therefore be discussed as part of the treatment of the spontaneous emission of radiations given later in this book. Natural radioactivity has, however, such a deep historical significance for nuclear physics, and has affected the whole development of the subject, including its terminology, to such a marked extent, that it seems proper to give at the outset a review of the steps that led to the nuclear hypothesis.

2.1 The discovery of radioactivity and the separation of radium

In 1896 Becquerel had been studying the luminescence of uranium salts excited by ordinary light. He had observed that the luminescent radiations were capable of casting shadows of opaque objects which could be recorded by a photographic plate wrapped in black paper. This phenomenon had also been demonstrated as one of the most dramatic and potentially valuable properties of the newly discovered X-rays and it seemed at first that the luminescent radiation from uranium might be similar to the X-rays themselves. The outstanding discovery† made by Becquerel was that the shadow-casting radiation from uranium persisted even when the exciting light was removed. He showed also that the radiation was found with all uranium compounds in proportion to their uranium content and that spontaneous emission of radiation, or *radioactivity*, was a property of the uranium atom itself in its normal state. It is now known that the radiations studied by Becquerel were the fast electrons emitted in the β-decay of daughter products of the nucleus ^{238}U; Becquerel himself showed that the radiations could be deviated by a magnetic field. He also found that the new radiations could discharge an electrified body and this discovery led quickly to the use of ionization chambers for quantitative assessment of the strength of the radiation, or *activity*.

After the announcement of Becquerel's work, Pierre and Marie Curie surveyed many other elements for radioactive emission and discovered that thorium showed a similar degree of activity. They also found that the uranium ore known as pitchblende contained more activity than was expected from

† H. Becquerel, *Comptes Rendus*, **122**, 501, 1896.

the chemical estimation of the uranium content and they rightly deduced the presence of other active elements. They then attacked the problem of separating these elements from the uranium, and from the other elements present in the pitchblende. Their chemical procedures first led them to concentrate a substance which seemed to behave chemically rather like bismuth, but which had a considerably greater activity than had uranium. This substance was separated from bismuth by volatilization, and, as the first new radioactive element to be isolated, was given the name *polonium*,† after Mme. Curie's native country. The activity of the separated polonium was later found to decrease slowly ($t_{1/2} = 138$ days) and by itself was insufficient to account for the excess activity of the pitchblende. The Curies therefore continued their work and next found another highly active substance which behaved chemically like barium and was named by its discoverers *radium*.‡ Neither polonium nor radium was obtained in sufficient quantity in the first extractions to permit a determination of atomic weight, and the Curies therefore set themselves the task of extracting the radium element from as much ore as they could obtain. The tireless patience of their work, which began with chemical separations and ended with repeated fractional crystallizations of radium and barium chlorides, is well known (Ref. 2.5). As fractionation proceeded, the specific activity of the product was found to have increased, and in the end about 0·1 g of radium, as chloride, was obtained from 10^3 kg of ore. Radium was shown to be several million times as active as uranium (mass for mass) and Mme. Curie was able to make the first chemical determination of the atomic weight of the new element as 225; it was shown also to have a characteristic optical spectrum. The activity of radium appeared to be constant, in contrast with the polonium activity, and was sufficiently strong to account for the original observations on pitchblende. It is now known that the mass number of radium is 226 and that its half-life is 1622 years.

In 1897 the radiations from uranium had excited the attention of Rutherford, who was then working under J. J. Thomson in the Cavendish Laboratory. After studying the conductivity of gases induced by the uranium radiations he turned his attention to thorium and continued work on this substance during his tenure of the Professorship of Physics at McGill University, Montreal (1898–1907). This was the heroic age of radioactivity, in which, mainly as a result of the work of Rutherford, Soddy and their collaborators, the subject was placed upon a quantitative basis and the stage was set for the appearance of the nuclear atom.

2.2 The radioactive radiations

The early experiments of Curie and of Rutherford showed that the radiations from radioactive substances contained components of different penetrating power, as assessed by their absorption in matter. The less penetrating rays,

† Pierre and Marie Curie, *Comptes Rendus*, **127**, 175, 1898.
‡ Pierre and Marie Curie, *Comptes Rendus*, **127**, 1215, 1898.

Plate 1 Expansion chamber tracks of α-rays from radium (C. T. R. Wilson, *Proc. roy. Soc.* A, **87**, 277, 1912).

Plate 4 Expansion chamber track of carbon ion, showing delta rays (J. C. Bower).

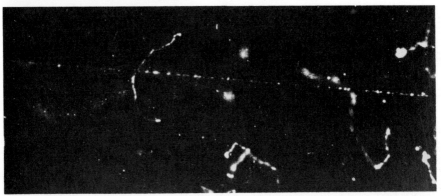

Plate 2 Expansion chamber tracks of fast and slow β-rays (electrons) produced by hard X-rays (C. T. R. Wilson, *Proc. roy. Soc.* A, **104**, 192, 1923).

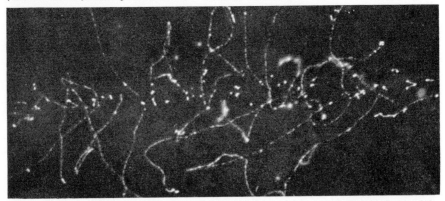

Plate 3 Expansion chamber tracks of recoil electrons due to beam of soft X-rays. Some of the short tracks are due to Auger electrons (C. T. R. Wilson, *Proc. roy. Soc.* A, **104**, 1, 1923).

which were completely absorbed by a few cm of air or by a thin sheet (about 0·1 mm) of metal were called α-*rays*. The more penetrating components, which were absorbed by about 1 mm of lead were named β-rays. Both the α- and β-rays were shown to be corpuscular in character by magnetic deflection methods. In 1900 Villard identified a third and even more penetrating type of ionizing radiation, capable of traversing as much as 10 cm of lead. These rays could not be deviated by a magnet and were described as γ-*radiation*; they are now known to be electromagnetic in nature. Figure 2.1 illustrates the difference between the radiations. The first decade of the study of radioactivity saw a great increase in knowledge of the nature and origin of these radiations and resulted in the development of several sensitive instruments (ch. 6) for their detection. Many experiments however were carried out with an ionization chamber connected to a Dolezalek electrometer; the active sample was placed on one of the electrodes inside the chamber (Fig. 2.7). This apparatus permitted the observation of decay rates and if absorbing foils were placed over the sample the radiations emitted could be distinguished according to their penetrating power. In this way specific radioactive

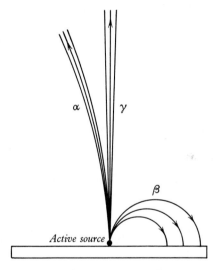

Fig. 2.1 Effect of a transverse magnetic field on radiations. The α-particles carry positive and the β-particles negative charge (Ref. 2.1).

bodies were classified according to their radiation properties. After 1908 the weaker sources of α-rays were often investigated by 'scintillation' counting. In this method the flashes of light produced when the α-rays struck a screen of zinc sulphide or other suitable material were observed and counted using a low power microscope.

The techniques of electric and magnetic deflection, which had already been used for the study of electrons and ions, were also applied extensively in analysis of the radioactive radiations. In a uniform magnetic field of flux density B a particle of mass m, charge e and velocity v perpendicular to B describes a circle of radius ρ in a plane perpendicular to the lines of force where

$$\rho = \frac{mv}{Be} \tag{2.1}$$

In a uniform electric field E the particle describes a parabola in a plane parallel to the lines of force and the deflection in a certain distance l in the field is given by

$$\frac{1}{2} \frac{Ee}{m} \left(\frac{l}{v}\right)^2 \tag{2.2}$$

assuming that v is initially at right-angles to E. From eqs. (2.1) and (2.2) both e/m and v can be obtained from the observed deflections in the two types of field.

The conclusions drawn about the nature of the radiations from ionization and deflection experiments were strikingly verified when the actual tracks of particles were made visible in the Wilson expansion chamber (ch. 6). Plates 1, 2 and 3 show the effects produced by the passage of α-rays and X-rays through such a chamber.

2.2.1 The alpha-rays

The properties of the α-rays were continuously and exhaustively investigated by Rutherford and his collaborators over the period 1898–1914. Deflection experiments on the α-rays from radium and its products showed to a first approximation that all the α-rays emitted by a single parent body had the same initial velocity. When thin sources of radioactive material were used it became possible to make accurate measurements and to obtain the apparent charge to mass ratio e_α/m_α for the radiation. The earliest measurements ranged between 43 and 64×10^6 C kg^{-1}. Although these values were not refined to the value $48 \cdot 20 \times 10^6$ until 1914 (Rutherford and Robinson) they were adequate to suggest that the actual value was probably half that ($e_H/m_H = 96 \cdot 52 \times 10^6$) obtained for the hydrogen ion from electrolysis and gaseous conduction experiments. The results indicated that the α-rays were actually behaving as *particles* of mass comparable with that of a light atom.

The velocity of emission of the α-particles was also determined in absolute units from the deflection experiments. Velocities (and thence energies) were also determined, once an absolute measurement had been made, by observing the relative ranges of different α-particle groups in air under standard conditions (15°C and 760 mmHg). All measurements confirmed the homogeneity in energy of the particles emitted from a given parent. Details of some of these techniques will be discussed in Chapter 7.

To determine the mass of the α-particles from the deflection experiments, it was necessary to measure the charge in absolute units. This was obtained by Rutherford and Geiger (1908) following their development of methods for counting single particles. Using a type of proportional counter (ch. 6) they found that the number of α-particles emitted per second from the product now known as RaC′ in equilibrium with 1 g of radium was $3{\cdot}4 \times 10^{10}$.† The total charge collected due to the α-particles from a source of RaC′ was then determined using the apparatus shown in Fig. 2.2, and the strength of the source was related to a standard source by comparison of γ-ray activities. The α-particle charge was thus found to be $+3{\cdot}1 \times 10^{-19}$ C; subsequent experiments have yielded a more accurate value which is just twice that of the electronic charge and of opposite sign. The α-particle therefore had a mass approximately four times that of the proton and appeared to be a helium atom with a double positive charge.

TABLE 2.1 The 'classical' constants of radioactivity

Half-life of radium	1622 years
Number of α-particles emitted per second from 1 g of radium	$3{\cdot}608 \times 10^{10}$
Number of α-particles emitted per second from 1 curie of radium	$3{\cdot}70 \times 10^{10}$
Volume of 1 curie of radon	$0{\cdot}64$ mm^3 at N.T.P.
Volume of helium produced per year by 1 g of radium in equilibrium with its products	169 mm^3 at N.T.P.
Heat emitted per hour by 1 g of radium in equilibrium with its products	619 J

Several important quantitative conclusions were drawn from the experiment of Rutherford and Geiger, including a value for the half-life of radium (Table 2.1) and figures for the rate of production of helium and for the evolution of heat by radioactive salts (Table 2.1 and Sect. **2.3**). The general agreement between those predictions which could be independently tested and actual observation confirmed the conclusion that the α-particle was a charged helium atom. This conclusion was further supported by the fact that radioactive ores were known to contain helium. The most elegant confirmation of the hypothesis was provided by an experiment of Rutherford and Royds in 1909 (Fig. 2.3). In this work α-particles from a large quantity of the active gas radon which had been compressed into a tube A, entered an evacuated space through the walls of the tube. After a period of a few days, the accumulated gases in the space were compressed into the capillary V and by exciting the contents of the capillary electrically, the characteristic optical spectrum of helium was observed. No such spectrum was seen in test experiments in which ordinary helium was allowed to stand for a few days in the tube A. It follows that on neutralization by collection of electrons (from residual gases in the apparatus) α-particles at rest give rise to helium gas.

This striking experiment helped to establish a remarkable conclusion that had begun to emerge, most clearly in the mind of Rutherford, from the study of radioactivity. This was that the atoms of matter were no longer to be regarded as immutable or indestructible, since some of them, notably uranium,

† This figure is now known more accurately (Table 2.1).

radium, radon and thorium could spontaneously emit atoms of helium. This idea now seems neither exciting nor improbable; in the early years of the century it constituted a scientific revolution. The development of the *transformation theory* is described in Sect. 2.4.1.

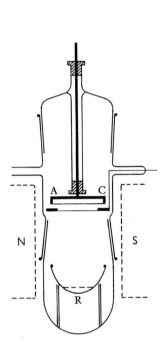

Fig. 2.2 Determination of charge of the α-particle (Rutherford and Geiger, *Proc. roy. Soc.* A, **81**, 162, 1908). A source of RaC was placed in the vessel R and the charge communicated to the screened plate AC by the α-particles from RaC′ was found from electrometer readings. The magnetic field was necessary in order to prevent the loss of secondary electrons from the plate AC.

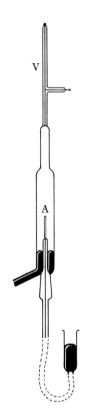

Fig. 2.3 Identification of the neutralized α-particle with the helium atom (Rutherford and Royds, *Phil. Mag.*, **17**, 281, 1909).

2.2.2 The beta rays

The electric and magnetic deflections obtained with β-rays corresponded both in direction and rough magnitude with those observed with the cathode rays in a discharge tube. The β-particles were thus negatively charged; their specific charge e_β/m_β, however, seemed to decrease with increasing velocity and the velocities themselves were distributed over a range of values up to a certain limit. For low velocities the value of e_β/m_β was 1.76×10^{11} C kg^{-1} which is almost exactly the value found for electrons. The decrease in the observed e/m for higher velocities is fully accounted for by the special theory of relativity, according to which the mass of a particle moving with velocity v is

$$m = \frac{m_0}{\sqrt{1 - v^2/c^2}} \tag{2.3}$$

where m_0 is the rest mass, effective only for very slow particles.†

The β-rays could thus be identified with ordinary electrons, endowed, because of their small mass, with velocities approaching that of light. They were distinguished from the α-particles not only by their specific charge, but also by their inhomogeneity in energy. This identification of the β-*particles* did not clarify the problems of their origin and energy distribution, for which a satisfactory solution had to wait for some thirty years (ch. 16).

Magnetic deflection experiments, as will be shown in more detail in Chapter 7, yield accurate momentum distributions of the β-particles from a given parent body. Detailed examination of this distribution shows that it contains a number of narrow, apparently homogeneous lines superimposed upon a general continuous distribution with an upper limit (Fig. 2.4). These lines (internal conversion lines) are associated with the emission of γ-radiation and not with the primary β-ray process; they do not appear in general when a source which does not emit γ-rays is studied. The substance RaE is of this latter type and its continuous spectrum of β-particles has been closely investigated. The decay process for this body is now known to be

$$\text{RaE} \xrightarrow{\ \beta\ } \text{RaF} \xrightarrow{\ \alpha\ } \text{Stable lead}$$

and by counting of individual α- and β-particles it was established that the continuous β-spectrum contained just one particle per transformation.

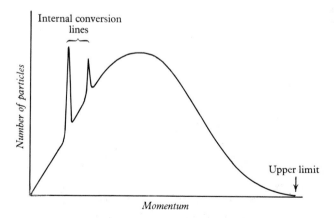

Fig. 2.4 Schematic representation of the momentum distribution of β-rays from a radioactive source.

† For $v = \frac{1}{2}c$, $m = 1\cdot15\,m_0$.

2.2.3 *The gamma rays*

The most penetrating component of the natural radiations is not deflected by magnetic fields, and interacts with matter as do X-rays. The γ-rays were therefore identified as electromagnetic radiation, and this conclusion was supported by the success of Rutherford and Andrade and of Frilley in diffracting these rays by crystals. The diffraction experiments indicated that the γ-rays probably had a line spectrum, and for some time it was considered that they arose as a result of the passage of β-particles, with which they were often associated, through the atomic structure. The electromagnetic interactions (photoelectric effect, Compton effect, ch. 5) were applied to determine the energy of the γ-rays and these energies were found to be commensurate with those observed in α- and β-emission. Relations between the energies of the γ-radiations themselves were extensively investigated by Ellis and as the accuracy of these experiments increased little doubt was left that the γ-rays formed the electromagnetic spectrum of a radiating system (now known to be the product nucleus) excited in the primary radioactive process of α- or β-emission.

2.3 General properties of the radiations

2.3.1 *Constancy of radioactivity with time*

Spontaneous emission of α-, β- and γ-rays was proved by Becquerel to be an atomic rather than a molecular phenomenon. The decay rate is not affected†
by the chemical environment of the active atom and has not been influenced by variations of temperature and pressure.

2.3.2 *Production of helium by radioactive materials*

The identification of the α-particle as a charged helium atom explained the occurrence of this gas in radioactive minerals. The counting experiments of Rutherford and Geiger permitted an accurate prediction of the rate of production of helium from radium (see Table 2.1) but the first measurements (Ramsey and Soddy 1903, Dewar 1908) of the evolution of helium from radium salts gave results differing slightly from the calculated value. The discrepancy was removed in 1911 by the more accurate work of Boltwood and Rutherford.

2.3.3 *Generation of heat in radioactive materials*

The amount of energy liberated in a single radioactive process is given by the kinetic energy of the emitted particle, together with that of the recoiling residual fragment and these energies (≈ 5 MeV) are very large compared with

† This is not quite true. A special case will be discussed in Sect. **13.2**.

the energies involved in the individual chemical reactions between molecules ($\approx$ eV). Even before the nuclear hypothesis, this aspect of radioactivity had convinced Rutherford that some completely new mechanism, not envisaged in early theories of atomic structure, was necessary to explain the radioactive process.

The heating effect due to the α-rays from 1 g of radium in equilibrium with its products was calculated from the counting experiment of Rutherford and Geiger, using the known mass and velocity of the α-rays. The result obtained (Table 2.1) was about 10 per cent less than the value observed by Rutherford and Robinson (1912), owing to heat developed by absorption of β- and γ-rays.

The actual amount of heat developed by 1 g of radium and its products per second (≈ 8 J) may seem small in comparison with the 34 kJ released in the oxidation of 1 g of carbon to carbon dioxide, but it must be remembered that the emission from radium continues with a half-life of 1600 years. Altogether, before its activity has reached negligible proportions the 1 g of radium will have emitted $1\cdot34 \times 10^{10}$ J. Despite this large value the energy released by radioactive materials has, as emphasized by Rutherford, little practical application because the amount of natural radioactive material which can be concentrated safely, allowing for biological hazards, is really very small. The release of nuclear energy on a large scale involves stimulation of a particular process (fission) in a large mass of material in a space of relatively small dimensions; it was first achieved nearly fifty years after the discovery of radioactivity. The total mass of radioactive material in the earth is nevertheless very large, and the total emission of heat from this constituent of the geological crust must be taken into account in estimates of the age of the earth based on thermal data.

2.3.4 The decay law

Soon after commencing his experiments in Montreal on the properties of thorium, Rutherford found that an active gas, now known to be an emanation isotope, could be swept away from thorium oxide into an ionization chamber

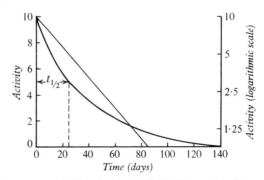

Fig. 2.5 Exponential decay curve (UX_1), showing half-life. The straight line is a logarithmic plot of the activity (Ref. 2.1).

by an air current. When the air current was stopped the conductivity of the chamber diminished with time according to a geometrical law, falling to half value in about 1 min (a time now known to be 51·5 s). This was the first observation of the *radioactive decay law*; all the other radioactive materials investigated until then had been long lived or in equilibrium with long-lived parents.

Further experiments of this type, on other radioactive bodies which in due course became available, showed that the decay law for a given mass of material was accurately exponential. Figure 2.5 illustrates the decay of activity observed in a typical case. If the activity is assumed to be due to an atom by atom process, it is of course proportional to the rate of decrease of the number of the atoms N_t present at time t, i.e.

$$\text{activity} = \frac{dN_t}{dt}$$

The observed exponential decay then requires further that

$$\text{activity} = \frac{dN_t}{dt} = -\lambda N_t \tag{2.4}$$

$$N_t = N_0 e^{-\lambda t} \tag{2.5}$$

where N_0 is the number of active atoms present at the beginning of the observations, $(t=0)$ and λ is the *radioactive decay constant*. The *mean life* of the atoms is

$$\tau = \frac{\int_{N_0}^{0} t \, dN_t}{\int_{N_0}^{0} dN_t} = \frac{\int_{0}^{\infty} t \frac{dN_t}{dt} \, dt}{\int_{0}^{\infty} \frac{dN_t}{dt} \, dt} = \frac{1}{\lambda} \tag{2.6}$$

The interval $t_{1/2}$ during which half the atoms disappear by decay (*half-value period* or *half-life*) is given by

$$N_0/2 = N_0 e^{-\lambda t_{1/2}}$$

from which

$$t_{1/2} = \frac{\log_e 2}{\lambda} = 0.693\tau \tag{2.7}$$

The exponential law of decay is characteristic of all radioactive processes and has been checked in vast numbers of experiments over extensive periods. It describes the disappearance by decay of short-lived unstable particles and excited states as well as of long-lived radioactive elements. A law of such general application must have some general basis, and it is simple to show that it may be deduced from the assumption that for a given atom the probability p of decay in a time Δ is independent of the previous life of the atom.

For if this is so then, for small enough Δ

$$p = c\Delta$$

where c is a constant and the probability that the atom shall not have disappeared in a time $t = k\Delta$ is

$$\begin{aligned} q &= (1 - c\Delta)^k \\ &= (1 - c\Delta)^{t/\Delta} \\ &= [(1 - c\Delta)^{-(1/c\Delta)}]^{-ct} \end{aligned}$$

and letting Δ become zero, while $k\Delta$ remains finite

$$q = e^{-ct}$$

and the decay law (2.5) follows, putting $c = \lambda$.

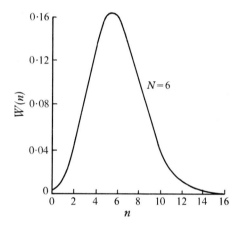

Fig. 2.6 Poisson distribution for $N=6$ (Ref. 2.4).

The random nature of radioactive emission was demonstrated by observation of fluctuations of the number of α-particles emitted by a given source in a given time interval. If N is the number of particles expected on the average in the interval, the probability of observing n particles in this interval is given by the Poisson distribution for random events:

$$W(n) = \frac{N^n e^{-N}}{n!} \tag{2.8}$$

This distribution is shown for $N=6$ in Fig. 2.6; it was checked by Rutherford and Geiger by recording on tape the time of appearance of individual α-particles. These and many similar experiments have abundantly verified the conclusion that α-particles are emitted at random in time.

2.4 Origin of the alpha- and beta-rays

2.4.1 *The transformation theory*

Our understanding of the nature of radioactive transformations dates generally from the arrival of Rutherford in Montreal in 1898 and especially from his subsequent association with Soddy. The years 1898–1911 saw a great increase in empirical knowledge of radioactive bodies, and the nature of the individual radiations as well as the general properties of radioactivity were established as described in Sects. **2.2** and **2.3**. In order to account for these facts, foremost among which was the appearance of new chemical substances, Rutherford and Soddy (1903)† proposed the *transformation theory* according to which atoms of a radioactive substance disintegrate spontaneously, with the emission of either an α- or a β-particle, and with the formation of a new chemical atom. This new atom may itself disintegrate similarly in succession and in this way a *radioactive series* of atoms, all genetically linked, arises. The transformations at the beginning of the thorium series, for instance, are

$$\text{Thorium} \xrightarrow{\alpha} \text{Mesothorium I} \xrightarrow{\beta} \text{Mesothorium II}$$
$$\xrightarrow{\beta} \text{Radiothorium} \xrightarrow{\alpha} \text{Thorium X} \xrightarrow{\alpha} \quad (2.9)$$

Each of these products has a definite chemical behaviour, which can be used in extracting it from a mixture of active elements.

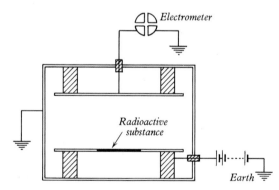

Fig. 2.7 Ionization chamber and electrometer used by Rutherford and Soddy in early studies of radioactivity.

The combination of chemical and physical techniques used in characterizing a series of radioactive transformations is well illustrated by the early work of Rutherford and Soddy‡ on thorium. In work with this element they found that the body described as thorium X (ThX) could be extracted by precipitating the thorium as hydroxide with ammonia and evaporating the filtrate

† E. Rutherford and F. Soddy, *Phil. Mag.*, **5**, 576, 1903.
‡ E. Rutherford and F. Soddy, *Phil. Mag.*, **4**, 370, 1902.

to dryness. When this operation was complete it was noted that (*a*) the thorium itself was reduced in activity temporarily, although it recovered its original strength in time, and (*b*) that the ThX activity in the filtrate decayed exponentially with a half-life of about four days. Figure 2.7 shows the ionization chamber used, and Fig. 2.8 the growth and decay curves obtained in this experiment. Such curves are now familiar in many other cases, and are wholly accounted for by assuming that thorium produces a chemically *dissimilar* body ThX as a result of its own decay, and that ThX decays, when separated, with its own characteristic half-life. The recovery of the initial thorium activity is due to creation of further ThX by decay until an equilibrium value is reached at which the production and decay rates of ThX are equal. That this equilibrium value is apparently constant indicates only that the half-life of the

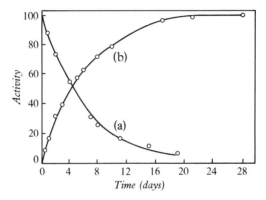

Fig. 2.8 (a) Decay of ThX extracted from thorium. (b) Recovery of ThX activity in a thorium sample after a chemical separation of ThX (Rutherford and Soddy).

original thorium is very long. The chemical and physical dissimilarity of Th and ThX (now known to be an isotope of radium) and their obvious genetic relationship proved that one type of atom must be decaying into another.

2.4.2 *The discovery of isotopes : the displacement laws*

An important discovery, closely linked with the transformation theory, was made by Soddy and others over the years 1906–11. It was that some bodies could be formed in radioactive decay which were physically distinct on the basis of their radioactive properties but were chemically *identical* since it was impossible to separate them by available chemical techniques. There are now many such examples, notably thorium and radiothorium, radium and mesothorium, uranium I and uranium II. These bodies were finally postulated by Soddy in 1911 to be *isotopes* of each other, that is to say they occupied the same place in the chemical system of the elements, but were nevertheless different physically. The property in which they differed became clear as soon as the nature of the radioactive radiations was established, and the *displacement laws* of Soddy, Russell and Fajans had been formulated (1913). These laws summarize a great deal of experience in the statements that

(i) the loss of an α-particle displaces an element two places to the left in the periodic table and lowers its mass by 4 units,

(ii) the loss of a β-particle displaces an element one place to the right in the periodic table but does not essentially alter the atomic mass.

It follows that in a sequence such as (2.9) there may be atoms of the same chemical nature but of different mass, e.g. thorium ($A = 232$) and radio-thorium ($A = 228$).

These laws showed clearly that the nineteenth-century concept of the immutability of atoms could not be maintained, as indeed had been already suggested strongly by the generation of helium in radioactive materials. Within a few years the *artificial* transmutation of atoms was to be demonstrated in Rutherford's laboratory at Manchester (1917–19), while the objective demonstration of the existence of the isotopes of neon by J. J. Thomson (1913) and Aston (1919) in a sense only supplemented what was already convincing evidence.

2.4.3 The radioactive series

It is not proposed to give in this book an account of the intricate procedures by which the physical and chemical nature of all the naturally occurring radioactive bodies were established. The unambiguous characterization of one of the natural series would be a not inconsiderable task even today with the highly developed resources of a radiochemical laboratory and the full support of nuclear theory. That a clear understanding of the basic phenomena could have been reached in only a few years after the discovery of radioactivity is a tribute to the genius of Rutherford and his fellow workers.

In general it was assumed that whenever a new half-life was discovered, a new radioelement might be involved and attempts were then made to separate it chemically and to identify it, if possible, by an atomic weight determination. Observation of genetic relationships between activities after chemical separations often indicated sequences of transmutations, and the application of the displacement laws permitted the atomic weights of new bodies to be deduced from those of known parents. The detection of a radioactive inert gas (emanation) and the identification of the stable end products of a decay chain provided important fixed points in the series.

In this way the radioactive bodies were identified, and their grouping into the three natural radioactive families was established. These families are the thorium series, the uranium–radium series and the uranium–actinium series. In addition another family, the neptunium series, has now been prepared artificially. In each series α- and β-emission are the main decay processes and the resulting sequences of elements and isotopes are generally similar between the families, but there are differences in detail. Since mass number changes are due only to the emission of α-particles the members of each series have masses which differ by multiples of four units. The naturally occurring series originate from nuclei with lives long compared with that of the earth. Neptunium has a relatively short life, but it is the longest lived member of its series. Each series terminates when the decay process results in the formation of a stable nucleus, which becomes the end product; the parents and end products of the four series are shown in Table 2.2.

TABLE 2.2 Parents and end products of radioactive series

Name of Series	Mass number	Parent	Half-life	End product
Thorium	$4n$	^{232}Th	$1 \cdot 4 \times 10^{10}$ yr	^{208}Pb
Neptunium	$4n+1$	^{237}Np	$2 \cdot 2 \times 10^{6}$ yr	^{209}Bi
Uranium–radium	$4n+2$	^{238}U	$4 \cdot 5 \times 10^{9}$ yr	^{206}Pb
Uranium–actinium	$4n+3$	^{235}U	$7 \cdot 2 \times 10^{8}$ yr	^{207}Pb

The naturally occurring series thus end in lead isotopes; this, of course, accounts for the presence of lead as well as helium in radioactive ores and for the difference between the isotopic constitution of this 'radiogenic' lead and the ordinary element.

The three naturally occurring families all include an emanation or inert gas (radon, thoron, actinon) as noticed by Rutherford (Sect. 2.3.4). This emanation engenders by recoil against α-decay an active deposit of products, which came to be known as the A, B, C members of each family, on surfaces to which it is exposed. The active deposit (formerly known as 'induced activity') is increased in the presence of an electric field, showing that the recoil 'A' particles are ionized.

In order to simplify the presentation of data on the radioactive series, the results of the nuclear theory will be assumed. The radioactive bodies can then be displayed on diagrams showing number of neutrons and number of protons in each nuclide as in Fig. 2.9. In this figure certain subsidiary decay chains have been omitted in the interests of clarity.

Most of the radioactive elements disintegrate in a definite manner with the ejection of an α- or β-particle. It will be seen, however, from Fig. 2.9 that a few undergo dual decay or *branching*. One of the best-known cases of this phenomenon is found in the thorium active deposit, in which ThB decays as follows

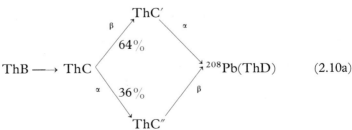

$$(2.10a)$$

The final product reached by each branch is the same and the energy release is also the same.

In the case of the body UX_2 there is a competition between β- and γ-decay, as shown below

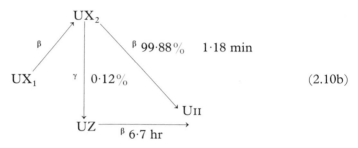

(2.10b)

The nuclei UX_2 and UZ both result from the decay of UX_1 and both decay by β-emission, but with different half-lives, into the same element UII. They are therefore essentially the same body and the difference between their decay properties is now explained by supposing that UX_2 is an *isomer* of UZ, that is to say it is the UZ nucleus excited to a metastable level of long life, from which β-decay to the ground state of the end product UII takes place. Such isomeric states are now well known in many lighter nuclei and are associated with nuclear levels between which there is a large spin difference (Sect. **13.5**). This first example of nuclear isomerism was established by Hahn (1921).

2.4.4 *The age of minerals: naturally occurring activities not connected with uranium or thorium*

It will be seen from the decay chains shown in Fig. 2.9 that the decay of the parent body and its products is accompanied (*a*) by the evolution of helium from the emitted α-particles, and (*b*) by the formation of lead of a particular atomic weight for each of the naturally occurring series. If, for example, neither helium nor lead of this atomic weight ($A = 206$) was present when a uranium bearing mineral was formed, and if no alteration of decay rate or loss of products has occurred, then determination of either total helium or total 'radiogenic' lead in the mineral, as a ratio to uranium content, should permit the age of the mineral to be calculated. Thus if the decay of uranium is assumed to take place at a constant rate the age t of a mineral is given simply by the equation

$$\lambda_U N_U t = N_{Pb} \tag{2.11}$$

where N_U and N_{Pb} are the present number of atoms of ^{238}U and ^{206}Pb in the mineral. Substituting for λ_U we find

$$\text{Age of mineral} = \frac{\text{Weight of } ^{206}Pb}{\text{Weight of } ^{238}U} \times 7 \cdot 5 \times 10^9 \text{ yr} \tag{2.12}$$

Many estimates of the age of geological strata have been made in this or in a similar way, using refined radiochemical and mass spectrometric techniques, and introducing corrections for the decay of the parent uranium activity. Results for pitchblende range from 640 to 1380×10^6 years.

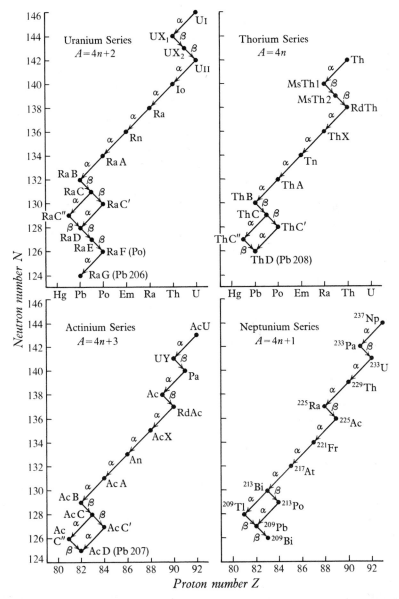

Fig. 2.9 The main decay chains of the four radioactive series. The $4n$ (thorium), $4n+2$ (uranium–radium), and $4n+3$ (uranium–actinium) series are found in nature but the $4n+1$ (neptunium) series must be prepared artificially. The historical names are shown on the diagram and the accepted chemical symbols on the axis (Ref. 2.4).

Radioactive data can also indicate the much longer time which has elapsed since the *formation of the atoms* as distinct from the consolidation of the minerals of the earth. A simple, but probably unreliable, estimate may be based on the assumption that the two isotopes ^{238}U and ^{235}U were created in comparable amounts; their present relative abundance (140/1) and their known half-lives then indicate a passage of 5×10^9 years since formation.

More recently methods of dating based on naturally occurring activities other than those of the main series have been developed. These will not be discussed here, since their interest is primarily geological, but a list of the active elements concerned is given below, Table 2.3. These are all low-energy β-emitters or electron capture (E.C.) bodies† with the exception of samarium and neodymium and their half-lives have been determined by careful observation of the activity of a known mass of material. In many cases detecting instruments of exceptionally low background were necessary in order to record the extremely weak activities.

TABLE 2.3 Naturally occurring active materials other than main families

Radioactive isotope	Half-life (years)	Radiation	Observed particle energy (MeV)
^{40}K	$1\cdot3 \times 10^9$	β-, (E.C.)	$1\cdot32$
^{50}V	5×10^{15}	β-, E.C.	$1\cdot19, 2\cdot39$
^{87}Rb	5×10^{10}	β-	$0\cdot273$
^{115}In	6×10^{14}	β-	$0\cdot6$
^{138}La	1×10^{11}	β-, (E.C.)	$0\cdot21$
^{144}Nd	3×10^{15}	α	$1\cdot8$
^{147}Sm	$1\cdot3 \times 10^{11}$	α	$2\cdot2$
^{176}Lu	$4\cdot5 \times 10^{10}$	β-, E.C.	$0\cdot43$
^{187}Re	4×10^{12}	β-	$0\cdot043$

2.4.5 *The transformation laws for successive changes*

The basic calculations of the rate of growth of radioactive bodies from radioactive parents and the conditions for radioactive equilibrium, were developed by Rutherford and his collaborators. They were crucial for the quantitative evaluation of the decay chains presented in Sect. 2.4.3. The general problem is that of a series of bodies transforming in sequence

$$A \longrightarrow B \longrightarrow C \longrightarrow D \ldots \qquad (2.13)$$

with different decay constants λ, and it is required to find the amount of a given product at any time after stated initial conditions. We consider for

† For electron capture see Chapter 16.

simplicity only the case in which the active element B grows from its parent A. The equations governing this process are:

$$\left.\begin{array}{ll} \text{for the decay of A} & \dfrac{dN_a}{dt} = -\lambda_a N_a \\[2ex] \text{for the growth and} & \dfrac{dN_b}{dt} = \lambda_a N_a - \lambda_b N_b \\ \text{decay of B} & \end{array}\right\} \qquad (2.14)$$

where N_a and N_b are the number of atoms of A and B present at time t. Eq. (2.14) may be solved to give N_a and N_b if the quantities of these two atoms present at time $t=0$ ($N_a(0)$ and $N_b(0)$) are known. For most purposes however it is the activity, or disintegration rate, which is experimentally required, and for these quantities we find

$$\left.\begin{array}{l} \text{Activity of A} = \lambda_a N_a = \lambda_a N_a(0)e^{-\lambda_a t} \\[2ex] \text{Activity of B} = \lambda_b N_b = \lambda_b N_b(0)e^{-\lambda_b t} + \dfrac{\lambda_a \lambda_b}{\lambda_b - \lambda_a} N_a(0)[e^{-\lambda_a t} - e^{-\lambda_b t}] \end{array}\right\} \quad (2.15)$$

If initially B is not present, $N_b(0)=0$ and then at any time

$$\frac{\text{activity of B}}{\text{activity of A}} = \frac{\lambda_b}{\lambda_b - \lambda_a}[1 - e^{-(\lambda_b - \lambda_a)t}] \qquad (2.16)$$

Two cases now arise:

(a) $\lambda_b > \lambda_a$ (daughter shorter lived than parent).

The activity ratio (2.16) tends to the constant value $\lambda_b/(\lambda_b - \lambda_a)$ and both activities decay ultimately with the half-life of the parent. When the ratio $\lambda_b/(\lambda_b - \lambda_a)$ has been established, a state of *transient equilibrium* exists. The variation with time of the activity of radiothorium ($\lambda_b = 1\cdot 2 \times 10^{-8}$ s^{-1}) growing via the short-lived MsTh2 from MsTh1 ($\lambda_a = 3\cdot 35 \times 10^{-9}$ s^{-1}) is shown in Fig. 2.10. The product activity reaches a maximum at a time given by $dN_b/dt = 0$ or, from (2.15) with $N_b(0)=0$

$$\lambda_a e^{-\lambda_a t} = \lambda_b e^{-\lambda_b t} \qquad (2.17)$$

and at this time, and at this time only

$$\lambda_b N_b = \lambda_a N_a \qquad (2.18)$$

In the regime of transient equilibrium, at later times, the activity of B is always greater than that of A.

If however $\lambda_b \gg \lambda_a$ then the ratio of activities tends to unity and we have the case of *secular equilibrium* in which B grows at a rate determined by its own decay constant until it reaches an activity equal to that of its parent. This case has already been illustrated (Fig. 2.8) in the production of ThX from radiothorium. In this case $\lambda_b = 2\cdot 2 \times 10^{-6}$ s^{-1} and $\lambda_a = 1\cdot 2 \times 10^{-8}$ s^{-1}. Many other examples, including the growth of radon from radium, could be cited.

In the case of a long-lived parent such as uranium or thorium it is clear that when secular equilibrium has been established the quantities of successive products present are given by the equations

$$\lambda_a N_a(0) = \lambda_b N_b = \lambda_c N_c = \cdots = \text{constant} \qquad (2.19)$$

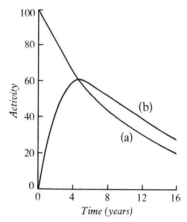

Fig. 2.10 (a) Decay of MsTh1. (b) Growth of RdTh in equilibrium with MsTh1. A state of transient equilibrium with an activity ratio (1·39) is established after about eight years. At 4·8 years the RdTh has a maximum activity equal to that of its parent (Evans, *The Atomic Nucleus*).

The activities are constant throughout and the relative amounts of different products are inversely proportional to their decay constants. The stable end product of the series must be excluded from this statement since it continually accumulates. The same equation (2.19) governs the equilibrium quantity N_b of an active material produced in an accelerator or reactor at a constant rate p, i.e.

$$\lambda_b N_b = p \qquad (2.20)$$

(b) $\lambda_b < \lambda_a$ (daughter longer lived than parent).

The activity ratio (2.16) increases continuously with time and there is no equilibrium in any significant sense. Transient equilibrium does indeed exist at the time given by (2.17) and the parent and daughter activities are then instantaneously equal.

The formulae used in this section have been elaborated by Bateman to deal with several successive decay products.

2.5 The scattering of α-particles by matter and the hypothesis of the nucleus

2.5.1 Large-angle scattering

During his work on the deflection of α-rays in a magnetic field (1906) Rutherford found that the presence of a small amount of air in the vacuum apparatus affected the path of the particles. The magnitude of this effect, in comparison with the similar effect that could be produced by externally applied fields, indicated that the atom must be 'the seat of very intense electrical forces'. With his genius for wresting the maximum information from the smallest effect, Rutherford saw in these deflections a possible method of studying the detailed structure of the atom. As soon as quantitative methods of counting α-rays had been established, as a result of his work with Geiger, he started the investigation of this structure, using α- and β-particles as probes, which was to lead to the nuclear hypothesis.

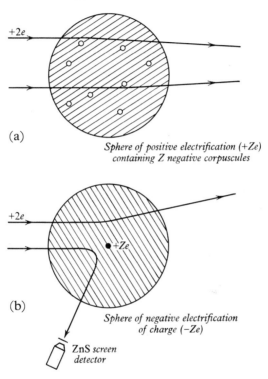

(a)

Sphere of positive electrification (+Ze) containing Z negative corpuscles

(b)

Sphere of negative electrification of charge (−Ze)

ZnS screen detector

Fig. 2.11 (a) The scattering of α-particles by a Thomson atom. (b) The scattering of α-particles by a nuclear atom.

At the time of the first experiment on the scattering of α-particles during their passage through thin foils of metal, the neutral atom was supposed, following the ideas of J. J. Thomson, to consist of a number of negatively charged electrons, accompanied by an equal quantity of positive electricity uniformly distributed throughout a sphere. The scattering of charged particles by matter was supposed to be a multiple effect, due to the statistical superposition of the results of a large number of atomic encounters, each with a small angle of deflection. Such a mechanism predicts that the angle of deflection of charged particles in passing through a thin foil should increase as

37

$\sqrt{t}$ where t is the thickness of the foil (Fig. 2.11(a)). This conclusion was supported by the experiments of Crowther (1910) on β-particle scattering, which also revealed however that some β-particles were deviated through very large angles (diffuse reflection). This was not considered surprising in view of the small mass of the β-particle. At about the same time Geiger, who was examining the deflection of α-particles by a gold foil using the scintillation method of detection found that the α-particles were also deviated through small angles in large numbers. This was in qualitative agreement with prediction based on the Thomson theory of the atom. Rutherford suggested that Marsden, a student working with Geiger, should look for large-angle scattering. The expectation was that such an effect would be very small, because of the large mass of the α-particle, if the main collisions were with the electrons of a diffuse charge distribution. The surprising result was found that a few α-particles were scattered through large angles, even greater than 90°, from even the thinnest foils that could be used. In Rutherford's words, long after this observation, 'It was quite the most incredible event that has ever happened to me in my life. It was almost as incredible as if you had fired a 15-inch shell at a piece of tissue paper and it came back and hit you.'

This 'incredible event' dominated Rutherford's thinking in the winter of 1910–11. The sequel is related in a letter from Geiger to Chadwick: 'One day, obviously in the best of spirits, he came into my room and told me that he now knew what the atom looked like and how the large deflections were to be understood. On the very same day I began an experiment to test the relation expected between the number of particles and the angle of scattering.' Rutherford's picture of the atom was simply that the electrons would fill a sphere of atomic dimensions (radius $\approx 10^{-10}$ m) but that their charge would be neutralized by a central positive charge on a *nucleus* of much smaller extent (radius $\approx 10^{-14}$ m). If the atom contained Z electrons, the nuclear charge would be $+Ze$ and since the electronic mass is small compared with that of an atom, most of the atomic mass, as well as the positive charge, would be concentrated in the nucleus. It was then shown by a straightforward and now celebrated calculation† that the Coulomb repulsion between the nucleus of a heavy atom and an incident α-particle would cause the α-particle to describe a hyperbolic orbit (Fig. 2.11(b)). Furthermore the sharp rise in the repulsive force, increasing as the inverse square of distance, as the α-particle approached the minute nucleus, would provide the strong fields necessary to cause the large-angle scattering.

Rutherford's calculations showed that the probability of deflection of an α-particle through an angle as a result of a single encounter with a nucleus was always greater than for multiple scattering in the electron distribution. He also deduced the statistical distribution of the singly scattered particles as a function of angle of scattering; the specific conclusions were that the number of α-particles detected at a certain point (Fig. 2.11(b)) after passage

† A derivation of the Rutherford scattering law is given in Sect. 5.1.4. See E. Rutherford, *Phil. Mag.*, **21**, 669, 1911.

through a thin scatterer of a heavy element would be proportional to

(a) cosec $^4\theta/2$; where θ was the angle of scattering,
(b) the thickness of the scatterer,
(c) the square of the central nuclear charge,
(d) the inverse square of the incident energy.

Each point was verified precisely in the careful experiments of Geiger and Marsden[†] over the years 1911–13. Darwin showed that any law of force other than that of the inverse square would be inconsistent with the data and the experiments indicated that for gold this law of force held down to less than 3×10^{-14} m. The validity of the hypothesis of a nucleus of minute dimensions, endowed with all the positive charge and most of the mass of the atom, seemed beyond dispute.

2.5.2 The nuclear atom

The nuclear atom proposed by Rutherford in 1911 transformed the appearance of the whole of atomic theory, and had profound implications throughout physics and chemistry. In the first place the concept effected an immediate separation between *nuclear* and *atomic* properties. The latter were clearly to be associated with the electrons, whose motions and mutual interactions were responsible for chemical binding, optical and X-ray spectra, and the macroscopic, observable properties of matter in bulk. The nucleus, on the other hand, was relatively remote and inaccessible to most influences except bombardment by fast particles; nuclear properties should be unaffected by the ordinary chemical and physical changes of their atoms and this was indeed a well-known characteristic of radioactivity. The energy changes in nuclear transformations were also clearly enormously larger (Sect. 2.3.3) than those involved in atomic rearrangements. The link between these two types of property was provided most convincingly by Moseley (1913), working in Rutherford's laboratory. He showed that the frequencies of the lines of the characteristic X-ray spectrum of the elements changed in a well-marked way from one element to another. The frequencies depended on the *atomic number* of the element concerned, defined as the ordinal number of the elements arranged in a sequence of increasing atomic weights. These results received an interpretation in the work of Bohr (1913) on the theory of spectra, for in this great achievement the nucleus played a fundamental role. The atomic number used by Moseley was in fact seen to be exactly the nuclear charge expressed in units of the electronic charge, or alternatively the number of electrons in the atom. The atomic number can thus in principle be determined from experiments of the type made by Geiger and Marsden and when these were sufficiently refined (Chadwick, 1920) it was found that the central nuclear charge number was indeed equal to Moseley's atomic number and to

[†] H. Geiger and E. Marsden, *Phil. Mag.*, **25**, 604, 1913.

the number of electrons per atom, as determined by X-ray scattering experiments. It is thus obvious that the properties of the Rutherford–Bohr atom are of special significance for the nuclear physicist.

The hypothesis of the nucleus immediately clarified many phenomena of radioactivity. The α-particle was recognized to be the nucleus of the helium atom and the displacement laws could be re-stated in terms of the alteration of nuclear charge due to α- or β-emission. The nuclear charge of an atom determines its electronic configuration and hence its chemical properties. Isotopes must have the same nuclear charge but a different mass. The emission of α- and β-radiation results from the energetic instability of one nucleus with respect to another; and the γ-radiation is essentially the electromagnetic spectrum of a nucleus (Fig. 2.12) having much the same relation to its structure as have optical radiations to that of the atom (cf. Sect. 2.2.3). The accumulation of helium in α-particle transmutations is an obvious consequence of the theory.

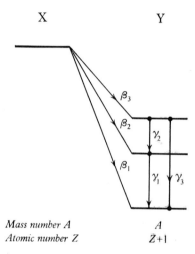

Fig. 2.12 Origin of radioactive γ-rays. A nucleus X is energetically unstable with respect to nucleus Y and decays with emission of an α- or β-particle. The nucleus Y is sometimes left in an excited state, from which γ-radiation arises. That the γ-radiation arises in Y rather than X is known from the energies of the 'internal conversion' electrons first studied systematically by Ellis (*Proc. roy. Soc.*, **99**, 261, 1921).

It will be surmised that Rutherford, having postulated the nucleus and defined its main properties of charge and mass was eager to enquire into its structure. The small size and strong force field of the nucleus clearly made such an investigation difficult from the outset. Despite this, the same technique which revealed the existence of the nucleus seemed capable of yielding further information, and after the First World War, the bombardment of a variety of elements with α-particles continued intensively under Rutherford in the Cavendish Laboratory. The successful outcome of this work in the production of artificial transmutations by accelerated ions and the enormous consequential expansion of nuclear physics forms the basic subject matter of Parts B, C, D and E of this book.

References

1. CHADWICK, J. *Radioactivity and Radioactive Substances*, revised J. Rotblat, Pitman, 1953.
2. CHADWICK, J. Rutherford Memorial Lecture, *Proc. roy. Soc.*, A, **224**, 435, 1954.
3. RUTHERFORD, E., CHADWICK, J. and ELLIS, C. D. *Radiations from Radioactive Substances*, Cambridge University Press, 1930.
4. *Experimental Nuclear Physics*, Vol. III, ed. Segrè, Wiley, 1959.
5. CURIE, EVE. *Madame Curie*, Heinemann, 1938.
6. EVE, A. S. *Rutherford*, Cambridge, 1939.
7. FEATHER, N. 'A History of Neutrons and Nuclei'. *Contemp. Phys.*, **1**, 191, 1960.
8. FEATHER, N. 'Radioactivity', *Chambers's Encyclopaedia*, 1963.

3 Atomic theory and radiation

Nuclear physics owes much to the development of atomic theory and to the extensive and precise measurements by spectroscopists of atomic transition frequencies. The nuclear model of the atom, as pointed out in Sect. 2.5.2, was quickly applied by Niels Bohr to elucidate the empirical regularities of the spectrum of atomic hydrogen and the existence round the nucleus of an electronic charge distribution exerts an important influence on many intrinsically nuclear phenomena. Apart from this, the interpretation of atomic structure offers a crucial test of the validity of the methods of wave mechanics. The interaction between the nucleus and the surrounding Z electrons is governed by the Coulomb law of force and to a very good approximation for many purposes the nucleus may be assumed to be a structureless point charge. The interactions between the individual electrons of the atom are also of the Coulomb type and all forces existing within the atom, as distinct from within the nucleus, are therefore known. The problem of predicting the energy levels of a simple atom is therefore in principle soluble. The methods of solution provided by the new mechanics are used continually in discussions of nuclear phenomena and are discussed in many books (Ref. 3.1). In this chapter only the briefest mention will be made of the main results, but, because interaction with radiation is a property of both atoms and nuclei, and has a direct connection with classical electromagnetic theory, this topic is treated at some length.

3.1 The Schrödinger equation and its solutions (Ref. 3.1)

According to wave mechanics, the motion of a particle in a field of force is to be obtained by solution of the Schrödinger equation:

$$\hat{H}\psi = i\hbar \frac{\partial \psi}{\partial t} \tag{3.1}$$

In this equation $\psi(x, y, z, t)$ is the wave function, whose absolute square $|\psi|^2$ yields the probability of observing the particle at the point (x, y, z) and at time t. The Hamiltonian operator $\hat{H}$ is constructed from the classical expression for the total energy E of the particle by replacing the linear momentum by the proper quantum mechanical operator, e.g.

$$p_x \rightarrow \hat{p}_x = -i\hbar \frac{\partial}{\partial x}$$

In many problems the field of force, as represented by an appropriate potential function $V(x, y, z)$, is constant in time and it is then possible to write ψ in the form

$$\psi(xyzt) = \psi(xyz)v(t)$$

Equation (3.1) can then be separated into two equations, in the independent variables (x, y, z) and t. Of these, the time-dependent part has a plane wave solution representing a particle with energy E. The spatial equation is

$$\hat{H}\psi = E\psi \tag{3.2}$$

and solutions of these equations yield the *eigenfunctions u_n* and the *eigenvalues E_n* of the problem. These might correspond with the energies permitted to a particle moving in a specific box, or potential well; the (negative) energies of the bound states of the hydrogen atom (or *eigenstates*) are just the eigenvalues of equation (3.2) for an electron moving in a Coulomb field. Other dynamical variables, e.g., angular momentum, may be expressed by operators and the observation of these quantities is limited by basic *commutation relations*. These indicate for instance that total energy, total angular momentum (squared) and a resolved part of the angular momentum of a system may be observed simultaneously. The eigenfunctions of an atomic or nuclear system contain vital information since by applying to them the necessary operators, observable properties of the system such as electric and magnetic moments, may be obtained.

The motion of an electron in an atom or of a nucleon in nucleus is a potential well problem. In many cases the potential well is spherically symmetric, i.e. specified by a single radial coordinate, and the eigenfunctions are then characterized by

 (i) a *principal quantum number n* related to the total number of nodes in the eigenfunction,
 (ii) an *azimuthal quantum number l* specifying the angular momentum† and
 (iii) a *magnetic quantum number m_l* specifying the orientation of the angular momentum with respect to a quantization axis.

In such potential fields, for which $V(x) = V(-x)$, the eigenfunctions have the important property of *parity* which is a quantum index based upon the behaviour of the wave function on inversion of all coordinates with respect to the origin. The parity of the orbital motion of a single particle is even (odd) if its orbital angular momentum quantum number is even (odd). In atomic spectroscopy, Laporte's rule for dipole radiation shows the importance of the parity concept. Elementary particles may also be ascribed an *intrinsic parity* and this will be discussed later (Chapters 19, 20, 21).

The potential well of the nucleus differs essentially from that of the atom because it is of *short range* in contrast with the long range of the Coulomb

---

† For $l = 0$ we speak of s-states, $l = 1$, p-states, $l = 2$ d-states etc.

potential. The difference is illustrated in Fig. 3.1 which shows that for the long range potential the spacing of energy levels tends to zero as the zero of energy (dissociation energy) is approached while for short range the level spacing is much more uniform and may not vary much as the dissociation

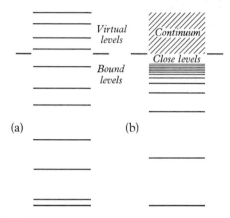

Fig. 3.1 Energy levels (schematic) for (a) short-range and (b) long-range potentials.

energy is passed. This is partly because of the reflection of the wave function at the abrupt potential discontinuity at the boundary of the well (Ref. 3.2, page 412). Levels which can dissociate with the emission of a particle are said to be *virtual* and those which cannot are *bound*.

3.2 Intrinsic spin

The necessity for the existence of a new, non-classical property of the electron was evident to Pauli in 1924 in order to explain the apparent existence of half-integral quantum numbers but it fell to Goudsmit and Uhlenbeck (1925) to suggest that this property was in fact an intrinsic angular momentum. This property is now embodied in Dirac's relativistic quantum mechanics which has proved a theory of extraordinary power and elegance. Intrinsic spin is also a property of nucleons, complex nuclei and of the wide variety of particles and states known in high energy physics.

Particles with intrinsic spin also exhibit magnetic behaviour, which is clearly seen in the Zeeman effect of atomic spectroscopy. In atoms, magnetic coupling between spin and orbital motion leads to a splitting of spectral terms which is expressed quantitatively in terms of the fine structure constant $\alpha = \mu_0 e^2 c / 2h$. The electron and muon, which can be described by Dirac's equations, exhibit a very definite ratio (gyromagnetic ratio) between the intrinsic spin and the associated magnetic moment. This is predictable with precision except for a minute anomalous part arising from coupling between the particles concerned and a field of virtual photons. In the case of nucleons and other massive particles a similar coupling, but to mesonic fields, is a major effect and a large part of the observable magnetic moment is anomalous.

3.3 Statistics (Ref. 3.3)

All the statistical concepts of atomic physics are required in a study of nuclear behaviour. The volume of phase space corresponding to a volume element $4\pi p^2 \, \mathrm{d}p$ in momentum space and a volume V in ordinary space is

$$4\pi p^2 \, \mathrm{d}p \, . \, V$$

and the number of allowed states of motion in this volume is

$$\frac{4\pi p^2 \, \mathrm{d}p \, . \, V}{h^3} \tag{3.3}$$

This formula is much used in discussing the breakup of nuclei or particles into simpler systems.

The occupation of these states by particles in a given problem, such as that of electrons in metals, is governed by the necessity for ensuring indistinguishability of identical particles in an assembly. By considering two such particles it may be shown that the wave function for the pair must be either totally symmetric or totally antisymmetric under interchange. There is no simple *a priori* way of knowing which of the particles of nature are described by a particular type of function but it may be settled by an appeal to experiment. Thus for electrons, the Pauli exclusion principle, as tested in the periodic system, indicates that the wavefunction for two particles with identical quantum numbers (including spin component) must vanish. It follows that assemblies of electrons are described by antisymmetric wave functions. Protons, neutrons, muons and generally all nuclei of half integral spin behave similarly. Such particles are known as *fermions* because their distribution between the allowed states obeys the *Fermi–Dirac statistics*; for spin $s = \frac{1}{2}$ only two particles $(2s + 1)$ can enter each spatial state.

Symmetric wave functions must be used to describe assemblies of photons, pions, deuterons, α-particles and nuclei with zero or integral spin. The Pauli principle does not limit the number of such identical particles which may enter a given state. The distribution is now determined by the *Einstein–Bose statistics* and the particles are known collectively as *bosons*. The type of statistics obeyed by a particle forming part of an assembly is a fundamental property with important applications in molecular phenomena and in nuclear models (ch. 9). It must be clearly distinguished from parity, which can be defined for single particles.

A significant example of quantum mechanical effects due to identity of particles is furnished by the helium atom. Two non-combining sets of spectral terms, usually described as belonging to ortho- and para-helium, arise.†

† The existence of two sets of terms is instructively shown by considering the classical motion of two coupled pendulums.

These are characterized by triplet $(s_1 + s_2 = 1)$ and singlet $(s_1 + s_2 = 0)$ spin systems respectively. Similarly, ortho- and para- forms of the hydrogen molecule exist because of the identity of the two protons. The same considerations are important in nuclear structure.

3.4 Molecules and superconductors

The explanation of the specific heats of gases and of their band spectra by quantum mechanics is an impressive example of the behaviour of the quantized oscillator and rotator. Since the early 1950s it has been realized that nuclei exhibit collective motions of a type which are susceptible to similar analysis, and this has developed into a major field of experimental and theoretical research in nuclear structure. In addition the special behaviour of homonuclear diatomic molecules has provided a powerful method of nuclear spin determination, as discussed in Chapter 4.

The development of the Bardeen–Cooper–Schrieffer theory of superconductivity of metals has also stimulated much development of nuclear theory. In the BCS theory correlations between electrons with equal and opposite momenta arise because of interaction with lattice phonons and relatively weak interactions can in this way profoundly affect the properties of a metal. In nuclear matter, correlations between nucleons with equal and opposite angular momenta can arise because of a 'pairing' force in the effective nucleon–nucleon interaction. When this analogy is pursued quantitatively, many of the features familiar in the theory of superconductivity, such as an 'energy gap', can be discerned in nuclear spectra (Ref. 12.11).

3.5 Semi-classical theory of radiation

The production, propagation and absorption of radiation are fundamental processes in atomic and nuclear physics. So far as propagation is concerned, the classical electromagnetic theory of Maxwell appears entirely adequate, but in the interaction of radiation with matter, in the processes which lead to the creation or disappearance of the radiation, it is necessary to introduce the concept of photons. A satisfactory and comprehensive description of all radiative processes has only been achieved by the development of quantum field theory. For many purposes, however, a simple semi-classical theory, which describes the radiation field by Maxwell's equations and the sources of radiation by quantum mechanics, is useful. Such a theory can in particular predict rough orders of magnitude for the lifetimes of excited states and can give the polarization of emitted radiation. It includes as a limit the correspondence principle of Bohr, according to which the radiation from a quantized system must agree in frequency, intensity and polarization with that from a classical oscillator for sufficiently large quantum numbers. In this section we indicate certain results of this theory which are of general use in atomic and nuclear physics.

It may be noted that the success of the semi-classical theory is due to the fact that the non-classical phenomena of quantum electrodynamics are really only very small effects. Recoil due to the emission of virtual photons by electrons enters formulae in powers of α, the fine structure constant, and the corrections (as seen in the magnetic moment of the electron and muon, in the Lamb shift, and in the theory of positronium) are negligible except when extremely accurate measurements are made.

3.5.1 *Spontaneous emission; line width*

According to Maxwell's theory an oscillating charge distribution emits radiation. In the case of a charge e, of mass m, oscillating simple harmonically with respect to a centre of force (Fig. 3.2) we define a dipole moment

$$D = ea \cos \omega_0 t = ea \cos 2\pi\nu_0 t \tag{3.4}$$

The instantaneous rate of radiation of energy into a solid angle at angle θ with the direction of a is given in Ref. 3.4 as

$$\frac{\mu_0 \ddot{D}^2}{(4\pi)^2 c} \sin^2 \theta \, d\Omega \tag{3.5}$$

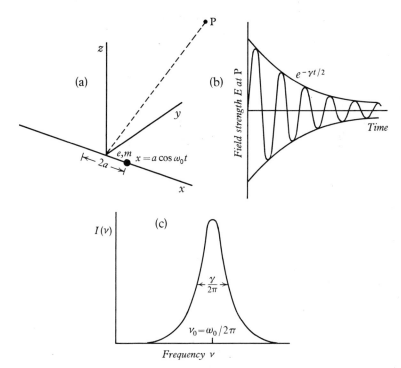

Fig. 3.2 (a) Classical dipole oscillator of moment $ea \cos \omega_0 t$. (b) Damped wave train radiated by classical oscillator. (c) Fourier analysis of damped wave train.

The total rate of emission of energy is obtained by integrating (3.5) over a sphere as

$$\frac{dW}{dt} = \frac{\mu_0 \ddot{D}^2}{6\pi c} = \frac{\mu_0 e^2 a^2 \omega_0^4}{6\pi c} \cos^2 \omega_0 t \ \text{J s}^{-1} \quad (3.6)$$

and the mean value of this over many cycles is

$$\frac{\overline{dW}}{dt} = \frac{\mu_0 e^2 a^2 \omega_0^4}{12\pi c} \quad (3.7)$$

Classically this energy must be supplied by the oscillator, whose amplitude will therefore decrease (unless it is driven). At the time when the amplitude is a, the total energy is

$$W = \tfrac{1}{2} m a^2 \omega_0^2 \quad (3.8)$$

and from (3.7) and (3.8), this can be expressed in the form

$$W = W_0 e^{-\gamma t} \quad (3.9)$$

where

$$\gamma = \frac{\mu_0 e^2 \omega_0^2}{6\pi m c} = \frac{2\pi \mu_0 e^2 v_0^2}{3mc} \quad (3.10)$$

The corresponding electric field vector at P may be written

$$\mathbf{E} = \mathbf{E}_0 e^{-\gamma t/2} \cos \omega_0 \left(t - \frac{r}{c} \right) \quad (3.11)$$

and this damped harmonic wave may in turn be represented by a superposition of undamped waves covering a range of frequencies of the order γ. If the radiation is examined with a spectroscope of high resolution it appears as a line of *finite spectral width*. A Fourier analysis of the expression (3.11) shows that the spectral line has an *intensity* distribution near the oscillator frequency v_0 given by

$$I(v) = \frac{\text{constant}}{(v - v_0)^2 + \tfrac{1}{4}(\gamma/2\pi)^2} \quad (3.12)$$

The damped wave train is shown in Fig. 3.2(b) and its frequency analysis in Fig. 3.2(c), from which it is seen that the full width of the distribution at half maximum intensity is just $\gamma/2\pi$. The formula (3.12) represents the *Lorentz shape* of a spectral line.

In taking these results into quantum theory, we assume that radiation arises as a result of a transition of a system such as an atom or a nucleus between two energy levels E_i and E_f as shown in Fig. 3.3 and that the frequency is given by the Bohr condition

$$h v_0 = E_i - E_f = E_0 \quad (3.13)$$

It is found that the line shape still has the Lorentz form, but the half-width is now interpreted as proportional to the transition probability per unit time for the system excited to the level E_i, i.e. to the reciprocal of the mean life τ of this level. By Heisenberg's uncertainty principle the mean life may also be expressed in terms of an energy width Γ of the level E_i given by

$$\Gamma\tau = \hbar = 6\cdot6 \times 10^{-16} \text{ eV s} \tag{3.14}$$

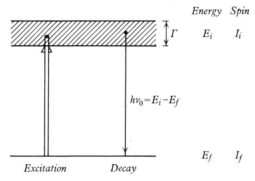

Fig. 3.3 Quantum picture of radiative process between levels of energy E_i and E_f.

In this representation the width of the spectral line is due to the level width Γ, or strictly to the sum of the widths of the two levels concerned in the transition, and the Lorentz shape of the line, in terms of energies, may be written

$$I(E) = \frac{\text{constant}}{(E - E_0)^2 + \frac{1}{4}\Gamma^2} \tag{3.15}$$

where E_0 is given by (3.13). For the D-lines of sodium the *classical* line width $\gamma/2\pi$ corresponds to an energy width of 4×10^{-8} eV and a lifetime, by (3.14), of $1\cdot7 \times 10^{-8}$ s. The remarkable precision of measurement offered by the Mössbauer effect (Sect. 13.6.4) has permitted the Lorentz shape for the 14·4 keV nuclear transition in ^{57}Fe ($\Gamma = 4\cdot6 \times 10^{-9}$ eV) to be examined and confirmed in detail.

It is not possible to calculate the quantum mechanical width or probability of spontaneous emission, from wave mechanics. This probability may indeed be deduced from the probability of the inverse process ($E_f + h\nu \rightarrow E_i$, Fig. 3.3) which is the excitation of a system by a light quantum, by application of a statistical argument. We prefer here, however, to use the classical approach starting with (3.6). If a suitable quantum mechanical form for the dipole moment D can be found and if the energy radiated from the excited system is to be carried away by photons of energy $h\nu_0$, the number emitted per unit time must be

$$\frac{\mu_0 \ddot{D}^2}{6\pi h\nu_0 c} = \frac{\mu_0 \ddot{D}^2}{6\pi\hbar\omega_0 c} \tag{3.16}$$

and the mean time τ_1 required for the emission of a photon is the reciprocal of the time-averaged value of this quantity. The dipole moment is a function of the two states of the radiating system and of the time and it is customary to express it as

$$D(r, t) = 2D(r) \cos \omega_0 t = D(r)\{e^{i\omega_0 t} + e^{-i\omega_0 t}\} \tag{3.17}$$

where $D(r)$ is a function of the states of motion only. If we consider the simple case in which the levels E_i, E_f are defined by the motion of a particle of charge e in a potential well, and if the corresponding wave functions are ψ_i, ψ_f then a suitable expression for $D(r)$ is

$$D(r) = e \int \psi_f^* \mathbf{r} \psi_i \, d\tau = e\bar{\mathbf{r}}_{if} \tag{3.18}$$

This effectively averages a vector quantity $e\mathbf{r}$ of the dimensions of the classical dipole moment, with respect to the charge distribution in the initial and final states.

From (3.16), (3.17) and (3.18) we obtain for the dipole mean lifetime τ_1

$$\frac{1}{\tau_1} = \frac{\mu_0 e^2 \omega_0^3}{3\pi\hbar c}|\mathbf{r}_{if}|^2 = \frac{\Gamma_1}{\hbar} \quad \text{from (3.14)} \tag{3.19}$$

This also may be written, remembering that $\omega_0 = 2\pi\nu_0 = E_0/\hbar$ as

$$\Gamma_1 = \frac{\mu_0 e^2}{3\pi c}\left(\frac{E_0}{\hbar}\right)^{3-}|\bar{\mathbf{r}}_{if}|^2 \tag{3.20}$$

so that the dipole width Γ_1 is proportional to the cube of the transition energy.

A classical charge distribution which may be completely described by a dipole moment is of course a very special case. More generally several electric moments are required, as explained in more detail in Appendix 2, and when current loops develop as a result of the motion of the charges, magnetic moments also arise. An electric quadrupole moment would have dimensions $\approx \Sigma er^2$ and the electric 2^L-pole moment is $\approx \Sigma er^L$. When the charge distribution oscillates, radiation may be associated with each of these moments, but because of their more symmetrical nature they do not radiate so strongly as the dipole moment. It can be shown that in the *long wavelength approximation*, when the wavelength of the emitted radiation is very much greater than the dimension a of the radiating system, the intensity of radiation from each successively higher moment decreases by a factor $(a/\lambda)^2$. In most cases of interest the yield of high multipole radiation is very small. Thus for a sodium atom emitting $\lambda = 600$ nm we have $(a/\lambda)^2 \approx 10^{-5}$ while for a nucleus emitting radiation of energy 0·5 MeV, and with $a = 10^{-14}$ m, $(a/\lambda)^2 = 10^{-3}$. Nuclei tend to show more higher order radiation than atoms because it may be shown that a system of particles in which the centre of charge coincides with the centre of mass has no dipole moment. Because of the strong forces binding neutrons to protons this may well be the case for nuclei.

These results may be carried over into the quantum mechanical treatment and formulae similar to (3.20) for the 2^L-pole width may be derived. For both electric and magnetic radiation for example

$$\Gamma_L \propto E_0^{2L+1} \tag{3.21}$$

Such formulae are always expressible as a product of an energy dependent factor and a factor characteristic of the source for the particular type of radiation. The former factor is quite general, since it arises unambiguously as a result of the long wavelength approximation, but the latter is specific to the radiating system and must be calculated on the basis of a model of that system.

The expression (3.5) contains the angular distribution factor for classical dipole radiation; the intensity along the axis of the dipole is zero. The classical analogy shows that the angular distribution of radiation from the 2^L-pole moment contains powers of $\cos \theta$ up to $\cos^{2L} \theta$ only. The polarization of the radiation is also predictable from the classical moments.

3.5.2 *Multipolarity; selection rules*

A radiative transition is in principle possible between any two levels of an atom or nucleus, providing that one of them at least has a finite spin I. Whether such a process will in practice be observable in competition with other types of de-excitation, or with transitions from the excited state to other lower states, depends on the radiative width Γ. Whether radiation of a particular *type*, e.g. dipole or quadrupole, occurs at all between the two levels depends on certain absolute *selection rules*. If for instance the dipole moment (3.18) vanishes, then radiation of a higher order may become important. Selection rules for radiative processes are based ultimately on the properties of the electromagnetic field, particularly with respect to angular momentum and parity. If these quantities are conserved in the electromagnetic field, and all evidence so far available suggests that they are, then the emission of a photon of known properties must alter the state of the emitting system correspondingly, not only in energy, but in symmetry character.

It may be shown directly from Maxwell's equations that the electromagnetic radiation from an oscillating charge distribution of small dimensions transports angular momentum as well as energy. The two are connected by the result (Ref. 3.2, p. 802) that: z-component of angular momentum $= (M/\omega) \times$ energy in field where M is an integer. If one quantum of energy $\hbar\omega$ is emitted the z-component of angular momentum is $M\hbar$. The total angular momentum of the quantum is the vector $\mathbf{L}$ (of absolute value $\sqrt{L(L+1)}\hbar$ just as in the case of a particle), and $|M| \leqslant L$. The L-value defines the *multipolarity* of the photon, or of the corresponding process, as the number of units of angular momentum removed by the quantum emission. For radiation of multipolarity L to be observed, the corresponding moment must be finite. The angular distribution and polarization are then those corresponding to the radiation from a classical 2^L-pole moment (Sect. 3.5.1). We shall write the angular distribution function for radiation of multipolarity L with z-component M as

$$F_L^M(\theta) \tag{3.22}$$

This gives the intensity observed at an angle θ with respect to the z-axis. For dipole radiation we have

$$F_1^0(\theta) = 3 \sin^2 \theta$$
$$F_1^{\pm 1}(\theta) = \tfrac{3}{2}(1 + \cos^2 \theta) \tag{3.23}$$

corresponding to dipoles lying in the z-axis and rotating in the x–y plane respectively. The numerical factors are introduced so that the angular factor in the mean rate of radiation over a sphere is unity.

In the case of motion of a single particle the angular momentum also determines the parity (Sect. **3.1**). This is not so for the electromagnetic interaction, which has fields of both even and odd parity. Figure 3.4 illustrates this for the

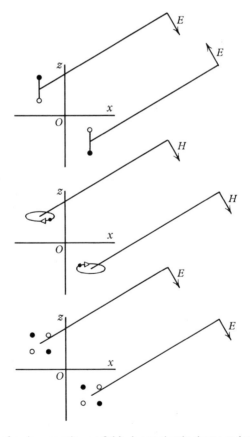

Fig. 3.4 Effect of parity operation on fields due to simple charge and current distributions.

inversion through the origin of simple radiating systems. We note that radiation of given multipolarity may be either even or odd and that in particular the field of an electric dipole has odd parity while electric quadrupole and magnetic dipole fields are even. In general radiative processes are classed as of electric or magnetic type (E or M) to indicate the parity change, as shown in the following table; the multipolarity is given as a number following the letter denoting the radiation type.

TABLE 3.1 Classification of electromagnetic radiation

Type of radiation	$E1$	$E2$, $M1$	$E3$, $M2$
Name	Electric dipole	Electric quadrupole Magnetic dipole	Electric octupole Magnetic quadrupole
Multipolarity	1	2, 1	3, 2
Parity change	Yes	No	Yes

These observations on the radiation field now permit the *selection rules* for the sources of the radiation to be formulated (Ref. 3.5). For angular momentum, we see that the multipolarity of the quantum emitted in the process shown in Fig. 3.3 is determined by the vector equation

$$\mathbf{I}_i - \mathbf{I}_f = \mathbf{L} \tag{3.24}$$

which implies for the quantum numbers

$$I_i + I_f \geqslant L \geqslant |I_i - I_f| \tag{3.25}$$

and for a specified axis Oz (Fig. 3.5(a))

$$m_i - m_f = M \tag{3.26}$$

The inequality (3.25) may permit several multipolarities, e.g. states of spin 3 and 2 could give $L = 1, 2, 3, 4, 5$. Some of these, however, are forbidden by the *parity selection rule* which may be obtained from Table 3.1. In the case 3^+ (even) $\to 2^-$ (odd) for instance $E1$, $M2$, $E3$, $M4$ and $E5$ radiation only would be allowed; in practice only the lowest multipoles, $E1$ and perhaps $M2$ would be effective, because of the factor $(a/\lambda)^2$ in the intensity formula (Sect. 3.5.1). If the spin change necessarily leads to radiation of high multipolarity, the excited state will have a long lifetime (Sect. **13.5**). The electromagnetic field does not provide any type of radiation with $L = 0$ (which would be a longitudinal wave) and for $I_i = I_f = 0$ radiation is *strictly forbidden*. Alternative processes for the nuclear $0 \to 0$ transition are discussed in Chapter 13.

In (3.24) and (3.25) I is the total angular momentum quantum number. If spin–orbit coupling is weak, the quantum numbers L_i and S_i are also useful and spin and orbital vectors provide separate selection rules. Thus for electric dipole radiation, there is no interaction with intrinsic spins and we have:

$\Delta I = 0, \pm 1$; $\Delta L_i = 0, \pm 1$; $\Delta S_i = 0$ and parity change, e.g. the transitions $^1D \to {}^1P$, or $^2P \to {}^2S$ (sodium D-lines). Magnetic dipole radiation can arise either from orbital motion, or because an intrinsic spin changes its direction in an internal field. The corresponding rules are

$$\Delta I = 0, \pm 1 \text{ and no parity change}$$
$$\Delta L_i = 0, \pm 1; \Delta S_i = 0 \text{ for orbital motion}$$
$$\Delta L_i = 0; \Delta S_i = 0, \pm 1 \text{ for spin-flip}$$

e.g. the transitions $^3S \to {}^1S$, $^3D \to {}^1D$.

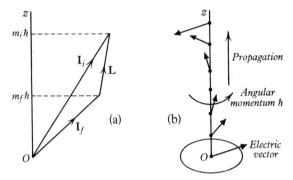

Fig. 3.5 (a) Vector diagram (schematic) for emission of radiation of multipolarity L. The component of $\mathbf{L}$ along the axis Oz is $(m_i - m_f)\hbar$. (b) If Oz is taken to be the direction of propagation, then $m_i - m_f = \pm 1$ and the radiation observed in the direction Oz is circularly polarized.

For some purposes it is useful to describe the angular momentum of the photon itself as composed of an orbital part and an intrinsic spin of unity. If the axis of quantization is taken along the direction of emission, the orbital part of the angular momentum has no component in this direction. The intrinsic spin can have components $\pm \hbar$ only, the $M = 0$ component being excluded by the requirement that the electromagnetic wave shall be transverse. We consequently arrive at the picture of a photon propagating along Oz as a circularly polarized plane wave, with two possible directions of rotation of the electric vector (Fig. 3.5(b)).

The transformation properties of the Maxwell field show that the intrinsic parity of the photon must be odd (Sect. 21.2.1). Electric dipole radiation $(L = l + s = 1)$ corresponds to the emission of a photon with even orbital momentum $(l = 0, \text{ or } 2, s = 1)$ and with parity change, while magnetic dipole radiation $(L = l + s = 1)$ corresponds to emission with one unit of orbital momentum $(l = 1, s = 1)$ and no parity change.

3.5.3 *The Zeeman effect*

The line-splitting observed when a radiating atom or nucleus is situated in a magnetic field is a good example of the application of selection rules. For the sodium D-lines $^2P_{3/2,\,1/2} \rightarrow\ ^2S_{1/2}$ the spin $\frac{3}{2}$ has four substates, $m_J = \pm\frac{3}{2}, \pm\frac{1}{2}$ and the spin $\frac{1}{2}$ has two substates, $m_J = \pm\frac{1}{2}$ and there is a parity change in the transitions. Application of the selection rule

$$\Delta J = \pm 1, 0; \qquad \Delta m_J = \pm 1, 0$$

predicts four components of electric dipole radiation for the D_1 line and six for D_2 (Fig. 3.6). If the magnetic field is taken to define the axis Oz, then

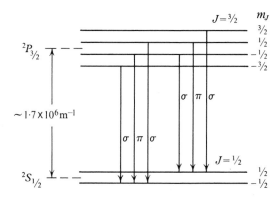

Fig. 3.6 Zeeman components of the D_2 line ($\lambda = 589$ nm) of the sodium atom, neglecting hyperfine structure. The separations of the substates are not shown to scale; in absolute units they are of the order of 100 m^{-1} T^{-1}.

components with $\Delta m_J = 0$ arise, in the classical analogy, from an electric dipole aligned with the axis of $\hat{z}$. The corresponding radiation vanishes in the direction $\pm Oz$ itself and is plane polarized (π) when viewed in any other direction. The components with $\Delta m_J = \pm 1$, on the other hand, are circularly polarized when viewed along the lines of force and plane polarized when seen in planes perpendicular to the lines of force; they are known as σ components.† Classically these correspond to radiation from a dipole rotating in the x–y plane.

The angular distribution of the Zeeman components is given by the classical analogy as proportional to $\sin^2 \theta$ for $\Delta m_J = 0$ and to $(1 + \cos^2 \theta)$ for $\Delta m_J = \pm 1$, i.e. by the distribution functions $F_1^0(\theta)$ and $F_1^{\pm 1}(\theta)$ (Sect. 3.5.2).

† In magnetic dipole transitions, which are frequently encountered in nuclear processes, the electric and magnetic vectors are 90° displaced with respect to the vectors for an electric transition if the moments have the same axis. The π, σ nomenclature is then reversed.

References

1. The following texts on quantum mechanics may be found useful:
   MATTHEWS, P. T. *Introduction to Quantum Mechanics* (2nd edn.), McGraw-Hill, 1969.
   DICKE, R. H. and WITTKE, J. P. *Introduction to Quantum Mechanics*, Addison Wesley, 1960.
   CASSELS, J. M. *Basic Quantum Mechanics*, McGraw-Hill, 1970.
   SCHARFF, M. *Elementary Quantum Mechanics*, Wiley-Interscience, 1969.

References 1.1–1.11 give accounts of the application of quantum theory to atomic and nuclear phenomena.

2. BLATT, J. M. and WEISSKOPF, V. F. *Theoretical Nuclear Physics*, Wiley, 1952.
3. POINTON, A. J. *Introduction to Statistical Physics*, Longmans, 1967.
4. HEITLER, W. *The Quantum Theory of Radiation* (3rd edn.), Oxford University Press, 1954.
5. KUHN, H. G. *Atomic Spectra* (2nd edn.), Longmans, 1969.
6. WOODGATE, G. K. *Elementary Atomic Structure*, McGraw-Hill, 1970.

4 Nuclear effects in spectroscopy

The availability of a large body of information on the static properties of nuclei, properties such as angular momentum, magnetic moment and size, has greatly advanced the development of nuclear theory. Such properties are not invoked in the simple theory of the atom outlined in Chapter 3, but much of the relevant information has been obtained by spectroscopy in the widest sense, in several regions of the electromagnetic spectrum. In ordinary optical spectroscopy, nuclear moment effects appear as small perturbations or splittings of spectral terms of much greater magnitude, and refined methods are necessary for their detection. Transitions between the hyperfine structure components of a spectral term are, however, also possible and the corresponding frequencies often lie in the microwave or radiofrequency region where accurate measurement is feasible. Microwave and radiofrequency methods, combined with the elegant techniques of the atomic and the molecular beam, have been responsible for a great increase in our knowledge of the electric and magnetic moments of the nuclear ground state; these and other methods will be briefly outlined in the present chapter.

4.1 Nuclear size and nuclear moments

The broad features of atomic spectra can be interpreted with the sole assumption of a nucleus of vanishingly small size and of infinite mass. The fine structure of spectral lines can be explained by the further addition of the concepts of electron spin and spin–orbit interaction, without the necessity for involving additional nuclear properties. The hyperfine structure of line spectra, on a much smaller scale, cannot, however, be so explained, and in 1927 Goudsmit and Back, following earlier ideas of Pauli (1924), suggested that this effect in the bismuth spectrum might be due to the existence of a nuclear angular momentum, with which would be associated a magnetic moment. If nuclear magnetic moments are considered to arise partly as a result of circulating charges in the nucleus, it is clear that the nucleus must have a finite size, and small effects in optical spectroscopy should then result from the penetration of electronic wave functions of s-states into the nuclear volume. Both the spin and magnetic moment enter into the quantitative treatment of the hyperfine structure. In analogy with the electronic case we define the following quantities:

(*a*) *the nuclear angular momentum vector* **I**, with absolute magnitude

$$|\mathbf{I}| = \sqrt{I(I+1)}\,\hbar \tag{4.1}$$

where I, the nuclear spin quantum number, is integral or half-integral. In accordance with the uncertainty principle the direction in space of the vector **I** cannot be determined, but if an axis Oz of quantization is defined, e.g. by an external magnetic field, the component of **I** along Oz will be observable and will have the value $m_I \hbar$ where m_I has one of the $2I+1$ values $I, (I-1), \ldots, -(I-1), -I$. In a vector diagram **I** is considered to lie on a cone described round the axis Oz with semi-angle β given by

$$\cos \beta = \frac{m_I \hbar}{|\mathbf{I}|} = \frac{m_I}{\sqrt{I(I+1)}} \tag{4.2}$$

The maximum value of the component of **I** is $I\hbar$ and this is usually known as the nuclear spin.

(*b*) *the nuclear magnetic moment vector* $\boldsymbol{\mu}_I$, which is taken to be a vector parallel or antiparallel to **I**. It may be expressed in absolute units or in nuclear magnetons, defined as $\mu_N = eh/4\pi m_p$ in analogy with the Bohr magneton, but smaller by a factor of $m/m_p = 1/1836$. The observable value of $\boldsymbol{\mu}_I$, usually known as magnetic moment μ_I, is the z-component of $\boldsymbol{\mu}_I$ taken when m_I has its maximum value of I.

(*c*) *the nuclear gyromagnetic ratio* γ_I, equal to the nuclear moment in absolute units divided by the nuclear spin also expressed in these units, i.e.

$$\gamma_I = \frac{\mu_I}{\mathbf{I}} = \frac{|\boldsymbol{\mu}_I|}{|\mathbf{I}|} = \left(\frac{\mu_I}{\mathbf{I}}\right)_z \quad \text{or simply} \quad \left(\frac{\mu_I}{I\hbar}\right) \tag{4.3}$$

The ratio of the nuclear moment expressed in nuclear magnetons to the nuclear spin expressed in units of $\hbar$, is defined to be the dimensionless nuclear *g-factor* g_I and is related to γ_I by the equation

$$g_I = \gamma_I \frac{2m_p}{e} \tag{4.4}$$

If the nuclear moment and spin are oppositely directed, as in the case of the neutron, μ_I, γ_I and g_I are negative. The g-factor defined in (4.4) has a value of the order of unity. For some purposes it has become customary to use nuclear g-factors defined in terms of the Bohr magneton and these g-factors are then of the order of $1/2000$. From (4.3)

$$\mu_I = \gamma_I I\hbar = g_I I\mu_N \quad (\text{or } g_I I\mu_B) \tag{4.5}$$

(*d*) *the nuclear electric quadrupole moment* Q_I, which measures the deviation of the nuclear charge distribution from a spherical shape; a discussion of this moment, which provides clear evidence for a finite nuclear size, is included in Appendix 2.

4.2 Nuclear effects in optical spectroscopy

4.2.1 Nuclear mass

The energy levels of the hydrogen atom are given in terms of the Rydberg constant R_H for hydrogen. This differs from the Rydberg constant R_∞ for an infinitely heavy nucleus by a factor which corrects for the relative motion of the electron and nucleus. We may write in general

$$R_M = R_\infty \frac{M}{m+M} \qquad (4.6)$$

where m is the electron mass and M the nuclear mass for a hydrogen-like atom. It is clear that the term energies and hence the frequencies of the associated spectral lines will depend on the nuclear mass. This will also be true for atoms not of hydrogen type and each separate isotope will lead to its own set of spectral lines. This isotope effect, purely dependent on nuclear mass, is most easily observable for the isotopes of light elements, and permitted the discovery of deuterium by Urey, Brickwedde and Murphy (1932). For the hydrogen isotopes the change in frequency expected for the H_α line ($\tilde{v} \approx 1.5240 \times 10^6 \text{ m}^{-1}$) observed by Urey and his co-workers, is $\Delta \tilde{v} \approx 400 \text{ m}^{-1}$ or $\Delta \lambda = 0.18$ nm. For heavy atoms this isotope effect becomes small as may be seen by differentiating (4.6):

$$\frac{\Delta \tilde{v}}{\tilde{v}} = \frac{\Delta R_M}{R_M} \approx \frac{m}{M^2} \Delta M \qquad (4.7)$$

The nuclear mass also affects the rotational and vibrational energies of diatomic molecules. The isotopes ^{13}C, ^{15}N, ^{17}O and ^{18}O were discovered by analysis of the electronic bands of carbon, nitrogen and oxygen. Pure rotational spectra, with which much lower frequencies are associated, have been studied by microwave methods (Sect. **4.3**).

4.2.2 Nuclear size

A further type of isotope effect depending on nuclear volume arises because the nuclear charge is spread over a finite volume in space. The potential energy of an electron in interaction with the nucleus decreases less rapidly than for a point charge when the wave function begins to penetrate the nucleus and the energies of the spectral terms are therefore shifted. The shift is seen as a difference between the term values of the several isotopes of a given element. This effect increases with increasing nuclear mass (in contrast with the isotope effect discussed in Sect. 4.2.1) and has been observed in many heavy elements. The results of such observations, like those on the similar phenomenon observed in muonic atoms (Sect. 11.2.3), provide important evidence on nuclear radii.

4.2.3 Nuclear spin and magnetic moment from hyperfine structure

The total energy of an atom with unpaired electrons, containing a nucleus with spin, includes a term dependent on the interaction of the nuclear magnetic moment with the magnetic moment of the electrons, and with the magnetic field due to the electronic motion in the atom. The effect of externally applied fields is neglected for the present. It is now easy to see, as in the case of electronic fine structure, that the energy levels of the atom are split into a group of states whose number may be determined by the nuclear spin and whose separation is dependent on the nuclear magnetic moment. Spectral lines originating between terms split in this way are seen under high resolution to form a multiplet, usually known as the *hyperfine structure*.

In addition to magnetic interaction, both internal and external electric fields contribute to the energy of the atom by coupling with the nuclear electric quadrupole moment Q_I. If the electric field gradient at the nucleus is q, the corresponding energy is written $b = eqQ_I$. Although quadrupole interactions contribute to many hyperfine structure patterns, they are not observable (App. 2) in the important cases in which $I = \frac{1}{2}$ or $\mathcal{J} = \frac{1}{2}$. In the present chapter we shall consider for simplicity mainly atomic states of the single electron type, with $\mathcal{J} = \frac{1}{2}$, for which $b = 0$.

If the total angular momentum due to electronic motion is $\mathbf{J}$ ($\neq 0$) and the nuclear spin is $\mathbf{I}$ ($\neq 0$) then the total mechanical angular momentum is $\mathbf{F}$ where

$$\mathbf{F} = \mathbf{I} + \mathbf{J} \qquad (4.8)$$

Quantum mechanically this vector sum can be performed in $2I + 1$ ways if $I < J$ and $2\mathcal{J} + 1$ ways if $J < I$ and each of these arrangements will correspond to a different relative orientation of the comparable vectors $\mathbf{I}$ and $\mathbf{J}$. Each such arrangement also corresponds to a different energy. The internal magnetic field at the nucleus, with a flux density B_i of 10–100 T, is proportional to $\mathbf{J}$ but opposite to this vector in a single electron atom because of the negative sign of the electronic charge. The nuclear magnetic moment $\boldsymbol{\mu}_I$ is parallel to $\mathbf{I}$, but may be positive or negative. The interaction energy

$$W = -(\boldsymbol{\mu}_I . \mathbf{B}_i) = -|\boldsymbol{\mu}_I| \, |\mathbf{B}_i| \cos (\boldsymbol{\mu}_I \mathbf{B}_i)$$

may therefore be written

$$W = +a|\mathbf{I}| \, |\mathbf{J}| \cos (\mathbf{IJ})$$

and introducing the quantum mechanical value for $\cos (\mathbf{IJ})$

$$W = \frac{a}{2} \{F(F+1) - I(I+1) - \mathcal{J}(\mathcal{J}+1)\} = \frac{aC}{2} \qquad (4.9)$$

The coupling constant a is proportional to the nuclear moment and if this is positive the hyperfine state with the lowest F has the lowest W. The coupling constant a also contains the internal atomic field, which must be calculated from the electronic wave function. If this is known, μ_I can be deduced from the observed hyperfine structure intervals. The nuclear spin can in principle be obtained by counting components of the optical line when $I < J$,† but if the components cannot be resolved then the ratio of total intensity associated with transitions to two states of different F may be observed. This ratio is equal to the ratio of the statistical weights $2F + 1$ of the two states under

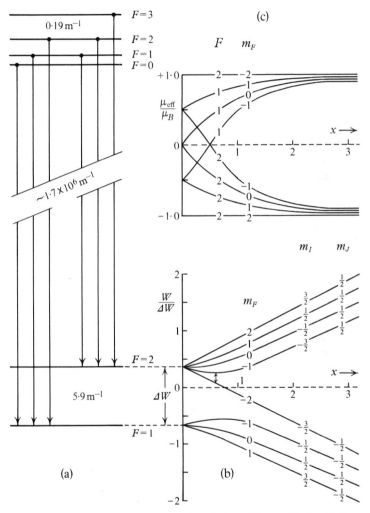

Fig. 4.1 (a) Hyperfine structure of one of the sodium D-lines ($^2P_{3/2} \rightarrow {}^2S_{1/2}$). (b) Energy levels of the sodium atom in weak and intermediate magnetic fields. The Zeeman splitting of the hyperfine levels of the ground state is shown in units of ΔW as a function of the quantity $x \approx 2\mu_B B_0 / \Delta W$. A radiofrequency transition used in the 'flop-in' technique of magnetic resonance (Sect. 4.4.4) is also shown. (c) Effective magnetic moments of the sodium atom in the states shown in 4.1(b).

† See for example, K. L. van der Sluis and J. R. McNally, *J. opt. Soc. Amer.*, **45**, 65, 1955.

normal circumstances when there is no reason for transitions to any particular substate m_F to be favoured.

The energy difference between the two states F and $F-1$ is, according to (4.9), equal to aF; this interval is known as the hyperfine structure separation or *hfs splitting* ΔW and the corresponding Δv is typically a few thousand MHz. The optical hyperfine structure of one of the sodium D-lines is shown in Fig. 4.1(a). In this case $I = \frac{3}{2}, \mathcal{J} = \frac{3}{2}$ for the initial state and $I = \frac{3}{2}, \mathcal{J} = \frac{1}{2}$ for the final state. The components of the line are determined by the selection rules

$$\Delta F = \pm 1, 0 \quad \Delta \mathcal{J} = \pm 1, 0 \quad \Delta l = \pm 1$$

and the magnitudes of the hfs splittings are shown. For the state with $I = \mathcal{J} = \frac{3}{2}$ the intervals should be in the ratio $1:2:3$; deviations from such a regular sequence in the spectrum of europium gave the first indication of the existence of the electric quadrupole moments and now provide one method for the determination of these quantities. The hfs splitting of the $^2S_{1/2}$ ground state of hydrogen ($I = \mathcal{J} = \frac{1}{2}$; $F = 0$, 1) is the source of the 21 cm 'galactic' $F = 1 \rightarrow F = 0$ radiation observed astronomically. The hyperfine separation in this case is simply the difference in energy between the states in which the nuclear moment and internal atomic field are parallel and antiparallel.

If an external magnetic field of flux density B_0 is applied to an atom with nuclear spin Zeeman effects occur in the hyperfine structure and may be discussed as in the electronic case. In weak fields the coupling between **I** and **J** persists and the resultant vector **F** orients in the field. Each of the $2F+1$ permitted orientations corresponds to a different interaction energy between the field and the resultant magnetic moment. If the field is increased the coupling between **I** and **J** may be broken and these vectors and their associated magnetic moments will orient independently. The fields required are considerably less than those necessary to uncouple the **LS** vectors composing **J** owing to the small value of the nuclear magnetic moment in comparison with the Bohr magneton. The Zeeman effect of the hyperfine structure of the ground state of sodium is illustrated in Fig. 4.1(b). In the strong field region, analogous to the Paschen–Back region in optical spectroscopy, no new levels arise (since $\Sigma(2F+1) = (2I+1)(2\mathcal{J}+1)$) but the spacing alters.

The total magnetic energy W of the atom in a field with flux density B_0 is given by a formula due to Breit and Rabi, valid for $I = \frac{1}{2}$ or $\mathcal{J} = \frac{1}{2}$ only. This is complicated except in the limiting cases of:

(a) *weak fields* for which F is a good quantum number and

$$W(F, m_F) = \frac{aC}{2} - m_F g_F \mu_B B_0 \tag{4.10}$$

where g_F is a factor of the order of g_J, the atomic Landé-factor (which is often negative). For atoms with $\mathcal{J} = \frac{1}{2}$ (Fig. 4.1(b)) and for $F = I + \frac{1}{2}$ the energy difference between the substates $m_F = F$ and $m_F = -F$ is mainly due to the orientation of the atomic moment μ_J parallel and antiparallel to the field and is therefore $2\mu_J B_0$. The energy difference between adjacent magnetic substates is easily seen to be approximately $\mu_J B_0 / F$ since there are $2F$ intervals between extreme values of m_F.

(b) *strong fields*, in which I and J orient independently, and

$$W(m_I, m_J) = a m_I m_J - m_I g_I \mu_B B_0 - m_J g_J \mu_B B_0 \tag{4.11}$$

in which the first term measures the interaction between nuclear and atomic moments and the other two give the energy of these moments individually in the external field. The nuclear term is usually negligible since $g_I \approx 1/2000\, g_J$.

The Breit–Rabi formula may be used to calculate the effective magnetic moment

$$\mu_{\text{eff}} = -\frac{\partial W}{\partial B} \tag{4.12}$$

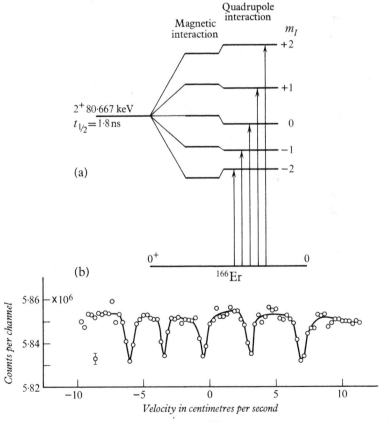

Fig. 4.2 (a) Hyperfine structure of the 2^+ level of ^{166}Er observed by the Mössbauer technique, showing both magnetic dipole and electric quadrupole interaction effects. (b) Resonance absorption pattern for ^{166}Er (R. L. Cohen and J. H. Wernick, *Phys. Rev.*, **134**, B503, 1964).

of an atom in a magnetic field. The result for $I = \frac{3}{2}, \mathscr{J} = \frac{1}{2}$ is shown in Fig. 4.1(c); at high fields all moments tend to μ_B since the nuclear and atomic moments are uncoupled and the former is negligible in comparison with the latter.

Experiments in which use is made of atomic moments usually give the interaction constants a and b in the first place. Evaluation of the moments μ_I and Q_I then necessitates adequate knowledge of electronic wave functions.

A very clear demonstration of the energy differences which arise when a nuclear moment orients in an internal field is provided by the Mössbauer effect (Sect. 13.6.4). The effect provides nuclear gamma radiation sources of very high homogeneity and of frequency controllable by movement of the source, through the Doppler effect. Resonant absorption of this mono-chromatic line in the magnetic substates of nuclei identical with the parent nucleus gives a direct survey of the hyperfine pattern. Figure 4.2 shows the level splitting of a nuclear excited state (^{166}Er*) and some experimental results leading to determinations of moments.

Electric quadrupole hyperfine structure can be seen in the X-ray spectra of muonic atoms (Sect. 11.2.3). In these atoms, which are all hydrogen-like, muonic wavefunctions can be calculated accurately and the transition energies are so large that they can be accurately observed by nuclear detectors of high resolution. Magnetic hyperfine effects exist but are small in comparison with magnetic hfs for electrons because of the small magnetic moment of the muon.

4.2.4 Nuclear spin and statistics from molecular spectra

Successive lines of the electronic (emission) band spectra of homonuclear diatomic molecules are found to show an alternation in intensity. This effect can be explained by symmetry considerations. If the nuclear spin is I, there are $2I + 1$ substates, with respect to an axis of quantization, for each nucleus. Combining these, there are $(2I + 1)^2$ spin functions of the type

$$\psi_s = \psi_{m'}(A)\psi_m(B) \tag{4.13}$$

with individual magnetic quantum numbers m, m' for the pair of nuclei. Since the nuclei are identical the probability function $|\psi_s|^2$ must not change on interchange of particles and the form (4.13) is therefore unsatisfactory. Instead we must construct a linear combination of pairs of functions of this type, as used in the case of the helium atom for electrons, and in the case of ortho- and para-hydrogen. The suitable combinations are

$$\psi_{ss} = \psi_m(A)\psi_{m'}(B) + \psi_m(B)\psi_{m'}(A)$$
and
$$\psi_{sa} = \psi_m(A)\psi_{m'}(B) - \psi_m(B)\psi_{m'}(A) \tag{4.14}$$

Since two functions of the type (4.13) combine to give two of type (4.14), there are still $(2I + 1)^2$ spin functions altogether. Of these $2I + 1$ have $m = m'$ and are obviously symmetrical in spin. Of the remaining $2I(2I + 1)$ functions with $m \neq m'$, half are of type ψ_{ss} and half of type ψ_{sa}. These are respectively symmetrical and antisymmetrical. In all, therefore there are

$$(I+1)(2I+1) \text{ symmetrical nuclear spin states}$$
$$I(2I+1) \text{ antisymmetrical nuclear spin states}$$

(4.15)

The symmetrical spin states belong to the ortho form of the molecule and the antisymmetrical states to the para form; for hydrogen $(I = \frac{1}{2})$ the ratio of ortho to para states is 3 to 1.

In a molecule, rotational motion is in general present and this causes the nuclei to change positions. The wave function for rotational motion is symmetric or antisymmetric according as $\mathcal{J}$ is even or odd. If the nuclei have half-integral spin (Fermi–Dirac statistics) the total *nuclear* wave function $\psi_r \psi_s$ must be antisymmetrical and therefore the ortho molecules have only antisymmetrical *rotational* states and the para molecules symmetrical rotations. For particles obeying the Einstein–Bose statistics the situation is reversed. From (4.15) we then obtain the following relative probabilities of even and odd rotational states

$$\text{Fermi–Dirac statistics} \qquad \frac{\text{even } \mathcal{J}}{\text{odd } \mathcal{J}} = \frac{I}{I+1} \qquad (4.16)$$

$$\text{Einstein–Bose statistics} \qquad \frac{\text{even } \mathcal{J}}{\text{odd } \mathcal{J}} = \frac{I+1}{I} \qquad (4.17)$$

These statistical weights modify the normal Boltzmann distribution of molecules between the rotational states in thermal equilibrium.

The selection rules for electric dipole radiative transitions, as given in Sect. 3.5.2, now imply that *homonuclear diatomic molecules exhibit no pure rotational or rotational-vibrational spectrum*. This is because the change of rotational quantum number $\Delta \mathcal{J} = \pm 1$ required in this case to provide the angular momentum removed by the radiation would result in a transition between symmetrical and antisymmetrical spin states. Such a transition is very improbable and the ortho and para forms should be regarded as non-combining. In an electronic emission band however, transitions take place between the rotational fine-structure levels of electronic energy states and the angular momentum and symmetry changes can in many cases be provided by the electronic change. Molecular symmetries will not be discussed here; we note only that many spectra indicate transitions with both $\Delta \mathcal{J} = 0$ and ± 1. The lines originating from a state of given $\mathcal{J}$ have an intensity proportional to the population, or statistical weight, of the state and from (4.16) and (4.17) it is apparent that the intensities of successive line will stand in the ratio $I+1/I$. In the case of the $^3\text{He}_2$ spectrum for instance the ratio is 3:1, indicating a spin $I = \frac{1}{2}$. It is also found, from knowledge of the molecular wave functions, that it is the odd $\mathcal{J}$ states in this case which are enhanced; the ^3He nuclei thus obey Fermi–Dirac statistics.

Band-spectrum analysis therefore determines both spin and statistics and it has been applied for several of the lighter nuclei, particularly when atomic hyperfine effects are difficult to observe. Strikingly clear evidence is obtained for nuclei such as ^{4}He, ^{12}C or ^{14}C and even A, even Z nuclei in general. In these cases the states of odd J and the corresponding lines are absent, which indicates that $I = 0$ and that the nuclei obey Einstein–Bose statistics.

4.3 Nuclear effects in microwave spectroscopy

4.3.1 *Gaseous absorption spectroscopy*

The microwave region of the electromagnetic spectrum (10–1000 m^{-1} approximately) is particularly suitable for studying rotational transitions in fairly heavy molecules. This is because of the availability of accurate frequency measuring apparatus and easily controllable klystron oscillators. The most useful technique is that of gaseous absorption spectroscopy, although similar effects may be found with solids and liquids; observations are made of the variation of the absorption coefficient of gas in a cell as the frequency of the incident radiation is varied. The structure visible in the absorption near the frequency of a rotational transition is a hyperfine effect but is not directly due to the nuclear magnetic moment. Molecular magnetic fields are small and the magnetic hyperfine structure is not observable because of collision broadening; the nuclear spins do however orient in the normal way in the internal field. Energy differences then arise because of the interaction of the nuclear electric quadrupole moment with the molecular electric field gradient. Since the axis of the quadrupole moment is fixed with respect to the nuclear spin axis, the number of hyperfine states is directly determined by the nuclear spin but their spacing is not determined by the nuclear magnetic moment. The electric quadrupole moment may be found if the molecular electric field can be calculated.

Nuclear spins have been determined by this method for ^{10}B, ^{33}S, ^{35}Cl and several other nuclei using suitable molecules consisting mainly of atoms with spin zero nuclei. The lighter, simpler molecules have rotational frequencies lying in the infra-red and cannot be studied by microwave methods since these techniques do not extend much above a frequency of 1000 m^{-1} ($300\,000$ MHz).

It is also possible to observe the Zeeman splitting of the hyperfine lines in a given rotational transition. The energy differences arising in an applied external field are due to the nuclear magnetic moment and the resulting shape of a given hyperfine component of the absorption spectrum (e.g. $J = 2$, $F = \frac{9}{2} \leftarrow J = 1$, $F = \frac{9}{2}$ in a particular case) leads to a value for μ_I.

There are distinct rotational absorption lines for each isotopic species and microwave spectroscopy can be used for nuclear mass determination as a method of high precision. The method can be developed experimentally into one of high sensitivity and observable hyperfine patterns are obtainable with quantities of material as small as 1 μg. This makes possible a study of many radioactive isotopes produced by cyclotron and reactor bombardments as well as ordinary stable nuclei.

4.3.2 *Electron paramagnetic resonance*

Figure 4.1(b) shows the energy levels of an atom with $\mathcal{J}=\frac{1}{2}$ and $I=\frac{3}{2}$ in a magnetic field of flux density B_0. If this diagram is continued to a flux density

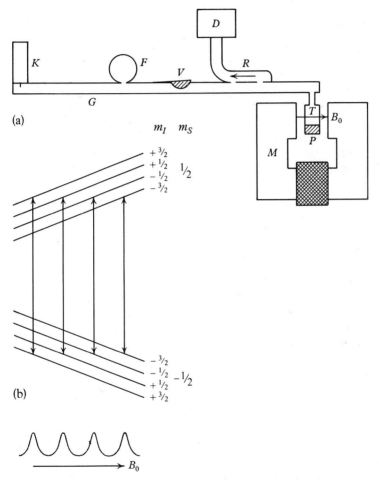

Fig. 4.3 (a) Paramagnetic resonance absorption apparatus. K is a klystron which generates a microwave of constant frequency in the waveguide G. The specimen P is placed in a resonant cavity T mounted in the field B_0 of a magnet M. R is a directional coupler which conducts the signal reflected from T to the crystal detector D. The signal level is controlled by the wavemeter F and attenuator V. The detector output is recorded at a fixed frequency as the magnetic field is varied. Frequencies of 3×10^4 MHz and fields B_0 of 0·5 T are typical (Van Wieringen, *Philips Tech. Review*, **19**, 301, 1958). (b) Energy levels of a paramagnetic ion with spin $S=\frac{1}{2}$ and nuclear spin $I=\frac{3}{2}$ in a field B_0. Absorption lines, corresponding to $\Delta m_I=0$, are indicated (Bleaney and Stevens, *Rep. progr. Phys.*, **16**, 108, 1953).

of 0.5 T $(x \approx 8)$ the spacing between the group of levels with $m_J = \frac{1}{2}$ and $m_J = -\frac{1}{2}$ becomes $W_{1/2} - W_{-1/2} \approx 7\Delta W \approx 40$ m^{-1}. This corresponds to a microwave quantum and absorption of energy in the microwave band, with reversal of the atomic moment, should be possible.

This possibility is realized most conveniently in the solid state. Many paramagnetic ions in crystalline solids behave as if their magnetic effect is due to a single electron spin and are characterized by a quantum number $S = \frac{1}{2}$. If their nuclei have a spin I, a hyperfine splitting arises, as in Fig. 4.1(b), due to the internal interaction between nuclear and electronic moments. In an external field B_0 each of the two electronic levels, $m_S = \pm \frac{1}{2}$, is split into $2I + 1$ hyperfine levels. Microwave absorption at a frequency $v_e \approx 2\mu_B B_0 / h$ reverses the electron spin direction with respect to the applied field and takes place between substates with the same value of m_I, because of the small interaction of the nuclear moment with the microwave field. For a given microwave frequency, $2I + 1$ absorption peaks are then found as the field B_0 is increased. This is shown in Fig. 4.3(b); a typical paramagnetic resonance absorption apparatus is shown in Fig. 4.3(a).

The nuclear spin I may be found by counting the absorption peaks and the nuclear moment μ_I may be obtained from the spacing between them if the internal field can be calculated. The method is extremely sensitive and has been widely used, not only for determinations of nuclear quantities but for investigations of the solid state. The mechanism by which spin reversal is achieved in electron paramagnetic resonance is the same as in the case of nuclear paramagnetic resonance and is discussed in Sect. 4.4.2.

4.4 Nuclear effects in radiofrequency spectroscopy

4.4.1 Orientation and precession of nuclear magnetic moments

An atom with a permanent magnetic moment in a free state will orient in a uniform magnetic field and the permitted orientations are specified by the quantum mechanical values of the magnetic quantum number m_J giving the component of total angular momentum along the magnetic field axis. This orientation was the basis of the Stern–Gerlach experiment on a collision free beam of atoms for the determination of magnetic moments of the order of μ_B, the Bohr magneton. Exactly similar orientation effects are observed with nuclei possessing a magnetic moment, and experiments of the Stern–Gerlach type can be adapted to measure nuclear moments and to indicate nuclear spins, although the deflections are scaled down by a factor of $\mu_N / \mu_B = m/m_p = 1/1836$ compared with those obtained in the electronic case.

In an assembly of atoms in which the magnetic effects of the electronic shells may be disregarded (e.g. for completed shells or sub-shells with $S = L = 0$) the nuclei, with spin I, will have $2I + 1$ orientations β with respect to the lines of force of an external magnetic field (Sect. **4.1**). Classically the magnetic field B_0 exerts a couple on the nucleus due to interaction with the dipole moment μ_I, but owing to the collinear angular momentum, the classically expected motion is a precession of the spin vector about the field direction (axis Oz) with an angular velocity

$$\omega_0 = \frac{\text{Couple}}{\text{Angular momentum}} = \frac{|\mu_I| \sin \beta B_0}{|\mathbf{I}| \sin \beta} = \gamma_I B_0$$
$$= \frac{g_I \mu_B B_0}{\hbar} \tag{4.18}$$

as may be seen from Fig. 4.4 and eqs. (4.3) and (4.5). This is the angular velocity of *Larmor precession*; it is independent of the orientation of the

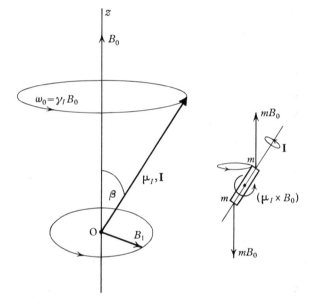

Fig. 4.4 Precession of nuclear moments. The inset diagram represents the classical case of a spinning magnet in a steady field. If a field B_1 is applied in the plane yOx and caused to rotate with angular velocity ω_0, a torque on μ_I is maintained and the angle β alters.

magnetic moment with respect to the z-axis, i.e. it is the same for any of the $2I+1$ orientations of $\mathbf{I}$ permitted quantum-mechanically. For a free proton the value of γ_p (see App. 9) leads to an angular velocity of precession

$$\omega_0 = 2 \cdot 6753 \times 10^8 B_0 \text{ rad s}^{-1}$$

or a frequency of precession

$$v_0 = \frac{\omega_0}{2\pi} = 4 \cdot 258 \times 10^7 B_0 \text{ Hz} \tag{4.19}$$

For magnetic fields of flux density about $0 \cdot 5$ T, v_0 lies in the radiofrequency range of about 20 MHz.

An interesting example of the use of nuclear precession is found in the work of Hillman, Stafford and Whitehead† on the polarization of high energy neutron beams. In these experiments it was at first thought necessary to move a rather complex counting apparatus round a target to investigate asymmetries of scattering. Later it was found much simpler to leave the counting apparatus in position and to rotate the direction of the neutron spin by passing the beam through a solenoid of length l producing a field B_0. For neutrons of 100 MeV energy the spin vector is rotated through an angle of $90°$ for $lB_0 = 1·33$ T m. The same method has been used (Sect. 21.2.2) in a direct determination of the gyromagnetic ratio of the free electron.

4.4.2 *Nuclear paramagnetism and nuclear resonance*

The energy associated with the nuclear precessional motion is essentially the potential energy of the nuclear dipole in the field B_0, i.e.

$$
\begin{aligned}
W = -(\boldsymbol{\mu}_I . \mathbf{B}_0) &= -|\mathbf{I}|\omega_0 \cos \beta \\
&= -m_I \hbar \omega_0 \\
&= -m_I h \nu_0 \quad \text{from (4.2) and (4.18)} \quad (4.20)
\end{aligned}
$$

The $(2I+1)$ allowed orientations of $\boldsymbol{\mu}_I$ and $\mathbf{I}$ thus define a set of energy levels of spacing

$$
h\nu_0 = g_I \mu_B B_0 = \frac{\mu_I B_0}{I} \quad \text{from (4.5) and (4.18)}
$$

If all these levels were equally populated in an assembly of nuclei, no resultant magnetization would be observed when the field was applied. In fact the populations are not equal because of the Boltzmann factor $e^{-W/kT}$ which, if there is some thermal contact between the nuclear spin system and its environment, increases the relative number of nuclei in the lower energy states. The observed magnetization can now be evaluated exactly as in the case of a paramagnetic gas and the susceptibility is given by the Langevin–Brillouin formula for the case $\mu_I B_0 \ll kT$, i.e.

$$
\chi_I = \frac{\mu_0 N |\boldsymbol{\mu}_I|^2}{3kT} = \frac{\mu_0 N \gamma_I^2 \hbar^2 I(I+1)}{3kT} \quad \text{from (4.3) and (4.1)} \quad (4.21)
$$

where N is the number of nuclei per unit volume, k is Boltzmann's constant, and T is the absolute temperature. Since $\mu_I \approx 10^{-3} \mu_B$ this susceptibility is only about 10^{-6} of the usual static susceptibilities observed for atoms and is much too small to be determined directly with accuracy at room temperature although the nuclear paramagnetic effect has been demonstrated for solid hydrogen. It was pointed out by Gorter‡ however that the nuclear suscepti-

† P. Hillman, G. H. Stafford and C. Whitehead, *Nuovo Cimento*, **4**, 67, 1956.
‡ C. J. Gorter, *Physica*, **3**, 995, 1936.

bility should show a resonance, fór oscillatory fields, at a frequency determined by the steady field B_0 in which the sample was placed. The development of magnetic resonance methods[†] has permitted the determination of gyromagnetic ratios with high accuracy.

To see how the resonance principle works, suppose that a small magnetic field of flux density B_1 is applied in the (x, y) plane (Fig. 4.4) and allowed to rotate with angular velocity ω, i.e.

$$B_x = B_1 \cos \omega t \qquad B_y = B_1 \sin \omega t \qquad (4.22)$$

The effect produced by this field can be described in two ways. Classically we can consider the interaction of B_1 with the resultant magnetization of a macroscopic sample of material. In such a sample the quantum mechanical conditions allow β to be considered as a continuous variable as may be seen from (4.2) if I is very large. The interaction is then a torque tending to alter β. If the rotation of the field is in the sense opposite to the precession of the resultant magnetic vector about B_0 or of a different frequency there is no recurrent effect, but if the sense is the same and $\omega = \omega_0$ (resonance), energy is exchanged between the field B_1 and the motion and the angle β alters. It may be shown (Ref. 4.6) that at resonance β becomes $\pi/2$ and the component of magnetization rotating in the (xy) plane thus rises to a maximum.[‡] Quantum mechanically we consider a single nucleus and envisage the absorption from the field of a quantum of resonance radiation which causes a transition (upward or downward) between the levels defined by (4.20). From (4.20) the transition between two adjacent levels corresponds to a unit change of magnetic quantum number, i.e.

$$\Delta m = \pm 1$$

Since the transition only alters the magnetic moment of the system, it may be classified as a magnetic dipole (M1) interaction.

On either picture of the process there is a reorientation of the nuclear moments which can be detected as a reaction in the external radiofrequency circuit supplying the field B_1 or by direct induction effects in a pickup coil. The result of the energy exchange is to reduce the magnetization given by (4.21), i.e. to equalize the populations of adjacent levels; alternatively we may regard the energy supplied by B_1 as raising the 'spin temperature' of the system. The observable effect therefore saturates and after removal of the resonance field B_1 a finite time (relaxation time) must elapse before the original magnetization is re-established by thermal contact with the normal temperature lattice in which the nuclei are immersed. A study of these relaxation times gives important information on the structure of solids and liquids.

[†] I. I. Rabi *et al.*, *Phys. Rev.*, **53**, 318, 1938; **55**, 526, 1939.

[‡] If the frequency ω or the field B_0 is altered through the resonance value, the angle β will pass from 0 to π approximately, so that the spin system is inverted. This so-called adiabatic fast passage is fully described in Ref. 4.6; it is used in sources of polarized ions for accelerators.

The saturation effect has in fact been used† to permit a direct determination of the nuclear paramagnetic susceptibility for the protons in water: application of the resonant frequency for a given static field destroys the nuclear magnetism and the effect can be seen by a sensitive balance technique.

Historically nuclear magnetic resonance was first observed by Rabi and his collaborators using an atomic beam. Gorter's earlier suggestion that nuclear reorientation effects should be detectable in ordinary material was however confirmed in 1946 by two groups of workers whose methods well illustrate the two ways of regarding the phenomenon. These macroscopic methods both depend upon the fact that although the nuclear magnetization is very small, a sufficiently large sample will provide a measurable effect at resonance. The first experiments were made with protons, which have the largest known g-factor and the known spin $I = \frac{1}{2}$. In this case magnetic resonance in a field B_0 corresponds to reversal of the spin direction and the frequency v for resonance is given directly by the energy equation

$$h v = 2 \mu_p B_0 \tag{4.23}$$

which also follows from (4.18).

4.4.3 Observation of nuclear magnetic resonance absorption (NMR)

(*a*) In the *nuclear induction experiment* of Bloch, Hansen and Packard‡ represented in Fig. 4.5 a sample of a suitable diamagnetic material such as water is placed in a steady field of flux density $B_z = B_0$. An inducing coil surrounds the sample and provides an oscillatory field

$$B_x = 2B_1 \cos \omega t$$

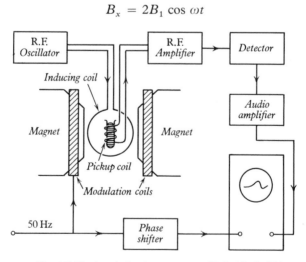

Fig. 4.5 Nuclear induction apparatus (Ref. 4.2, Smith).

† J. A. Poulis *et al.*, *Proc. Phys. Soc.*, **82**, 611, 1963.

‡ F. Bloch, W. W. Hansen and M. Packard, *Phys. Rev.*, **69**, 127, 1946; **70**, 474, 1946.

This field is equivalent to two oppositely rotating fields of the form (4.22), and the field rotating in the direction opposite to that of the precession about B_0 may be disregarded. A pickup coil with its axis at right angles to both B_0 and B_1 (and therefore relatively insensitive to direct pickup from the field B_1) is used to detect reorientation of the rotating magnetization. The output voltage of the pickup coil is amplified and displayed as the y-deflection on an oscilloscope. It is convenient in practice to modulate the field B_0 by a small 50 Hz field so that the resultant magnetic field sweeps through the resonance value $B_0 = \omega/\gamma_I$; the 50 Hz modulation can then provide the time base for the oscilloscope.

The observation of a signal in a nuclear induction experiment shows that the applied frequency ω is in resonance with the Larmor frequency of the protons in the solution in the applied field B_0 and that reorientation of the macroscopic nuclear magnetic moment is taking place. The shape of the signal depends on the relaxation time for the protons in the particular solution used and on the internal fields in the specimen. In the original experiment a sample of volume about 1·5 cm^3 was used and with a frequency of 7·765 MHz the resonance field was $B = 0·1826$ T and signals of about a millivolt were obtained on a 10-turn receiver coil.

(b) In the *nuclear resonance absorption experiment* of Purcell, Torrey and Pound,[†] the sample forms part of a resonant circuit and the transmission of radiofrequency power through the circuit is measured as a function of a steady magnetic field B_0 applied to the sample. The radiofrequency oscillatory field B_1 developed by the circuit is at right angles to B_0. At the nuclear resonance field $B_0 = \omega/\gamma_I$ the extra absorption of power by the precessing nuclei lowers the Q-value of the resonant circuit, and a sharp change in the output power may be detected.

In the original experiment a cavity resonator filled with solid paraffin was used, and the resonant frequency was 29·8 MHz. The output from the cavity was balanced against a signal derived from the input in a bridge circuit, so that the resonant absorption could be clearly displayed. As the field B_0 was varied through the nuclear resonance value of 0·7100 T (conveniently by 50 Hz modulation), a 50 per cent change in output power occurred. A typical signal is shown in Plate 5.

The elegant methods outlined in this section essentially measure nuclear gyromagnetic ratios, usually in terms of the accurately known value for the proton. If the spin number I is known, then the nuclear magnetic moment μ_I may be obtained in absolute units. If I is not known then the gyromagnetic ratio yields alternatively the nuclear g-factor by eq. (4.4). The spin I may in fact be found from the resonance experiment itself if there are internal electric fields which disturb the even spacing of the magnetic substates. There is then a fine structure of $2I$ lines instead of a single resonance frequency. The induction method is able to give the sign of the nuclear magnetic moment if a

† E. M. Purcell, H. Torrey and R. V. Pound, *Phys. Rev.*, **69**, 37, 1946; **73**, 679, 1948.

rotating field, with a known sense of rotation, is applied from two coils with their axes set at 90°.

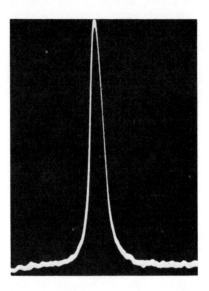

Plate 5 Nuclear resonance signal from protons in ferric nitrate solution (N. Bloembergen *et al., Phys. Rev.*, **73**, 686, 1948).

The relative simplicity of the apparatus required makes nuclear resonance very suitable for precision measurements of magnetic fields and many magnetometers have been based on the principle. Conventional methods require fairly large quantities of material, but for radioactive nuclei polarized by low temperature or nuclear reaction techniques the sensitivity of NMR may be improved by a very large factor by observing the emission of beta† or gamma radiation.‡ Anisotropy of the radiation pattern is reduced or destroyed at the resonant frequency. Magnetic resonance is of course important for the study of the relaxation effects due to the interaction of the nuclear spin system and a solid lattice (Refs. 4.1, 4.6). This forms a large subject in itself: for the purposes of nuclear physics, rather more information has been obtained from nuclear magnetic resonance in matter in such a state that relaxation effects may be entirely disregarded, i.e. in molecular and atomic beams.

4.4.4 *Nuclear moments from molecular and atomic beam experiments*

Figure 4.6(a) shows a typical molecular beam apparatus of the type used by Stern and Gerlach to determine atomic magnetic moments. The essential

† T. Tsang and D. Connor, *Phys. Rev.*, **132**, 1141, 1963; K. Sugimoto *et al., J. Phys. soc. Japan*, **21**, 213, 1966.
‡ E. Matthias and R. J. Holliday, *Phys. Rev. Lett.*, **17**, 897, 1966.

parts are a source of atoms or molecules, a magnet providing an inhomogeneous deflecting field, a detector and a vacuum envelope maintained at such a low pressure that the beam is essentially collision free. If molecules such as H_2 or D_2, with no resultant magnetic moment due to the electronic motion are used, the deflection observed on the detector when the magnetic field is switched on is due either to the nuclear moments or to moments associated with the rotation of the molecule. Both effects are small and slow molecules and long magnetic fields are essential, but Stern† and his collaborators were able in 1933 to make the first measurements of the proton moment by this method.

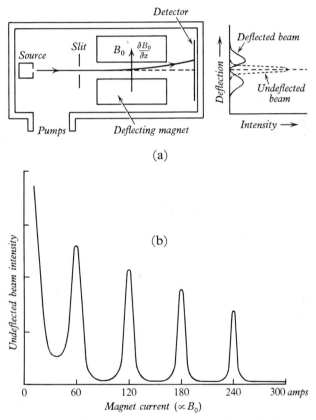

Fig. 4.6 (a) Direct determination of atomic moments by the Stern–Gerlach method. (b) Zero-moment method for a nucleus with spin 9/2. The undeflected beam is observed with a narrow detector as a function of deflecting field B_0 (Ref. 4.2, Smith).

Deflection patterns dependent on nuclear spin may also be observed in some cases with atomic beams, when electronic moments mainly determine the displacement (since $\mu_B \approx 10^3 \mu_N$). The most successful method of experiment, known as the *zero moment technique* depends on the fact that the

† R. Frisch and O. Stern, *Zeits. fur Physik*, **85**, 4, 1933.

effective magnetic moment μ_{eff} vanishes at certain values of the applied field B_0 for nuclear spins greater than $I=\frac{1}{2}$, as may be seen in Fig. 4.1(c). In an apparatus of the sort shown in Fig. 4.6(a), equipped with a detector of extent comparable with the beam width, the observed beam will have maximum intensity for the zero-moment values of the magnetic field B_0. From these values of B_0 the nuclear moment can be found; typical results are shown in Fig. 4.6(b).

The most important development of the technique of molecular beams is the application of the nuclear magnetic resonance principle of Rabi, Zacharias, Millman and Kusch, as illustrated in Fig. 4.7(a). In this apparatus neutral

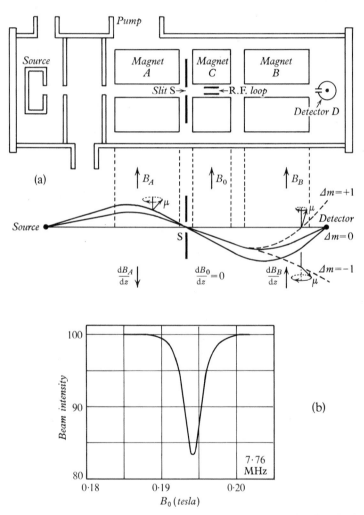

Fig. 4.7 (a) Molecular beam magnetic resonance apparatus (Ref. 4.2, Smith). (b) Resonance curve for the ^{19}F nucleus observed in NaF (Rabi *et al.*, *Phys. Rev.*, **55**, 526, 1939).

atoms or molecules with zero electronic angular momentum (atomic state $\mathscr{J} = 0$ or molecular Σ state), but with a magnetic moment μ_I associated with nuclear spin, leave a source as shown and enter an inhomogeneous magnetic field of density B_A. The magnetic moment vector orients with respect to the lines of force of the field B_A, and because the field is inhomogeneous the particles are deflected to an extent depending on the resolved part of μ_I. Those with certain suitable initial angles of travel will be able to pass through the collimating slit S and be brought to a focus on a detector D by a second field B_B parallel to B_A, whose gradient is reversed. If a steady homogeneous field B_0 is applied at C, parallel to the fields B_A, B_B there is no resulting change in the magnetic state of the beam but a set of energy levels spaced by $h\nu_0$, where ν_0 is the Larmor frequency, is defined, as discussed in Sect. 4.4.2. A field B_0 of flux density about 0.5 T is sufficient to decouple the nuclei of a molecule so that they may be regarded as free. If now a radio-frequency magnetic field B_1 is produced by a coil parallel or perpendicular to B_0 there will be transitions between the energy levels when the resonance condition $\nu = \nu_0$ is fulfilled. If B_1 is perpendicular to B_0 transitions with $\Delta m_I = \pm 1$ may take place and the magnetic state of the beam is altered, i.e. the effective moment changes. The molecules which have suffered transitions enter the field B_B with an effective moment different from its value at field B_A and are not refocussed at the detector (see Fig. 4.7(a)) so that the intensity drops. At resonance a large fraction of the molecules may undergo transitions with a suitably designed apparatus. The transmitted beam intensity is measured as a function of B_0 or the radiofrequency ν and resonance dips of the form shown in Fig. 4.7(b) are obtained. The detector D may be a surface ionization instrument in which incident molecules (or atoms) with low ionization potentials are ionized by impact on a heated tungsten wire and are then detected electrically. In the important case of hydrogen, a less sensitive detector depending on the change of temperature of a fine wire due to accumulation of gas molecules in a surrounding chamber was used.

Molecular beam experiments have yielded many nuclear g-values. The sign of the magnetic moment may be obtained if a magnetic field rotating in one direction is used instead of an oscillatory field. Deviations of the observed pattern of resonances from expectation may be due to electric quadrupole interactions and the first evidence for the existence of an electric quadrupole moment of the deuteron was obtained in this way.

The magnetic resonance method has also been extensively used with atomic beams, for which much larger deflections are obtained since the magnetic moments are of the order of μ_B. In certain forms of the apparatus the magnet B and the detector may be set to respond only to atoms which have undergone a particular type of transition, which reverses the electron spin, in the radio-frequency field (*'flop in' method*). Good intensity is achieved because for a given transition, all source velocities are focussed. The nuclear spin may easily be deduced from weak field observations with such an apparatus for atoms in a $^2S_{1/2}$ state. In this case the Zeeman splitting, from (4.10), is $\mu_B B_0 / hF =$

$1{\cdot}40 \times 10^4 B_0/F$ MHz. For a given field the frequency is set to a sequence of values indicated by this method until 'flop-in' transitions with $\Delta F = 0$, $\Delta m_F = \pm 1$ are observed; F is then determined and thence $I(=F-\frac{1}{2})$. A suitable transition is shown in Fig. 4.1(b). When the nuclear spin has been determined the hyperfine structure splitting ΔW may be found by observations at higher fields. The nuclear magnetic moment (and electric quadrupole moment) may be deduced from the observed splitting if electronic wave functions are well enough known. This general method has been widely applied to the measurements of spins and moments of radioactive isotopes, for which detection methods are especially sensitive. A mixture of several active isotopes can be studied by setting the apparatus for a particular spin I and observing the decay curve of the activity collected at this setting. Figure 4.8 shows the results of an investigation of the light caesium isotopes by this method. The observed nuclear spins can also be associated with a particular mass number by including an ionizer and a mass spectrometer in the detection equipment.

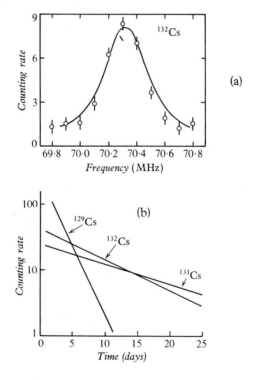

Fig. 4.8 Determination of spins and moments of radioactive caesium isotopes by the 'flop-in' method. (a) Activity of ^{132}Cs collected as a function of frequency for a fixed magnetic field. (b) Decay of activity of Cs isotopes collected at frequencies set for the appropriate spins. (From W. A. Nierenberg *et al.*, *Phys. Rev.*, **112**, 186, 1958; see also E. H. Bellamy and K. F. Smith, *Phil. Mag.*, **44**, 33, 1953.)

4.4.5 *Neutron moment measurement*

The molecular beam technique was adapted by Alvarez and Bloch† to measure the magnetic moment of the neutron. Use was made of the magnetic scattering of slow neutrons, as a result of which a slow neutron beam in passage through a strongly magnetized iron plate becomes partially polarized. The resulting spin distribution is then favourable for easy passage through a second similarly magnetized plate. If however a radiofrequency field is

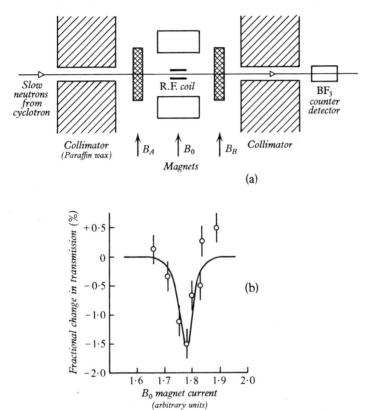

Fig. 4.9 Determination of magnetic moment of neutron by magnetic resonance method. (a) Schematic of apparatus. (b) Transmission dip at resonance obtained by Alvarez and Bloch at a frequency of 1·84 MHz with $B_0 = 0·0622$ T.

applied in a steady field of flux density B_0 between the two magnetized plates the beam will tend to become depolarized at the resonance field, since the number of 'upward' and 'downward' transitions will be proportional to the number of neutrons with spins in the corresponding states. Any reduction in polarization will result in a drop of transmission of the beam through the two magnetized plates, and observation of such a drop may be used to determine

† L. Alvarez and F. Bloch, *Phys. Rev.*, **57**, 111, 1940.

the resonance field B_0 for a given radiofrequency. The neutron moment then follows from an equation similar to that already given for proton resonance,

$$hv = 2\mu_n B_0 \qquad (4.24)$$

which gives the energy necessary to reverse the neutron spin in the field B_0 (i.e. to cause a transition between the states with $m_I = \pm\frac{1}{2}$).

The apparatus of Alvarez and Bloch is sketched in Fig. 4.9(a). The source of particles was the ^{9}Be (d, n reaction in a cyclotron, with suitable moderators to slow down the fast neutrons to thermal energies, and collimating channels in hydrogenous material to define a beam. The iron plates A and B, which may be regarded as polarizer and analyser, had their magnetization parallel to the steady field B_0 in which the radiofrequency transitions were induced. Figure 4.9(b) shows the small but quite definite resonance dip in transmission observed.

By this important experiment and later improvements, in particular the use of magnetized mirrors as polarizer and analyser and of two separated oscillatory fields,[†] the neutron moment was shown to be negative, and of value

$$\mu_n = -1\cdot913148 \pm 0\cdot000066 \text{ nuclear magnetons}$$

In separate experiments[‡] a strong electric field was applied parallel to the steady magnetic field B_0 and an upper limit of $|d_n/e| < 5 \times 10^{-25}$ m was set to the *electric dipole moment* of the neutron. The proton electric dipole moment is also known, from a molecular beam resonance experiment,[§] to be $|d_p/e| < (7 \pm 9) \times 10^{-23}$ m. An elementary particle can have no electric dipole moment if parity and time reversal invariance (Chapter 21) are valid for the electromagnetic interaction.

4.4.6 The double resonance technique

An interesting method of investigating excited atomic states, which can provide nuclear information, was introduced by Kastler and Brossel.[||] The method is essentially a paramagnetic resonance experiment on electrons in excited states, and as in ordinary paramagnetic resonance, hyperfine effects may be observed. The technique may be illustrated for the sodium atom, in which the optical line of wavelength 589 nm arises as a result of transitions between the $^2P_{3/2}$ and $^2S_{1/2}$ levels. If the atom is placed in a weak magnetic field of flux density B_0, the Zeeman states of these levels are as shown in Fig. 3.6, in which hyperfine structure is neglected. Absorption of unpolarized resonance radiation by the atom creates equal populations of the substates of

[†] V. W. Cohen, N. R. Corngold and N. F. Ramsey, *Phys. Rev.*, **104**, 283, 1956.
[‡] J. V. Baird, P. D. Miller, W. B. Dress and N. F. Ramsey, *Phys. Rev.*, **179**, 1285, 1969.
[§] G. E. Harrison, P. G. H. Sanders and S. J. Wright, *Phys. Rev. Lett.*, **22**, 1263, 1969.
[||] Reviews are given by A. Kastler, *Proc. phys. Soc.*, **67**, 853, 1954 and by G. W. Series, *Rep. progr. Phys.*, **22**, 280, 1959.

the level† but if plane polarized radiation, with an electric vector parallel to the magnetic field, is used the selection rule (Sect. 3.5.3)

$$\Delta m_J = 0$$

permits only the formation of the states with $m_J = \pm\frac{1}{2}$. This is shown in Fig. 4.10(a), which displays the Zeeman states in a convenient way. This

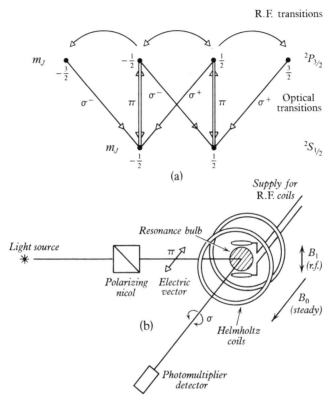

Fig. 4.10 (a) Optical and radiofrequency transitions between Zeeman states of the $^2P_{1/2} \rightarrow \ ^2S_{1/2}$ (589 nm) line of the sodium atom. The diagram shows excitation in plane polarized light. (b) Apparatus for optical double resonance experiment, showing excitation by π light and observation of σ light (Series).

figure also shows that these states decay by emission of a mixture of plane polarized (π) radiation and circularly polarized (σ) radiation corresponding to downward transitions with $\Delta m_J = 0$ and ± 1 respectively. If the scattered radiation is observed along the lines of force of the field B_0 only circularly polarized components will be seen. Figure 4.10(b) shows the experimental arrangement for detection of this radiation. Other combinations of magnetic field directions and polarization may be used.

† In a weak field the Zeeman splitting of both the P and S states is small compared with usual (Doppler) line widths.

Suppose now that a radiofrequency field B_1 is applied in the xy plane. Then, at a frequency v given approximately by

$$hv = g_J \mu_B B_0$$

where g_J is the electronic g factor for the $^2P_{3/2}$ state, transitions with $\Delta m_J = \pm 1$ can take place, transferring excited atoms with $m_J = \pm\frac{1}{2}$ to the $m_J = \pm\frac{3}{2}$ substates (Fig. 4.10(a)). These latter states can decay only by emission of σ radiation, and the proportion of this radiation in the scattered light therefore increases. The intensity of radiation observed along the field axis can therefore be used to determine when electron paramagnetic resonance absorption is taking place in the excited state. In a typical experiment a fixed radiofrequency of about 40 MHz is used and the field B_0 is varied in the neighbourhood of 2 mT until an increase in σ radiation is detected. The resonance line shows structure as already discussed for paramagnetic resonance (Sect. 4.3.2) from which the hyperfine constants a and b may be deduced. From these, in turn the nuclear moments μ_I and Q_I may be calculated. The method is particularly valuable in cases where the atomic ground state has $\mathcal{J} = \frac{1}{2}$, since the quadrupole moment cannot be found by observation on such states alone. The technique is sufficiently sensitive to allow the use[†] of radio-active atoms e.g., ^{109}Cd.

4.5 Summary

The orders of magnitude of the effects of nuclear magnetic moments in spectroscopy are indicated in Sects. 4.2 and 4.4. In zero magnetic field, the hyperfine splitting of atomic ground states (Fig. 4.1(a)) is about 6 m^{-1} (1800 MHz). For paramagnetic ions in crystals the hyperfine splitting is rather less, $\approx 0\cdot 5 \text{ m}^{-1}$ (150 MHz).

Hyperfine structure cannot usually be fully resolved in optical lines which arise as $E1$ transitions between multiplets but optical spectroscopy is important historically and also for studies of the isotope shift.

An external magnetic field of flux density about $0\cdot 1$ T may decouple I and $\mathcal{J}$ almost completely, but internal fields of 10–100 T still exist and hyperfine structure is still observable. Magnetic resonance transitions ($M1$) within multiplets are observable in zero field but their main application is in externally applied fields in which case the resonance frequencies are given by

$$v_e/B_0 \text{ (paramagnetic resonance)} = 2\cdot 8 \times 10^4 \text{ MHz T}^{-1}$$

$$v_0/B_0 \text{ (proton resonance)} \qquad = 42\cdot 6 \text{ MHz T}^{-1}$$

Both optical and magnetic resonance measurements of the hyperfine structure give the coupling constants a, b from which nuclear moments may be obtained if internal atomic fields are known. Optical spectroscopy may give I directly. Magnetic resonance in systems in which strong internal fields do

† M. N. McDermott and R. Novick, *Phys. Rev.*, **131**, 707, 1963.

not arise, such as molecular beams and many solids and liquids, give g_I, from which μ_I may be calculated if I is known.

The following table lists the main quantities determined by the various techniques and indicates their general sensitivity (Refs. 4.2–4.4).

TABLE 4.1 Determination of nuclear constants by spectroscopic methods

Method	Quantity determined	Sensitivity
1. Optical isotope shift	Nuclear size	10^{-9} kg
2. Optical hfs	I, a, b	—
3. Molecular spectra	I, statistics	—
4. Microwave absorption	I, b, mass	10^{-12} kg
5. Microwave paramagnetic resonance	I, g_I (ratio)	10^{-15} kg
6. Deflection of atomic and molecular beams	I, a, b	—
7. Nuclear magnetic resonance	g_I, I	10^{-12} kg (beams) 10^{-6} kg (liquids and solids)
8. Double resonance	a, b	—
9. Mössbauer effect (special cases)	B_i, μ_I, b	—
10. Muonic atoms	a, b	—

References

1. ANDREW, E. R. *Nuclear Magnetic Resonance*, Cambridge University Press, 1955; D. J. E. Ingram, 'Nuclear Magnetic Resonance', *Contemp. Phys.*, **7**, 13 and 103, 1965.
2. BLIN-STOYLE, R. J. *Theories of Nuclear Moments*, Oxford University Press, 1957; K. F. Smith, 'Nuclear Moments and Spins', *Progr. nucl. Phys.*, **6**, 52, 1957; H. Kopferman, *Nuclear Moments*, Academic Press, 1958.
3. RAMSEY, N. F. *Molecular Beams*, Oxford University Press, 1956; K. F. Smith and P. J. Unsworth, 'Molecular Beam Spectroscopy', *Sci. progr.*, **53**, 45, 1965.
4. INGRAM, D. J. E. *Spectroscopy at Radio and Microwave Frequencies* (2nd edn.), Butterworth, 1967.
5. JOHNSON, R. C. *Molecular Spectra*, Methuen, 1949.
6. ABRAGAM, A. *The Principles of Nuclear Magnetism*, Oxford University Press, 1961.
7. GOUDSMIT, S. A. 'Pauli and Nuclear Spin', *Physics Today*, **14**, No. 6, 18, 1961.

B

Experimental techniques of nuclear physics

5 General properties of ionizing radiations

The liberation of charge, or ionization, in ordinary neutral matter due to the passage of radiation has proved a property of the utmost importance for the development of nuclear physics. Ionizing radiations have long been familiar, and as a result of the many beautiful early experiments on the gaseous discharge the following types of ionizing radiation were distinguished:

(a) *the positive rays* (or canal rays) which are now known as positive ions and which emerge through a hole in the cathode of a discharge tube,

(b) *the cathode rays*, which are electrons and which were used for many striking demonstrations because of the ease with which they could be deflected by a magnet,

(c) *X-rays*, which could not be deflected by electromagnetic fields at all.

These radiations were all produced by the application of an electric field to a gas at a low pressure. The discovery of naturally occurring radioactivity provided a spontaneous source of similar radiations which were classified, as has been seen in Chapter 2, into corresponding groups, namely the α-particles, β-particles and γ-rays.

In low energy nuclear physics a general division of radiations into heavy particles, light particles and electromagnetic radiation is familiar and clear. In high energy physics, where energies of many times the rest mass of the proton are encountered, the distinction is rather between radiations which interact strongly with nuclei and those which do not but the ionization produced by such particles is still an important experimental property.

The general types of ionization to be discussed are illustrated in Plates 1, 2 and 3 which show the tracks of α-particles and of fast and slow electrons in a cloud chamber. There are other forms of energy loss which contribute to the absorption or retardation of radiations in matter and these will also be outlined. Many of the phenomena to be discussed are atomic, in the sense that the interaction is between an incident particle or photon and a free or bound electron; but some, including the celebrated α-particle scattering process which led to the discovery of the nucleus, yield specifically nuclear information, e.g. the charge and size of the scattering centre. Most of the interactions described have been applied in the design of nuclear detectors (ch. 6).

and opposite momenta

$$\frac{M_1 M_2}{M_1 + M_2} v_1 \quad \text{or} \quad M_0 v_1$$

where

$$M_0 = \frac{M_1 M_2}{M_1 + M_2}$$

is the reduced mass of the pair. In elastic scattering the particles are each turned through the same angle θ in the centre-of-mass system, since the total momentum must remain zero, and the relation between θ, which is required for analysis, and θ_L, which is actually observed, may be obtained as shown in the vector diagram 5.1(c), i.e.

$$\tan \theta_L = \frac{\sin \theta}{\cos \theta + M_1/M_2} \tag{5.3}$$

From the diagram it is also seen that

$$\phi_L = \tfrac{1}{2}\phi = \frac{\pi}{2} - \frac{\theta}{2} \tag{5.4}$$

In all collision processes an energy of $\tfrac{1}{2}(M_1 + M_2)v_c^2$ is associated with the centre-of-mass motion and is not available for producing internal effects such as nuclear excitation. The energy remaining from the initial kinetic energy is thus

$$\tfrac{1}{2}M_1 v_1^2 - \tfrac{1}{2}(M_1 + M_2)\frac{M_1^2 v_1^2}{(M_1 + M_2)^2} = \frac{1}{2}\frac{M_1 M_2}{M_1 + M_2} v_1^2$$
$$= \tfrac{1}{2}M_0 v_1^2 \tag{5.5}$$

which shows that the behaviour of the initial system in centre-of-mass co-ordinates may be described in terms of that of a particle of the reduced mass M_0 moving with the laboratory system incident velocity v_1. It is this energy† which can be used to initiate a nuclear reaction (Sect. 14.2.1).

5.1.3 Cross-sections; beam attenuation

The probability of occurrence of a particular collision process, such as the elastic scattering of a particle through a certain angle, is conveniently expressed as a *cross-section*. The origin of this concept may be understood from Fig. 5.2.

Consider a parallel beam of n_0 particles per second incident on a thin slice of material of thickness t containing N scattering or absorbing centres per

---

† The energy associated with the centre-of-mass motion may also be written $\tfrac{1}{2}p^2/M$, where p is the incident momentum and M is the mass of the compound system. This expression may be used in calculation of the available energy for an incident photon (Sect. 13.6.4).

unit volume. If the probability of collision of any sort is determined by an area σ ascribed to each centre, then for a single incident particle the chance of collision in passing through the thin lamina dx of area A is

$$\frac{NA\,dx\,\sigma}{A} = N\sigma\,dx = \frac{dx}{\lambda}$$

where $\lambda = 1/N\sigma$ is the 'mean free path' for collision. If each interaction removes the particle from the beam the attenuation is then described by the equation

$$dn = -nN\sigma\,dx$$

which gives

$$n_x = n_0 e^{-N\sigma x} = n_0 e^{-\mu x} \tag{5.6}$$

where $\mu = N\sigma = 1/\lambda$, measured in m^{-1}, is known as the *linear attenuation coefficient* of the material for the incident beam. The simple exponential formula is only valid if the collision area is essentially constant for the range of particle energies in the beam at all points within the lamina.

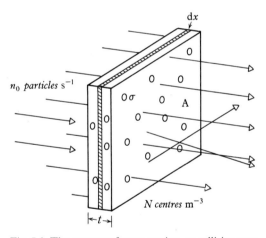

Fig. 5.2 The concept of cross-section, or collision area.

The quantity $\mu_m = \mu/\rho$, where ρ is the density, is the *mass attenuation coefficient* and is measured in $m^2\,kg^{-1}$. It is useful to note that

$$\mu_m = \frac{\mu}{\rho} = \frac{N\sigma}{\rho} = \frac{\sigma}{m_A}$$

where m_A is the mass of a scattering centre.

In the thickness t of material

$$n_t = n_0 e^{-N\sigma t} \tag{5.7}$$

and if now attention is directed to the number of collisions which have actually taken place, namely the yield Y of the process, we find

$$Y = n_0 - n_t = n_0(1 - e^{-Nσt})$$
$$\approx n_0 Ntσ \tag{5.8}$$

if the attenuation is small. Now Nt is the number of scattering centres per unit area of the material perpendicular to the incident beam and since from (5.8)

$$σ = \frac{Y}{n_0 Nt} \tag{5.9}$$

we have the definition of the *total cross-section* $σ$ as numerically equal to the probability of an event for one particle incident on a sample of material containing one scattering centre per m^2 of projected area. From (5.9) $σ$ is measured in m^2 per atom or nucleus; for atoms, units of $πa_0^2$, where a_0 is the Bohr radius, are often used and for nuclei the *barn*, equal to 10^{-28} m^2, is appropriate. The ratio n_t/n_0 is usually known as the *transmission* of the sample.

If we consider only scatterings or other events which are observed in a small solid angle $dΩ$ at a certain specified angle $θ$ with the incident beam then we may define a *differential cross-section*

$$dσ = σ(θ)\, dΩ \quad \text{(m}^2 \text{ per atom or nucleus)} \tag{5.10}$$

which is the collision area associated with this type of event for each centre. If the phenomenon is independent of azimuth about the incident beam direction then

$$dΩ = 2π \sin θ\, dθ \tag{5.11}$$

and

$$dσ = 2π \sin θ\, σ(θ)\, dθ \tag{5.12}$$

from which the total cross-section is obtained by integration

$$σ = 2π \int_0^π σ(θ) \sin θ\, dθ \tag{5.13}$$

The quantity $σ(θ) = dσ/dΩ$ gives the *angular distribution* of the particular events; it is sometimes itself described as the differential cross-section, but strictly it is the differential cross-section per unit solid angle and is measured in m^2 (atom or nucleus)$^{-1}$ steradian^{-1}.

The angular distribution $σ_L(θ_L)$ observed in the laboratory system is related to the centre-of-mass angular distribution by the equation

$$σ(θ) \sin θ\, dθ = σ_L(θ_L) \sin θ_L\, dθ_L$$

which states that the same number of events is observed in each coordinate system for corresponding angles and solid angles. For *elastic scattering*, this gives, using (5.3),

$$\sigma_L(\theta_L) = \frac{(1+2\gamma\cos\theta+\gamma^2)^{3/2}}{1+\gamma\cos\theta}\,\sigma(\theta) \tag{5.14}$$

where $\gamma = M_1/M_2$. For *inelastic processes*, in which the particles M_1 and M_2 transform into particles M_3 and M_4 with an energy release Q, formulae (5.3) and (5.14) also hold but with

$$\gamma = \sqrt{\frac{M_1 M_3}{M_2 M_4}\frac{E}{E+Q}}$$

where $E = \frac{1}{2}M_0 v_1^2$.

In this book we shall assume, unless specifically stated to the contrary, that all experimental quantities are converted to the centre-of-mass system (θ, ϕ) for comparison with theory.

5.1.4 Collision between charged particles—the Rutherford scattering law

If the force of interaction between two particles is known, a detailed description of the collision process, subject still, however, to the general laws of Sect. 5.1.1, may be given.

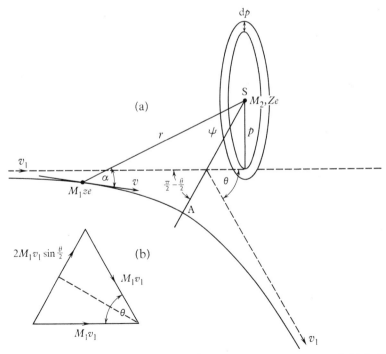

Fig. 5.3 Rutherford scattering: (a) Trajectory of a charged particle M_1 scattered elastically by a nucleus M_2 fixed at S. (b) Vector diagram for initial and final momenta.

We consider a particle of mass M_1, charge ze, and velocity v_1 approaching a stationary particle of mass M_2 and charge Ze as shown in Fig. 5.3(a) and again assume that the incident velocity is small compared with that of light. If the force between particles at distance r is accurately given by the law of the inverse square, i.e.

$$F = \frac{zZe^2}{4\pi\varepsilon_0 r^2} \tag{5.15}$$

then it is known from the general theory of central orbits that both for attractive and repulsive forces the particles describe hyperbolic orbits with respect to their centre-of-mass and with respect to one another. This solution was applied to the scattering of α-particles by matter by Rutherford (Sect. 2.5.1) and led to the theory of the nuclear atom. Rutherford's original calculation made use of the geometrical properties of the hyperbola (see Ref. 5.1) but the basic result can be obtained perhaps more simply as follows:

For simplicity we consider the case of an infinitely heavy centre of force M_2 and a repulsive interaction. The orbit of M_1 is then a hyperbola with M_2 in the outer focus S (Fig. 5.3(a)). Let the perpendicular distance from S to the direction of incidence, known as the *impact parameter*, be p.

The effect of the collision is to deviate the incident particle through an angle θ without changing its energy. The momentum change necessary results from the Coulomb force acting between the two charges and can be seen from Fig. 5.3(b) to be

$$\Delta q = 2M_1 v_1 \sin \theta/2 \tag{5.16}$$

in the direction of SA. The impulse due to the Coulomb force, resolved along SA, is

$$\int_0^\infty \frac{zZe^2}{4\pi\varepsilon_0 r^2} \cos \psi \, dt = \int_{-\infty}^\infty \frac{zZe^2}{4\pi\varepsilon_0 r^2} \cos \psi \, \frac{ds}{v} \tag{5.17}$$

$$= \frac{zZe^2}{4\pi\varepsilon_0} \int_{-(\pi/2 - \theta/2)}^{\pi/2 - \theta/2} \frac{\cos \psi}{r^2 v} \frac{r \, d\psi}{\sin \alpha} \tag{5.18}$$

where α is the angle between the radius vector and the trajectory. The law of conservation of angular momentum requires that

$$pv_1 = vr \sin \alpha \tag{5.19}$$

so that (5.18) becomes

$$\Delta q = \frac{zZe^2}{4\pi\varepsilon_0} \int_{-(\pi/2 - \theta/2)}^{\pi/2 - \theta/2} \frac{\cos \psi \, d\psi}{pv_1} = \frac{zZe^2}{2\pi\varepsilon_0 pv_1} \cos \frac{\theta}{2} \tag{5.20}$$

Equating this to the expression (5.16) we obtain

$$p = \frac{zZe^2}{4\pi\varepsilon_0 M_1 v_1^2} \cot \frac{\theta}{2} = \frac{b}{2} \cot \frac{\theta}{2} \tag{5.21}$$

where

$$b = \frac{zZe^2}{2\pi\varepsilon_0 M_1 v_1^2} \tag{5.22}$$

The quantity b is known as the *collision diameter* and is easily seen to be the distance of closest approach for a repulsive Coulomb interaction; it is a useful parameter also for the attractive interaction.

The assumption of infinite mass for M_2 may be removed by replacing M_1 in formula (5.16) by the reduced mass M_0. The angle of deflection in formula (5.21) then becomes the angle of deflection in the centre-of-mass system and

$$b = \frac{zZe^2}{2\pi\varepsilon_0 M_0 v_1^2} \tag{5.23}$$

These formulae may now be used to calculate the probability of scattering of the incident particle through an angle between θ and $\theta+d\theta$. Since this scattering corresponds to an impact parameter p given by (5.21) the differential cross-section, or collision area for this process is simply (Fig. 5.3)

$$d\sigma = 2\pi p \, dp$$

Substituting for p we obtain (cf. Sect. 5.1.3)

$$|d\sigma| = \sigma(\theta) \, d\Omega = 2\pi \frac{b^2}{8} \cot \frac{\theta}{2} \mathrm{cosec}^2 \frac{\theta}{2} \, d\theta$$

whence, using (5.11),

$$\sigma(\theta) = \frac{b^2}{16} \mathrm{cosec}^4 \frac{\theta}{2}$$

$$= \left(\frac{zZe^2}{8\pi\varepsilon_0 M_0 v_1^2}\right)^2 \mathrm{cosec}^4 \frac{\theta}{2} \tag{5.24}$$

which is the well-known scattering law of Rutherford, written here in centre-of-mass coordinates. From (5.8), (5.10) and (5.24), the actual number dY of particles observed in a solid angle $d\Omega$ is

$$dY = n_0 Nt \left(\frac{zZe^2}{8\pi\varepsilon_0 M_0 v_1^2}\right)^2 \mathrm{cosec}^4 \frac{\theta}{2} \, d\Omega \tag{5.25}$$

For 40 MeV α-particles incident on uranium the squared term has the value 0·02 barns; the $\mathrm{cosec}^4 \theta/2$ factor exhibits a rapid rise towards small angles, characteristic of the Coulomb interaction assumed. This rise however does not continue indefinitely because small angular deflections correspond with large impact parameters and for such distant collisions the nuclear charge will be screened by atomic electrons.

The derivation of the Rutherford formula given here assumes that the law of force is accurately of the Coulomb type to a distance of approach less than the collision diameter b. If this is not so, there will be deviations from the angular distribution and energy dependence predicted by (5.25). Quite independently of the law of force, the classical method of discussing collisions in terms of definite trajectories will fail if these trajectories cannot be defined wave mechanically. From (5.20), assuming θ to be small we see that the momentum transfer in the interaction for impact parameter p and velocity v is

$$\Delta q = \frac{zZe^2}{2\pi\varepsilon_0 pv}$$

The uncertainty in the impact parameter then cannot be less than

$$\Delta p \approx \frac{\hbar}{\Delta q} = \frac{2\pi\varepsilon_0 \hbar pv}{zZe^2}$$

and for classical orbits in the Coulomb field to be meaningful we must have $\Delta p \ll p$. This condition therefore requires that

$$\frac{zZe^2}{2\pi\varepsilon_0 \hbar v} \gg 1 \tag{5.26}$$

This inequality is independent of p, and is valid for both attractive and repulsive Coulomb forces. For 40 MeV α-particles incident on uranium the quantity on the left-hand side of this equation is approximately equal to 20 and the classical condition is well satisfied. It will usually be so for slow particles and for high nuclear charges. For fast particles incident on electrons or light nuclei, the condition is not fulfilled and the full wave-mechanical treatment of scattering (ch. 14) must be used. The wave mechanical solution agrees exactly with the classical theory in the case of scattering by a pure inverse square law field and thus leads to Rutherford's scattering law. An essential modification, peculiar to wave mechanics, enters when the incident and target particles are identical, since there is then coherence between the waves representing the scattered and recoil particles and interference effects are observed (Sects. 5.3.1 and 14.2.2).

The inequality (5.26) for the Coulomb field is equivalent, as may be seen from (5.22), to the statement that the de Broglie wavelength λ of the incident particle at infinite distance is much less than the collision diameter. For a field of force of finite range a the classical condition requires $\lambda \ll a$ and this is satisfied at *high* velocities rather than low.

5.2 Passage of charged particles through matter

5.2.1 General

The main process by which a charged particle loses energy in passing through matter is interaction with atomic electrons, through the Coulomb force. When the work necessary to excite these electrons to new levels or to remove them from the atom is small compared with the incident particle energy, this process may be regarded as elastic. There is also a similar loss of energy to the nuclei of the stopping medium but this is small in comparison with the energy loss to electrons except in the special case of ions such as fission fragments carrying several electrons with them. Nuclear encounters are, however, of prime importance in determining the scattering of charged particles. The direct removal of electrons from neutral atoms by the incident particle is the *primary ionization*; the electrons so produced may have an energy up to $4m/M$ times the kinetic energy of the incident particle (mass M)† and their tracks may be seen in expansion chamber photographs (Plate 4) or in nuclear emulsions. The ionization produced by these *delta-rays* is known as the *secondary ionization* and is difficult to calculate theoretically. If the δ-rays have such a short range that they do not move an observable distance from the primary track, the only measurable quantity is the *total ionization*. When the incident particle is no longer able to ionize it has reached the end of its *range* in the stopping medium and will revert to a neutral atom.

Nuclear excitation and transformation processes may remove particles from a beam and therefore contribute to the effective absorption. If such mechanisms become important in comparison with ionization losses it is no longer possible to define a range for a particle in matter and the intensity of an incident beam as a function of thickness of matter traversed is given by the exponential law (Sect. 5.1.3) where the attenuation coefficient μ is determined by nuclear cross-sections. Such a law was originally proposed to account for the behaviour of both α-particles and β-particles in matter. When observations with thin sources were made it became clear that α-rays were characterized by a definite maximum range and that β-rays had a similar property when account was taken of their considerable scattering.

The close encounters between a fast charged particle and nuclei may lead to sudden accelerations. These in turn result in radiation according to the laws of electrodynamics and the '*bremsstrahlung*'‡ thus excited is an important mechanism of energy loss for electrons. Another form of radiative loss, essentially different from bremsstrahlung, is the *Cherenkov radiation* which arises from longitudinal polarization of a transparent medium when a charged particle traverses it at a velocity exceeding the phase velocity of light in the medium.

If T is the kinetic energy of a particle moving through an absorbing medium we write, neglecting nuclear transformations

$$-\frac{dT}{dx} = \textit{specific energy loss} \ (\text{Jm}^{-1}) \ \text{or} \ \textit{absolute stopping power}$$

$$= \left(-\frac{dT}{dx}\right)_{\text{collision}} + \left(-\frac{dT}{dx}\right)_{\text{radiation}} \tag{5.27}$$

† The maximum electron velocity is obtained from (5.2), setting $M_1 = M$, $M_2 = m$ ($\ll M$) and $\cos \phi_L = 1$.

‡ Literally 'braking' or 'deceleration' radiation.

Table 5.1 gives approximate values of the energy loss of electrons and protons in water and Fig. 5.4 shows the variation of dT/dx with energy for these particles in their passage through lead.

TABLE 5.1 Energy loss of a charged particle in 10 mm of water

Particle	Energy (MeV)	$\beta = \dfrac{v}{c}$	Collision loss (MeV)	Bremsstrahlung loss	Cherenkov loss (keV)
Electron	100	1	2	2 MeV	2·7
Proton	1000	0·87	2	0·01 keV	1·65

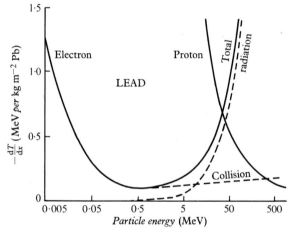

Fig. 5.4 Energy loss of electrons and protons in lead as a function of kinetic energy of particle (Ref. 5.12).

Energy loss by collision (Sect. 5.2.2) is to a good approximation the same for all particles of the same charge and velocity and it reaches a minimum at relativistic energies. At lower energies the collision loss varies as $1/v^2$, where v is the velocity of the particle, and above the minimum bremsstrahlung becomes important for electrons. For heavy particles radiative losses are negligible, but nuclear reactions become significant for protons even before the minimum in the collision loss is reached. The region of applicability of the formula for collision loss thus depends on the nature of the particle; it is most extensive for muons, which are too heavy to radiate and have a very weak interaction with nuclei.

It is useful in discussing the passage of electrons through matter to define a *critical energy* ε, at which the energy loss by collision per unit path is equal to that lost by radiation. The path length of absorber from which an electron of energy much greater than ε (i.e. when radiation loss is the dominant process) emerges with a fraction $1/e$ of its initial energy is known as the *radiation length* X_0. Table 5.2 gives values of these quantities for certain materials.

For energies well above the critical energy, radiation losses are proportional to T and to Z^2 (Sect. 5.2.3) while collision losses are mainly proportional to Z (Sect. 5.2.2) and nearly independent of T.

TABLE 5.2 Radiation lengths and critical energies (Ref. 5.2)

Material	Radiation length X_0/kg m^{-2}	Critical energy/MeV
H	580	400
C	446	102
Fe	141	24·3
Pb	65	7·8
Air	377	84·2
Water	371	83·8
Glass	100–110	13·0–16·0

5.2.2 *Classical theory of energy loss by collision*

When a heavy charged particle of mass M_2 ($=M$) and velocity v, collides with an electron of mass M_1 ($=m$, $\ll M$), the incident particle is essentially undeviated, i.e. θ may be set equal to zero in (5.20). Putting also $Z=1$ for the electron the momentum transfer is $\Delta q = ze^2/2\pi\varepsilon_0 pv$, and for non-relativistic motion the corresponding energy loss of the incident heavy particle is

$$Q = \frac{1}{2}\frac{(\Delta q)^2}{m} = \frac{z^2 e^4}{8\pi^2 \varepsilon_0^2 m p^2 v^2} \tag{5.28}$$

As the heavy particle passes through a thickness of absorber in which there are N atoms of atomic number Z per m^3 the number of collisions transferring energy between Q and $Q+dQ$ is the number of electrons within an annulus of area $2\pi p\,dp$, i.e. $NZ\,dx\,.2\pi p\,dp$. The total energy transferred, for all impact parameters, is then

$$-dT = 2\pi NZ\,dx \int_{p_{min}}^{p_{max}} Qp\,dp$$

$$= \frac{NZ}{4\pi\varepsilon_0^2}\,dx\,\frac{z^2 e^4}{mv^2}\int_{p_{min}}^{p_{max}}\frac{dp}{p}$$

whence

$$-\frac{dT}{dx} = \frac{z^2 e^4}{4\pi\varepsilon_0^2 mv^2}\,NZ\,\log\frac{p_{max}}{p_{min}} \tag{5.29}$$

This formula is incorrect, because the energy loss becomes infinite for $p_{min}=0$. In a more accurate derivation using eqs. (5.2), (5.4) and (5.23),

(5.28) is replaced by

$$Q = Q_0 \frac{b^2/4}{p^2 + b^2/4} \tag{5.30}$$

where b^2 is given by (5.23) with $M_0 = m$ the electron mass and Q_0 is the maximum energy transferred which is easily seen to be

$$Q_0 = \tfrac{1}{2}mv_2^2 \approx \tfrac{1}{2}m(2v)^2$$
$$= 2mv^2 = \frac{4m}{M}T \tag{5.31}$$

with $T = \tfrac{1}{2}Mv^2$ the initial kinetic energy of the heavy particle. The argument of the logarithm then becomes

$$\frac{p_{max}^2 + b^2/4}{p_{min}^2 + b^2/4} \tag{5.32}$$

In this p_{min} may be set equal to 0, corresponding to a head-on collision with the energy transfer Q_0, but p_{max} is more difficult to define. It cannot be made infinite since this leads to an infinite energy loss owing to the large number of distant collisions with small energy transfers. For collisions with these large impact parameters, however, the electron can no longer be considered as free and interaction with such distant electrons is therefore approximately adiabatic, i.e. without net energy transfer. Bohr pointed out that this would be so if the collision time ($\approx p/v$) were about equal to the period of vibration of the electron concerned in its parent atom. We thus put

$$p_{max} = \frac{v}{\omega} \tag{5.33}$$

where ω is a characteristic frequency which must be calculated from an atomic model. Assuming that $p_{max} \gg b/2$, the stopping power formula becomes

$$-\frac{dT}{dx} = \frac{NZ}{8\pi\varepsilon_0^2} \frac{z^2 e^4}{mv^2} \log \frac{4p_{max}^2}{b^2} \tag{5.34}$$

$$= \frac{z^2 e^4}{mv^2} \frac{NZ}{4\pi\varepsilon_0^2} \log \frac{4\pi\varepsilon_0 mv^3}{\omega z e^2} \tag{5.35}$$

using (5.23) and (5.33).

An expression of this form was first given by Bohr. Later Bethe and Bloch gave a quantum mechanical treatment of stopping power based on the Born approximation method of wave mechanics and valid subject to the condition (cf. 5.26)

$$\frac{zZe^2}{2\pi\varepsilon_0 \hbar v} \ll 1 \tag{5.36}$$

which means physically that the perturbation due to the incident particle does not seriously disturb the electronic motion for large impact parameters. The Bethe–Bloch formula is, to a good approximation

$$-\frac{dT}{dx} = \frac{z^2 e^4}{4\pi\varepsilon_0^2 mv^2} NZ \log \frac{2mv^2}{I}$$
(5.37)

where I is a parameter interpreted as a mean atomic excitation potential. This formula is extended to relativistic incident particles as follows:

$$-\frac{dT}{dx} = \frac{z^2 e^4}{4\pi\varepsilon_0^2 mv^2} NZ \left\{\log \frac{2mv^2}{I} - \log(1-\beta^2) - \beta^2\right\}$$
(5.38)

where $\beta = v/c$. The quantity I is not the first ionization potential of the atom but can be calculated from the Thomas–Fermi electron distribution function for an atom in the form

$$I = kZ$$
(5.39)

where $k \approx 11\cdot5$ eV; in practice I is deduced from experimental results.

It will be noted that formulae (5.37) and (5.38) (and similar formulae for ionization and δ-ray production) show that to a good approximation non-relativistic collision loss of energy is (*a*) proportional to the square of the charge of the particle, (*b*) inversely proportional the the square of its velocity and (*c*) *independent of the mass of the particle* (note that m is the mass of the electron), i.e.

$$-\frac{dT}{dx} \propto \frac{z^2}{v^2}$$
(5.40)

The variation of collision loss with particle energy is shown in Figs. 5.4 and 5.8.

The *primary ionization* may be estimated by evaluating the number of collisions for which the energy transfer to an electron lies between Q_0, the maximum possible ($=2mv^2$), and I_0, the first ionization potential. This number is

$$NZ \, dx \int_{Q=2mv^2}^{Q=I_0} 2\pi p \, dp$$
(5.41)

$$= \frac{NZ \, dx \, z^2 e^4}{8\pi\varepsilon_0^2 mv^2} \left(\frac{1}{I_0} - \frac{1}{2mv^2}\right)$$
(5.42)

using limiting values of p derived from (5.28).
Since $I_0 \ll 2mv^2$ this is approximately

$$\frac{z^2 e^4 NZ}{8\pi\varepsilon_0^2 mv^2 I_0} \text{ ion pairs m}^{-1}$$
(5.43)

For protons of 5 MeV energy passing through nitrogen at atmospheric pressure this formula indicates a primary ionization of the order 60 ion pairs

mm^{-1}. The observed ionization however is 240 ions pairs mm^{-1} and the difference is to be accounted for by secondary ion production by the δ-rays ejected in the primary process. No simple formula analogous to eq. (5.43) can be given for the total specific ionization but empirically at least it appears to follow roughly the same law of variation with velocity as the total energy loss.

The distribution of delta rays (Plate 4) along the track of a fast particle is obtained directly from (5.41) and (5.28) as

$$dn = \frac{z^2 e^4}{8 \pi \varepsilon_0^2 m v^2} NZ \frac{dT}{T^2} \tag{5.44}$$

where dn is the number of delta rays with kinetic energy between $T(=Q)$ and $T+dT$ per unit length of track. The total number of such rays of energy greater than T is then given by

$$n = \frac{z^2 e^4}{8 \pi \varepsilon_0^2 m v^2} \left\{ \frac{1}{T} - \frac{1}{2 m v^2} \right\} \tag{5.45}$$

5.2.3 Radiative loss of energy

If a particle of mass M and charge ze enters the field of a nucleus of charge Z in an absorber the acceleration produced is proportional to zZ/M. According to classical electrodynamics the resulting radiation would have an intensity proportional to $z^2 Z^2 / M^2$. This radiation, which is the bremsstrahlung referred to in Sect. 5.2.1, is therefore much smaller for protons and mesons than for electrons (Table 5.1 and Fig. 5.4). For the latter particles it is an important mechanism of energy loss for energies above a few MeV, particularly in absorbers of high Z. At lower energies the bremsstrahlung produced by the impact of an intense beam of electrons on a massive target furnishes the continuous X-ray spectrum discovered by Röntgen and at extreme relativistic energies bremsstrahlung participates in the development of cosmic-ray showers.

Electrons of kinetic energy T may in passing through a thin absorber give rise in each radiative collision to a bremsstrahlung photon of any energy between 0 and T. Since small deflections are more probable than large, low energy quanta are emitted preferentially, but when the *energy loss* per unit frequency interval is calculated it is found that this is to a first approximation independent of the quantum energy. Bremsstrahlung therefore has an *equienergy spectrum*, as shown in Fig. 5.5(a).

The total radiative loss of energy per unit path is obtained by integrating over the bremsstrahlung spectrum and may be written

$$-\left(\frac{dT}{dx}\right)_{rad} = NEZ^2 f(Z, T) \tag{5.46}$$

where N is the number of atoms per m³ of the stopping material, $E (= T + mc^2)$ is the total energy of the incident electron and $f(Z, T)$ is a slowly-varying function of Z and T.

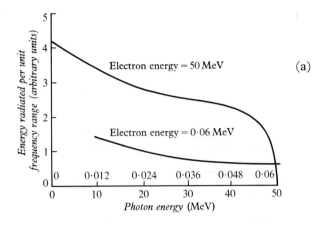

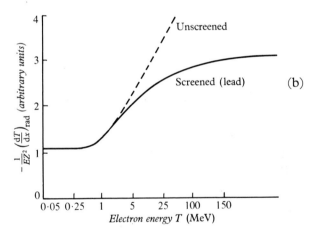

Fig. 5.5 Bremsstrahlung. (a) Radiation from 50 MeV electrons passing through lead and from 0·06 MeV electrons passing through aluminium. If the ordinate is divided at each point by the energy $h\nu$, the result gives the number distribution of the bremsstrahlung photons ($\propto 1/h\nu$). (b) Energy variation of radiative loss for electrons of kinetic energy T in lead, showing the effect of screening of nuclear charge by atomic electrons. Note that the ordinates must be multiplied by $E (= T + mc^2)$ to give the form of the energy loss, which rises very steeply with T (Ref. 5.12).

The variation of the radiative energy loss with incident energy T is shown in Fig. 5.5(b) for electrons in lead. For a bare nucleus the radiative loss increases indefinitely with energy, corresponding to production of photons at greater and greater distances from the nucleus. For an actual atom, however, a limit to the radiative loss is imposed owing to the screening of the nuclear charge by the atomic electrons. The bremsstrahlung radiation is emitted into a cone of semi-angle

$$\theta \approx \frac{mc^2}{mc^2 + T} \qquad (5.47)$$

about the direction of the incident electrons and is thus constrained more and more into the forward direction as the electron energy increases.

Although bremsstrahlung can occur in electron–electron collisions the Z^2 factor in (5.46) means that as a practical source of energy loss it is a nuclear process. It differs essentially from the radiation discovered by Cherenkov which depends on the gross structure of the medium through which an incident particle passes. The electric field of a charged particle passing through matter produces a macroscopic polarization due to displacement of bound electrons. The time variation of this polarization can in principle lead to a radiation field at a point P (Fig. 5.6(a)). For slowly moving particles the

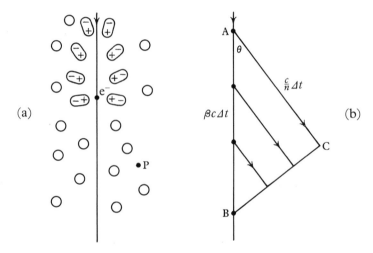

Fig. 5.6 Cherenkov radiation (Ref. 5.2). (a) Polarization of atoms of a transparent medium by passage of a charged particle. (b) Formation of coherent wavefront.

polarization has complete symmetry about the position of the electron, and no resultant field, or radiation, is observed at P. For fast moving particles however the polarization is axially asymmetric as shown in Fig. 5.6(a) and a resultant time-varying field exists at P. In general the contributions to the radiation field at P from different parts of the track of the particle cancel out. If however the particle velocity exceeds the velocity of light in the medium, c/n, where n is the refractive index, coherence is possible between the contributions at P from the polarizations at different points along the path of the particle. This is because the radiation from point B (Fig. 5.6(b)) under these conditions is in phase with the radiation from point A, and a coherent wavefront can be propagated through the medium, if it is transparent, in the direction AC for which

$$\cos \theta = \frac{c}{n} \Big/ \beta c = \frac{1}{\beta n} \tag{5.48}$$

The Cherenkov radiation is thus due essentially to longitudinal polarization of the medium, it is emitted from all points along the path of the incident particle in directions lying on the surface of a cone with semi-angle given by (5.48), and unlike bremsstrahlung, it is independent of the mass of the moving particle. The electric vector of the Cherenkov radiation is perpendicular to the surface of the cone. The angle of emission increases with the particle velocity in contrast with the angular distribution of bremsstrahlung. Although the basic mechanism for Cherenkov energy loss resides in small energy transfers to atomic electrons, and is therefore present for all incident electron energies, coherence does not appear until the electron velocity reaches the critical value given by

$$\beta = \frac{1}{n} \tag{5.49}$$

and below this threshold there is effectively no radiation of this type.

The theory of the Cherenkov effect given by Frank and Tamm (see Ref. 5.2) shows that the radiative loss is

$$-\left(\frac{dT}{dx}\right)_c = \mu_0 \pi z^2 e^2 \int \left(1 - \frac{1}{\beta^2 n^2}\right) v \, dv \tag{5.50}$$

where the integration extends over all frequencies of emission v for which the refractive index is such that $\beta n > 1$. This confines the radiation to wavelengths greater than those corresponding to the ultraviolet absorption bands of the medium, and above this limit the spectral distribution of energy loss is proportional to $v \, dv$, in contrast with the equi-energy distribution, proportional to dv, for bremsstrahlung. The Cherenkov light is therefore found predominantly at the blue end of the visible spectrum.

The Cherenkov effect is usually small in comparison with both ionization and other radiative losses (cf. Sect. 5.2.1) but it is thought to play an important part at energies greater than the energy of minimum ionization for a relativistic particle. Technically the phenomenon has become of considerable importance because of the production threshold and the highly directional angular distribution. These properties, together with the sharpness of the light pulse ($\ll 10^{-10}$ s) and the concentration of energy into the visible region, have led to the development of energy-selective, directional counters for use in high energy physics (see Sect. 6.1.6) particularly with protons and mesons, for which the Cherenkov energy loss is as large as with electrons.

The orders of magnitude of radiative losses are shown in Table 5.1. The Cherenkov radiation loss is of the order of 0·1 per cent of the energy loss by ionization for a relativistic particle in a typical transparent medium.

From (5.50) the number of quanta N emitted within a narrow spectral range λ_1 to λ_2 by a particle passing through a length l of medium can be obtained as

$$N \equiv \frac{\mu_0 \pi z^2 e^2 c}{h} l \left(1 - \frac{1}{\beta^2 n^2}\right)\left(\frac{1}{\lambda_1} - \frac{1}{\lambda_2}\right) \qquad (5.51)$$

This gives typically about 10 photons per mm between 400 and 600 nm for a relativistic electron passing through water.

5.3 Experimental results

From the formulae developed in Sects. **5.1** and **5.2** it is clear that the slowing down of a charged particle passing through matter is mainly due to collisions with electrons, while the large angle scattering is mainly due to collisions with nuclei.

5.3.1 Scattering of charged particles

The scattering of α-*particles* passing through matter due to collisions with nuclei does not exceed a few degrees even for absorbers of thickness comparable with the particle range. Large angle scattering, due to close nuclear collisions, is described by the Rutherford scattering law, eq. (5.24), and the first detailed verification of this law by Geiger and Marsden has already been described (Sect. 2.5.1). These early experiments were made with α-particles from radon and radium B + C, of about 7 MeV energy, and with fairly heavy scattering nuclei such as Ag and Au. In such cases the criterion of the validity of the classical theory (eq. (5.26)) is well satisfied and no deviations from the Rutherford scattering law were found. In particular it seemed that the Coulomb law of force between an α-particle and a nucleus was valid down to a distance of less than 3×10^{-14} m. When the α-particle experiments were extended to lighter atoms two new effects were found which may be seen in Fig. 5.7, showing the results of Chadwick[†] for the scattering of α-particles in helium:

(*a*) the intensity at about 45° in the laboratory system (90° centre-of-mass) for slow α-particles is just twice that predicted by (5.24), even allowing for the fact that scattered and recoil particles cannot be distinguished. This is a quantum mechanical effect arising from the identity of the particles and consequent interference between the waves representing the scattered and recoil particles (Sect. 14.2.2).

(*b*) The intensity at 45° as a function of increasing velocity of the α-particle at first falls off and then increases largely above the value expected from (5.24). This 'anomalous' scattering was also found with other light elements and was interpreted as evidence for a finite nuclear size or for

† J. Chadwick, *Proc. roy. Soc.*, A, **128**, 114, 1930.

non-Coulomb forces. Since the scattering at first decreases, the non-Coulomb nuclear force must be first attractive as the α-particle approaches the nucleus, but for close distances of approach all that could be concluded from the early experiments was that the force was stronger than the Coulomb repulsion.

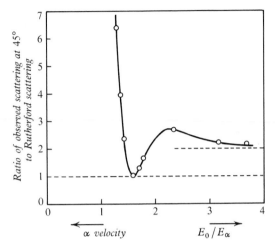

Fig. 5.7 Scattering of α-particles in helium at 45° laboratory angle (Chadwick); E_0 is a reference energy.

Scattering of protons and α-particles by nuclei under conditions such that $b < \lambda$, so that the classical theory is inadequate, is now an important source of information on nuclear sizes. Interpretation must be based on a nuclear model and on the full wave mechanical theory of scattering, which is outlined in Chapter 14. For the case $b > \lambda$, in which Rutherford scattering is expected both classically and quantum mechanically, many accurate verifications of the angular distribution (5.24) are now available for artificially accelerated particles and a wide variety of nuclei.

The scattering of *electrons* by nuclei and by other electrons was also studied in the early days of radioactivity with naturally occurring β-particle emitters as sources. Such electrons, and indeed all electrons of more than 100 keV energy, must be considered to be relativistic particles. This means that the Rutherford scattering law must be used, when applicable, in relativistic form but most cases of the scattering of electrons will not in fact be suitable for classical treatment because of the low value of the quantity $zZe^2/2\pi\varepsilon_0\hbar v$. Suitable formulae based on the Born approximation of wave mechanics, which in fact requires $zZe^2/2\pi\varepsilon_0\hbar v \ll 1$, have been given by Mott and others. Beta particles suffer large deflections in both nuclear and electronic collisions and the verification of the laws of deflection is experimentally difficult because of multiple scattering. Early experiments by Chadwick, using annular geometry, and by Schonland, together with a rough analysis of deflections observed in cloud chamber photographs showed that it was possible to obtain conditions of effectively single scattering, which merged into multiple scattering as the thickness of the scattering foil increased. The main effect is due to nuclear scattering (Mott scattering) because of the factor Z^2 in (5.24) but for the lightest elements such as hydrogen and helium, electronic scattering proportional to Z may be observed. The theoretical formula for electron–electron collisions (Möller scattering) includes an interference term due to identity of the two particles. Formulae have also been developed for positron–electron scattering which differ from the electron case because of the different sign of charge (which reduces nuclear scattering of positrons in comparison with that of electrons) and because of the possibility of annihilation in e^+–e^- collisions. Experimental results on electron-nuclear scattering at about 2 MeV obtained by Van de Graaff and others† verify the Mott formula for distances of approach between electron and nucleus down to 7×10^{-15} m. As the electron energy is increased the point-charge scattering of electrons becomes a worse approximation and the scattering of electrons of several hundred MeV has become the most important method of studying the size of the nuclear charge distribution.

The multiple scattering angle of charged particles in passing through absorbers is large for electrons and small for heavy particles. This dependence on particle mass has proved particularly useful in the identification of mesons and other new particles by observation of the deviations from linearity of their tracks in nuclear emulsions, in which elastic collisions with silver and bromine nuclei give rise to the effect. From (5.20) it can be seen that the change in momentum in a small angle collision between M_1 and M_2 $(\theta \approx 0)$ is $zZe^2/2\pi\varepsilon_0 pv_1$ and the angle θ is thus given by

$$\theta = \frac{\Delta q}{q} = \frac{zZe^2}{2\pi\varepsilon_0 pvq}$$

By integrating θ^2 over all impact parameters p an estimate of the mean square scattering angle is obtained (Ref. 1.6, page 39). This is often given, for a path length x of material, in the form

$$\overline{\theta^2} = \frac{z^2 E_s^2}{q^2 \beta^2 c^2} \frac{x}{X_0} \tag{5.52}$$

where E_s is a constant ($=21$ MeV), independent of the mass of the particle and of the nature of the medium, q is the momentum, $\beta = v/c$ and X_0 is the radiation length (Sect. 5.2.1) for the medium. For the example given in Table 5.1 the root mean square deflection of the proton is about 8′ of arc.

5.3.2. *Stopping power of matter for charged particles*

The *absolute stopping power* of a material for a heavy charged particle is the quantity $-dT/dx$ given in (5.38) and expressed in J m^{-1} or in an equivalent

† R. J. Van de Graaff, W. W. Buechner and H. Feshbach, *Phys. Rev.*, **69**, 452, 1946.

unit such as MeV per kg m^{-2}. It is usual to write

$$-\frac{dT}{dx} = \frac{z^2 e^4}{4\pi\varepsilon_0^2 m v^2} NB \qquad (5.53)$$

where N is the number of stopping atoms per m^3 and B the *atomic stopping number* is given, for a non-relativistic particle by the expression

$$B = Z \log \frac{2mv^2}{I} \qquad (5.54)$$

The variation of the stopping power, or energy loss by collision, for air is shown in Fig. 5.8. The curve may be placed on an absolute scale of energy

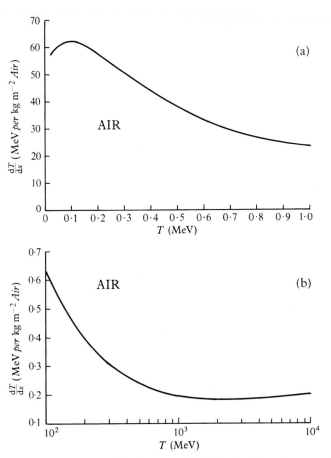

Fig. 5.8 Absolute stopping power of air for protons (collision loss), based on experimental observations (Ref. 5.6). (a) Low energy region ($\beta \ll 1$). The main part of the curve is described by (5.38). Below $T = 0\cdot1$ MeV the collision loss falls off because capture and loss of electrons reduces the effective charge of the proton. (b) High energy region ($\beta > 0\cdot4$). The curve is given by (5.38) using an experimental value for I. 'Minimum ionization' occurs for $T = 1500$ MeV.

loss once a value for I in (5.38) is calculated or observed. For all but the lightest elements it is a good approximation to assume that

$$I = kZ \text{ eV} \qquad (5.55)$$

as predicted by Bloch, and extensive calculations of energy loss based on the value of $k = 11 \cdot 5$ eV have been made (Ref. 5.8).

Formula (5.38) is valid in its simple form only so long as the velocity of the incident particle remains large compared with the velocities of the atomic electrons and for slow particles corrections must be made for the binding of the K-electrons which can no longer be considered as free. For a heavy ion especially, and for light ions to some extent, the energy loss to electrons decreases at low energies because of reduction of average charge by the capture and loss process (Sect. 5.3.6). In the end the average charge of the ion is so near zero that the main part of the energy loss is due to interaction between the nuclei of the ion and of a target atom. The relative importance of nuclear and electronic stopping is discussed in Ref. 5.4 (Northcliffe); it is strikingly illustrated by the behaviour of fission fragments in matter. Nuclear collisions involve large energy losses and appreciable angular deflections of the projectile together with the ejection of recoil nuclei from the absorber atoms. These nuclear delta rays may be seen along the tracks of fission fragments and other heavy ions in expansion chamber photographs.

At relativistic energies the stopping power as shown in Figs. 5.4 and 5.8 reaches a minimum value, known usually as 'minimum ionization' and then begins slowly to increase. The relativistic increase is in practice diminished for condensed media by a polarization of the medium which tends to screen distant atoms from the field of the incident particle (density effect), but the average ionization still increases with increasing energy owing to increase of the maximum possible energy transfer. If attention is confined to energy transfers of less than a certain specified amount as may be necessary when useful quantitative measurements are to be made on the tracks of charged particles in nuclear emulsions or expansion chambers, the ionization is found to reach a constant value at high energies (Fermi plateau).

Experimental determinations of absolute stopping power, leading to values of I, have been made by measuring the energy of a beam of particles before and after the interposition of an absorber of known thickness in the path of the beam. Relative stopping powers have also been determined for many elements for 340 MeV protons by observing the change of range of the particles when an absorber is placed in the beam. From these observations the stopping power per electron relative to aluminium was deduced and the I values for the elements were calculated with the assumption of a standard value of I for aluminium. Typical results are shown in Table 5.3, together with some older values of atomic stopping power relative to air based on magnetic deflection experiments with 6 MeV α-particles.

It will be noted that the effectiveness of an electron in retarding a charged particle is greater for the lighter elements; this is because I increases with Z

in the expression (5.54) for B/Z and arises because of the increased nuclear binding of inner electronic shells of heavy atoms.

TABLE 5.3 Relative stopping powers

Element	Stopping power per electron relative to Al (340 MeV protons)[a]	Stopping power per atom relative to air (6 MeV α-particles)[b]
H	1·280	0·21
C	1·084	0·93
Al	1·000	1·45
Cu	0·924	2·43
Pb	0·804	4·35

[a] C. J. Bakker and E. Segrè, *Phys. Rev.*, **81**, 489, 1951.
[b] Ref. 5.7, Table XLVIII.

We may conclude that present experimental results support the theory of energy loss of heavy charged particles by collision, according to which we may write, in the non-relativistic case (eqs. (5.37), (5.39))

$$-\frac{dT}{dx} = \frac{z^2}{v^2} f(v) \tag{5.56}$$

The fall-off in dT/dx approximately as $1/v^2$, the dependence on z^2 and independence of mass are all verified by observation over the range of energy from about 50 keV to 300 MeV. The difficulties still remaining are chiefly concerned with calculations of I values for which a suitable atomic model is necessary. The additional complication of capture and loss effects at energies below about 1 MeV per nucleon is beyond the simple theory of stopping power and is discussed in Sect. 5.3.6.

It is less easy to make precise comparison between theory and experiment for the stopping power of matter for electrons. Although for light absorbers and electrons of a few MeV energy radiative losses may be neglected, difficulties still arise because of increased nuclear scattering and because of large straggling effects. These result from the large possible energy loss in a single collision, and will be discussed in Sect. 5.3.5.

5.3.3 Range–energy curves

The theoretical range of a heavy charged particle is equal to its path length in matter because scattering is negligible. The range may formally be obtained by integration of the expression for energy loss, giving

$$R = \int_0^T \frac{dT}{dT/dx} \tag{5.57}$$

but in practice this cannot be carried out for the full range of the particle because of the corrections necessary to the explicit formula for dT/dx at low energies. Range–energy curves are therefore constructed semi-empirically by combining observations of the range of particles of known energy with integrations of the energy–loss formula over some particular region; for high energies the formula should be valid.

The first range–energy curve to be established was for α-particles in air. Energies of radioactive α-particles were obtained from the classical magnetic deflection experiments of Briggs, Rosenblum and Rutherford with a precision of 1 part in 10^5 in favourable cases. Ranges were measured by observing tracks in an expansion chamber or by allowing the α-particles to pass through a shallow ionization chamber which was then moved along until the particle terminated its range in the chamber. The ranges displayed in range–energy curves are by convention mean ranges, i.e. the average range of a group of

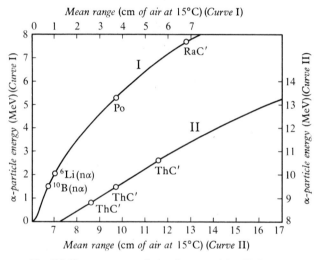

Fig. 5.9 Range–energy relation for α-particles (Ref. 5.1).

particles initially of homogeneous velocity and observed ranges must be reduced to mean values by correction for fluctuations, or straggling, as described in Sect. 5.3.5. It is possible to obtain fixed points on the α-particle range–energy relation by using observations on the ranges, in an expansion chamber, of the particles from nuclear reactions with accurately known energy release, e.g. $^{10}B(n, \alpha)^7Li$ and $^6Li(n, \alpha)^3H$. The present α-particle range–energy curve, based on a conventional value of 8·57 cm of air at 15°C and 760 mmHg pressure for α-particles of energy 8·776 MeV (ThC′), is shown in Fig. 5.9.

If the energy loss equation is written in the form (5.56) then by substitution of $\frac{1}{2}Mv^2$ for the kinetic energy T we obtain

$$Mv\frac{dv}{dx} = \frac{z^2}{v^2}f(v) \tag{5.58}$$

from which a range–velocity relation may in principle be obtained by integration. If the energy dependence of the function $f(v)$ could be neglected this would immediately give

$$R \propto \frac{M}{z^2} v^4 = \text{const.} \times v^4 \tag{5.59}$$

for a given particle; empirically it was early found by Geiger that the range–velocity relation for α-particles is more nearly

$$R = \text{const.} \times v^3 \quad \text{(Geiger's rule)} \tag{5.60}$$

Equation (5.59) also shows that for a given velocity, for which $f(v)$ has the same value for all particles independently of mass and charge in a given stopping medium, the ranges of different particles are proportional to M/z^2. This is a particularly useful result since it enables a great many range–energy curves to be derived from one standard curve such as that for α-particles, providing that the necessary small corrections for different charge exchange effects can be applied. For particles of the same charge and velocity, the ranges are directly proportional to their mass, e.g. the ranges of protons, deuterons and tritons of energies 10, 20 and 30 MeV stand accurately in the ratio $1:2:3$. Range–energy tables are given in Ref. 5.8.

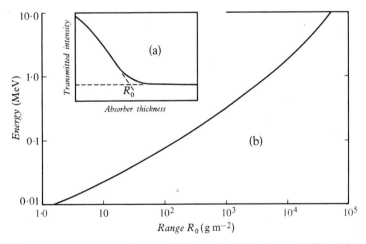

Fig. 5.10 Range–energy curve for electrons (Katz and Penfold, *Rev. mod. Phys.*, **24**, 28, 1952).

The range of a slow electron in matter, as may be seen in expansion chamber photographs (Plates 2, 3) may be much less than its path length because of large angle scattering. For the same reason the range is not sharply defined as it is for a heavy particle and depends on the method of measurement used. If a beam of electrons of homogeneous initial velocity is passed through an absorber the transmitted intensity as a function of absorber thickness is as shown in Fig. 5.10(a). By extrapolation of the sloping part of the curve to the axis, or to the level of background, an extrapolated range R_0 may be defined and this procedure is sufficiently reproducible to permit such ranges to be used in establishing a range–energy curve. To a good approximation ranges are independent of the material of the absorber. The relation $R_{kg\,m^{-2}\,Al} = 5\cdot43\ (E_0/MeV) - 1\cdot60$, first given by Feather, has been much used for finding the maximum energy in a β-ray spectrum. Figure 5.10(b) shows an empirical curve based partly on measurements with homogeneous electrons and partly on β-ray spectra; in all cases the energies are determined by magnetic analysis. The slowing down of electrons in matter inevitably introduces a large straggling which is not only considerably greater in proportion than that for heavy particles but is also asymmetric, because of the high probability of loss of a large amount of energy in a single collision.

5.3.4 *Ionization and associated effects*

Knowledge of the ionization produced by charged particles is of importance not only because it forms the basis of many methods of charged particle detection but also because it is the process by which radiation causes biological damage. All modern methods of radiation dosimetry are ultimately based on ionization measurements (Sect. 6.1.1). We consider here, however, only effects in gases.

The intense and continuous ionization along the path through a gas of a heavy charged particle (perhaps 3000 ion pairs mm^{-1}) is well shown in expansion chamber photographs (Plate 1). The expansion chamber was in fact the first instrument used, by Feather and Nimmo, to study the variation of ionization along the path of a single α-particle, and it was possible to show that the density fell off rapidly in the last few mm of the range. This conclusion was consistent with earlier measurements on *beams* of α-particles passing through shallow ionization chambers, but such observations must be corrected for straggling before they yield the true single particle curve. The shallow ionization chamber has been used by Holloway and Livingston as a pulse instrument to give not only the relative ionization near the end of the range of a single α-particle, but also the absolute *specific ionization*, i.e. the total number of ion pairs per unit path. The results are shown in Fig. 5.11, which also gives the specific ionization for protons obtained by Jentschke by differentiating a curve relating total ionization of single protons to their residual range (i.e. the range remaining to a particle of given velocity).

The proton and α-particle curves in Fig. 5.11 are displaced by a distance of 2 mm which takes account of the different effective charge of the two particles, as a fraction of their full charge, in the last few mm of range. At greater residual ranges the ionizations tend to stand in the ratio 4:1 expected theoretically for equal velocities on the basis of a z^2 dependence of energy loss. Experimental observations practically always give total ionization, which includes in addition to the primary ions (Sect. 5.2.1) all additional secondary ions due to the absorption of δ-rays and radiation resulting from the primary events. From the specific ionization and the specific energy loss dT/dx, the average expenditure of energy to form one ion pair (ω) is obtained. The

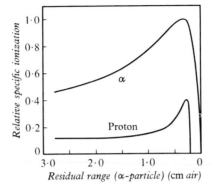

Fig. 5.11 Specific ionization of a single α-particle and a single proton in air at 15°C and 760 mm Hg. The maximum ionization is 6600 ion pairs mm^{-1} for the α-particle and 2750 ion pairs mm^{-1} for the proton (Ref. 5.1).

general result of a great many measurements of ω is that it varies little with type of particle, incident velocity or type of gas in which the ionization is produced. Table 5.4 shows typical results:

TABLE 5.4 Energy in eV required to form one ion pair[a]

Gas	Slow electrons (5 keV)	Polonium α-particles (5·3 MeV)	Protons (340 MeV)
Argon	27·0	25·9	25·5
Helium	32·5	31·7	29·9
Hydrogen	38·0	37·0	35·3
Nitrogen	35·8	36·0	33·6
Air	35·0	35·2	33·3
Oxygen	32·2	32·2	31·5
Methane	30·2	29·0	—

[a] Taken in part from Ref. 5.5, p. 57.

The measurements with α-particles are made by collection of ions, under conditions when recombination is small, either from the whole of the track of an α-particle or from a short section of it. The figures given in Table 5.4 for 340 MeV protons are due to Bakker and Segrè, who combined measurements of relative ionization in a particular gas and in argon with a separate observation that one 340 MeV proton produced 16·6 ion pairs per mm in argon at atmospheric pressure and 0°C. In this last experiment the ionization produced by a measured beam current of particles was recorded.

The existence of an ionization minimum and the subsequent rise in ionization predicted by (5.38) has been tested mainly for cosmic ray muons, for which radiative losses may be neglected. The momentum of the incident particle is found by magnetic deflection, and the ionization is recorded by making coincident observations in a proportional counter or pressure ionization chamber, or expansion chamber. The incident particles may be selected for momentum by adjusting the magnetic field. The results of Ghosh, Jones and Wilson,† shown in Fig. 5.12 illustrates the rise from the ionization minimum of about 4 ion pairs mm^{-1} of air and the 'Fermi plateau' (Sect. 5.3.2) for muons passing through oxygen.

Ionization in condensed media has been studied mainly in connection with the grain density resulting along the track of a charged particle passing through

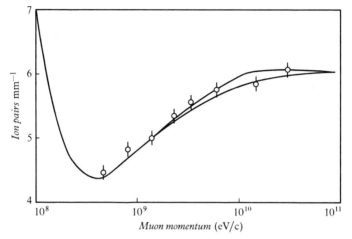

Fig. 5.12 Ionization by relativistic muons in oxygen excluding energy transfers of more than 1 keV. The curves are calculated theoretically.

a photographic emulsion. The existence of a minimum density and the rise to a 'plateau' are well established, and the correlation of grain density with multiple-scattering angle forms an important method for identification of charged particles.

5.3.5 Straggling and fluctuation phenomena

An average α-particle from a radioactive substance makes about 10^6 collisions resulting in small discrete energy transfers to electrons before coming to rest in an absorber. A group of such particles of initially uniform velocity, after passage through a certain thickness of matter, will show a distribution of velocities about a mean value owing to the statistical nature of the energy loss. The ranges of the particles will therefore be grouped about a mean value and since the number of collisions is large, and since also they are independent

† S. K. Ghosh, G. M. D. B. Jones and J. G. Wilson, *Proc. phys. Soc. Lond.*, **67**, 331, 1954; see also A. Crispin and G. N. Fowler, *Rev. mod. Phys.*, **42**, 290, 1970.

processes, the range (or energy) distribution may be expected to be approximately Gaussian. The *mean range* of the group has a definite value and it is this range which is always given in range–energy tables.

The range straggling is estimated quantitatively in Ref. 5.1, pp. 661. It is there shown that the standard deviation σ_R of the range distribution of heavy particles of energy T slowed down in an absorber is given by

$$\sigma_R^2 = \frac{z^2 e^2}{4\pi\varepsilon_0^2} NZ \int_0^T \left(\frac{dT}{dx}\right)^{-3} dT \tag{5.61}$$

where the symbols are those used in Sect. 5.2.2. By substitution from (5.37) for dT/dx and integration, neglecting the energy variation of the logarithmic term, we find approximately

$$\frac{\sigma_R^2}{R^2} = \frac{2m}{M} \frac{1}{\log (2mv^2/I)} \tag{5.62}$$

For the α-particles of polonium ($E = 5.3$ MeV, $R = 3.84$ cm of air) we find $\sigma_R/R = 0.9$ per cent. For protons of the same initial velocity as these α-particles, (5.62) shows that the straggling σ_R/R is just double that for the α-particles. This is because the α-particle dT/dx is roughly four times greater than that for protons so that the straggling effect is relatively smaller.

The number of particles in a group of n_0 with ranges between x and $x + dx$ can now be written, assuming a normal distribution,

$$dn = \frac{n_0}{\sigma_R \sqrt{2\pi}} \exp -\frac{(x-R)^2}{2\sigma_R^2} dx$$

$$= \frac{n_0}{\alpha \sqrt{\pi}} \exp -\left(\frac{x-R}{\alpha}\right)^2 dx \tag{5.63}$$

in which R is the mean range and $\alpha = \sqrt{2}\,\sigma_R$ is the *range straggling parameter*. The *number–distance curve* follows immediately by integration; in this curve† (Fig. 5.13) the mean range is at the point of inflection of the curve, and is the range reached by half of the particles. The intercept on the range axis of the linear part of the number–distance curve gives the *extrapolated number range* (ENR) which is readily determined in counting experiments. It is easily shown that this range exceeds the mean range by $\sqrt{\pi/2}\,\sigma_R$. Figure 5.13 also shows the specific ionization curve of a single α-particle, which is drawn to terminate at the mean range of the straggled group, and by multiplying this by the straggling distribution for an initially homogeneous group, a curve known as the *Bragg* or *average ionization curve* is obtained. This is essentially the ionization measured in a very shallow chamber drawn along the path of a beam of α-particles passing through a gas. If the linear part of the Bragg curve is extrapolated to the range axis we obtain the *extrapolated ionization range* (EIR).

† M. G. Holloway and M. S. Livingston, *Phys. Rev.*, **54**, 18, 1938.

Straggling is normally shown and measured by the experiments which determine energy loss. It may be seen in the energy distribution of protons after passage through an absorbing foil and in any observation of a Bragg curve by counters or an expansion chamber.

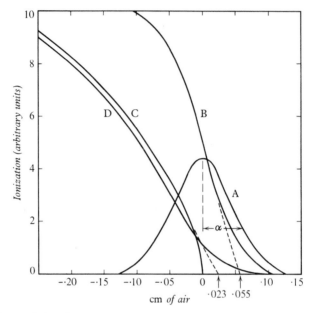

Fig. 5.13 Characteristics of range curves for polonium α-particles in air. A, range straggling distribution about the mean range; B, number–distance curve showing extrapolated range; C, differential specific ionization of a single particle of mean range; D, average ionization (Bragg) curve for a group of α-particles with the range distribution of curve A, showing the extrapolated ionization range. The straggling parameter α is shown (Holloway and Livingston).

The assumption of a Gaussian distribution for the energy loss, although in approximate agreement with experiment for non-relativistic particles, neglects the effect of the small number of collisions with high energy loss, which contribute a 'tail' to the distribution. This effect is particularly marked for electrons because of the large energy loss which is possible in a single collision and a detailed theory has been given by Williams, by Landau and by Symon. The type of distribution found is shown in Fig. 5.14 which gives results of Goldwasser, Mills and Hanson† on the energy loss of 15·7 MeV electrons in a thin metal foil, determined by magnetic analysis, following the original studies of White and Millington. The characteristic asymmetrical shape of the 'Landau' distribution is clear; it is also necessary to take account of the density effect in condensed media to obtain agreement with experiment. The width of the Landau distribution has been measured and compared with theory for protons, electrons and muons (Ref. 5.5); it must always be taken into account in computing the most probable energy loss of a relativistic

† E. L. Goldwasser, F. E. Mills and A. O. Hanson, *Phys. Rev.*, **88**, 1137, 1952.

particle. As the velocity of the particle decreases the relative importance of small energy transfers increases and the straggling tends to the symmetrical Gaussian form. The Landau–Symon distribution has been verified for proton energies as low as 1 MeV.†

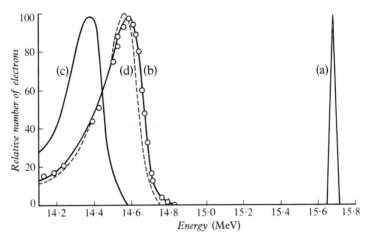

Fig. 5.14 Energy distribution of electron beam before and after passing through 8·6 kg m^{-2} of aluminium: (a) incident beam, (b) transmitted beam, (c) Landau theory, (d) Landau theory with density correction (Goldwasser *et al.*).

5.3.6 *Capture and loss (charge exchange)*

It was found by Henderson in 1922 that a beam of α-particles passing through matter always consisted of a mixture of singly and doubly charged particles. The proportion of singly charged particles increases markedly towards the end of the range of the particles and a small component of neutral helium atoms also develops. These phenomena are due to rapid changes in the charge state of the helium ion resulting from pick-up of atomic electrons and to subsequent re-ionization by collision. The existence of the three beams may easily be demonstrated by a magnetic deflection experiment as shown in Fig. 5.15(a); in general if the α-particles have passed through absorber on their way to the detecting screen, there will be three distinct traces. If gas is admitted to the deflection chamber the variation in the number of He$^{\circ}$, He^{+} and He^{++} particles with pressure may be studied and a mean free path for capture of an electron by a He^{++} ion of given velocity may be found.

Let λ_c be the mean free path for the capture process

$$He^{++} + e^{-} \rightarrow He^{+} \tag{5.64}$$

and λ_l the corresponding mean free path for the loss

$$He^{+} \rightarrow He^{++} + e^{-} \tag{5.65}$$

† T. R. Ophel and J. M. Morris, *Phys. Lett.*, **19**, 245, 1965.

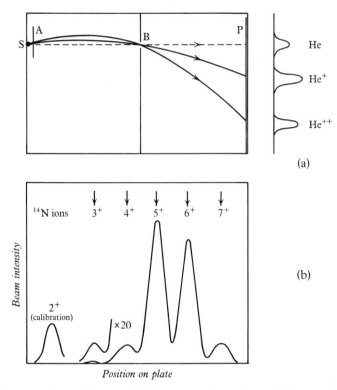

Fig. 5.15 Capture and loss of electrons by heavy charged particles. (a) Principle of apparatus; α-particles from a source pass through an absorber A and collimating slit B in a vacuum box. They are deflected in a transverse magnetic field and the detector shows traces due to neutral, singly-charged and doubly-charged particles. (b) Results for nitrogen ions accelerated in a cyclotron (Stephens and Walker).

If these mean free paths are small compared with the thickness of absorber through which the α-particle has passed the beam reaches an equilibrium in which the charge ratio, is

$$\frac{\text{He}^+}{\text{He}^{++}} = \frac{\lambda_l}{\lambda_c} = \frac{\sigma_c}{\sigma_l} \tag{5.66}$$

where σ_l, σ_c are the cross-sections per atom of the absorber for the charge exchange processes. Experimentally determined mean free paths λ and average charges ez_{av} for α-particles of different velocities are shown in Table 5.5, based on the work of Rutherford; the effect of the He° particles is neglected in calculating z_{av}.

The table shows that the interchange of charge near the end of the range of the α-particle is very rapid and may take place 100 times in the last m of air before the particle is brought to rest. The mean free paths seem to be independent of the nature of the absorber and are so short for low velocity particles that charge equilibrium is reached for extremely thin layers of matter. Thus α-particles of about 1 MeV emerging from a target of thickness of only a few nm will already be partly in the He^+ state.

TABLE 5.5 Mean free paths for capture and loss of α-particles in air at $15°C$ and 760 mmHg. (From Ref. 5.1, p. 634.)

E_α/MeV	λ_c/mm	λ_l/mm	*Average charge number* z_{av}
6·78	2·2	0·011	1·995
4·43	0·52	0·0078	1·985
1·70	0·037	0·0050	1·883
0·65	0·003	0·003	1·500

The phenomenon of charge exchange has been extensively studied for protons and α-particles and also for fission fragments (App. 4) which carry a large number of electrons because of their high nuclear charge. Many of the phenomena are more clearly exhibited by the behaviour of accelerated heavy ions such as carbon or nitrogen passing through gases or through foils thick enough to produce charge equilibrium but thin enough to give little reduction of velocity. In the work of Stephens and Walker† 15·0 MeV nitrogen ions emerging from a cyclotron were passed through an organic film of thickness 10^{-7} m and analysed magnetically (Fig. 5.15(b)), using a photographic emulsion as a detector. The population of the possible ionic states of the beam is clearly shown. From experiments such as this the average charge of the ion for a given velocity can be found as ez_{av} where

$$z_{av} = \sum_0^z Zq_Z \tag{5.67}$$

where q_Z is the fraction of ions with charge $Z (\leqslant z)$. Since stopping effects vary as charge squared, the effective charge for use in heavy ion stopping power formula is not $(z_{av})^2$ but is nearer to a root mean square value (Ref. 5.4, Northcliffe).

5.3.7 *Passage of charged particles through crystals* (Ref. 5.9)

In Sect. 5.2.2 the theory of energy loss by collision was developed on the assumption of an isotropic medium in which the distribution of impact parameters p for collision with electrons could be written in the form $2\pi p \, dp$. If a charged particle moves through a single crystal however the crystal structure defines certain directions or planes for which the electron density may be especially low, so that within a small range of angles ($\approx 1/10°$ about these directions in typical cases) the impact parameter distribution may differ considerably from the isotropic form. In these directions, e.g. directions of high symmetry, there is increased penetration and finely-collimated incident ions emerge from a thin crystal with an energy loss smaller than that which they

† K. G. Stephens and D. Walker, *Proc. roy. Soc.*, A, **229**, 376, 1955.

experience in other directions.† This phenomenon is known as *channelling*; for random angles of incidence and in amorphous solids the energy loss corresponds to the average electron density normally used in stopping power calculations.

The occurrence of channelling in a simple two-dimensional case is illustrated in Fig. 5.16(a). An ion travelling in approximately the direction of a

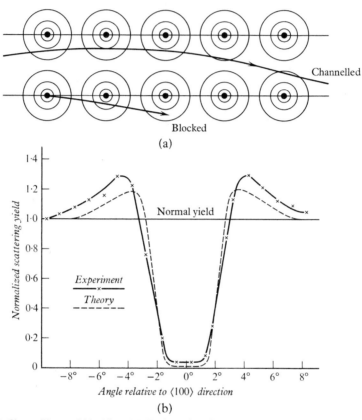

Fig. 5.16 Channelling and blocking: (a) Trajectories of ions between rows of atoms. A particle incident nearly parallel to the row is channelled; a particle emitted from a lattice site is scattered away from the channelling direction. The circles represent the atomic electron density. (b) Blocking dip in 135° Rutherford scattering of 480 keV protons incident, with an angular spread of 0·1°, along the ⟨100⟩ direction of a tungsten crystal (J. U. Andersen and E. Uggerhoj, *Can. jnl. Phys.*, **46**, 517, 1968).

crystal axis approaches an atom and finds itself in a net repulsive field deriving from the resultant of the ion–nucleus and ion–electron interactions. This deflects the incident particle slightly towards the region of low electron density between the atomic rows. At the next atom of the row a similar effect occurs and as a result of the correlated series of Coulomb scatterings the ion

† G. Dearnaley, *Trans. I.E.E.(N.S.)*, **11**, No. 3, 249, 1964.

moves from side to side between the rows of atoms. It is therefore steered through the crystal mainly in a region of low electron density and its specific energy loss is correspondingly reduced. It also tends to avoid close nuclear collisions. Evidently such effects will be enhanced for low energy heavy ions† for which nuclear as distinct from electronic stopping is in all cases important. Channelling however is also observed for light particles of an energy of several MeV and for electrons.

An effect closely allied to channelling is the *blocking* of the emission of charged particles from lattice sites. Because of the strong Coulomb deflections particles will not be able to enter a direction of crystal symmetry over a small range of angles. Nuclear reactions and large angle Rutherford scattering provide convenient sources of particles emerging from lattice points, and yields show a marked dip as a crystal target is rotated round a beam direction because close collisions are inhibited. Figure 5.16(b) shows results obtained for back-scattered protons.

Channelling and blocking effects in crystal absorbers yield striking emergence patterns which may be recorded photographically. The sharp angle dependence of these phenomena may be used for crystal alignment, for the location of impurity atoms (which if present at lattice sites will not interact with a channelled beam), for the probing of interatomic potentials, and for nuclear lifetime determinations (Sect. 13.6.1).

5.4 Interaction of electromagnetic radiation with matter

5.4.1 *General*

The passage of electromagnetic radiation through matter is characterized by the exponential law of absorption, eq. (5.6). This is because radiation, in its interaction with matter as distinct from its propagation, must be considered as a stream of photons and absorption processes remove individual photons from a beam to a degree directly proportional to the incident number. In this respect, radiation resembles protons of about 1000 MeV energy, which are 'absorbed' by nuclear collisions, and electrons of a few MeV energy which are heavily scattered in nearly all collisions in matter, rather than radioactive α-particles, for which large angle collisions are rare. Although at high energies (say > 100 MeV) the development of cascade showers may be described in terms of a radiation length, electromagnetic radiation of lower energy is not considered to have a range, as in the case of non-relativistic heavy particles, but only an attenuation coefficient μ.

The processes resulting in attenuation are (*a*) absorption, in which there is a direct conversion of photon energy in whole or in part into kinetic energy of easily absorbed particles, and (*b*) scattering, in which a photon is deflected out of the beam. There is a connection between these two types of process since in any given interaction there is usually an immediate appearance both

† R. S. Nelson and M. W. Thompson, *Phil. Mag.*, **8**, 1677, 1963.

of kinetic energy and of scattered quanta. Since however scattered quanta may not be absorbed again in the scatterer it is useful to make a distinction between the two processes and to write

$$\mu = \mu_a + \mu_s \tag{5.68}$$

The coefficient μ_a then gives the *deposition of energy* in a medium, as distinct from the diminution in the number of photons (given by μ) and is important in assessing radiation hazards.

The coefficients μ, μ_a and μ_s depend on the atomic number of the absorbing material and on the energy of the incident quanta in a way determined by the details of the interaction processes. These are:

(*a*) elastic scattering (Rayleigh, Thomson and nuclear resonant),
(*b*) photoelectric effect,
(*c*) Compton scattering,
(*d*) pair production for photon energies above 1 MeV,

of which the last three are usually stated to be inelastic. The Compton effect is actually elastic in the sense of Sect. 5.1.1 since there is no loss of kinetic energy, but there is a wavelength change and the scattered radiation is essentially incoherent with the incident beam. In addition nuclear inelastic scattering, the nuclear photoeffect and photomeson production are absorption processes which are important and interesting in detail but do not normally contribute essentially to the observed values of attenuation coefficients. Of the elastic processes Rayleigh scattering (Sect. 5.4.2) is usually by far the most important, but in special circumstances (Mössbauer effect, Sect. 13.6.4) nuclear attenuation may exceed that due to electronic processes.

If cross-sections σ_R, σ_{PE}, σ_C and σ_{PP} are assigned to the individual processes the linear attenuation coefficient for removal of photons from a homogeneous beam may be written†

$$\mu = N(\sigma_R + \sigma_{PE} + \sigma_{PP}) + ZN\sigma_C \tag{5.69}$$

where N is the number of atoms of absorber per m³. The atomic number Z multiplies the cross-section σ_C because the Compton effect takes place with individual electrons rather than with atoms as a whole. This is in fact true only when the momentum transferred to an electron in the incoherent scattering process considerably exceeds $\sqrt{2m_e B}$ where B is the electron binding energy, so that the electron may be treated as free. If this is not so the Compton scattering cross-section per atom $Z\sigma_C$ is reduced. From μ, the mass attenuation coefficient μ_m is obtained by dividing by the density for a particular absorber. Figure 5.17 shows the variation of μ_m with energy for lead; clearly the energy is not a single-valued function of μ_m because the pair production cross-section increases with energy while the photoelectric and Compton cross-sections decrease. The relative importance of these three main pro-

† In most practical cases the term σ_R may be omitted (Sect. 5.4.2).

cesses as a function of energy and atomic number of absorber is shown in Fig. 5.18. Measurement of attenuation coefficient near minimum absorption is therefore not an unambiguous method of determining photon energy in this region of the spectrum.

The wavelength of a photon of energy 10 MeV is $0 \cdot 12 \times 10^{-12}$ m. For energies below this it is a good approximation to assume that the instantaneous electric field of the photon does not vary over a typical nucleus. Variations

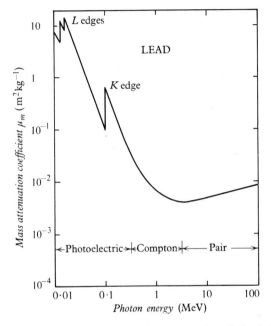

Fig. 5.17 Mass attenuation coefficient for electromagnetic radiation in lead (Ref. 5.1).

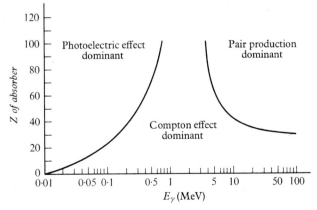

Fig. 5.18 Relative importance of the three major types of γ-ray interaction (Ref. 5.1).

125

over the *atom* however must occur, and can be used for radiation of X-ray wavelength to reveal details of atomic structure.

5.4.2 Elastic scattering of radiation

If a plane-polarized electromagnetic wave falls upon a free electron (Fig. 5.19(a)), the charge receives an acceleration

$$\ddot{z} = \frac{eE_0}{m} \cos \omega t \tag{5.70}$$

where the electric field of the incident wave is taken to be

$$E_z = E_0 \cos \omega t \tag{5.71}$$

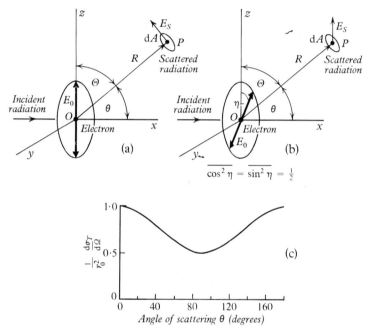

$$\cos^2 \eta = \sin^2 \eta = \tfrac{1}{2}$$

Fig. 5.19 Thomson scattering of radiation from a free electron, observed in plane xOz. (a) Plane polarized wave. (b) Unpolarized wave. (c) Angular distribution of scattered radiation from un-polarized wave.

The solution of eq. (5.70) is

$$z = -\frac{eE_0}{m\omega^2} \cos \omega t \tag{5.72}$$

so that the electron moves in anti-phase with the electric vector. This accelerated motion creates a time-varying dipole moment $D = ez$, and according to classical electrodynamics, a radiation field results, in which the amplitude of the electric vector at a point P distant R from the dipole is

$$E_s = -\frac{\mu_0 e^2}{4\pi m} \frac{E_0 \sin \Theta}{R} = \frac{f(\Theta)E_0}{R} \qquad (5.73)$$

where Θ is the angle between the polarization vector and the direction of observation and $f(\Theta)$ is known as the *scattering amplitude*. The time-averaged flux of radiation across a small area dA perpendicular to R is (eq. 3.5).

$$dW = \frac{1}{2} \frac{\mu_0}{(4\pi)^2} \frac{\ddot{D}^2}{c} \sin^2 \Theta \frac{dA}{R^2}$$

$$= \frac{\mu_0 e^4 E_0^2 \sin^2 \Theta}{32\pi^2 m^2 c} \, d\Omega \, \text{Js}^{-1} \qquad (5.74)$$

The incident intensity of radiation is

$$I = \frac{1}{2} \frac{E_0^2}{\mu_0 c} \, \text{J s}^{-1} \, \text{m}^{-2} \qquad (5.75)$$

From the considerations of Sect. 5.1.3 (see also Ref. 5.1, p. 821) we may relate the radiation dW scattered by a *single electron* to the incident radiation intensity through the concept of cross-section, and write

$$dW = I \, d\sigma_T = I\sigma_T(\Theta) \, d\Omega \qquad (5.76)$$

whence

$$\sigma_T(\Theta) = \left(\frac{\mu_0 e^2}{4\pi m}\right)^2 \sin^2 \Theta = |f(\Theta)|^2 \qquad (5.77)$$

If the incident radiation is unpolarized an average over the possible orientations of the incident electric vector must be made, and by resolving this vector parallel and perpendicular to the scattering plane xOz (Fig. 5.19(b)) we obtain

$$dW = \frac{\mu_0 e^4 E_0^2}{32\pi^2 m^2 c} \, d\Omega \, [\sin^2 \Theta \, \overline{\cos^2 \eta} + \overline{\sin^2 \eta}]$$

$$= \frac{\mu_0 e^4 E_0^2}{32\pi^2 m^2 c} \, d\Omega \, \frac{1 + \sin^2 \Theta}{2}$$

$$= I \left(\frac{\mu_0 e^2}{4\pi m}\right)^2 \, d\Omega \, \frac{1 + \cos^2 \theta}{2} \qquad (5.78)$$

where θ is the angle of scattering. The differential cross-section for scattering of unpolarized radiation through angle θ by a *single electron* is thus

$$d\sigma_T = \left(\frac{\mu_0 e^2}{4\pi m}\right)^2 \frac{1 + \cos^2 \theta}{2} \, d\Omega \qquad (5.79)$$

127

and the total cross-section, obtained by writing $d\Omega = 2\pi \sin\theta \, d\theta$ and integrating between the limits $\theta = 0$ and $\theta = \pi$ is

$$\sigma_T = \frac{8\pi}{3}\left(\frac{\mu_0 e^2}{4\pi m}\right)^2 = \frac{8\pi}{3} r_0^2 \tag{5.80}$$

where $r_0 = \mu_0 e^2/4\pi m$ is the classical electron radius. This cross-section, first calculated by J. J. Thomson and now known after him, is independent of primary frequency and enters into all more accurate expressions for the scattering of electromagnetic waves. It is also valid quantum mechanically in the non-relativistic limit ($h\nu \ll mc^2$) for scattering of radiation by free electrons. Figure 5.19(c) shows the angular distribution of the *Thomson scattering* of unpolarized radiation, as given by (5.79).

When radiation is incident on an atom containing Z bound electrons and a nucleus of charge Ze, elastic scattering can take place:

(a) from the bound electrons (*Rayleigh scattering*), providing that the electrons do not receive sufficient energy to eject them from the atom,
(b) from the nuclear charge (Thomson scattering).

Nuclear resonant scattering (Sect. 13.6.4) and scattering of the photons by the electromagnetic field of the nucleus (potential or Delbruck scattering) can also take place but will be disregarded in this chapter. The Rayleigh scattering by the bound electrons acting together is calculated by assuming that each electron scatters a wave with amplitude at unit distance given by the Thomson factor $(\mu_0 e^2/4\pi m) \sin\Theta$ and that a coherent superposition of these amplitudes for a given direction of scattering must be made. An integration over the electron distribution then leads to an *atomic scattering factor* f_θ in terms of which the differential cross-section for Rayleigh scattering may be written

$$d\sigma_R = \left(\frac{\mu_0 e^2}{4\pi m}\right)^2 |f_\theta|^2 \frac{1 + \cos^2\theta}{2} \, d\Omega \tag{5.81}$$

In the limit of very long wavelengths $f_\theta \to Z$, but as the wavelength decreases (and the non-relativistic approximation becomes worse) the Rayleigh scattering from an atom is concentrated into forward angles as expected from simple diffraction theory. The Rayleigh scattering from atoms arranged in a regular crystal structure is responsible for X-ray diffraction phenomena.

The Thomson scattering by the nuclear charge Ze is well described by the classical theory because the nuclear mass M must replace the electron mass in eq. (5.70) and the condition $h\nu \ll Mc^2$ is amply fulfilled. Thomson and Rayleigh scattering are coherent since they arise from the motions of complementary charges in the same electric field, although this may vary over the atom for the shorter X-rays and gamma rays. The theoretical values of Rayleigh and Thomson scattering for 411 keV gamma rays are shown in Fig. 5.20(a), given by Moon.† This figure also gives for comparison the cross-

† P. B. Moon, *Proc. phys. Soc. Lond.*, **63**, 1189, 1950.

section for incoherent scattering due to the Compton effect (Sect. 5.4.4) which gradually replaces the coherent Rayleigh scattering as the photon energy increases and electrons are increasingly removed from their bound states. For very small angles the Rayleigh scattering, then proportional to Z^2, exceeds the Compton scattering, but since it is so strongly peaked forward it contributes little to the attenuation of a beam, and is not included in Fig. 5.17. The variation of the Rayleigh cross-section for lead with energy is shown in Fig. 5.20(b).

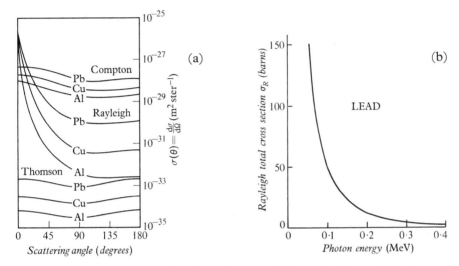

Fig. 5.20 (a) Scattering of 411 keV radiation by atoms of Al, Cu, Pb (Moon). (b) Variation of Rayleigh total cross-section σ_R with energy for lead (Ref. 5.1).

5.4.3 Photoelectric effect; Auger effect

A free electron cannot wholly absorb a photon because it is impossible simultaneously to conserve both energy and momentum. Quanta of energy greater than an atomic ionization potential may however eject a bound electron from a free atom (Fig. 5.21(a)) because the residual ion is able to take up the balance of momentum. The probability of this photoelectric absorption is greater the more tightly bound the electron and hence chiefly involves the K-shell, if sufficient energy is available, so that the energy of the emitted photoelectron is

$$T = h\nu - E_K \qquad (5.82)$$

where E_K is the ionization energy of the K-electron.

The theoretical treatment of this effect is only simple if the electron energy is small compared with mc^2, so that non-relativistic wave functions may be used. It is then found that most of the photoelectrons are emitted in the direction of the electric vector of the incident radiation, so that the angular

129

distribution about the beam is as shown in Fig. 5.21(c). The cross-section, and hence the absorption coefficient, is roughly given by

$$\sigma_{PE} \approx Z^5 \lambda^{7/2} \text{ (sometimes given as } Z^4 \lambda^3) \tag{5.83}$$

As the primary quantum energy increases the direction of maximum photoelectric emission is displaced towards the forward direction (Fig. 5.21(c)) and in the extreme relativistic region the cross-section varies more nearly as

$$\sigma_{PE} \approx Z^5 \lambda \tag{5.84}$$

The theoretical variation of σ_{PE} with energy for lead is shown in Fig. 5.21(b); there is a sharp increase in cross-section with increasing quantum energy at the ionization potentials of successive atomic shells. The process is important for absorption at energies below 0·5 MeV.

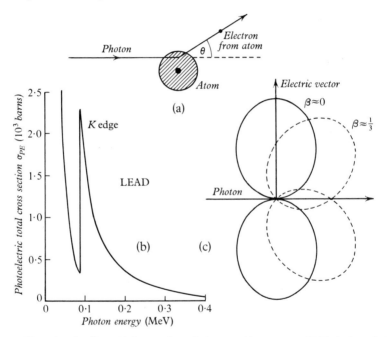

Fig. 5.21 Photoelectric effect. (a) Interaction of photon with an atom. (b) Variation of photoelectric total cross-section σ_{PE} with energy for lead (Ref. 5.1 and 5.11). (c) Polar diagram showing angular distribution of photoelectrons with respect to electric vector of incident radiation.

The photoelectric effect essentially gives no scattering of the incident radiation in the sense discussed in Sect. 5.4.1 because the small amount of energy which does not appear directly in the photoelectron is emitted as X-radiation and Auger electrons resulting from the atomic vacancy and these radiations are usually easily absorbed.

The Auger effect is the emission of low energy photoelectrons as an alternative to X-rays after the creation of a vacancy in one of the atomic shells by photoelectric effect, nuclear electron capture or other transition. If the K-shell is excited, the Auger electron may originate in the L-shell, and if the K-vacancy is filled by another electron from the L-shell the energy of the L-Auger electron is

$$T = h\nu_K - E_L = E_K - 2E_L \tag{5.85}$$

where ν_K is the frequency of the K X-ray line, and E_L is the energy of the L-edge. The Auger process is sometimes described as internal conversion of X-rays although it is strictly an alternative process for removal of energy. The relative probability of Auger emission and X-radiation is measured by the *fluorescence yield*

$$W_K = \frac{\text{number of } K\text{-quanta}}{\text{number of } K\text{-shell vacancies}} \text{ (for the } K\text{-shell)} \tag{5.86}$$

and the empirical variation of this quantity with Z shown in Fig. 5.22. For heavy atoms radiation is the more probable because of the high nuclear charge which effectively increases the coupling of the system to the radiation

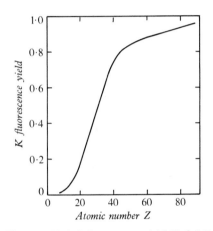

Fig. 5.22 *K*-shell fluorescence yield (Ref. 5.1).

field; for light atoms the Auger effect is predominant. The effect was discovered by the observation of short electron tracks associated with the emission of longer photoelectrons along the path of a beam of X-rays in an expansion chamber, cf. Plate 3. It has been possible to give an accurate theoretical treatment of the process and in particular to predict the curve shown in Fig. 5.22. The Auger effect is also important in transitions between the levels of muonic atoms (Sect. 11.2.3).

5.4.4 Compton effect

Neither the Rayleigh elastic scattering nor the photoelectric absorption cuts off sharply at any particular quantum energy, and both depend upon the reaction of an atom as a whole. As the quantum energy increases the wavelength becomes shorter and there is a greater tendency for interaction to take place with individual electrons, provided always that the condition of large enough momentum transfer (Sect. 5.4.1) is satisfied. Since, however, relativistic effects also become important as the wavelength decreases, the simple Thomson theory of free electron scattering is inadequate and the interaction must be regarded as the collision of a photon with an electron. This process

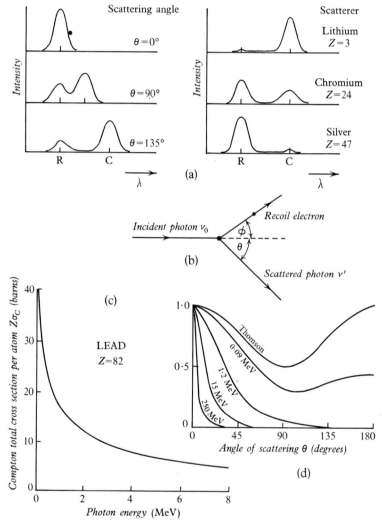

Fig. 5.23 Compton effect. (a) Qualitative dependence of Compton scattering (C) on angle and atomic number of scatterer, in relation to Rayleigh (unmodified) scattering (R). The intensity scales are arbitrary. (b) Interaction of photon with free electron. (c) Variation of Compton total cross-section per atom of lead ($Z\sigma_C$) with energy (Ref. 5.11, Table XIV). (d) Relative intensity of Compton scattering from free electrons as a function of angle (Ref. 1.3).

with free electrons is known as the Compton effect and is the most important mechanism of energy absorption for radiation of energy between, say, 0·5 MeV and 10 MeV. It differs most markedly from the Thomson scattering because the scattered radiation is of longer wavelength than the incident beam, and its wavelength depends on the angle of scattering. At each angle the Compton-scattered (modified) radiation is accompanied by a Rayleigh-scattered (unmodified) component, and the ratio between the two depends on the atomic number of the scatterer, which determines the ratio of effectively free to effectively bound electrons. These effects are illustrated in Fig. 5.23(a).

The energy and momentum relations for the Compton effect for an electron initially free and at rest follow from Fig. 5.23(b). For conservation of linear momentum

$$\frac{h v_0}{c} = \frac{h v'}{c} \cos \theta + p \cos \phi$$

$$0 = \frac{h v'}{c} \sin \theta - p \sin \phi \tag{5.87}$$

where p is the (relativistic) momentum of the recoil electron and for conservation of energy

$$h v_0 = h v' + T \tag{5.88}$$

where T is the kinetic energy of the electron, related to its momentum p by the equation

$$p^2 c^2 = T(T + 2mc^2) \tag{5.89}$$

From these equations, which must hold whatever the details of the scattering process, by elimination of ϕ and p the change in frequency or wavelength of the scattered photon is found to be

$$\frac{c}{v'} - \frac{c}{v_0} = \lambda' - \lambda_0 = \frac{h}{mc}(1 - \cos \theta) \tag{5.90}$$

The Compton shift in wavelength $\lambda' - \lambda_0$ is thus

(a) independent of the incident wavelength, so that photons of high energy (short wavelength) lose a large amount of energy in the scattering,
(b) independent of the material of the scatterer,
(c) dependent on angle in a way which is found to agree with experiment if m is taken to be the mass of an electron,
(d) expressible in terms of a fundamental constant h/mc, the Compton wavelength of a free electron, which is the wavelength of a photon of energy equal to mc^2, the electron rest energy.

The energy of the scattered quantum is

$$h v' = \frac{h v_0}{1 + h v_0 / mc^2 (1 - \cos \theta)} \tag{5.91}$$

and the kinetic energy of the recoil electron is then

$$T = h(v_0 - v') = hv_0 \frac{hv_0/mc^2(1-\cos\theta)}{1+hv_0/mc^2(1-\cos\theta)} \tag{5.92}$$

which is zero for $\theta = 0$ and a maximum for $\theta = \pi$, as might be expected. For $hv_0 \gg mc^2$ the Compton scattered photons at $\theta = 90°$ always have the energy mc^2 (511 keV).

These consequences of the photon hypothesis and of the laws of collision have been verified with great quantitative detail. The simultaneity of the emission of the scattered quantum and recoil electron, which now tends to be implicitly assumed, is of great importance for the photon hypothesis and has not remained unquestioned. The early experiments of Geiger and Bothe, in which the quantum and recoil electron were recorded in counters, and of Compton and Simon, in which tracks of recoil electrons and of photoelectrons from scattered γ-rays were observed in an expansion chamber, have, however, been abundantly confirmed by more recent work.

The probability of Compton scattering cannot be calculated simply, since it depends on details of the photon–electron interaction, e.g. the relation between the direction of polarization of the radiation and the direction of spin momentum of the scattering electron. A full discussion of the Klein–Nishina formula, which gives the differential cross-section for Compton scattering, is given by Evans (Ref. 5.1, p. 677). The fundamental expression for the differential collision cross-section per electron for polarized radiation is

$$d\sigma_c = \frac{r_0^2}{4}\, d\Omega \left(\frac{v'}{v_0}\right)^2 \left(\frac{v_0}{v'}+\frac{v'}{v_0}-2+4\cos^2\Theta\right) \tag{5.93}$$

in which r_0 is the classical electron radius and Θ is the angle between the polarization vectors of the incident and scattered radiation. This formula gives the probability of the scattering of a photon into the solid angle $d\Omega$ with energy hv'; if the solid angle element is $2\pi \sin\theta\, d\theta$, corresponding to scattering through angle θ, then θ is given in terms of v' by eq. (5.91). It may also be shown (Ref. 5.1) that the scattered photon and recoil electron tend to lie in a plane perpendicular to the electric vector of the incident radiation. This fact is utilized in the design of polarimeters for studying nuclear gamma radiation. Circularly polarized radiation would have no preferential scattering plane, but the scattering of this type of radiation is sensitive to the direction of the electron spins in the scatterer, and iron magnetized to saturation can be used as a polarimeter in this case. For unpolarized incident radiation averaging of (5.93) over all incident and scattered polarizations gives for the differential collision cross-section for deflection θ

$$d\sigma_c = \frac{r_0^2}{2}\, d\Omega \left(\frac{v'}{v_0}\right)^2 \left(\frac{v_0}{v'}+\frac{v'}{v_0}-\sin^2\theta\right) \tag{5.94}$$

This formula can be used to calculate the *number* of photons observed at a scattering angle θ. Since the energy of the scattered photon varies rapidly with θ, the intensity of radiation or energy scattered through the angle θ is proportional to $v'\,d\sigma_C$. Figure 5.23(d) shows the relative intensity of Compton scattering as a function of angle; for small angles and for small incident energies it approaches the value predicted by the Thomson formula. In practice the incoherent scattering cross-section will fall as the 'Thomson' limit is approached because of the binding of the inner shell electrons.

The *total collision* cross-section σ_C is obtained by integrating (5.94) over all angles θ from 0 to π. This gives the probability of removal of a photon from the beam by a Compton process and is shown for a particular case in Fig. 5.23(c); it also appears directly in the attenuation coefficient μ_m shown in Fig. 5.17. The Klein–Nishina theory has been exhaustively and successfully tested by measurements of Compton total and differential cross-section for a wide range of atomic numbers and energies; as shown in Fig. 5.23(c) the cross-section decreases with quantum energy.

5.4.5 Pair production

Dirac's relativistic wave equation, which gives such an excellent description of the properties of the electron and of the fine structure of the hydrogen spectrum, leads to the conclusion that electrons possess states of negative energy. From the relativistic formula connecting energy and momentum of a particle of rest mass m we have for the total energy

$$E^2 = (T+mc^2)^2 = p^2c^2 + m^2c^4$$

whence

$$E = \pm\sqrt{p^2c^2 + m^2c^4}. \tag{5.95}$$

This means that the states of energy for an electron are as shown in Fig. 5.24(a), extending positively from mc^2 to $+\infty$ and negatively from $-mc^2$ to $-\infty$. Classically the negative energy states† could be excluded as having no physical meaning, but quantum mechanically an external field can cause transitions to these states which must therefore be accommodated in the theory. The derivation of the Klein–Nishina formula for the Compton effect is based on Dirac's theory and explicitly involves transitions of this kind in which momentum, but not energy, is conserved. Such processes are permitted within the limits of the uncertainty principle; energy need not be conserved in an intermediate state if it is sufficiently short-lived. The exact quantitative success of the Klein–Nishina formula is strong evidence for the existence of negative energy states.

In order to avoid the difficulty that the ordinary electrons of experience should all make transitions to negative energy states Dirac put forward his theory of 'holes' according to which in the absence of an external field all

† In a negative energy state acceleration would be in a direction opposite to the applied force.

negative energy states are filled with electrons and there can be no transitions to these occupied states. This sea of electrons is supposed to give no contribution to the total energy and momentum of a system. If now an electron is removed from the distribution by the action of an external field, creating a hole, the system acquires an energy $-(-E)$, a momentum $-(-p)$ and a charge $-(-e)$, that is to say it behaves as an electron of ordinary momentum

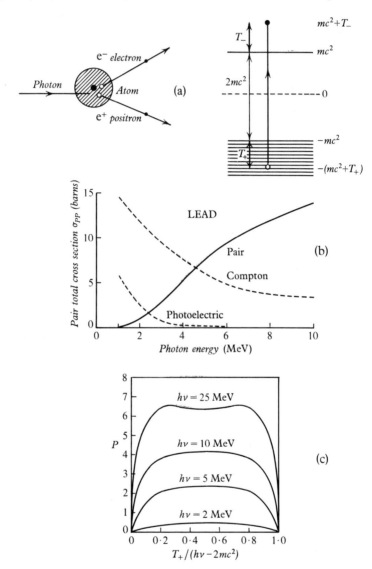

Fig. 5.24 Pair production. (a) Mechanism; the energy absorbed in the process is $2mc^2 + T_+ + T_-$. (b) Variation of pair total cross-section σ_{PP} with energy for lead. The Compton and photoelectric cross-sections are indicated for comparison (Ref. 5.11, Table XIV). (c) Distribution in energy between electrons and positrons (Ref. 5.1).

and energy but with positive charge. Since the electron elevated by this transition must finally occupy a state of total energy greater than mc^2, it is clear from Fig. 5.24(a) that the external field, which may be that of γ-radiation or that of a charged particle, must supply an energy greater than $2mc^2$ to induce the transition. The elevated electron and the 'hole' together form an electron–positron pair and the process, which obviously has a threshold energy of $2mc^2$, is known as pair production. In order to satisfy the conservation laws, pair production must take place in the field of an electron or a nucleus, which can absorb linear momentum. Positrons were discovered in cloud chamber photographs of cosmic radiation by Anderson[†] in 1932 and are now familiar particles, of common occurrence in nuclear β-processes (ch. 16); a picture of a positron track, distinguished by its curvature, is given in Plate 6.

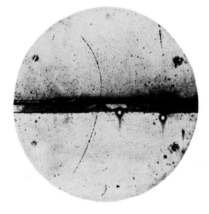

Plate 6 A positron of energy 63 MeV passes through a lead plate and emerges with an energy of 23 MeV (C. D. Anderson, *Phys. Rev.*, **43**, 491, 1933).

The interaction between electrons and positrons includes, in addition to ordinary scattering, the possibility of annihilation. This is the process inverse to pair production, and in it the electron makes a transition to the vacant negative energy state corresponding to the positron. Both particles disappear and there is an appearance of an energy of $2mc^2$ as electromagnetic radiation. In the annihilation of free positrons at rest by free electrons conservation of momentum requires that this energy shall appear as two oppositely directed quanta of energy $mc^2 = 0.511$ MeV, the so-called *annihilation radiation*, but in the field of a nucleus one quantum annihilation is also possible, with absorption of an electron. If a positron is annihilated by collision with a moving electron the angle between the two annihilation quanta becomes less than $180°$ and observation of these angles is an important means of learning about the distribution of electron momenta in matter. The high resolution of the germanium detector (Sect. 6.1.2) has made it possible to see the broadening

[†] C. D. Anderson, *Science*, **76**, 238, 1932.

of the annihilation line due to the electron motion.† Annihilation of positrons in flight is possible, and has been observed.

Annihilation of positrons is formally equivalent to the production of bremsstrahlung in the collisions of fast electrons with nuclei. The only difference is that in the former case the electron makes a transition to a state of negative, instead of positive, energy. The theory of both processes was given by Bethe and Heitler, who also deduced the cross-section for the inverse process of pair production from the same calculation. For creation of pairs by a gamma ray in the field of a nucleus the differential cross-section for the production of a positron of kinetic energy T_+ (and an electron of energy $h\nu - 2mc^2 - T_+$) may be written (Ref. 5.1, p. 703)

$$\mathrm{d}\sigma_{PP} = \frac{\sigma_0 Z^2 P}{h\nu - 2mc^2} \, \mathrm{d}T_+ \tag{5.96}$$

where

$$\sigma_0 = \frac{1}{137} \left(\frac{\mu_0 e^2}{4\pi m}\right)^2 = 5\cdot8 \times 10^{-32} \text{ m}^2$$

and P, shown in Fig. 5.24(c), is a slowly varying function of $h\nu$ and Z. The figure implies symmetry in the distribution of energy between electrons and positrons in pair production; in fact this is disturbed by the nuclear charge, which slightly accelerates the positrons and retards the electrons.

The variation of the total cross-section for pair production in lead with quantum energy is shown in Fig. 5.24(b). This is obtained by integrating the expression (5.96) over the energy spectrum. The cross-section increases at first with energy, as the pair production takes place at larger and larger distances from the nucleus, but a limit is set to the increase by the screening of the nuclear charge by the atomic electrons. In the absence of screening the pair cross-section is proportional to Z^2; it is therefore, as shown in Fig. 5.18, the predominant effect at high energies and for heavy nuclei in comparison with the Compton scattering and photoelectric effect. The electron pairs are mainly confined at high energies within a small forward cone of semi-angle mc^2/E, and play an essential role in the building up of the cascade showers of cosmic radiation.

The pair production discussed in this section is not a nuclear phenomenon, although a nucleus may be involved in the momentum balance. It is therefore known as *external* pair production to distinguish it from *internal* pair production (Sect. 13.3) which is a process alternative to radiative emission from an excited nucleus. Theoretically external production is treated as a *plane* wave phenomenon; internal production however depends markedly on the multipolarity (Sect. 3.5.2) of the radiative transition and hence on the *spherical* wave representation of the radiative field.

† G. Murray, *Phys. Lett.*, **24B**, 268, 1967. See also Example (5.23).

References

1. EVANS, R. D. *The Atomic Nucleus*, McGraw-Hill, 1955.
2. JELLEY, J. V. *Cherenkov Radiation and its Applications*, Pergamon Press, 1958.
3. RUTHERFORD, E., CHADWICK, J. and ELLIS, C. D. *Radiations from Radioactive Substances*, Camb. Univ. Press, 1930.
4. ALLISON, S. K. and WARSHAW, S. D. 'Passage of Heavy Particles through Matter', *Rev. mod. Phys.*, **25**, 779, 1953; FANO, U. 'Penetration of protons, α-particles and mesons', *Ann. rev. nucl. Sci.*, **13**, 1, 1963; NORTHCLIFFE, L. C. 'Passage of Heavy Ions through Matter', *Ann. rev. nucl. Sci.*, **13**, 67, 1963.
5. PRICE, B. T. 'Ionization by Relativistic Particles', *Rep. progr. Phys.*, **18**, 52, 1955.
6. BETHE, H. A. and ASHKIN, J. 'The Passage of Radiations through Matter', in *Experimental Nuclear Physics*, ed. E. Segrè, Vol. I, Wiley, 1953.
7. LIVINGSTON, M. S. and BETHE, H. A. 'Nuclear Physics (Part C)', *Rev. mod. Phys.*, **9**, 245, 1937.
8. WHALING, W. 'The Energy Loss of Charged Particles in Matter', *Encyclopedia of Physics*, **34**, 193, Springer, 1958.
9. MORGAN, D. V. and POATE, J. M. 'Channelling in Physics', *Sci. Jnl.*, August 1970, page 47; DATZ, S., ERGINSOY, C., LIEBFRIED, G. and LUTZ, H. O. 'Motion of Energetic Particles in Crystals', *Ann. rev. nucl. Sci.*, **17**, 129, 1967.
10. EVANS, R. D. 'Compton Effect', *Encyclopedia of Physics*, **34**, 218, Springer, 1958.
11. DAVISSON, C. M. and EVANS, R. D. 'Gamma Ray Absorption Coefficients', *Rev. mod. Phys.*, **24**, 79, 1952.
12. HEITLER, W. *The Quantum Theory of Radiation* (3rd edn.), Clarendon Press, 1954.
13. ROCHESTER, G. D. and WILSON, J. G. *Cloud Chamber Photographs of the Cosmic Radiation*, Pergamon Press, 1952.

6 Nuclear detectors

One of the most important applications of knowledge of the passage of radiations through matter has been in the design of detecting instruments. The availability of a new type of detector, such as the Geiger–Müller counter (Sect. 6.1.4) or the expansion chamber (Sect. 6.2.1) or the sodium iodide crystal (Sect. 6.1.5) has sometimes greatly enlarged experimental possibilities in nuclear physics, and much work is continually being undertaken to improve existing instruments. In the present chapter the main types of nuclear detectors are surveyed, with particular attention to the specific applications for which they are best suited; in the following chapter the general problem of the measurement of the energy and intensity of ionizing radiations will be discussed.

6.1 Detecting instruments (electrical)

6.1.1 Ionization chambers

In its simplest form the ionization chamber is a gas-filled metal vessel into which is inserted an insulated electrode (Fig. 6.1(a)); under the influence of ionizing radiation the application of a potential difference of a few hundred volts between the electrode and the wall causes a current to flow through the gas. If the insulated electrode is charged and then disconnected from the voltage source, the ionization current discharges the capacity of the electrode, and the fall in potential may be displayed as motion of a leaf or fibre as in an electroscope, rather than as continuous current. In both cases the chamber is responding smoothly to the integrated effect of a large number of ionizing events, and such *current chambers* are extensively used for flux determination or source intensity measurements for all types of radiation. In some applications the parallel plate type of chamber (Fig. 6.1(b)) is more convenient.

It has been long known from experiments with X-rays, that for a steady rate of primary ionization the current reaches a saturation value when all ion pairs are collected before recombination. If q ionizing events each producing n ion pairs take place per second in the chamber, the saturation current is

$$i_s = qne \tag{6.1}$$

where e is the electronic charge, and under these conditions the chamber can give an indication which accurately determines the primary ionization. The ionization current is usually measured electronically by observing the voltage developed by the current across a resistance of value 10^{10}–10^{13} ohms; alternatively the output may be converted into a.c. by a vibrating-reed electrometer, which is effectively a condenser whose capacitance is varied at a frequency of about 300 Hz. This provides a stable, high impedance input and permits measurement of currents of the order of 10^{-14}–10^{-16} A. Such ionization

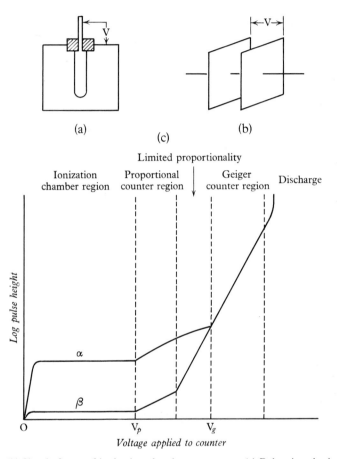

Fig. 6.1 (a), (b) Simple forms of ionization chamber or counter. (c) Pulse size obtained from a counting chamber as a function of voltage applied. Curves are drawn for a heavy initial ionization, e.g. from an α-particle and for a light initial ionization, e.g. from a fast electron. For voltages less than V_p only the primary ion pairs are collected; for $V > V_p$ ionization by collision takes place and the counter ceases to be a conventional ionization chamber. Very roughly the electric field in argon at atmospheric pressure corresponding to V_p is 10^6 V m^{-1} (from Korff, *Electron and Nuclear Counters*, van Nostrand, 1946).

chambers for detection of γ-radiation are usually filled with high pressure argon, for fast neutron detection with hydrogen, and for slow neutron detection with boron trifluoride (BF_3). In all these cases a secondary effect is detected, i.e. either electron conversion of photons, or recoils from nuclear elastic scattering or transmutation in the case of neutrons.

The development of high-gain pulse amplifiers has made it possible to use ionization chambers as *counting instruments* for single particles. An α-particle of energy 6 MeV wholly absorbed in a gas will produce about 2×10^5 ion pairs, equivalent to positive and negative charges of $3 \cdot 2 \times 10^{-14}$ C. In a chamber of capacitance 10pF such a charge causes a change of potential of $3 \cdot 2$ millivolts which is much larger than the input noise voltage of a typical amplifier. It is in fact possible to distinguish the pulse due to as few as 1000 ion pairs and hence, by means of a shallow chamber, to study in detail the variation of ionization along the track of a single α-particle.

The output pulse height from a counting chamber varies with applied voltage as shown in Fig. 6.1(c). In the region OV_p all the primary ions are collected and the pulse height has a saturation value. For higher voltages, ionization by collision occurs and the pulse height increases; counters operating in these regions will be described later (Sects. 6.1.3, 6.1.4). In the ionization chamber region OV_p the operation of a pulse chamber may be understood from Fig. 6.2(a), (b), in which a charged particle is supposed to traverse

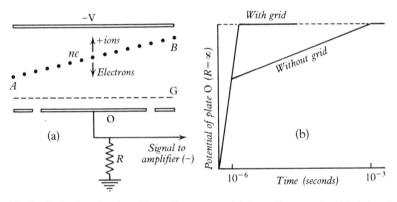

Fig. 6.2 The ionization chamber. The collector potential for ordinary and gridded chambers is shown as a function of time, neglecting increased collector capacity due to the grid.

the chamber along the path AB in a time of perhaps 10^{-9} s, leaving a track containing about 15×10^5 ion pairs m^{-1}. In gases such as pure hydrogen, nitrogen and argon, which do not readily form negative ions, the initial effect is the production of positive ions and electrons. If an electric field of about 10^4V m^{-1} acts across a chamber 10 mm deep and if the argon is at atmospheric pressure, the electrons will move towards plate O with a velocity of 5×10^3 m s^{-1} and the positive ions will move towards plate $-V$ with a velocity of about $1 \cdot 4$ m s^{-1}. If the plate is insulated from earth its potential immediately following the burst of ionization varies as shown in Fig. 6.2(b). The sharp rise is due to collection of the fast moving electrons in a time of the order of 10^{-6} s; the positive ions move much more slowly, and until they have all reached the plate $-V$ the potential of O continues to rise since the ions induce a negative charge on O which is not released until they are collected. The pulse is not complete until a time which may be considerably greater than 10^{-3} s and the magnitude of the initial sharp rise, for a given primary ionization, is dependent on the position of the ionizing track AB in the chamber. These undesirable features of 'slow' ionization chambers are removed in the *gridded chamber* (Fig. 6.2(a)) in which the plate O is shielded by a mesh of wires G held at a potential intermediate between zero and $-V$. The capacity between a track of ions and the collector O is then very much reduced and only the fast electron pulse, of size independent of the position of the track in the chamber (neglecting wall effects), is obtained. The gas filling such 'fast' ionization chambers must not form negative ions and in order to sharpen the electron pulse as much as possible the unidirectional drift velocity of the electrons through the gas in a given field E must be high. The drift velocity for *ions* is found by an elementary calculation to be

$$\frac{eE\lambda}{Mc} \tag{6.2}$$

where M is the ionic mass, λ the mean free path and c the root mean square thermal velocity. This shows that the drift velocity may be increased if the random velocity is reduced. For electrons the calculation of drift velocity is somewhat more complicated, since electrons lose very little energy by collision with gas molecules and the application of the electric field E raises the electron temperature. It has been found useful to check the increase of c by adding a small amount (2–10 per cent) of carbon dioxide or methane to the argon usually employed in ionization chambers. These molecules have many low-lying excited states and inelastic collisions continually remove the energy gained by the electrons from the field E, so that the drift velocity is increased. It is customary to allow a suitable mixture of gases to flow through the ionization chamber, so that electron-capturing impurities do not accumulate.

The pulse from an ionization chamber is detected by connecting the collector plate O (Fig. 6.2(a)) to earth through a high resistance R (10^{10}–10^{13} ohms) which together with the capacity C of the chamber forms part of the input circuit of a low-noise amplifier. If an argon filling is used in the chamber the amplifier time constants may be set to reproduce only the sharp rise of voltage resulting from electron collection. Fast counting of particles is then possible, and, in addition, low frequency microphonic disturbances are entirely eliminated.

Counting ionization chambers are not used for electrons because of the low primary ionization of these particles; for lightly ionizing events a counter providing either gaseous multiplication or a long path length is required. The parallel plate chamber has been much used for determination of the energies of radioactive α-particles, and offers a moderate energy resolution.† Figure 6.3(a) shows a spectrum obtained using an α-particle source in a gridded parallel plate chamber filled to an argon pressure sufficient to stop the α-particles within the chamber. If the ionizing events are independent, then

† The term 'energy resolution' is defined in Sect. 7.1.2.

for a homogeneous group of α-particles the number of ion pairs follows a Poisson distribution with standard deviation $\sigma = \sqrt{N}$ where N is the average number. This indicates that for 6 MeV α-particles the group width at half-height would be about $\frac{1}{2}$ per cent of its energy. In fact, as pointed out by Fano, the ionizing events are not independent because of the fixed total energy that can be lost, and the resulting correlation narrows the distribution so that $\sigma = \sqrt{FN}$ where F, the *Fano factor*, is of the order of 0·3.

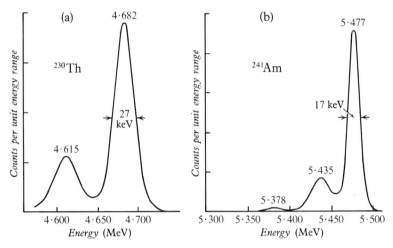

Fig. 6.3 Alpha-particle groups. (a) ^{230}Th in gridded ionization chamber (Engelkemeir and Magnusson, *Rev. sci. Instrum.*, **26**, 295, 1955). (b) ^{241}Am in semiconductor counter (Dearnaley and Whitehead, *Nucl. instrum. Meth.*, **12**, 205, 1961).

6.1.2 Semiconductor counters

(*a*) *General.* Gas-filling is not an essential part of an ionization chamber. It has been known since 1945 that semiconducting crystalline materials such as diamond, zinc sulphide and silver chloride will respond to the ionization produced by particles, and the conduction mechanism is well understood; Fig. 6.4(a) illustrates the general principle. Ideally the crystal has a low conductivity because the conduction band is empty and a high collecting field can then be maintained across a block of material without causing excessive current. Electrons released by an ionization process can enter the conduction band and move through the crystal under the applied field. In practice it is found that for most materials the number of electrons released by an ionizing process is considerably less than the number of charged carriers (electrons or positive holes) already present in the semiconductor. These arise because of thermal excitation of electrons from the filled band across the band gap and because of the presence of impurities which provide electron-donating levels near the conduction band (Fig. 6.4(a)). The presence of impurities also creates trapping centres which may indeed impede the motion of carriers and lower the conductivity of the crystal, and ultimately the storage of charge in the traps polarizes the crystal and prevents proper collection of electrons from an ionizing event. These effects can be avoided in a limited volume of semiconductor by making use of the properties of the p–n junction (or rectifying contact) and in large volumes by compensation techniques.

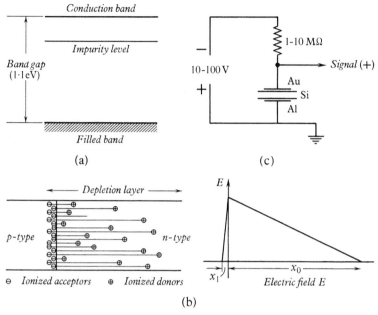

Fig. 6.4 The semiconductor counter (Ref. 6.6): (a) Bands and impurity level in semiconductor. (b) Charge and electric field distributions in the depletion layer of a p–n junction. (c) Circuit arrangement.

(b) Junction detectors for heavy charged particles. In an n-type semiconductor, conduction is due to the motion of electrons in the conduction band, and in p-type material the process is effectively a motion of positive holes resulting from the rearrangement of electrons between atoms in the crystal. The two materials are prepared from the intrinsic (i) semiconductor by the controlled addition of electron-donating or electron-accepting elements. If a junction between p- and n-type regions is formed in a single crystal (Fig. 6.4(b)) electrons migrate from the n region to the p region until a potential barrier sufficient to prevent further motion is established. A shallow layer near the junction, known as the *depletion layer*, is thus cleared of carriers. This is a layer of low conductivity and high field and can operate as an ionization chamber for a charged particle creating electron-hole pairs within it.

The depth of the depletion layer depends on the density of impurity centres and on any applied potential difference V (reverse bias). If ρ is the resistivity of the n-type region then the depth x_0 (Fig. 6.4(b)) is given by

$$x_0 \doteq \sqrt{\rho V} \quad \text{(Ref. 6.8)} \tag{6.3}$$

For silicon of the highest practicable resistivity a depletion depth of about 3 mm may be obtained.

If a charged particle enters the depletion layer the electron-hole pairs produced are swept away by the field existing across the layer and a pulse may be detected in an external circuit. For silicon the energy required per electron-hole pair is about 3·6 eV (i.e. a few times the band gap) compared with about 30 eV for a gaseous detector. When allowance is made for the Fano factor a line width of perhaps 5 keV may be predicted for a 6 MeV α-particle. In practice heavy particles suffer nuclear collisions towards the end of their range and because of the high ionization density, recombination effects are important. A certain statistical fluctuation, which limits overall performance, arises in this way, so that a resolution of only about 0·2 per cent is obtainable; Fig. 6.3(b) shows the spectrum of α-particles from ^{241}Am observed with a silicon detector of resistivity 30 Ω m.

Two types of p–n detector have been developed: (a) *the diffused junction* detector in which a donor impurity, usually phosphorus, is introduced into a p-type (boron doped) silicon single crystal to form a depletion layer at the diffusion depth, and (b) *the surface barrier* detector (Fig. 6.4(c)) in which a p-type layer is formed on the surface of n-type silicon by oxidation. Contact is made to the detectors through a layer of gold, which apparently plays some part in the formation of the rectifying contact. The counter base is often an ohmic (non-rectifying) layer of aluminium. Signals are derived from the surface barrier detector by the connections shown in Fig. 6.4(c). Since the depth of the depletion layer, and consequently the interelectrode capacitance, depends on the applied voltage, a charge-sensitive rather than a voltage-sensitive pre-amplifier is used.

The advantages of the silicon p–n junction as a detector of heavy particles are its excellent resolution and linearity, its small size and consequent fast response time, permitting a high counting rate, and the fairly simple nature of the necessary electronic circuits. It is of course a windowless counter, although in the diffused junction type an insensitive layer is generally present, and its simplicity permits the use of multi-detector arrays. The handling of data from such complex systems increasingly demands the availability of an on-line computer for storage of information and data processing. A special facility of high value in the use of solid state detectors for nuclear reaction experiments is the *particle-identification* 'telescope' in which a thin transmission (ΔE) counter through which particles pass is combined with a thicker (E) counter in which the particles stop. For a particle of velocity v and charge ze the signals from these counters are proportional (eq. 5.40) to $z^2 e^2/v^2$ and $\frac{1}{2}Mv^2$ respectively so that the product $E\Delta E$ measures Mz^2 independently of v.

A further property of increasing importance is *position-sensitivity*, which is achieved by the arrangement shown in Fig. 6.5. For a particle of given energy incident on a strip detector, a full height pulse V_C is obtained at the front contact C. The back contacts however are made via a resistive strip evaporated on to the detector surface and the pulse height at B then depends on the lateral coordinate x according to the relation

$$V_B = V_C \left(1 - \frac{x}{l}\right) \tag{6.4}$$

Such counters find an application in the focal plane of particle spectrometers.

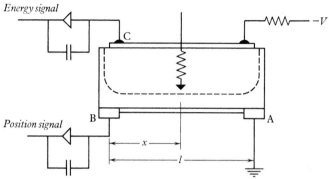

Fig. 6.5 Position-sensitive counter (from Fig. 2.4, p. 189 in: G. Bertolini and A. Coche (eds.), *Semiconductor Detectors*, North Holland, Amsterdam, 1968).

Junction detectors are used successfully for the detection of protons of energy up to 25 MeV but for higher energy particles the depth of the sensitive layer must be increased beyond the approximate 3 mm to which this figure corresponds. This can be done by compensating the acceptor centres in a p-type semiconductor by the controlled introduction of donor impurities. The resulting material, although not intrinsic, is of very low conductivity and can be used to detect ionization pulses. The most suitable donor is lithium, because this atom has a very low ionization potential in a semiconductor and also a high mobility. The lithium is 'drifted' into the bulk material from a surface layer by applying an electric field and raising the temperature to about 150°C. Compensated depths of about 15 mm can be obtained and by drifting into a block of semiconductor from several directions, or from an axial hole, a large volume detector (up to 100 cm^3 in germanium at present) can be prepared. Lithium-drifted silicon detectors can be used, singly or in $(E . \Delta E)$ combination, as spectrometers for energetic charged particles from nuclear reactions. Unfortunately the acceptor-donor balance can fairly easily be destroyed by radiation damage effects which do not anneal out; the counters then draw a large current and the resolution becomes very poor. The Si(Li) counter is also very successfully used for conversion electron spectroscopy.

Silicon detectors must be used in such a way that channelling effects (Sect. 5.3.7) are not significant.

(*c*) *Gamma ray detectors.* The major application of the lithium-drift technique is to the germanium counter,† which has transformed the subject of

† G. T. Ewan and A. J. Tavendale, *Nucl. instr. Meth.*, **26**, 183, 1964; *Can. jnl. Phys.*, **42**, 2286, 1964.

gamma ray spectroscopy. Germanium has a band gap of only 0·67 eV and must be cooled to liquid nitrogen temperature when used as a counter in order to minimize thermal excitation of electrons to the conduction band. This is an inconvenience which results in a preference for silicon counters (although these are also improved by cooling) for charged particle spectrometry. For gamma ray detection however the higher atomic number of germanium ($Z_{Ge} = 32$, $Z_{Si} = 14$) leads to a marked improvement in efficiency as

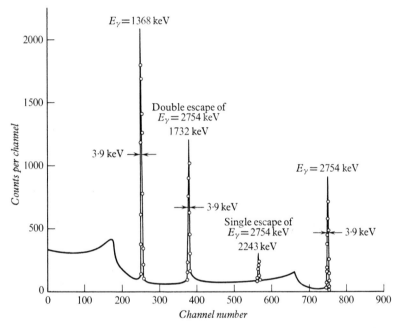

Fig. 6.6 ^{24}Na γ-ray spectrum taken with 30 cm^3 Ge(Li) detector (V. J. Orphan and N. C. Rasmussen, *Nucl. instrum. Meth.*, **48**, 282, 1967).

may be seen from the expressions for the basic processes of photoelectric, Compton, and pair-production absorption given in Chapter 5. The difference between Ge(Li) and Si(Li) counters is especially marked in the energy ranges in which the photoeffect and pair production predominate because of their particular Z-dependence. The Ge(Li) detector cannot yet be made as large as a sodium iodide scintillator (Sect. 6.1.5) and is less suitable than the latter when the highest detection efficiency is required, but its resolution is better by more than an order of magnitude. The resolution in the case of gamma ray detection via photo-electrons, Compton electrons or electron–positron pairs is not limited by the nuclear straggling effects noticed with heavy particles, and line widths of less than 1 keV have been reported.[†] This inherent performance imposes very high demands on the electronic circuits to which the detector is coupled and the most satisfactory low-noise head amplifying ele-

† H. R. Bilger and I. S. Sherman, *Phys. Lett.*, **20**, 513, 1966.

ment is now felt to be a field-effect transistor (FET) operated at a low temperature.

Figure 6.6 shows a spectrum of ^{24}Na gamma rays taken with a typical Ge(Li) detector. The ^{24}Na source emits two gamma ray lines of energies $E_\gamma = 2754$ and 1368 keV, and the figure shows, for the former, peaks at E_γ, $E_\gamma - mc^2$ and $E_\gamma - 2mc^2$. The full energy peak corresponds to complete retention of all secondary radiations by the detector; when however the photon is absorbed by a pair production process the positron usually slows down and ultimately annihilates with the production of two quanta of energy mc^2. If one of these escapes from the detector a pulse in the single escape peak $E_\gamma - mc^2$ is produced; if both escape the signal falls in the double escape peak $E_\gamma - 2mc^2$. The figure also shows, especially below the full energy peak for the 1368 keV gamma ray, the characteristic shape of the distribution of Compton electron energies. This is normally an acceptable background in Ge(Li) spectra but it may in fact be cut down by surrounding the Ge counter by a large sodium iodide detector to detect the scattered Compton photon. If Ge-pulses are only recorded when there is no coincident signal in the anti-Compton annulus, only the pair production and photoelectric events are in principle detected.

The absolute efficiency of a 20 cm^3 Ge (Li) counter as a function of energy is shown in Fig. 6.7. The rise time of the pulse from a semiconductor detector

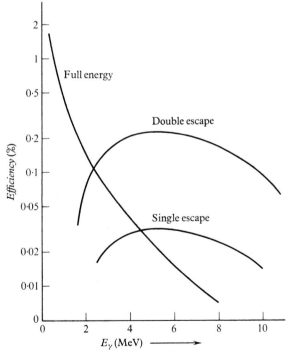

Fig. 6.7 Efficiency curve for a 20 cm^3 Ge(Li) detector (C. van der Leun *et al.*, *Nuclear Physics*, **100**, 316, 1967).

is a few nanoseconds. In a large volume Ge(Li) detector a spread of rise times may be observed because of the spatial distribution of ionizing events (cf. the ionization chamber, Sect. 6.1.1). This may be improved by the use of co-axial geometry.

6.1.3 *Proportional counters*

If the voltage applied to a gas-filled ionization chamber is increased beyond the point V_p in Fig. 6.1(c), the current begins to increase owing to ionization by collision. The field E necessary for ionization by electrons starting from rest in argon at atmospheric pressure is of the order 10^6 V m^{-1}; if the pressure p of the gas is reduced the mean free path increases and the field required decreases, in such a way that E/p is approximately constant.

The field required is most conveniently, although not necessarily, developed in a counter of the form shown in Fig. 6.8(a), employing cylindrical geometry. The field at radius r is given by the formula (Fig. 6.8(b))

$$E = \frac{V}{r \log_e b/a} \qquad (6.5)$$

where a is the radius of the central wire and b is the inner radius of the cathode cylinder. For a counter with $a = 0.1$ mm, $b = 10$ mm and $V = 1000$ volts the field required for ionization by collision in argon at atmospheric pressure is reached at a radius $r = 0.2$ mm. The electrons from ion pairs produced anywhere in the counter volume move towards the positively charged wire and produce further electrons, which build up into an 'avalanche' of ionization in the central region. The gas multiplication m is the number of electrons reaching the wire for each initial electron and is usually between 10 and 10^3. Although these avalanche electrons are collected rapidly, the electrical pulse in an external circuit obtained from the collector wire is due to the drift of the residual positive ions away from the wire to the outer cylinder, since while the positive ions are close to the wire they neutralize the effect of the electrons.

The shape of the pulse from a proportional counter is independent of the position of the ionizing track in the chamber because the main multiplication always occurs in the high field region. Figure 6.8(c) shows the pulse shape calculated for the counter specified earlier in the section; the pulse rises to half its final height in 2.5 μs, although before this rise commences there may be a time-lag of perhaps 1 μs while the initial electrons drift towards the wire. This lag may be reduced by the addition of carbon dioxide to the counter filling. Fast pulses may be obtained by using a short time constant of differentiation in a following stage of amplification.

The high multiplication available in proportional counters renders them suitable for the study of lightly ionizing particles, such as β-particles, mesons, and fast protons. Slow heavy particles are also readily detected and external amplification in such cases need only be very small. For this reason the proportional counter was historically the first type of counting tube to be used, before the development of pulse amplifiers permitted the application of ionization chambers to particle detection. It is however desirable whenever possible to keep the gas multiplication of a proportional counter low in the interests of stability even at the expense of loss of resolution. The proportional counter loses its proportionality for high initial ionizations or high multiplications owing to space-charge effects in the avalanche and in the limit of high multiplications the output pulse height becomes independent of the primary ionization; this is the so-called *Geiger* region (Fig. 6.1(c)).

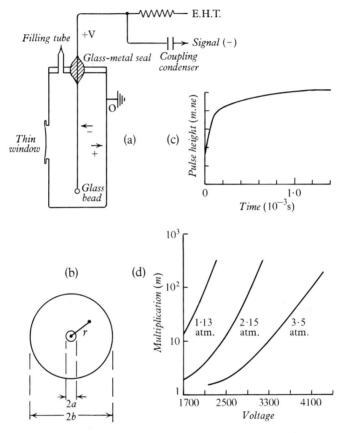

Fig. 6.8 The proportional counter. (a) Construction. The glass bead prevents sparking from points at the end of the wire. (b) Cross-section, cylindrical geometry. (c) Pulse shape (Ref. 6.10). (d) Dependence of multiplication on voltage for a cylindrical counter filled to the pressures indicated (Ref. 6.10).

Methane is better than argon as a filling for proportional counters to be used at high multiplication because of its ability to absorb photons which might provoke catastrophic breakdown. The gas is usually allowed to flow at atmospheric pressure through the counter in order to preserve purity and to facilitate the introduction of samples for counting. Typical proportional counter characteristics are shown in Fig. 6.8(d).

Proportional counters are of particular service in measuring the energy of low energy radiations which can be absorbed wholly in the gas filling of the counter. Gaseous β-emitters of low energy such as ^{3}H (18 keV), ^{14}C (155 keV) or ^{35}S (168 keV) have been studied extensively in this way and the counter is also widely applied in the determination of the energies of X-rays and low energy γ-rays by pulse height analysis. Figure 6.9 shows the spectrum of low energy radiations from a source of ^{57}Co, decaying into ^{57}Fe, obtained in a krypton-filled proportional counter. The energy scale may be calibrated using X-rays of known energy.

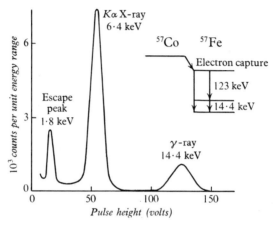

Fig. 6.9 Pulse height distribution for 14·4 keV γ-radiation and iron K X-radiation from a source of ^{57}Co (decay scheme inset) observed in a krypton-filled proportional counter. When a 14·4 keV photon produces a photoelectron in the krypton a K-shell vacancy usually results. If a K X-ray quantum is emitted it may escape from the counter. The total energy deposited in the counter is then 14·4 keV − 12·6 keV (the krypton K_α-energy) = 1·8 keV (Lemmer *et al.*, *Proc. phys. Soc.*, **68**, 701, 1955).

A special form of proportional counter in which many thin parallel wires, at a spacing of a few mm, are mounted between mesh electrodes has been used by Charpak *et al.*† for particle location in high energy experiments. Each individual wire effectively behaves as a cylindrical counter with a detection efficiency of 100 per cent for particles of minimum ionization passing within about 1 mm of the wire. Signals from the wires must be separately amplified and displayed.

6.1.4 *Geiger–Müller counters*

When the discharge characteristics of a proportional counter are so altered by reducing the gas pressure or increasing the applied voltage, that the output

† G. Charpak *et al.*, *Nucl. instr. Meth.*, **62**, 262, 1968; see also *Ann. Rev. nucl. Sci.*, **20**, 195, 1970.

pulse is independent of initial ionization the tube operates as a Geiger–Müller, or GM counter. The construction of a typical GM counter (Fig. 6.10(a)) may be almost identical with that of the proportional counter described in Sect. 6.1.3; a gas filling of 100 mmHg pressure of argon and 10 mmHg of ethyl alcohol vapour may be used. If an ion pair is produced in a GM counter, the electron moves towards the central wire, as in proportional counter action, and the positive ion of argon moves more slowly to the

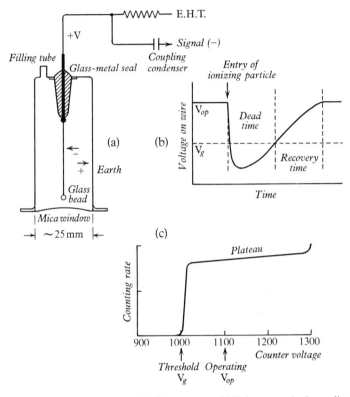

Fig. 6.10 The Geiger–Müller counter. (a) Construction. (b) Voltage on wire immediately following entry of an ionizing particle. The dead time and recovery time are each approximately 100–200 μs. (c) Counting rate of GM tube exposed to constant source, as a function of applied voltage.

cathode. The avalanche produced by the electron in the high field near the wire is much more violent than in a proportional counter and secondary ultraviolet photons are produced in sufficient number to convey the discharge down the whole length of the wire of the counter. The velocity of propagation is about 10^4–10^5 m s^{-1} and as a result of this spread of the avalanches the charge available for collection by the wire has a constant value, independent of the magnitude and location of the initial ionization. When the electrons from the avalanches have been collected the slowly moving sheath of positive ions acts as a partial electrostatic screen and reduces the field at the wire

153

below the value necessary for ionization by collision so that the discharge should cease. This simple expectation is however modified by the fact that the positive ions can eject electrons from the cathode when they reach it and since the field at the wire is then no longer reduced, avalanches can once more occur and a single ionizing event can therefore lead to multiple or continuous discharges. It is the function of the alcohol in the gas filling to 'quench' the discharge and this it does because of its low ionization potential (11·3 eV). The argon ions (ionization potential 15·7 eV) on their journey to the cathode are practically all neutralized by acquiring an electron from the alcohol molecules. The ions reaching the cathode are then alcohol ions and although they are themselves neutralized at the cathode, the energy available is absorbed in dissociating the alcohol molecule rather than in producing further electrons from the cathode. The discharge thus ceases when the central field has fallen sufficiently. The alcohol vapour will also absorb the ultra-violet photons emitted during the avalanche stage, and prevent them from ejecting photoelectrons from the cathode although not from causing propagation of the discharge down the central wire.

The main features of a GM counter resulting from this mechanism are:

(*a*) constant output pulse size, independent of initial ionization,
(*b*) sensitivity to the production of a single ion pair,
(*c*) a fairly long insensitive time following the entry of each particle (Fig. 6.10(b)). This is made up of a *dead time* during which the counter voltage has dropped below the counting threshold, and a *recovery time* during which pulses of reduced size are produced. The insensitive time is usually made definite by suitable design of the external electrical circuit; it is then known as the *paralysis time*.

The operating characteristics of a GM counter exposed to a source of ionizing radiation (Fig. 6.10(c)) shows:

(*a*) *a threshold voltage* V_g,
(*b*) *a plateau* of small slope, over which the counting rate only increases slightly with operating voltage,
(*c*) *a background counting rate* due to contamination of the materials, to cosmic radiation and to spurious discharges. This may be reduced by screening and anti-coincidence counters to perhaps 2 counts min^{-1} but is normally 7–20 counts min^{-1}.

The pulse size obtainable from a GM counter may be many volts, depending on the operating voltage. The pulse rise time is slower than that of a proportional counter and the dead time is far longer owing to the propagation of the discharge along the wire and to the large reduction of field by the positive ion sheath. The lifetime of a self-quenched counter is limited by the existence of the quenching agent, which in the case of argon–alcohol mixtures is dissociated after about 10^9 counts. Counters filled wholly with argon–halogen mixtures in which the halogen acts as quencher, have an indefinite lifetime. The characteristics of two commercial GM tubes are shown in Table 6.1.

TABLE 6.1 Characteristics of GM counters (1962)

| Type | Plateau | | Operating voltage V_{op} | Life (no. of counts) | Signal developed on 1 MΩ | Dead time (μs) | Recovery time (μs) |
	Length (volts)	Slope (per volt)					
GM4 Argon–Alcohol Type	250	0·05%	1300	6×10^8	2·5 V	100	250
MX123 Halogen Type	150	0·07%	675	$> 10^{11}$	5 V	85	250

There is obviously no energy sensitivity in the response of a GM counter. The tube responds to every charged particle which enters its sensitive volume and produces an ion pair there. The efficiency for counting photons depends upon the conversion of the radiation into electrons in the counter walls and is normally about 1 per cent for photons of energy 1 MeV for a counter of the type shown in Fig. 6.10, so that the GM counter is useful for counting charged particles in the presence of γ-radiation. Geiger–Müller counters, like proportional counters, are available with a wide range of physical dimensions.

6.1.5 Scintillation counters

The scintillation screen, a thin layer of zinc sulphide or barium platinocyanide viewed by a low-power microscope for the observation of light flashes, is historically important as the first detector of individual nuclear particles. It was used in many of the classical researches of Rutherford and was employed as late as 1932 by Cockcroft and Walton to demonstrate the disintegration of lithium by protons. The present extensive use of the scintillation method is due partly to the application of the photo-electron multiplier (photomultiplier) to the detection of the light flashes (Curran and Baker, 1945) and partly to the discovery of new scintillating materials or phosphors such as anthracene (Kallmann) and the alkali halides (Hofstadter) which are transparent to their own radiations and can therefore be used in the form of thick slabs.

In both types of phosphor the first result of the passage of a charged particle is the production of a trail of ionization in a time of the order of 10^{-10}–10^{-9} s. If we consider a particle of energy 1 MeV which stops in the scintillator, about $10^6/30 \approx 30,000$ ion pairs are produced. The next stage depends on the type of phosphor concerned. In inorganic crystals containing impurity (or activator) ions it is known that luminescence centres exist; these are atomic or molecular groups in which energy of excitation is liberated by a radiative transition rather than by direct mechanical interaction with the crystal lattice.

The initial ionization process releases electrons from the various electron bands of the solid and some of the resulting holes are filled by electrons from the luminescence centres; when this happens, a photon is emitted. The efficiency of conversion of the initial energy deposited in the scintillator into energy of radiation thus depends on the density of luminescent centres in the crystal in comparison with other non-radiative centres, and is usually 5–10 per cent. About 20,000 photons may thus result from the absorption of a 1 MeV particle, and these will be emitted over a time (≈ 1 μs) determined by the delay in transfer of energy to the luminescence centres and the decay time of the centres themselves. In organic scintillators similar processes take place but the energy transfer mechanisms are different (Ref. 6.12).

The scintillators which have been chiefly used for particle and photon detection are listed in Table 6.2. The particular properties of these phosphors which may be emphasized are:

(a) *Sodium iodide (thallium-activated).* This is the most versatile of all the phosphors, and is of outstanding importance for the study of γ-radiation, but it has the disadvantage that it is hygroscopic and must be sealed in an aluminium can (Fig. 6.12) with reflecting or diffusing walls. The efficiency for γ-ray detection is many times greater than that of a Geiger counter because of the effective thickness of the converter. The efficiency is also much greater than that of a germanium counter of the same volume because of the difference in atomic number, but the resolution is much poorer.

(b) *Zinc sulphide.* This is an excellent phosphor for particles of short range, but cannot be used in thick layers since it rapidly becomes opaque to its own radiation.

(c) *Anthracene and stilbene.* Organic phosphors have a faster decay time than the inorganic phosphors but a poorer efficiency, especially for heavy particles. Since they contain only light elements they are useful for counting β-particles in the presence of γ-radiation. For fast counting they have been largely superseded by

(d) *Plastic and liquid scintillators.* These are readily obtainable in very large volumes and can be adapted to many different geometrical arrangements. Energy of excitation in this case is transferred from a solvent to a solute, which then re-emits radiation in a wavelength range for which the solvent is transparent. Plastic scintillators form the detectors in the multiple counter 'telescopes' used in high energy physics to define the path of a particle, or to trigger the operation of a visual detector. Because of their high hydrogen content and fast response, organic scintillators are much used for detection of neutrons by observation of recoil protons from (np) collisions.

(e) *Gases.* Xenon is particularly useful when heavy charged particles are to be counted in the presence of γ-radiation. The main emission of light takes place in the ultra-violet and a wavelength shifter (often a film of grease) is necessary to convert the energy to a wavelength suitable for detection by a standard photomultiplier.

The linearity of response of a scintillator depends on the density of initial ionization in comparison with the number of luminescent centres and differs between organic and inorganic materials. High density of ionization impedes the processes of energy transfer in organic phosphors and the light output for heavy particles is suppressed with respect to that for electrons. The inorganic crystals suffer less from this disadvantage, and it does not occur at all in gaseous scintillators. Figure 6.11 shows the relative scintillation responses as a function of incident energy for two widely used phosphors and for

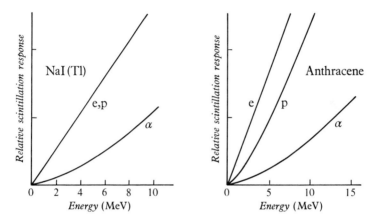

Fig. 6.11 Scintillation response of phosphors for electrons, protons and α-particles.

TABLE 6.2 Characteristics of scintillators

Material	Form	Density (kg m⁻³ × 10⁻³)	Refractive index	Main emission wavelength (nm)	Intrinsic efficiency (per cent)[b]		Decay time (μs)
					protons	electrons	
ZnS–Ag[a]	Powder	4·1	2·4	450	20	13·5	>10
NaI–Tl[a]	Crystal	3·7	1·7	410	20	20	0·25
KI–Tl[a]	Crystal	3·2	1·65	400	5	5	>10
CsI–Tl[a]	Crystal	4·5	1·75	white	—	—	>1
Anthracene	Crystal	1·25	1·62	445	—	10	0·03
Stilbene	Crystal	1·16	—	400–420	0·8	6	0·008
Terphenyl in xylene	Solution	0·86	1·5	320–400	—	5–10	0·002
'Organic'	Plastic	1·03	—	425	—	5–10	0·002
Xenon gas						10	0·01–0·1

[a] The concentration of activator is usually about 0·1 per cent (molar).
[b] The intrinsic efficiency is the ratio of energy emitted as radiation to the energy absorbed in the scintillator. The figures depend considerably on the preparation of the phosphor and should be taken as a rough guide only.

different types of particle. The linearity frequently obtainable and the high intrinsic efficiency are responsible for the wide use of scintillation spectrometers for determining gamma ray energies when detection efficiency is more important than resolution.

In such instruments the light output of the scintillator must be conveyed to the photo-sensitive cathode surface of a sealed-off photomultiplier tube. Many problems arise in obtaining efficient optical transfer; whenever possible the scintillator is mounted directly on the photomultiplier envelope with a suitable sealing liquid to ensure good optical contact, but in many applications

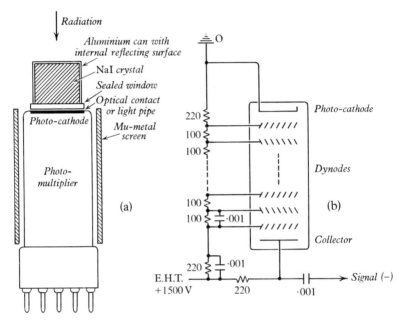

Fig. 6.12 Scintillation counter. (a) Arrangement of sodium iodide crystal on photomultiplier. (b) Circuit diagram, showing potentiometer chain for supplying dynode potentials. The resistance values are given in kilo-ohms and the capacitance values in microfarads.

some kind of Perspex light guide is necessary, with consequent reduction in efficiency owing to absorption in the Perspex. Figure 6.12 shows a typical photomultiplier arrangement. The photo-cathode of the multiplier tube should have a good response over the wavelength of the luminescence emission and a low 'dark current', i.e. spontaneous noise generation due to random emission of electrons from the cathode surface. The most frequently used surface is an antimony–caesium coating, which may have a quantum efficiency (number of photo-electrons produced per photon incident) of up to 20 per cent. High quantum efficiency is important for the study of low energy radiations because the spread in output pulse height is generally determined by the statistical fluctuation in the electron emission from the photo-cathode. Since the number of these electrons is proportional to the energy E

of the particle incident on the scintillator, the fluctuation is proportional to $E^{1/2}$ and the percentage spread of pulse height therefore varies as $E^{-1/2}$. All photomultipliers are to a greater or less degree sensitive to magnetic fields and must normally be heavily screened with iron or mu-metal if magnetic fields are likely to be present. The circuit arrangements for supplying dynode potentials and for developing the output signal are shown in Fig. 6.12(b).

The use of the NaI(Tl) spectrometer for energy measurements of γ-radiation is illustrated in Fig. 6.13. The efficiency of detection varies with energy in accordance with the variation of the linear attenuation coefficient for γ-radiation. For a 40 mm diameter block of sodium iodide 25 mm thick the efficiency is about 50 per cent for 500 keV radiation.

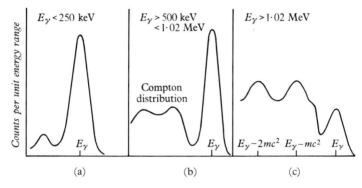

Fig. 6.13 Scintillation spectrometry. (a), (b), (c) Pulse spectrum of homogeneous γ-radiation of energy corresponding to interaction mainly by photoelectric effect, Compton effect and pair production. In each case there is a full energy peak due to capture of all radiations by the crystal but the height depends markedly on the relation between the crystal size and the γ-ray energy.

The pulse height distribution for homogeneous γ-radiation observed in a single crystal scintillation spectrometer depends on the quantum energy. For $E_\gamma < 250$ keV (Fig. 6.13(a)) the photoelectric effect predominates, and a full energy peak is observed (with a lower energy 'escape' corresponding to events in which the K X-ray photon of iodine has emerged from the crystal). For $E_\gamma > 500$ keV (Fig. 6.13(b)) the Compton effect is also important, and the full energy photoelectric peak is accompanied by a broad distribution of recoil electrons, with a well marked 'Compton edge'. For $E_\gamma > 1\cdot02$ MeV (Fig. 6.13(c)) pair production is possible and in addition to the Compton distribution peaks are found at the energies E_γ, $E_\gamma - mc^2$, and $E_\gamma - 2mc^2$. The lower energy peaks arise as discussed for the Ge(Li) spectrometer, because of escape of one or two annihilation quanta from the crystal. A resolution of about 8 per cent at 600 keV is obtainable in a crystal of the size quoted.

Organic scintillators used for neutron detection are also sensitive to gamma radiation and it may be desirable to distinguish between the two types of radiation. This can be done by making use of the characteristics of the (optical) decay time of the scintillator for pulses due to recoil protons (from neutrons) and electrons (from gamma rays). The decay curves show a long period component (≈ 300 ns) and a short period (≈ 10 ns) and the ratio of intensities of these components depends on the nature of the initiating particle. Electronic circuits which sense this difference can therefore be used to provide *pulse shape discrimination* (Ref. 6.12).

6.1.6 *Cherenkov counters*

The conditions for the production of Cherenkov radiation were given in Chapter 5, Sect. 5.2.3. Equation (5.51) shows that a relativistic particle of unit charge passing through 100 mm of glass or Perspex gives rise to about 2500 photons in the visible wavelength range. If these are all collected and focussed on to the photocathode of a multiplier tube a few hundred electrons are produced and pulses with a statistical spread in size of about 10 per cent may be detected. The light output of this type of counter is very much less than that of a *scintillating* material for the same total energy loss but the Cherenkov counter has the following important advantages:

(*a*) rapid rise and decay of pulse (intrinsically $\leqslant 10^{-10}$ s),
(*b*) directional emission of light, with an angle dependent on particle velocity,
(*c*) no emission of light for particles of velocity less than a critical value, and above this value a dependence of intensity on particle velocity.

In practical Cherenkov counters (*b*) and (*c*) have been chiefly exploited. Liquid and solid counters of large volume can be used as threshold detectors and another class of instruments with special optical systems can be employed to select particles with a narrow spread of velocity or of direction. In high energy physics differential counters of this type may be used in a beam of particles of known momentum, defined by magnetic analysis. The Cherenkov counter can then be used to select particles of given mass; a typical experimental application is discussed in Sect. 20.3.2.

Cherenkov counters have also been used for electron and photon energy measurement. In a high energy electromagnetic cascade in a transparent material, such as glass or thallium chloride, in which a narrow forward electron–photon shower develops, the electrons produce Cherenkov light. If the cascade is completely absorbed the light output will be proportional to the energy of the primary cascade-producing photon and may be determined, after calibration with known-energy particles, by pulse height analysis.

6.2 Detecting instruments (visual)

Electrical detecting systems have the advantage of rapid accumulation and display of information in an experiment. They are especially suitable in well-defined situations, as when the number of particles scattered from a beam at a given angle has to be found as a function of beam energy. Such counting systems are less suitable for the analysis of a complicated process in which several particles may be emitted and tests of momentum balance and co-planarity have to be applied in the analysis. In such cases visual techniques are necessary; in many of the simpler experiments such methods provide important confirmation of suggested mechanisms and at the very least, a considerable aesthetic satisfaction to the physicist.

6.2.1 The Wilson expansion chamber

The cloud chamber owes its existence to the interest of C. T. R. Wilson in meteorological phenomena on the mountain of Ben Nevis in 1894. The present form of the instrument differs hardly at all in essentials from the earliest models and tracks obtained by Wilson in 1911 are still among the best examples of performance of the chamber. The dramatic quality of the pictures obtainable with the Wilson expansion chamber is well illustrated by the examples of events shown in Plates 1–4 and practically every nuclear phenomenon has come under the scrutiny of this powerful technique in the half century following its invention.

The expansion chamber, whose essential features are illustrated in Fig. 6.14(a), causes condensation of a vapour into drops along the path of an ionizing particle passing through a gas. The gas and vapour are contained in a cylinder with suitable transparent windows, and the visible tracks are illuminated and photographed through the walls or top of the chamber as shown. Condensation occurs as a result of an adiabatic expansion, and this is conveniently produced by the controlled motion of a rubber diaphragm (Fig. 6.14(a)) which forms one boundary of the gas–vapour volume.

The principle of operation of the chamber may be understood from Fig. 6.14(b). The curve in this figure shows the relation between the vapour pressure p_0 in equilibrium with a liquid with a plane surface and the vapour pressure p near a convex surface of radius r of the same liquid at the same temperature. The curve divides an upper region of vapour pressure in which drops grow by condensation from a lower region in which they evaporate. If now a drop carries an electrostatic charge q curve (1) passes over into curve (2) which is given approximately by the equation

$$\frac{RT\rho}{M} \log_e \frac{p}{p_0} = \frac{2\gamma}{r} - \frac{q^2}{32\pi^2 \varepsilon_0 r^4} \tag{6.6}$$

where R is the gas constant, T the absolute temperature, γ the surface tension, M the molecular weight, ρ the liquid density and r the radius of the drop. The upper region of vapour pressures for which drops grow is now much larger and in particular minute drops formed with a radius corresponding to the point B will continue to grow and will reach visible size.

If the chamber initially contains a gas and a saturated vapour at temperature T_1, then after an adiabatic expansion from volume V_1 to V_2 the temperature is given by

$$T_2 = T_1 \left(\frac{V_1}{V_2}\right)^{\kappa - 1} \tag{6.7}$$

where κ is the ratio of the specific heats of the gas–vapour mixture. Initially the vapour pressure is p_1, corresponding to saturation at temperature T_1, but after the expansion it is

$$p = p_1 \left(\frac{V_1}{V_2}\right)^{\kappa} \tag{6.8}$$

which is considerably greater than the saturation vapour pressure p_2 for temperature T_2. If the ratio p/p_2, known as the *supersaturation* S, exceeds the value indicated by the maximum of curve 2 in Fig. 6.14(b) condensation occurs on drops carrying a charge and these grow. For smaller values of S, small drops (A_1, A_2) grow or evaporate until they reach conditions specified by curve (2) for which they are stable, but too small to be seen. For water vapour at approximately $0°C$ the critical supersaturation is $S = 4·2$ and the drop radius a_1 corresponding to the maximum of curve (2) is 6×10^{-10} m.

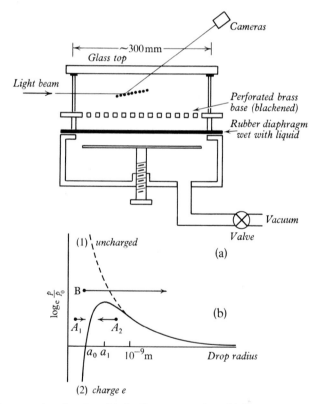

Fig. 6.14 The expansion chamber. (a) Outline construction. (b) Vapour pressure equilibrium curves for small droplets. Curve (1) is for uncharged drops and curve (2) for drops carrying one electronic charge (Ref. 6.14).

Condensation centres produced by ionization probably have a radius of the order of a_0 ($\approx 3 \times 10^{-10}$ m) in the first instance. If a sufficient degree of supersaturation is existing at the time of ionization the drops grow to a diameter of about 3×10^{-5} m in about 0·5 s and during this time of growth may be photographed with a suitably phased flash of light. It is found that positive ions require a somewhat greater degree of supersaturation, and hence *expansion ratio* V_2/V_1, than negative ions. The interval after production of supersaturation during which ionization leads to track formation is known as the *sensitive time*; it is primarily determined by the warming up of the chamber from the temperature T_2 with consequent reduction of supersaturation. The expansion ratios used in practice are limited by the appearance of background condensation on uncharged centres.

A typical sequence of operations for the photography of charged particle tracks is as follows:

At zero time a fast adiabatic expansion occurs (in about 0·01 s) with an expansion ratio of perhaps 1·30. The resulting temperature drop is about 28° C for a mixture of air and water vapour and a supersaturation of ≈ 6 is produced and persists for perhaps $\frac{1}{2}$ s. During this time a burst of particles is admitted by a shutter mechanism and after about 0·2 s for drop growth the tracks are illuminated by a short burst of light and photographed. The expansion ratio and timing sequence are adjusted for maximum clarity. The ions formed in the burst of particles must be removed after the photograph is taken and a potential of a few hundred volts supplies a clearing field to sweep the ions to the bottom of the chamber; the clearing field is removed during the expansion to minimize track broadening. Even with this field some centres for condensation may remain after the adiabatic expansion but many of these may be removed by a subsequent slow expansion. The ideal timing sequence cannot always be achieved, particularly in the important case of a *counter-controlled chamber*. In this application the expansion is triggered by a particle which has passed into or through the chamber, and in addition through one or more counters, and there is of necessity some diffusion of ions in the track before it is recorded. The broadening is minimized by making the expansion as rapid as possible. The development of this technique by Blackett and Occhialini in the 1930s led to many detailed studies of the cosmic radiation, and, in due course, to the discovery of strange particles by Rochester and Butler (Sect. **20.1**).

Expansion chambers have been made in many different sizes and they operate with many different gases and vapours with pressures up to tens of atmospheres. Rapid chambers, in which the resetting time is reduced by particular forms of pressure cycle, have been constructed to operate as frequently as once in every 10 seconds. Large sensitive volumes are available and stereoscopic photography with twin cameras is usually possible. Tracks are examined in detail by reprojection onto a white surface through the cameras used for the photography.

The expansion chamber has the limitation of a short sensitive time and the evolution of continuously sensitive diffusion chambers (Ref. 6.15) by Langs-dorf in 1936 was a considerable step forward. The stopping power of the gas is however rather low even at high pressures and after a brief period of use with high energy accelerators, this type of chamber (together with the expansion chamber) has been discarded in favour of the rapid-cycling bubble chamber.

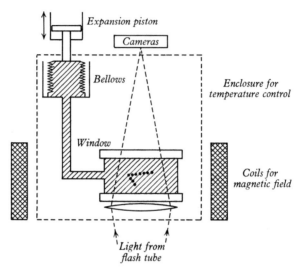

Fig. 6.15 Outline construction of a bubble chamber with glass windows and metal walls, of a type suitable for use with liquid hydrogen.

6.2.2 The bubble chamber

The limitation of track chambers was very considerably alleviated in 1952 by the invention by Glaser of the bubble chamber, in which the supersaturated gas–vapour mixture of the cloud chamber is replaced by a superheated liquid. The passage of an ionizing particle through such a superheated liquid may cause bubbles to grow along the track to a size at which they may be photographed without first precipitating uncontrolled boiling throughout the volume of the fluid. Figure 6.15 illustrates the construction and operation of the apparatus; a suitable liquid is heated above its normal boiling point and is maintained in the liquid phase by the application of a pressure above the saturation vapour pressure at the operating temperature. The pressure is then reduced by about 10 per cent by movement of a piston or diaphragm and the liquid becomes superheated and sensitive to the presence of ionization; after some seconds general boiling will occur unless a recompression prevents it. The first theory of bubble formation followed the electrostatic theory of condensation in the expansion chamber (Sect. 6.2.1) very closely, and led to curves of the type shown in Fig. 6.14(b) for bubble radius. The charges required for this mechanism are however unrealistic and it is now considered that bubble development is a thermal effect due to the deposition of energy by short delta rays originating in the initial ionizing process.

Unless special care is taken to avoid roughness on surfaces, the sensitive time of bubble chambers after expansion (before uncontrolled boiling) is only a few milliseconds and entry of particles and photography must take place during this time. In this sense the bubble chamber has apparently lost the advantage of continuous sensitivity enjoyed by the diffusion chamber but this is not in practice so because (*a*) the continuity of the diffusion chamber is of limited use owing to vapour depletion, (*b*) the bubble chamber accumulates very few background tracks, (*c*) rapid cycling of the bubble chamber is possible because the whole sequence of events from expansion to recompression occupies only about 25 ms, and (*d*) the sensitive depth of a bubble chamber is not limited. When the increased yield of events due to increased density is taken into account the bubble chamber is operationally much the more economic. Bubble growth occurs in about 10^{-3} s in an average chamber (cf. $\approx 10^{-1}$ s in a cloud chamber) but the actual centres on which the bubbles form seem to last less than 10^{-6} s (Ref. 6.15). It is therefore impossible to adapt the chamber for counter control since the centres produced by the triggering particle would have disappeared before the chamber could be expanded. Expansion must take place before the particles enter the chamber, as a result, for instance, of an early signal from an accelerator.

If the delta ray theory of the origin of bubbles is correct the density of bubbles along a track should be determined by a formula such as (5.44) and observation of the density may indicate the particle velocity $\beta = v/c$. In practice it is often easier to determine *gap* density along the track rather than bubble density.

Practical chambers have used pentane, propane, xenon and above all, because of their significance in high energy physics, liquid hydrogen, deuterium and helium. Hydrogen chambers give the most readily interpreted information, although in the case of deuterium, which is used for studying interactions with neutrons, corrections for the internal motion of the particles in the deuteron must be made. Heavy liquid chambers suffer from the disadvantage that the primary collision in a complex nucleus is not uniquely defined, but they have good sensitivity for detection of gamma rays, e.g. from π° decay, and therefore complement the hydrogen chambers. Table 6.3 gives typical operating conditions for some of these liquids; the heavy liquid chambers now tend to be based on a 75–25 (by volume) ratio of propane to CF_3Br mixture operating at about $43°C$. Composite chambers containing both neon and hydrogen, in separate compartments, offer good definition of the primary interaction (in H), reasonable gamma detection efficiency (in Ne) and good momentum determination (in H, in which scattering is minimized).

The only limit to the size of bubble chambers is their cost and a 2·5 m deep liquid hydrogen chamber 3·75 m in diameter has been designed for the Argonne Laboratory, U.S.A. A magnetic field is now vital in bubble chamber work for determination of momentum and of sign of charge for particles of near-minimum ionization and especially in the case of liquid hydrogen chambers, superconducting magnets offer considerable technical advantage. The successful analysis of bubble chamber information is critically dependent on track quality, particularly in respect of spatial location (≈ 3 μm) and of momentum definition, i.e. on the maximum magnetic field available (≈ 7T). The measurement of bubble chamber tracks is now a highly specialized activity which is increasingly based on automatic machines under computer control. A summary of some of the current procedures is given in Ref. 6.15. Plates 7, 17, 19, 20 show events detected in a liquid hydrogen chamber, and the use of such a chamber in high energy physics is described in Sect. 20.3.1.

TABLE 6.3 Operating conditions for bubble chamber fluids.

Fluid	*Temperature* °C	*Pressure* (atmospheres)	*Mean free path for* 100 MeV γ-*rays* (m)	*Density* (kg m^{-3})
Hydrogen	-246	5	27	60
Deuterium	-241	7	20	130
Helium	-269	1	18	130
Propane	58	21	2·2	430
Pentane	157	23	—	500
Xenon	-20	26	0·07	2300

6.2.3 The nuclear emulsion

Photographic methods have exerted a profound influence on the development of nuclear physics ever since the discovery of radioactivity. The passage of ionizing radiation through a photographic emulsion renders grains of silver bromide developable and this fact has been applied for many years in the detection of X-rays and γ-rays, and in the recording of beams of charged particles deflected by magnetic spectrometers. In this application the small grain size of the emulsion provides in principle extremely high resolution of recorded spectra. The use of the emulsion for the detection of single particles also dates from the early days of radioactivity and was slowly developed, by gradual improvement of the emulsion characteristics, until in the mid-1930s it was possible to distinguish under suitable magnification the tracks of both α-particles and protons as rows of individual grains. At about this time the rapidly increasing interest in nuclear transmutations and cosmic rays led to more detailed study of all types of nuclear detector and the Ilford and Kodak companies were soon successful, in collaboration with workers in Cambridge and elsewhere, in producing emulsions which, although relatively slow photographically, had a small grain size and a tolerable density of 'background' grains on development. Techniques for processing and subsequent examination of the tracks by the binocular microscope were also evolved and by 1940 the nuclear emulsion was available as a new and promising detector of heavy charged particles, and of neutrons through the projection of recoil protons.

Although the simplicity and economy of the nuclear emulsion detector won early recognition, its reliability as a quantitative method was not widely accepted until the publication of the early work of Powell in which he demonstrated very convincingly that measurement of the length of tracks in Ilford emulsions could give reliable values for the energy of particle groups.

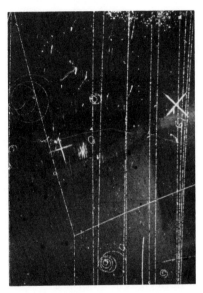

Plate 7 Hydrogen bubble chamber tracks of 1 GeV protons, showing a proton–proton collision (J. B. Kinson).

The emission of protons and neutrons in the reactions

$$^{10}\text{B} + \text{d} \rightarrow\ ^{11}\text{B} + \text{p} + 9{\cdot}22\ \text{MeV} \tag{6.9}$$

$$^{10}\text{B} + \text{d} \rightarrow\ ^{11}\text{C} + \text{n} + 6{\cdot}46\ \text{MeV} \tag{6.10}$$

was studied and for the former reaction number–range curves agreeing well with those previously obtained by a counter method were found. The results for the neutron-producing reaction were even more promising. In an exposure of a few minutes made by placing a small photographic plate coated with nuclear emulsion near a boron target in a Cockcroft–Walton apparatus (Sect. 8.1.1), some 3000 recoil proton tracks were obtained in a few square centimetres of emulsion area. Subsequent microscope work led to a neutron spectrum of higher quality than that obtained from a similar experiment in which 20 000 expansion chamber photographs were taken.

This remarkable work, and parallel investigations in which the easily portable plates were exposed at mountain altitudes in a search for cosmic ray

167

events, clearly demonstrated the great power of the emulsion method. The major limitations at this time seemed to be (i) the time consumed in scanning the emulsions and measuring tracks, (ii) the fact that owing to the very small length of most tracks no magnetic curvatures could be measured in the emulsion, and (iii) the fact that the emulsions were not sensitive to particles of minimum ionization. The first two points remain as limitations for some experiments at present but as a result of further work by Powell and his collaborators, and by the Kodak and Ilford companies, the Nuclear Research Emulsion, which can be made electron-sensitive, became available. In this emulsion the visibility of tracks is much improved over those obtained in earlier (halftone) plates by an increase in the concentration of silver halide; proton and α-particle tracks are dense, continuous lines and particles of minimum ionization including the readily scattered electrons also leave easily distinguishable tracks. Many beautiful examples of tracks in modern emulsions are now available; Plate 8 may suffice to show the improvement in technique obtained over a few years, and the potentialities of the present emulsions. The discovery† in 1947 of the π-meson by the exposure of concentrated emulsions to cosmic radiation at mountain altitudes stems directly from the technical improvements made in 1946 and from the simplicity and visual property of the method. The ability of the emulsion to store information was also vital in this discovery, and the important advantage of integration has been much deployed in subsequent years.

The composition of Ilford Nuclear Research Emulsions is approximately as shown in Table 6.4 (from Ref. 6.18).

TABLE 6.4 Composition of dry Ilford emulsion

Element	H	C	N	O	S	Br	Ag	I
atoms m$^{-3} \times 10^{-28}$	2·93	1·51	0·31	0·75	0·02	1·15	1·17	0·03

Mean density $3·9 \times 10^3$ kg m^{-3}

The size of the developed grains when examined under the microscope depends on the method of preparation; so too does the probability of a grain being rendered developable by the passage through it of a charged particle of given velocity, and the grain size and sensitivity (which may be defined quantitatively as the number of electrons required to render a grain developable) are interdependent. Table 6.5 (from Ref. 6.18) gives the characteristics of a few types of Ilford emulsion.

Of these emulsions the type C2 is most used for non-relativistic particles and the type G5 for studies of electron and high energy tracks in general. All emulsions are sensitive in greater or less degree to the presence of γ-radiation, which causes an increase of background grains, or even the appearance of distinct tracks due to photoelectrons or Compton electrons.

† C. M. G. Lattes, G. P. S. Occhialini and C. F. Powell, *Nature*, **160**, 453 and 486, 1947.

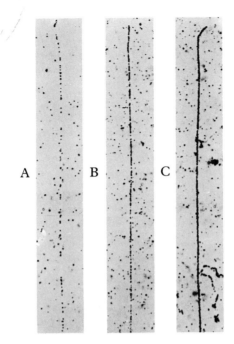

Plate 8 Tracks of protons in different emulsions.
A Ilford Half-Tone, B Ilford C2, C Kodak NT4
(C. F. Powell, taken from Ref. 6.18).

TABLE 6.5 Characteristics of nuclear emulsions (1962)

Ilford Emulsion Type	D1	E1(K1)	C2(K2)	B2	G5
Mean grain diameter (μm)	0·12	0·14	0·10	0·21	0·18
Highest velocity detectable (v/c)		0·2	0·31	0·46	all
Highest detectable energy of protons (MeV)		20	50	120	all

Nuclear emulsion is usually supplied in the form of coated glass plates with emulsion thicknesses from 20 μm to 600 μm. Since the range of a proton of 10 MeV in emulsion is about 580 μm, and since in any case the tracks under investigation may often be produced tangentially in the plates, such thicknesses are generally adequate for nuclear reaction studies. In high energy, cosmic ray and meson physics, however, it is desirable to follow tracks of particles over very large distances and it is now possible to use large blocks of emulsion sheets which form effectively a solid track chamber. After exposure to radiation these blocks are suitably marked with fiducial lines, divided up, processed, and mounted on glass backing plates for observation. Several laboratories can then participate in the examination of selected tracks.

The processing of nuclear emulsion is similar to that of ordinary photographic material, but is determined by fixed routines which ensure that the lower layers of thick glass-backed emulsions are not under-developed, or the top layers over-developed. Wet emulsion has about 2·3 times the volume of dry emulsion and distortion of tracks during shrinkage has to be avoided. The examination of the developed emulsion under the microscope is usually carried out with oil-immersion objectives (×45 or ×95) and a total magnification up to ×1500 dependent on the particular application; in 'area scanning' for outstanding events a relatively small magnification is used, but for 'along the track scanning' small deviations may be important and high magnification and precision stage movements are required.

Plate 9 Disintegration star produced by a high-energy proton in nuclear emulsion (C. F. Powell *et al.*, Ref. 6.19).

The many applications of the nuclear emulsion technique fall into two broad classes: (*a*) the uses in which interactions originating in the emulsion are studied, and (*b*) the uses in which the emulsion is employed simply as a detector of charged particles. The study of interactions is somewhat complicated by the complex composition of the emulsion (Table 6.4) which makes it difficult to identify the type of nucleus with which an incident particle, such as a proton or neutron, has interacted. An exception arises in the case of the hydrogen content when high energy protons are being used; the resulting collisions may be identified fairly easily because of the simple kinematics of the process. For collisions between high energy particles and the other emulsion constituents, assignment of observed stars (such as that shown in Plate 9) to the light or heavy elements of the emulsion must be based on detailed consideration of the prongs of the stars. In some cases specific effects may be observed when other elements are introduced into the emulsion either in manufacture or by impregnation; the best known example is the boron-loaded emulsion in which short α-particle tracks due to the reaction

$$^{10}B + n \rightarrow {}^7Li + \alpha + 2\cdot79 \text{ MeV} \tag{6.11}$$

may be seen after irradiating with thermal neutrons. Deuterium loading and uranium loading have also frequently been used in studies of photodisintegration and fission respectively. A further important feature of the emulsion for the study of interactions is the high resolution of distance provided by the small grain size. Distances of recoil or travel from a particular point, such as the location of a star, of the order of 1 μm or less can be estimated, and the possibility of the decay of unstable particles in this sort of distance can be assessed. This property, coupled with the high stopping power of the emulsion, has led to some evidence on the lifetime of the π°-meson ($\approx 10^{-16}$ s).

The identification of the nature and energy of charged particles from observed tracks in nuclear emulsions depends on knowledge of (a) range-energy relations, (b) grain density, (c) delta ray production and (d) multiple scattering properties. For heavily ionizing particles 'gap' density rather than grain density is used. In some cases, momentum may be known by deflection or selection in external magnetic fields. A single observed quantity usually depends on both e/m for the primary particle and on its velocity v and two independent observations are necessary to give full information. The methods used are fully described by Powell (Ref. 6.21) with particular reference to the π- and μ-mesons (muons).

The second main application of the emulsion technique is the integral recording of the number of charged particles originating in a given experiment, together with identification or selection by special characteristics, such as grain density or residual range. Nuclear plates are widely used for the recording of nuclear reaction products after high resolution magnetic analysis; the plates are examined only to yield the number of tracks per mm² as a function of angle to a primary beam, or as a function of some coordinate along the plate related to momentum or energy. The emulsion itself may be used to yield an energy spectrum of incident particles if ranges can be measured (Fig. 6.16); absolute track counts in experiments of this sort yield reaction cross-sections.

It is clear that the nuclear emulsion provides a powerful and versatile detection technique, in which many of the properties of other detectors are combined. It records continuously and stores the recorded information; it has energy and velocity discrimination properties and high stopping power. Although large lengths of track can if necessary be followed by means of the emulsion stack the nuclear plate requires no auxiliary apparatus and can therefore if necessary be used in difficult or remote situations, where small size and microscopic observation technique provide high resolution both in distance and angle. It will never be possible to obtain the immediate response provided by the counter systems and some phenomena (e.g. variations of cross-section with energy) are not conveniently studied by this technique. Only minor adjustments can be envisaged in the chemical composition of the emulsion, but this is not important when the plates are used solely as detectors. The tedious process of scanning and measurement may be alleviated by developments in automatic scanning and analysis (Refs. 6.15, 6.16).

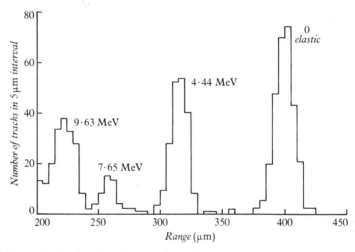

Fig. 6.16 Range distribution in nuclear emulsion of 38·1 MeV α-particles scattered by carbon nuclei. The groups marked correspond to elastic scattering and to inelastic scattering with excitation of the carbon nucleus to the levels indicated (Aguilar *et al.*, *Proc. roy. Soc.*, A, **254**, 395, 1960).

6.2.4 Solid state track detectors†

The photographic principle of the nuclear emulsion is not a necessary requirement for charged particle detection. In many solids the trail of damage left by an ionizing particle may be rendered visible under the microscope by suitable chemical etching, providing that the rate of energy loss dT/dx has exceeded a certain critical value. For mica and for argon ions, track registration takes place for all energies from a few MeV to about 200 MeV, but for neon ions the energy loss is never sufficient for production of a recognizable track. The solid state track detector thus has the advantage, in addition to simplicity, of complete insensitivity to particles with an energy loss rate below threshold. Applications to the recording of fission fragments and of heavy nuclei of the cosmic radiation have been made. In addition to mica, polymers such as cellulose nitrate and polycarbonate may be used.

† R. L. Fleischer, P. B. Price, R. M. Walker and E. L. Hubbard, *Phys. Rev.*, **A133**, 1443, 1964.

6.3 The triggered spark chamber

The visual techniques described in Sect. **6.2** provide excellent spatial resolution for particle studies, but poor or non-existenf time determination. The counter techniques (Sect. **6.1**) on the other hand offer highly accurate time resolution but are unsuitable for the study of three-dimensional spatial distribution except in well-defined circumstances, such as a simple scattering or transmutation experiment.† The first successful attempt to combine the two desirable features of high resolution both in space and time resulted in the counter-controlled cloud chamber, but the re-cycling time of chambers limits the efficiency of this type of detector.

Triggered visual detectors have been designed in which a bank of neon tubes is pulsed with a high voltage immediately after the passage of a primary ionizing particle through a counter telescope. The neon tubes through which the particle has passed then discharge and their glow may be photographed.

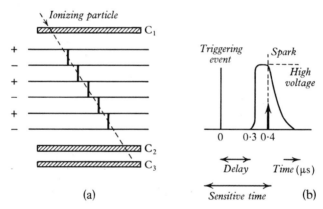

Fig. 6.17 The triggered spark chamber. (a) Construction; C_1, C_2, C_3 are counters forming a telescope. (b) High voltage pulse.

Such arrays have been extensively used in sea-level cosmic ray experiments. This detector however is two-dimensional only and is an array of some complexity and of low resolution. A simpler three-dimensional form of detector operating in the same way is the triggered multiplate spark chamber, as used by Cranshaw and de Beer.‡ In later forms of this instrument (Fig. 6.17) an ionizing particle passes through a stack of thin metal plates with a spacing of 10–20 mm contained in a vessel filled with a noble gas at about 1 atmosphere pressure. A signal from a counter telescope is used to trigger a voltage impulse of 10–15 kV which is applied to the plates within about 0·3 μs of the primary event. The rise time of the voltage pulse is about

† The use of position-sensitive solid state detectors, or multiwire proportional counters provide respectively one- or two-dimensional position information. An array of scintillating crystals coupled to photomultipliers through light guides may also be used.
‡ T. E. Cranshaw and J. F. de Beer, *Nuovo Cimento*, **5**, 1107, 1957.

0·025 µs and within 0·1 µs of its appearance on the plates an electron avalanche builds up through a channel defined by what remains of the initial ionization. This gives a bright, well-defined spark, which may be photographed. Tracks due to unwanted particles are minimized by the application of a constant clearing field (≈ 10 kV m^{-1}) to the chamber to remove residual ionization and a sensitive time of about 0·5 µs can be obtained. The high voltage pulse is cut off abruptly by the discharge and terminates in 0·1 µs; there is a dead time of 1–10 ms while the positive ions are being cleared from the spark channel. This detector is clearly capable of rapid operation.

In a multiplate chamber it is desirable that the general direction of the particle trajectories shall be approximately perpendicular to the plates. The positional information is not as good (≈ 0.5 mm) as is obtained with the bubble chamber, but the absence of unwanted tracks due to counter control and to the short sensitive time is an important advantage for operation under conditions of high background. The spark chamber can be made fairly cheaply in a very large size and can be built for a specific experiment; Plate 10 shows the quality of the information obtainable with the instrument.

The use of a rare gas as the spark chamber filling permits the registration of multiple tracks, as first shown by Fukui and Miyamoto, and encourages the spark discharge to follow the actual particle trajectory.

The track in a multiplate chamber is of course fragmented but complete tracks including vertices of events may be seen in the *wide gap* chamber in which by careful control of conditions electrode spacings of up to 0·4 m have been used. The performance of both narrow and wide gap chambers for multiple tracks is improved by operating the chambers in the *streamer* mode which relies on a stage of gaseous breakdown preceding the full spark; in this mode an especially short voltage pulse with current limitation is used and tracks are viewed through transparent electrodes in the direction of the field. Both spark and streamer chamber information, recorded photographically, may be processed by the techniques devised for bubble chamber work, although the accuracy is not so high.

If a simple two-dimensional location device only is required, e.g. to define a beam line, or to record particles analysed by a magnetic spectrometer, single narrow-gap chambers may be used. In such chambers sonic detectors (sound-ranging transducers) are very appropriate. These record the acoustic wave from the spark in the chamber gas and for a single incident particle immediately yield x and y coordinates from the time-of-flight information. This can be rapidly processed and displayed by an on-line computer so that with an appropriately arrayed series of such chambers the fate of each incident particle may at once be traced. An obvious development of this type of chamber suitable for digital data handling is to replace one of the thin plates of the chamber by a series of parallel wires. The spark to a given wire, determining its x or y coordinate, may be made to alter the magnetic state of a ferrite core which may later be interrogated. Alternatively the current pulse from the spark can influence a magnetostrictive wire and the time of passage of the resulting acoustic pulse along this wire will give positional information.

The spark chamber is well adapted to the detection of high energy gamma radiation by characteristic pair production in the electrodes, followed by the development of a small electromagnetic shower. In a further development

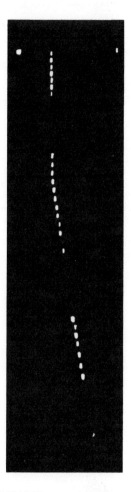

Plate 10 Track of 170 MeV proton, including a nuclear scattering, recorded in a spark chamber (J. G. Rutherglen).

of the technique spark chambers may be placed between the poles of a magnet so that momentum information is obtainable. The read-out from such chambers may use the *vidicon* system in which the spark image is projected on to a photosensitive surface where it is scanned by an electron beam as in a television camera.

References

1. SHARPE, J. *Nuclear Radiation Detectors*, Methuen, 1960.
2. STAUB, H. 'Detection Methods', in *Experimental Nuclear Physics*, ed. E. Segrè, Vol. 1, Wiley, 1953.
3. SIEGBAHN, K. (ed.) *Alpha, Beta and Gamma Ray Spectroscopy*, North Holland, 1965.
4. WILKINSON, D. H. *Ionization Chambers and Counters*, Cambridge University Press, 1950.
5. DEARNALEY, G. and NORTHROP, D. *Semiconductor Counters for Nuclear Radiations* (2nd edn.), Spon, 1966.
6. DEARNALEY, G. 'Solid State Radiation Detectors', *Contemp. Phys.*, **8**, 607, 1967.
7. GUNNERSEN, E. M. 'Recent applications of semiconductor techniques in the study of nuclear radiations', *Rep. progr. Phys.*, **30**, 27, 1967.
8. BERTOLINI, G. and COCHE, A. (eds.) *Semiconductor Detectors*, North Holland Publ. Co., 1968.
9. EWAN, G. T. 'Semiconductor Spectrometers', *Prog. nucl. techn. Instr.*, **3**, 67, 1968.
10. WEST, D. 'Energy Measurements with Proportional Counters', *Progr. nucl. Phys.*, **3**, 18, 1953.
11. GARLICK, G. F. 'Luminescent Materials for Scintillation Counters', *Progr. nucl. Phys.*, **2**, 51, 1952; NEILER, J. H. and BELL, P. R. Ref. 6.3, page 245.
12. BROOKS, F. D. 'Organic Scintillators', *Nucl. instrum. Meth.*, **4**, 151, 1959; OWEN R. B. *I.R.E. Trans. in Nucl. Sci.*, **NS8**, 255, 1961.
13. JELLEY, J. V. *Cherenkov Radiation*, Pergamon Press, 1958.
14. WILSON, J. G. *Principles of Cloud Chamber Technique*, Cambridge University Press, 1951.
15. HENDERSON, C. *Cloud and Bubble Chambers*, Methuen, 1970.
16. SHUTT, R. P. (ed.) *Bubble and Spark Chambers*, Academic Press, 1967.
17. NEWTH, J. A. 'Devices for the detection of energetic particles', *Rep. progr. Phys.*, **27**, 93, 1964.
18. ROTBLAT, J. 'Photographic Emulsion Technique', *Progr. nucl. Phys.*, **1**, 37, 1950.
19. POWELL, C. F. *et al*. *The Study of Elementary Particles by the Photographic Method*, Pergamon Press, 1959.
20. ZACHAROV, B. *Digital Systems, Logic and Circuits*, Allen and Unwin, 1968.
21. POWELL, C. F. 'Mesons', *Rep. progr. Phys.*, **13**, 350, 1950.

7 Measurement of energy and intensity of ionizing radiations

The radiations of nuclear physics are encountered as *single particles* or *photons*, as *collimated beams* of particles or photons with an intensity such that single members cannot be identified, or as *fluxes* in which the single particles or photons travel with a wide range of angles and the usual measure of intensity is the number crossing unit area in unit time. All observations of radiation rely on some kind of detector, which, except in the case of a current-measuring instrument, will usually be one of those described in Chapter 6. Energy measurements often involve the combination of an analysing instrument such as a magnetic spectrometer or a time-of-flight spectrometer with a detector for plotting a line profile, as in optical spectrometry, but many detectors are themselves energy sensitive and furnish information on energy and intensity simultaneously.

The object of intensity determinations in nuclear physics is usually to obtain the absolute yield of a particular process and from it the corresponding cross-section (Sect. 5.1.3). Absolute intensity determinations can be avoided in simple cases when a collimated beam is available and only one type of process is significant, since a transmission method may then be used to find the linear attenuation coefficient (Sect. 5.1.3) and the cross-section follows immediately. More usually, however, a beam or flux of particles or photons bombards a target and the intensity of a particular radiation, often in a particular direction, is estimated.

The present chapter briefly surveys some of the more important techniques used in nuclear structure studies. In high energy physics, similar information is sought but techniques are determined by the necessity for clearly defining the particular process under study or by the desire to examine the whole of a complex event. Examples of the methods used in this field are given in Chapter 20.

7.1 Measurement of energy and intensity of charged particles

7.1.1 Semiconductor detectors

Silicon surface barrier or diffused junction detectors, or lithium-drifted silicon detectors, are now used almost exclusively for charged particle detection in all experiments in which the highest possible resolution is not required. The operation and performance of these detectors is described in Sect. 6.1.2.

The semiconductor detector has essentially replaced many traditional methods such as range measurements using shallow ionization chambers or nuclear emulsions, and light-output methods based on the absorption of a particle in a scintillator.

Low energy electrons however (e.g. from tritium decay, $O < T_\beta < 18$ keV) are still best detected in a gaseous proportional counter. High energy electrons (say of 20 MeV and above) have too great a range to be stopped in any economically possible type of counter except a block of transparent material, in which the total light output due to Cherenkov radiation provides energy information.

7.1.2 *Electrostatic and magnetic spectrometry (general)*

The deflection of charged particles by electrostatic and magnetic fields has been of outstanding importance in the development of physics since the first studies of the gaseous discharge. At the present time the theory of such deflections finds application in

(*a*) energy or momentum measurements,
(*b*) monochromators, steering and focusing magnets for particle beams,
(*c*) mass spectrometry, and
(*d*) accelerator design.

For heavy particles of energy above about 50 MeV accurate electrostatic and magnetic spectrometers are comparable in size with accelerators. Large magnets with or without double focusing properties and especially quadrupole magnets (Sect. **8.4**) are widely used for deflecting and concentrating particle beams from synchrocyclotrons and proton synchrotons and are an essential part of the beam engineering installations. They usually have a long focal length and are not suitable for anything better than a crude estimate of particle energies. For such purposes deflecting magnets may be calibrated by the 'floating wire' technique,† but the information thus obtained is likely to be much inferior to that provided by knowledge of the properties of the accelerator itself.

In low energy nuclear physics magnetic spectrometers give the most precise information available on nuclear energy changes in transmutations, and lead to a scale of atomic masses comparable with that based on mass spectrographic measurements (which are also made in essence with magnetic spectrometers). The development of β-ray spectrometers originated in the study of naturally occurring radioactive elements as a method of refining the crude data provided by absorption methods. Such instruments are now used for practically all β-spectroscopic measurements except in cases in which the low disintegration rate of the source under investigation demands the use of a large angle of collection.

† It is easy to show that a flexible wire under tension T and carrying current i takes up the path of a singly charged particle of momentum p in a magnetic field of flux density B where $p = Te/i$.

The response of a spectrometer in which a radiation of homogeneous energy is being examined may often be represented as in Fig. 7.1, which also applies to an energy-sensitive nuclear detector. In both cases we distinguish between ΔE, the full width of the response curve at half intensity and δE, the accuracy with which the energy E is known.† The width ΔE depends jointly on the properties of the particular instrument and on the size of the source and detector of particles used in the spectrometer; it determines the resolution $R = \Delta E/E$ of the apparatus. The accuracy δE depends on calibrations. In general, resolution can be improved by special attention to magnetic field design and also by some sacrifice of transmission (T), which is defined to be the fraction of particles of a given energy or momentum emitted by the source which is collected and focused on a detector by the spectrometer.

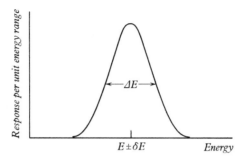

Fig. 7.1 Response of spectrometer, or energy sensitive nuclear detector, corresponding to homogeneous radiation of energy E.

Heavy particle spectrometers have generally followed existing designs of β-ray instruments but certain new features connected with their use in nuclear reaction work have been introduced.

7.1.3 Beta and electron spectrometers

(a) Deflection or prismatic type. Magnetic deflection through 180° in a uniform field was first applied to beta rays by von Baeyer, Hahn and Meitner. If the electrons move in a plane perpendicular to the lines of force of the field there is first order focusing; particles of the same velocity passing through the slit AB from a point source (Fig. 7.2(a)) describe circles of the same radius and converge to a focus F. The trace of such particles on a photographic plate in the focal plane has a sharp edge on the high energy side corresponding to rays emitted at right-angles to the diameter SF, and this permits accurate measurements to be made (Fig. 7.2(b)). If the slit AB (which may be at any point on the trajectory) defines an angle of acceptance of 2α

† In many cases it is more convenient to express the response in terms of momentum p, rather than energy E.

then the width of the image of a point source at the focus is approximately

$$s = 2\rho(1 - \cos\alpha) \approx \rho\alpha^2 \qquad (7.1)$$

where ρ is the radius of curvature for a given (relativistic) momentum p, i.e.

$$\rho = \frac{p}{eB} \qquad (7.2)$$

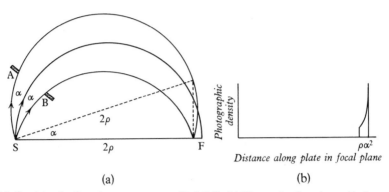

Fig. 7.2 Semicircular focusing spectrometer (Ref. 7.5). (a) First order focusing at F of particles of uniform energy from a point source S after 180° deflection in the plane of the diagram. (b) Microphotometer trace of line on a photographic plate placed in the focal plane.

Proportionality of the image width to the square of the angle α is the characteristic of a first-order focus. The resolution in momentum, i.e. the relative change in momentum necessary to shift the line by its width s, is

$$R = \frac{dp}{p} = \frac{d\rho}{\rho} = \frac{s}{2\rho} = \frac{\alpha^2}{2} \quad \text{from (7.1)} \qquad (7.3)$$

The dispersion D, defined as the change in line position per unit relative momentum change is

$$D = \frac{d(2\rho)}{dp/p} = 2p\frac{d\rho}{dp} = 2\rho$$

If now trajectories in other planes are considered, it is found that over a narrow range of angles α there is still first order focusing at F. The transmission factor in the case when AB is a circular hole subtending a solid angle Ω is

$$T = \frac{\Omega}{4\pi} = \frac{1}{4\pi} \cdot 2\pi(1 - \cos\alpha) \approx \frac{\alpha^2}{4} \qquad (7.4)$$

In practice the source is deposited on a fine wire, whose diameter must be taken into account in calculating the actual resolution, which is then inversely proportional to the dispersion of the instrument. A resolution of 1 per cent and a transmission of 0·1 per cent are typical figures for semicircular spectrometers of modest performance. The accuracy of energy measurement is usually determined by the measurement of B.

The detector in an electron spectrograph may be a photographic film or a nuclear emulsion placed in the focal plane or a counter set at a fixed radius. In the latter case the magnetic field is varied and a momentum spectrum is obtained by plotting a corrected counting rate as a function of magnetic field. Since according to (7.3) the detector aperture will embrace a momentum interval dp proportional to p at any given setting, the corrected counting rate is obtained by dividing the observed rate by p.

Improved transmission, which is important in coincidence experiments in nuclear spectroscopy, may be obtained by use of a magnetic *lens spectrometer* in which electrons pass from source to detector in helical paths through a long or short solenoid. Figure 7.4(a) shows a typical β-spectrum observed in a lens type spectrometer.

(*b*) *Inhomogeneous field (double focusing) spectrometers.* The good transmission property of the lens spectrometer and the potentially high resolution of the constant field 180° instrument are combined in spectrometers based on the double focusing property of an inhomogeneous field (Fig. 7.3). If the

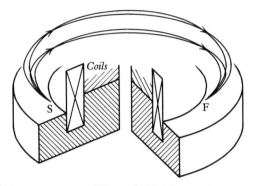

Fig. 7.3 Double focusing spectrometer. The top half of the annular magnet is largely cut away to show the orbits (Ref. 7.4).

magnetic flux density B between the poles of a spectrometer with axial symmetry is written

$$B = B_0 \left(\frac{r_0}{r}\right)^n \tag{7.5}$$

where r is the distance from the axis and B_0, r_0 and n (the field index) are parameters, then for $n = \frac{1}{2}$ electrons diverging from a point will converge back to a point after deflection through an angle $\sqrt{2}\,\pi(=254 \cdot 6°)$ from the source (App. 3). This type of spectrometer with a field dependence

$$B \propto r^{-1/2} \tag{7.6}$$

was suggested by Svartholm and Siegbahn. An instrument with a radius of 500 mm and a pole gap of 28 mm had a transmission of 1 per cent and a resolution of 0·9 per cent with a source of 5 mm wide. The source and detector must be within the field if the instrument is to be used for absolute measurements. The spectrometer is non-linear because of the iron in the magnetic circuit, but shaped fields can also be obtained without iron by suitably arranged conductors if sufficient power is available.

Figure 7.4(b) shows a spectrum of electron lines due to internal conversion (Sect. 2.2.2) obtained in a double focusing spectrometer. At a resolution setting of 0·1 per cent a transmission of 0·26 per cent could be used.

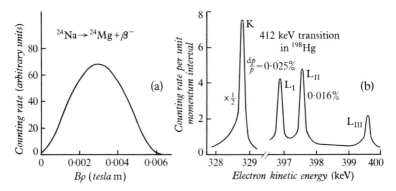

Fig. 7.4 (a) Beta-spectrum of ^{24}Na observed in a lens-type instrument (Siegbahn, *Phys. Rev.*, **70**, 127, 1946). (b) Internal conversion lines observed in a double focusing spectrometer (Graham *et al.*, *Nucl. instrum. Meth.*, **9**, 245, 1960).

7.1.4 Heavy particle spectrometers

The design of heavy particle spectrometers has generally followed that of β-spectrometers with the limitation that iron-free magnetic instruments cannot be used because of the high momenta involved.

The first *uniform field spectrometer* as distinct from magnetic deflection apparatus was designed by Cockcroft for the analysis of naturally occurring α-particle groups by Rutherford and his collaborators (1933). The magnet was of the 180° focusing type, and in order to use only a minimum of iron an annular gap was employed. The thin radioactive source and an ionization chamber or proportional counter as detector were immersed in the field at opposite ends of a diameter. For absolute measurements of the highest precision† the diameter must be compared directly with a standard metre by optical means, and the magnetic field must be measured by the nuclear resonance method (Sect. 4.4.3) and corrected (Hartree correction) for azimuthal variation. Under these circumstances the precision of energy measurement is about 1 part in 5000 and the resolution of the instrument (for mean radius of about 350 mm) can be about 0·1 per cent. This performance is however combined with a small transmission of 0·005 per cent and in the type of measurement in which the magnetic field is varied while a line profile is

† See for example, E. R. Collins, C. D. McKenzie and C. A. Ramm, *Proc. roy. Soc.*, A, **216**, 219, 1953; and the review by G. H. Briggs, *Rev. mod. Phys.*, **26**, 1, 1954.

plotted from counter readings the exploration of the spectrum can be extremely tedious.

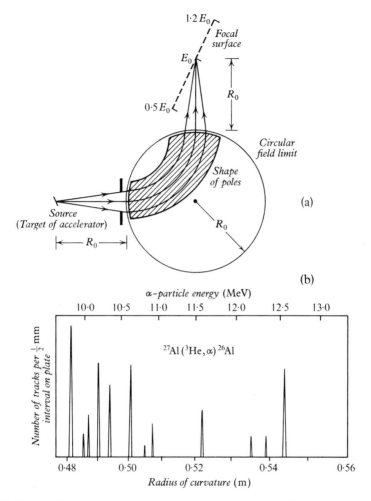

Fig. 7.5 The broad range spectrograph. (a) Geometrical arrangement. There is adequate focusing for particles of a range of energy about E_0, the energy for $90°$ deflection (Ref. 7.6). (b) Results for the $^{27}\text{Al}(^3\text{He}, \alpha)^{26}\text{Al}$ reaction (Hinds and Middleton, *Proc. phys. Soc.*, **73**, 501, 1959).

For experiments in which only a momentum distribution is required it is better to convert the spectrometer into a spectrograph by using a nuclear emulsion as a detector. Particles with a considerable range of momenta can then be recorded simultaneously, and immediate discrimination between singly and doubly charged particles is obtained from grain density on examination. If a sector-shaped magnetic field with deflection angle less than $180°$ is used both source and detector can be some distance outside the magnetic field, and the arrangement is very convenient for transmutation work

reaching a detector. The detector output may be displayed on a time base which is synchronized with the source modulation pattern and if the source is monokinetic, signals are obtained only at the definite time delay l/v after the source pulse. If the source is not monokinetic the signals displayed on the time base give a velocity spectrum of the neutrons. Alternatively signals may be accepted from the detector only through a 'gate' delayed by a variable time from the source pulser. All such signals are then due to neutrons of a given velocity, which may thus be selected from a continuous distribution of source velocities. If a sample of material is placed in the flight path the absorption for neutrons of the selected velocity may be studied.

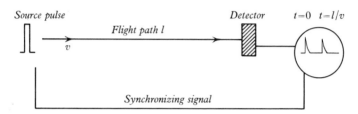

Fig. 7.6 Principle of time-of-flight technique using modulated source.

In the first successful application of this method at the University of Columbia the source was a cyclotron in which neutrons were produced by the deuteron bombardment of beryllium as a result of the nuclear reaction

$$^9\text{Be} + \text{d} \longrightarrow {}^{10}\text{B} + \text{n} + 4 \cdot 36 \text{ MeV} \tag{7.8}$$

The ion source of the cyclotron was modulated and the fast neutrons emerging from the target were slowed down in a block of paraffin wax near by. The slow neutrons traversed a flight path to a boron-trifluoride ionization chamber. In experiments made with a path of 5·4 metres, pulses of length between 10 and 1000 μs were used with repetition rates of 100–1000 Hz; the resolution was determined by the pulse lengths. The method has recently been developed at the University of Columbia by use of the 385 MeV proton synchrocyclotron to produce neutron bursts of 0·1 μs duration at a rate of 60 Hz. A flight path of 35 metres was used and the yield of gamma radiation from neutrons interacting with a sample at the detector position was observed. Similar velocity spectrometers for slow neutrons have been based on electron linear accelerators as source, the neutrons being produced by (γ, n) reactions in beryllium as a result of bremsstrahlung pulses. The accelerator pulse length of about 1 μs is very suitable for time-of-flight work.

Slow neutron velocity spectrometry has also been applied to the neutron beams from nuclear reactors. The beam of mixed, but near thermal, energies emerging from a channel in the reactor shielding is interrupted rapidly by a rotating shutter of a strongly absorbing material such as boron or cadmium. Such 'fast choppers' can provide pulses of 1 μs duration for analysis by the time-of-flight method or for use in transmission experiments. The general improvement in resolution in slow neutron spectrometry over a few years is illustrated in Fig. 7.7 for a number of well-known instruments; in this figure the transmission of a sample at a given energy has been converted into a cross-section (eq. 5.8).

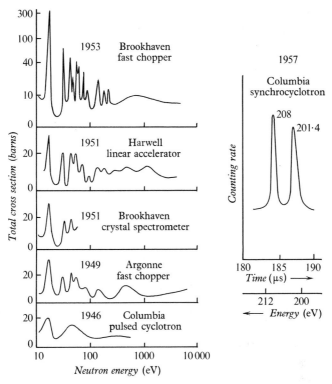

Fig. 7.7 Determination of total neutron cross-section for silver by time-of-flight methods, 1946–1957 (Rainwater, *Encyclopedia of Physics*, **40**, 373, 1957).

(b) *Fast and high energy neutrons.* The time-of-flight method can be used in the energy range 0·5–10 MeV but a neutron of energy 1 MeV has a velocity of $1·4 \times 10^7$ m s^{-1} and nanosecond circuit techniques become necessary for timing. It is possible to pulse the beam of an electrostatic accelerator producing neutrons by the reaction

$$^3\text{H} + {}^2\text{H} \longrightarrow {}^4\text{He} + \text{n} + 17·58 \text{ MeV} \tag{7.9}$$

with a duration of about 2×10^{-9} s at a frequency of 3·7 MHz and a flight path of 2 m is then convenient. The neutrons must be detected by fast counters using organic scintillators. In another variant of this technique a particle or photon associated with the neutron production, e.g. the α-particle in the reaction (7.9), provides the timing pulse and neutrons are observed in delayed coincidence with this signal. Cyclotrons and linear accelerators are also particularly suitable for this work because the beam is bunched during the acceleration process and arrives at the neutron producing target as a succession of short pulses in any case. Figure 7.8 shows a neutron spectrum obtained with a pulsed electrostatic accelerator.

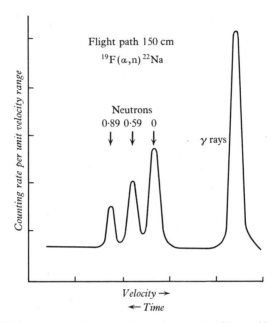

Fig. 7.8 Time-of-flight spectrum of neutrons from the reaction ^{19}F(α, n)^{22}Na. The large peak corresponds to γ-radiation (travelling between target and detector with the velocity of light). Three neutron peaks are shown, corresponding to formation of ^{22}Na in its ground state and in excited states of energy 0·59 and 0·89 MeV, giving neutrons of lower energy and longer flight time (Batchelor and Towle, *Proc. phys. Soc.*, **73**, 307, 1959).

(*c*) *General.* The resolution of time-of-flight methods of energy measurement is usually expressed in terms of a time spread per metre of flight path. For the Columbia (1957) results shown in Fig. 7.7 a resolution of 0·01 µs m^{-1} is indicated corresponding to an energy resolution of about 0·5 per cent at 150 eV. This has been improved to better than 0·1 per cent in electron linear accelerator installations in which neutrons are produced by photo-reactions in beryllium of electron bremsstrahlung. The time-of-flight method has been extensively applied to the study of spectra of inelastically scattered neutrons, which are detected with increased delay in the spectrometer.

7.2.2 *Measurements based on recoil protons*

The energy of a proton recoiling at a laboratory angle ϕ_L with the direction of incidence of a neutron of energy E_0 is

$$E_r = E_0 \cos^2 \phi_L \qquad (7.10)$$

from (5.2). If the neutron energy is less than about 10 MeV it is a good approximation to assume that the neutron–proton scattering has an isotropic angular distribution in the centre-of-mass system. The differential cross-section for scattering of the neutron through an angle θ is then

$$d\sigma = \frac{\sigma}{4\pi} d\Omega$$

$$= \frac{\sigma}{2} \sin \theta \, d\theta \tag{7.11}$$

By (5.4) the angles ϕ_L and θ are connected by the relation

$$\theta = \pi - 2\phi_L$$

so that

$$d\sigma = -2\sigma \sin \phi_L \cos \phi_L \, d\phi_L$$

$$= \frac{\sigma}{E_0} dE_r \tag{7.12}$$

It follows that the energy distribution of protons projected from a thin hydrogenous target by a homogeneous beam of neutrons is uniform up to the maximum available energy E_0, as shown in Fig. 7.9. Neutron energies may

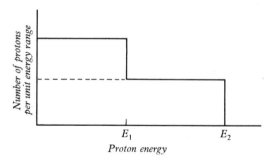

Fig. 7.9 Distribution of energies of protons projected from a thin hydrogenous target (or in a hydrogen-filled ionization chamber) by a neutron beam containing two homogeneous groups of energy E_1 and E_2 giving isotropic scattering.

be determined by observing this distribution, or more usually, by observing the energy distribution of the protons projected within a small angular range near $\phi_L = 0$.

Following early experiments with hydrogen or helium filled cloud chambers visual observations of recoil protons have usually employed the nuclear emulsion. A more flexible technique, limited mainly by its low efficiency, is that of the counter telescope, in which the range distribution of forward-going protons from a hydrogenous radiator is determined by means of absorbers. The low efficiency is a consequence of reduction in stopping power of the radiator in order to reduce the spread of energy for the range measurement. This technique has been mainly used in the energy region above 20 MeV, in which accurate energy measurement may be less important than discrimination between neutrons and other radiations. In high energy physics, an anticoincidence counter is generally used to veto pulses in the telescope arising from incident charged particles.

Intensity measurements using the techniques described in this section depend on knowledge of the neutron–proton collision cross-section at the energy concerned. This is easily determined by a transmission method, and is known for a wide range of energies.† Observations made with nuclear emulsions must also be corrected for escape of recoil protons.

Energy determinations may also be obtained from pulse-size distributions in ionization chambers, proportional counters or organic scintillators. These are as predicted by equation (7.12) so long as the scattering is isotropic in the c.m. system. Large liquid scintillators offer efficient detection for fast neutrons. In high energy physics protons recoiling from neutrons of energies of a few hundred MeV may be able to produce light by the Cherenkov effect (Sect. 5.2.3) in a transparent medium. The special conditions necessary for this phenomenon provide a low energy cut-off and in practice counters of this sort are operated in conjunction with anti-coincidence counters to discriminate against primary charged particles.

7.2.3 Reaction methods

The most satisfactory method of determining a flux of neutrons of mixed energies is that of *thermalization*; the neutron source is placed, if possible, in a large volume of water containing a suitable absorber such as $MnSO_4$. The neutrons are slowed down by the water to thermal energies and are then captured with high probability by the manganese yielding active ^{55}Mn by the reaction

$$^{55}\text{Mn} + \text{n} \longrightarrow {}^{56}\text{Mn} + \gamma + 7 \cdot 27 \text{ MeV} \tag{7.13}$$

In equilibrium the rate of absorption by the ^{55}Mn, which may be assessed by taking samples of the stirred solution and finding their activity, is equal to the rate of production by the source.

Slow neutron flux measurements are conveniently made with ionization chambers or scintillators containing boron in which neutrons induce the reaction

$$^{10}\text{B} + \text{n} \longrightarrow {}^7\text{Li} + {}^4\text{He} + 2 \cdot 79 \text{ MeV} \tag{7.14}$$

which may be detected as an electrical pulse. The total cross-section for this reaction has been accurately measured as a function of energy by the crystal spectrometer method and has the value

$$\sigma \, (^{10}\text{B}) = 3820 \pm 10 \text{ barns}$$

† This cross-section and many others will be found in 'Neutron Cross-Sections', by D. J. Hughes and R. B. Schwartz, Brookhaven Report BNL 325.

for a neutron velocity of 2200 m s^{-1}. A $1/v$ dependence has been shown to hold in this region and probably up to energies of about 1000 eV. The boron chamber is thus extremely suitable for flux measurement since it effectively reduces an inhomogeneous neutron flux to the equivalent flux at standard energy.

7.3 Measurement of energy and intensity of photons

The advent of the lithium-drifted germanium counter (Sect. 6.1.2) leaves many of the earlier techniques of gamma ray measurement with mainly a historical interest, at least in the field of low energy nuclear reactions. The germanium counter is not absolute, although it may easily be calibrated with standard sources, and it has a low efficiency, but its resolution is excellent. In this section only methods offering absolute energy determination or especially suitable for use in high energy experiments will be mentioned.

7.3.1 *The curved crystal spectrometer*

Crystal diffraction has been applied to the spectroscopy of γ-radiation by DuMond and his collaborators. The main instrument is the transmission-type of bent crystal spectrometer first used by Cauchois and illustrated in principle in Fig. 7.10(a), (b). A thin cylindrical lamina of crystalline quartz, initially flat, is bent to have a radius $2R$ so that diffracting planes intersect at distance $2R$ from the centre of the crystal. The source of radiation is placed

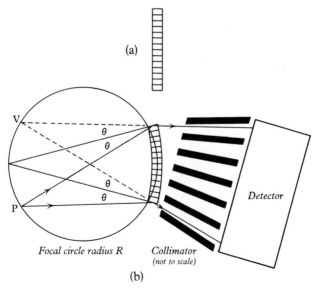

Fig. 7.10 Curved crystal spectrometer (Cauchois–DuMond); (a) crystal before bending, (b) ray diagram (DuMond, *Ann. rev. nucl. Sci.*, **8**, 163, 1958).

either on the convex side of the crystal (Cauchois arrangement) or on the focal circle (DuMond arrangement). In each case the crystal picks out the wavelength λ appropriate for reflection at the Bragg angle θ given by

$$\lambda = 2d \sin \theta \tag{7.15}$$

where d is the lattice spacing, and a real or virtual image is formed at the point V. In the DuMond arrangement the reflected beam diverging from the

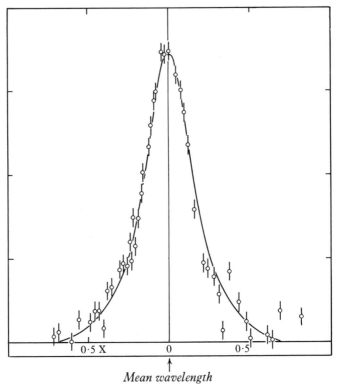

Mean wavelength

Fig. 7.11 Spectrum of annihilation radiation ($E = 511$ keV) obtained with curved-crystal spectrometer $1X = 10^{-13}$ m (DuMond *et al.*, *Phys. Rev.*, **75**, 1226, 1949).

virtual image V is received by a scintillation counter shielded from the direct radiation from the source by a system of baffles or collimator. The source is moved mechanically with high accuracy along the focal circle to explore a range of wavelengths, and the crystal is automatically rotated as the source moves so that the detector and collimator can remain in a fixed position. The grating constant d for quartz is 0·1178 nm and the Bragg angle θ for $E_\gamma = 511$ keV is 34 minutes of arc.

Figure 7.11 shows the profile of the annihilation quantum line at 511 keV obtained with a curved crystal spectrometer (DuMond arrangement) using a quartz crystal 1 mm thick, 70 mm high and 50 mm wide, a radius of curvature R of 2 metres and a source strength of 2·5 curies of positron activity. The wavelength was given as

$$\lambda = 0\cdot0024271 \pm 0\cdot0000016 \text{ nm}$$

The resolution was approximately 1 per cent, the accuracy of measurement 0·04 per cent and the efficiency of the instrument about 5×10^{-8} counts per photon emitted from the source. The Cauchois arrangement has been used by DuMond in a study of the γ-radiation following inelastic scattering of protons from a cyclotron in a variety of targets; exposures of many hours are necessary to obtain film records even using extremely high (≈ 5 mA) proton currents at 3·7 MeV.

The curved-crystal spectrometer has been used with great success for measurements of energies up to 500 keV and provides standards of γ-ray energy over this range (Ref. 7.2). Its main limitations arise from the fall of crystal reflecting power and the small angles of reflection at high quantum energies. The latter limitation has been overcome in a new type of two (flat) crystal spectrometer developed in the Chalk River laboratories in which neutron capture radiations of energy up to 2 MeV have been measured with a resolution of 0·4 per cent at 1 MeV.

7.3.2 Magnetic spectrometers

The cloud chamber, with a magnetic field, was one of the early methods of determining the momentum of electrons produced by photon interactions in thin radiators. In nuclear reaction experiments it has been entirely superseded, but in the study of radiative processes in elementary particle physics, the heavy liquid bubble chamber and the spark chamber are both used in this way for measuring the momenta of secondary electron–positron pairs.

The photoelectric effect, the Compton effect and pair production have all been made the basis of high resolution magnetic spectrometers for photon measurements. The design of such instruments for γ-radiation is obviously essentially the same as for β-spectrometers of a similar energy range but additional emphasis on transmission is desirable because of the loss of overall efficiency in the conversion of the photons into electrons. Gamma radiation arising from nuclear reactions may be accompanied by internal conversion electrons (Sect. **13.2**) or by internal conversion pairs (Sect. **13.3**) or both, and in such cases the photon energy is obtained directly from observations of the conversion spectrum (Figs. 7.4 and 7.13). The electron energies are

$$E_\gamma - E_K, E_L \ldots$$

where E_K, $E_L \ldots$ are automatic ionization energies.

Deflection (prismatic) spectrometers, as described in Sect. 7.1.3, were the first precision instruments used in the study of γ-radiation. Ellis (1932) used a $180°$ semicircular focusing permanent magnet spectrometer with photographic recording for accurate measurements of internal conversion lines in the Th B+C spectrum. In such work the resolution and transmission achieved are as given for β-spectrometers. If internal conversion lines are of low intensity, external conversion in a radiator foil may be employed; the resolution and sensitivity then depend on foil thickness in addition to instrumental factors. Photoelectrons from heavy radiators (Pb) and Compton electrons from light foils (Be) have frequently been used, usually with counter

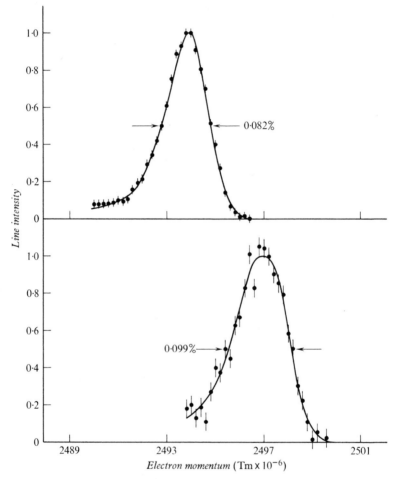

Fig. 7.12 Precise determination of γ-ray energy by external conversion method, using a thin uranium radiator. The upper diagram is for the 412 keV radiation of ^{198}Hg; the lower is for 511 keV annihilation radiation (G. Murray, R. L. Graham and J. S. Geiger, *Nucl. Phys.*, **45**, 177, 1963).

detection. Figure 7.12 shows results obtained by Murray, Graham and Geiger using the iron-free double-focusing electron spectrometer already referred to in Fig. 7.4(b). This instrument is not absolute because energies cannot be deduced precisely from orbit radii and the magnetic field. It was however possible to compare the momentum of L_{III}-photoelectrons (Fig. 7.4(b) ejected from a thin uranium foil by the 412 keV gamma ray of ^{198}Hg with K-electrons ejected from the same foil by 511 keV annihilation radiation ($E_\gamma = mc^2$) of thermalized positrons (as positronium). The nuclear gamma ray energy was thus obtained directly in terms of the electron mass. A value of $411 \cdot 799 \pm 0 \cdot 007$ keV (1961 constants) was quoted.

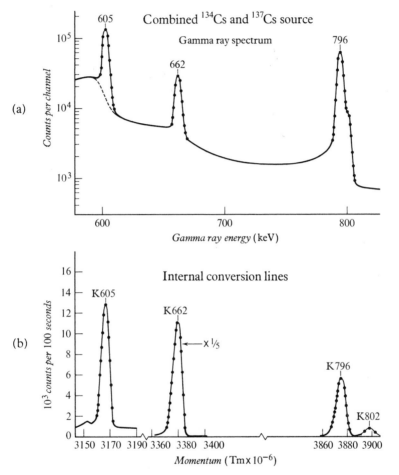

Fig. 7.13 Comparison of response of Ge(Li) detector (a) with that of a high resolution electron spectrometer (b) (R. A. Brown and G. T. Ewan, *Nucl. Phys.*, **68**, 325, 1965).

Figure 7.13 compares the response of a high resolution electron spectrometer with that of a $3 \cdot 5$ mm deep Ge(Li) counter.

Before the development of the germanium detector, the *magnetic pair spectrometer* was often used for the study of gamma rays of energy above about 3 MeV. The principle of this instrument, which has found application in high energy physics, is illustrated in. Fig. 7.14(a). A thin radiator is used and positrons and electrons are detected in coincidence by fixed counters after opposite deflections through 180°. The electron pairs from high energy radiation (say $E \approx 5$ MeV) go predominantly forward, and, again for high energies,

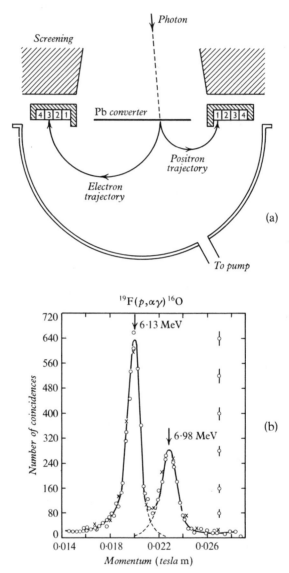

Fig. 7.14 Prismatic pair spectrometer. (a) Section through gap of magnet. (b) Pair spectrum for γ-radiation from ^{19}F (p, αγ)^{16}O reaction (Walker and McDaniel, *Phys.Rev.*, **74**, 315, 1948).

the sum of the momenta of the two particles is constant since the sum of the energies is constant and equal to $E_\gamma - 2mc^2$. The sum of the radii of the two trajectories is therefore also constant and pairs from all parts of the radiator (with the correct energy ratio) may be detected. Figure 7.14(b) shows the results obtained by Walker and McDaniel in a study of (p, γ) capture reactions; the resolution was 6 per cent. The efficiency of such instruments is low (≈ 1 coincidence count per 10^7 photons at 7 MeV) although it increases rapidly with energy as the pair production cross-section rises. The intensity of the incident radiation is obtained from calculations based on this cross-section and on the geometrical properties of the spectrometer.

In all measurements with magnetic spectrometers a stability of 1 in 50 000 in magnetic field should be aimed at. The γ-ray energies obtained from the instruments may require correction because of Doppler shifts due to recoil motion of the radiating source. If the nucleus emitting the radiation moves with velocity v, the energy observed at angle θ with the direction of motion is shifted by an amount

$$E_\gamma \frac{v}{c} \cos \theta \qquad (7.16)$$

where E_γ is the transition energy in the centre-of-mass system. This can give a shift of as much as 1 part in 200 and it is not always clear whether such a correction should be applied. This depends on the ratio between the lifetime of the radiating nuclei for γ-emission and the slowing down time for these nuclei in the source material; observation of the Doppler effect has in fact been made the basis for lifetime measurements (Sect. 13.6.1).

7.3.3 Total absorption spectrometers

In cases in which the highest possible efficiency in the MeV range of energy is required, and in which modest resolution is satisfactory, the sodium iodide scintillation crystal is a useful detector. It has the disadvantage that a single

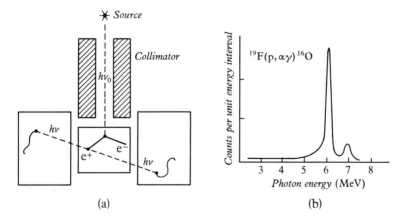

Fig. 7.15 Three-crystal pair spectrometer. (a) General arrangement of crystals. (b) Energy spectrum of radiation from ^{19}F(p, $\alpha\gamma$)^{16}O reaction (Bent and Kruse, *Phys. Rev.*, **108**, 802, 1957).

homogeneous line may give more than one peak in a pulse height spectrum from a typical crystal (Fig. 6.13(c)). The analysis of a complex spectrum then becomes difficult. This trouble has been overcome in the three-crystal spectrometer, which is arranged as shown in Fig. 7.15(a). The incident radiation produces a pair in the centre crystal, and the positron and electron both stop in this crystal. The two oppositely directed annihilation quanta from the positron have a chance of escaping from the centre crystal and being detected, one in each side crystal. Triple coincidence counts are then taken between the crystals and the output is used to gate a pulse analyser recording the size of pulses from the centre crystal. A single peak, representing the kinetic energy of the pair, is then obtained for each quantum energy (Fig. 7.15(b)). With crystals of a size the order of a 25 mm cube, a resolution of 5 per cent at 4·4 MeV has been obtained with a detection efficiency (which may be calculated from the cross-sections of sodium iodide) of one in 10^4 of the quanta from a suitably collimated source. This is several hundred times better than the performance of a magnetic pair spectrometer.

Three crystal arrangements with a germanium centre counter have been used at lower energies and offer advantages in resolution at the expense of efficiency. Two-crystal spectrometers based on the Compton effect (one crystal detecting the recoil electron and the other the scattered photon) have been operated with both Ge(Li) and NaI(Tl) detectors. Because of the sensitivity of the Compton effect to the direction of the incident electric vector they can be used as gamma ray polarimeters.

Other forms of total absorption gamma-ray spectrometer are:

(a) the *Si(Li) semiconductor counter* (Sect. 6.1.2) and the *proportional counter*, (Sect. 6.1.3) which have high resolution and efficiency for X-ray photons,
(b) the *Cherenkov total absorption spectrometer* (Sect. 6.1.6).

Intensity calculations in all these cases must be based on known geometry and elementary cross-sections.

7.3.4 *Flux determination by ionization chamber*

Ionization chambers have three principal applications in determining photon flux:

(a) *for measurement of the relative yield* of γ-radiation in nuclear reactions as a function of bombarding energy,
(b) *for radiation monitoring* in health physics,
(c) *for monitoring the output* of electron accelerators such as betatrons, synchrotrons and linear accelerators.

The intensity of the photon beam from such machines is often measured by the 'quantameter'† in which an electromagnetic shower is produced in an

† R. R. Wilson, *Nucl. instrum. and Methods*, **1**, 101, 1957.

ionization chamber containing absorbing plates. The intensity may be expressed either as an energy content U, deduced calorimetrically or directly from the observed ionization, or as a total number of 'equivalent quanta' Q given by

$$Q = \frac{U}{E_0} = \frac{1}{E_0} \int_0^{E_0} kn(k) \, dk \qquad (7.17)$$

where E_0 is the maximum bremsstrahlung energy and $n(k) \, dk$ is the number of photons with energy between k and $k + dk$ in the spectrum.

References

1. FRIEDLANDER, G. and KENNEDY, J. W. *Nuclear and Radiochemistry*, Wiley, 1955.
2. SEGRÈ, E. (ed). *Experimental Nuclear Physics*, Vol. III, Wiley, 1959; WAPSTRA, A. H. *Nuclear Physics*, **57**, 48, 1964; MARION, J. B. *Rev. mod. Phys.*, **38**, 660, 1966.
3. SIEGBAHN, K. (ed.). *Alpha, Beta and Gamma Spectroscopy*, North Holland Publishing Co., 1965.
4. CAVANAGH, P. E. 'Spectroscopy of β- and γ-rays', *Prog. nucl. Phys.*, **1**, 140, 1950.
5. VERSTER, N. F. 'The Electron Optical Properties of Magnetic β-ray Spectrometers', *Prog. nucl. Phys.*, **2**, 1, 1952; SIEGBAHN, K. in Ref. 7.3, page 79.
6. BUECHNER, W. W. 'The Determination of Nuclear Reaction Energies by Deflection Measurements', *Prog. nucl. Phys.*, **5**, 1, 1956.
7. CRANBERG, L. and ROSEN, L. 'Measurement of Fast Neutron Spectra', in *Nuclear Spectroscopy*, Part A, ed. AJZENBERG-SELOVE, F. Academic Press, 1960.
8. ALLEN, W. D. *Neutron Detection*, Newnes, 1960.

8 The acceleration of charged particles high energies

In the ten years following Rutherford's discovery of transmutation (1919) efforts were increasingly devoted to attempts to improve the intensity of the observed effects. A naturally radioactive α-particle source of 100 mCi strength emits $3 \cdot 7 \times 10^9$ particles per second into a solid angle of 4π and therefore provides a flux density of 3×10^{10} particles $m^{-2} s^{-1}$ at a distance of $0 \cdot 1$ m. A positive ion beam of 1 μA, which may easily be collimated to a cross-sectional area of $10^{-4} m^2$, delivers 6×10^{12} singly charged particles per second. On the other hand the α-particles from radium and its products have velocities† corresponding to an energy of 5–8×10^6 eV and the production of such energies by direct acceleration of positive ion beams entailed high-voltage apparatus of a type quite unknown at the time. Despite this, high voltage development proceeded and by 1930, Cockcroft and Walton, in the Cavendish Laboratory, Cambridge, were able to announce the production of a beam of 10 μA of 280 keV hydrogen ions. Later the accelerating voltage was increased to 700 kV, and the successful transmutation of lithium in 1932, with protons of this energy (Sect. 14.1.2), gave great impetus to the rapidly developing field of accelerator technology.

Present-day accelerator design and construction is a major field of scientific effort in which vast sums of money are available. The enormous growth of this branch of physics in the last thirty years is due partly to a new awareness of the types of problem awaiting attack, and partly to technical progress in providing the necessary means. The problems are, briefly, the study of nuclear structure by means of nuclear reactions, and the study of the properties of elementary particles. The former may be investigated with relatively modest machines, whose output energies match the potential barriers surrounding nuclei; the latter continues to tax experimental ingenuity to the utmost by demanding the highest possible energy or intensity, or both. It is in the latter field especially that progress depends mainly on new concepts and principles.

The division of the problems into two types also suggests a corresponding division of accelerators into direct current (d.c.) or pulsed (r.f.) machines. The latter class may be further subdivided into linear and orbital accelerators. The only exception from this division is the betatron, an orbital accelerator whose output appears in bursts, but which does not employ radiofrequency

† A singly charged particle after falling through a potential difference V (volts) acquires an additional energy of V electron-volts.

acceleration. The present account will adopt this classification, although it should not be inferred that the study of nuclear reactions is confined to d.c. machines.

8.1 High-voltage d.c. accelerators

Installations of this type always include an ion source, an accelerating tube, a method of generating high voltages for application to the electrodes of the accelerating tube, and an analysing system for selecting ions of a particular type after acceleration or for improving the energy homogeneity of the beams.

8.1.1 The cascade generator

The circuit used by Cockcroft and Walton† for production of a steady terminal voltage of 700–800 kV was a two-stage doubling arrangement of transformer, rectifiers and condensers; Fig. 8.1 illustrates the principle. If the

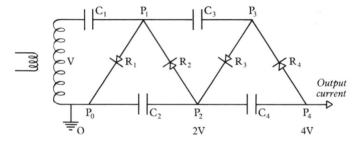

Fig. 8.1 Cascade generator circuit (two stages of voltage doubling).

transformer develops a peak secondary voltage V, then condenser C_1 charges up to this potential difference through rectifier R_1 (load currents and rectifier voltage drops are neglected). The voltage across R_1 then varies from O to 2V sinusoidally during each cycle. This voltage is effectively applied to the circuit R_2C_2 and charges up condenser C_2 to a potential difference 2V, thus completing the first stage of voltage doubling. The voltage across R_2 also varies sinusoidally from O to 2V and by similar arguments it can be seen that C_3 and C_4 will also charge up to a potential difference 2V. In the steady state condenser C_1 is charged to potential V and C_2, C_3, C_4 to potential 2V; C_1, C_3 are in series with the transformer secondary and their voltage *to ground* varies by $\pm V$ during each cycle, but C_2, C_4 have potentials fixed with respect to ground and the final d.c. terminal voltage at P_4 is 4V. The operation of the circuit may also be understood by regarding the rectifiers as switches through which the charge on C_1 is shared in successive a.c. cycles with the other condensers of the circuit. The number of stages of voltage doubling may be increased by adding more rectifiers and condensers and for a number $2n$ of

† J. D. Cockcroft and E. T. S. Walton, *Proc. roy. Soc.*, A, **136**, 619, 1932.

each of these components, the circuit is a *n*-stage voltage doubler providing a no-load output voltage of 2*n*V. Each condenser must be rated for a voltage of 2V, and each rectifier must withstand a reverse peak voltage of the same magnitude.

If a current *i* is drawn from the high-voltage terminal P_4 and if the frequency of the transformer mains supply is *f* Hz then in one cycle the condenser C_4 drops in voltage by i/fC. This has to be made good by charge supplied by the sharing process from the rest of the circuit during the cycle and as a result both a *voltage drop* ΔV from the no-load value and a *ripple* δV of supply mains frequency appear in the output. It may be shown that, for a *n-stage* voltage doubler, in which all condensers have the same capacitance *C*, and stray capacities are neglected

$$\delta V = \frac{n(n+1)}{2} \frac{i}{fC} \tag{8.1}$$

and

$$\Delta V = (\tfrac{2}{3}n^3 + \tfrac{1}{2}n^2 + \tfrac{1}{3}n) \frac{i}{fC} \tag{8.2}$$

so that it is advantageous to use the smallest possible number of stages and the highest possible capacities and charging frequencies. Limits arise because of the voltage rating and frequency characteristics of the components.

Cockcroft and Walton used continuously pumped thermionic rectifiers erected in a tall glass column sealed by Apiezon-Q compound. Modern cascade generators use high-voltage selenium rectifiers with a peak inverse rating up to about 200 kV.

Air-insulated cascade generators are reasonably convenient for voltages up to about 1000 kV, but beyond this, size is a major limitation. This can be overcome by enclosing the generator in a pressure vessel particularly if the cascade structure is simplified by high frequency operation. In one recent development† the condenser stacks have been eliminated by employing direct capacitative coupling from the electrodes of a radiofrequency power circuit ($f = 300$ *k*Hz) to a rectifier column. In this way a.c. power is fed to the rectifiers in parallel but a d.c. potential is developed across the rectifiers in series, and may be applied to an accelerating tube. In such a generator, operating in the range 300 kV to 4 MV, the ripple voltage at the output terminal is of the order of 0·1 per cent. Ripple reduction for this generator requires only a small R.F. choke in the output lead; for the condenser-stack generator a RC filter may be connected to the output as shown in Fig. 8.2.

The cascade generator is connected to the continuously pumped accelerating tube (in which a pressure of 10^{-4}–10^{-5} mmHg is maintained), by high resistances (Fig. 8.2) through which the electrodes of the tube are kept at constant potential. Typical accelerating gaps are shown in the figure; the gaps are mounted in series and in operation constitute a succession of electro-

† The Dynamitron generator, made by Messrs. Radiation Dynamics Inc.; insulated with sulphur hexafluoride gas.

static lenses of long focal length. A focused beam of ions is delivered from an ion source (Sect. 8.1.3) through a small canal at the high voltage (positive) end of the accelerating tube and emerges at ground potential with a velocity corresponding to the accelerating voltage. The canal has a very low pumping speed and permits a much higher pressure in the ion source than in the accelerating tube. Power for the ion source is conveyed to the high-voltage terminal either by a belt driving a generator in the terminal, or by high-frequency currents circulating in the condenser stacks.

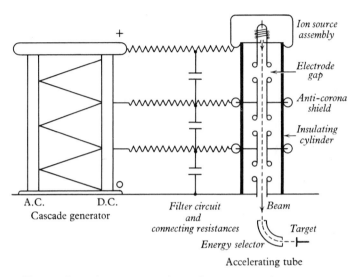

Fig. 8.2 General arrangement of cascade generator and accelerating tube.

The main advantages of the cascade generator are that it uses standard components and that it has a large output; currents of up to 10 mA of positive ions (or electrons) are obtainable with an energy up to 4·0 MeV. Many more modest generators in the 200–300 keV range are in use for neutron production through the (d–T) reaction (see eq. 7.9).

8.1.2 *The electrostatic generator*

This generator is a direct illustration of the definition of the potential of a conductor in elementary electrostatics as the work done in bringing unit charge from a standard reference point to the conductor. In the machine developed by Van de Graaff (1931) charge is sprayed from sharp corona points at a voltage of about 100 kV on to a moving insulating belt (Fig. 8.3). The belt conveys the charge to an insulated terminal electrode within which the charge is removed by collector points and allowed to flow to the surface of the electrode through a resistance R. If the capacity of the terminal is C

to ground, the potential at any instant is

$$V = \frac{q}{C} \tag{8.3}$$

where q is the stored charge. If the belt delivers a current i to the terminal the rate of rise of terminal potential is

$$\frac{\mathrm{d}V}{\mathrm{d}t} = \frac{i}{C} \tag{8.4}$$

and this may be as much as 10^6 volts s^{-1}. As the terminal voltage increases,

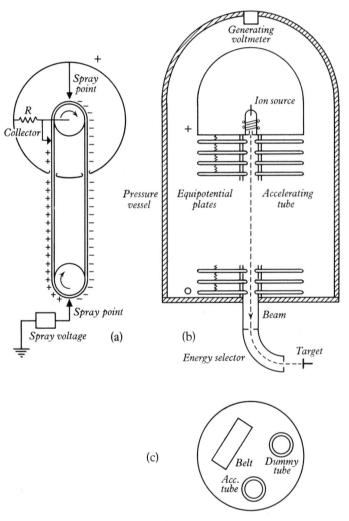

Fig. 8.3 Electrostatic generator. (a) Principle. (b) General construction of vertical generator showing accelerating tube and equipotential plates. (c) Cross-section of generator stack.

the current drain due to corona, losses through supporting insulators, and possible drain due to acceleration of ions or electrons, increases until an equilibrium between load current and charging current is established. The effective charging current may be increased by insulating the belt pulley in the upper terminal so that it reaches a potential higher than that of the terminal by an amount equal to the resistive voltage drop in R (Fig. 8.3(a)). A spray point mounted above the belt then deposits charge of opposite sign on the receding belt surface. The processes of charge collection and charge spray are unaffected by the potential difference between the terminal and ground.

The amount of charge which may be placed on the belt is limited by breakdown of the surrounding gas at some critical field strength E_c; the charge density is then

$$\sigma = 2\varepsilon_0 E_c \tag{8.5}$$

For air at atmospheric pressure E_c is about 3 MV m^{-1}, and this breakdown field should increase proportionately with pressure, but in practice irregularities on the surface of the terminal prevent this. Practically all electrostatic generators are enclosed in a pressure vessel and operated at a pressure of up to 25 atm of nitrogen or some other insulating gas in order to permit high terminal voltages. Multiple belt systems are also used to increase the charging current; speeds up to 40 m s^{-1} can be employed, with charging currents of about 400 μA per belt.

A typical construction of a vertical electrostatic generator is shown in Fig. 8.3(b). The high-tension terminal is mounted at the top of a column of metal equipotential plates, separated by insulators. The potentials are determined by corona currents between the plates or by conduction through a chain of high resistances. This arrangement gives a uniform field of electric force throughout the column. The field surrounding the terminal may also be controlled by intermediate shields anchored to appropriate points of the column, and the greater the uniformity achieved in the design, the greater will be the ultimate breakdown potential. The belts run through slots cut in the equipotential plates and further holes are provided for accelerating tubes. The radius of curvature of all high-potential surfaces, including the terminal, is kept as large as possible to reduce field gradients. Although the weight of the top terminal is most easily supported in a vertical arrangement, many horizontally-mounted machines have been made.

Accelerating tubes used for ion beams in electrostatic generators are usually alternate sections of insulator and metal spaced to match the spacing of the equipotential planes. Such a tube has good breakdown properties, providing that the electrode structure is designed to prevent internal multiplication of small electron currents. The tube should not affect the convergence or parallelism of an ion beam and should be equally efficient at all terminal voltages. A vacuum of the order of 10^{-5} mmHg or better is maintained by pumps at ground potential. Ions enter the tube through a narrow canal from a compact source situated in the high-potential terminal and in order to prevent too much gas from the ion source entering the main accelerating tube, extra pumping near the canal may be provided via a dummy tube running through the column. For the more efficient ion sources, this differential pumping is unnecessary. Ion source power supplies are obtained from a small generator driven from the belt pulley; controls may be operated photoelectrically from ground potential.

The terminal voltage of an electrostatic generator (or the voltage of the outer corona shield) is usually measured by a generating voltmeter mounted at the top of the pressure vessel. This instrument has an insulated probe which is alternately exposed to and screened from the electrostatic field of the machine by a rotating electrode at ground potential. The currents induced in the probe circuit are rectified and the resulting steady current measures the generator potential on a scale which may be calibrated by observation of well-known nuclear resonance levels. The voltmeter output may be used to stabilize the generator voltage; control of the corona spray current is usually used. Another method which provides extremely fine control is to adjust the load current by means of a small and variable electron current travelling up the accelerating tube or the differential pumping tube; a stability of ± 150 volts in 10^6 volts has been claimed for this method.

The electrostatic generator has the great advantages of stability, ease of voltage control, and high voltage rating, which offset the apparent disadvantage of small output current in comparison with that given by cascade generators. The beams are highly homogeneous since there is no ripple voltage, and the residual spread of energy, due perhaps to the ion source, may be further reduced by magnetic or electrostatic analysis after acceleration. Such beams are admirable for nuclear resonance level work (Sect. **15.2**). By scrupulous attention to all points of design, the High Voltage Engineering Corporation has been able to raise the rating of recent generator columns to about 3 MV m^{-1} and a standard generator for 6–7 MeV, with a current output of 100 μA of hydrogen ions, is now available.

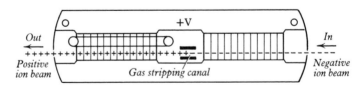

Fig. 8.4 Principle of tandem electrostatic generator.

A development of outstanding importance in the evolution of electrostatic generators is the introduction of the *tandem principle* (Fig. 8.4 and Ref. 8.7). Two insulating columns, mounted horizontally, are contained within one pressure tank, and the high voltage terminal at their junction is charged positively by a normal belt system within one column. There is no positive ion source in the terminal, but a *negative ion source* outside the pressure vessel supplies a beam of negative ions to the grounded end of the first accelerating tube. These ions are accelerated to about 7 MeV in their passage to the central terminal where they are stripped of electrons and converted to *positive ions*. The stripping is achieved by passage through a canal in which there is an increased gas pressure or through a thin carbon foil. The positive ions then enter the second accelerating tube in which they receive a second increment of velocity, so that for a single positive charge they emerge finally at a ground potential with a velocity corresponding to 14 MeV, i.e. twice the terminal voltage. The arrangement has most of the advantages of the normal electro-static generator, together with the manifest improvement of having the ion source readily available. Beam currents of 1·5 μA of protons have been obtained and although this is smaller than can be obtained from lower-energy machines it is ample for most experiments of high resolution at 14 MeV. The machine has also been used to provide beams of heavy ions such as ^{12}C and ^{16}O.

The ultimate limiting energy which will be achieved by use of the tandem principle depends not only on the maximum terminal voltage and on the insulating properties of the accelerating tube structure, but also on the nature of the installation. For a single tandem machine a terminal voltage of 10 MV is available commercially and in the negative-positive mode a proton energy of 20 MeV with a spread of ±1 keV is obtainable. If however such a machine is used in the neutral-negative mode, in which a beam of neutral atoms travels from an earth potential source to the terminal and there forms negative ions in an electron-adding canal, the resulting 10 MeV negative beam may be injected into a second accelerator. The final output energy for protons is then 30 MeV (±5 keV) and a current of the order of 1 μA is available. Such an installation, in which many types of heavy ion may also be accelerated, repre-sents an extremely advanced level of technique in nuclear spectroscopy both for its high resolution and easy variability of energy.

8.1.3 Ion sources

Practically all sources of positive ions used in accelerators rely on some form of gaseous discharge for ion production. The only exceptions are sources of lithium ions, which may be obtained thermionically from a heated salt. All discharge sources allow neutral gas molecules to stream into the accelerator vacuum together with the ions, and it is desirable to make the ratio of ion current to gas current as large as possible, particularly in the case of highly rated d.c. accelerating tubes, in which slight deterioration of vacuum may seriously lower the breakdown voltage. Most sources are designed for the production of beams of hydrogen ions, but there is also considerable interest in the use of heavier ions.

The two types of source now most widely used in high voltage d.c. accelera-tors are the PIG (Philips ionization gauge) source, and the radiofrequency source introduced in 1946 by Thonemann. *The PIG source*† (Fig. 8.5(a)) has

† See for example J. D. Gow and J. S. Foster, *Rev. sci. Instrum.*, **24**, 606, 1953.

a cylindrical anode and a twin cathode and the whole assembly is mounted in a chamber within a solenoid providing an axial magnetic field. At a pressure of 10^{-2} mmHg of hydrogen, and with a magnetic field of about 0·05 T, a discharge takes place between the anode and cathode at a potential difference of about 500 volts. Electrons produced near one of the cathodes are accelerated towards the anode but are constrained by the magnetic field to move helically in the direction of the lines of force and therefore pass through the (electric) field-free volume of the anode and out towards the other cathode, where they

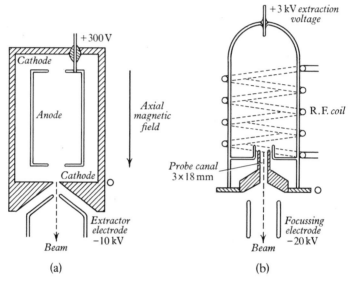

Fig. 8.5 Ion sources. (a) PIG type, the containing vessel forming the cathode. (b) Radiofrequency type.

are decelerated and reflected by the anode–cathode field. This process will continue until electrons have lost so much energy in collisions that they do not emerge from the anode volume and are finally collected. Before this happens, each electron may have created about 10 ion pairs in its long path, and these ions, together with the associated electrons, contribute to the formation of a plasma within the anode and out towards the cathodes. Ions emerge from this plasma under the influence of the field due to an extractor electrode which draws them down a canal into an electrostatic lens system (Fig. 8.5).

In the *radiofrequency ion source*† a plasma is produced in hydrogen or some other gas in a glass vessel (Fig. 8.5(b)). The electric field in the discharge is maintained by surrounding the glass tube by a solenoid excited from an oscillator with an output of about 300 watts at a frequency of 20 MHz. If the discharge volume, with the exception of the outlet to the probe canal (Fig. 8.5(b)), is entirely enclosed by glass, it is found that a high percentage of protons is obtained in the emergent beam; it is known that recombination of

† P. C. Thonemann, J. Moffat, D. Roaf and J. H. Sanders, *Proc. phys. Soc., Lond.*, A, **61**, 483, 1948.

atomic ions to molecular ions or molecules is promoted by metal surfaces. The operation of this widely used source is characterized by the bright red glow of the Balmer series of atomic hydrogen and by positive ion densities in the plasma of about 10^{16} ions m^{-3}. Optimum values for gas pressure and radiofrequency power input are usually established empirically. Notable improvements in radiofrequency ion sources have resulted from concentration of the beam on the canal by means of an auxiliary magnetic field.

Special sources have been developed for the efficient production of the negative ions required in tandem electrostatic generators. The charge state of atoms and ions can be altered by passing the beam through a narrow canal containing gas. For negative ion production, collisions in the vapour of an alkali metal have been found to be very efficient and have even permitted the formation of beams of negative helium ions.† Conversion of negative ions after first stage acceleration to the positive state is achieved in gas or thin foil strippers in the tandem centre terminal and the final beam is of singly charged hydrogen (H^+), doubly charged helium (He^{++}) or multiply charged heavy ions (O^{6+} etc.).

8.2 Linear accelerators

In a linear accelerator, charged particles move down a vacuum tube under the influence of an electric field which either accompanies the particles as a travelling wave, or appears regularly, in correct phase, at a series of electrode gaps. In this way high velocities can be attained without the application of correspondingly high voltages and serious problems of insulation are avoided. Of the two types of accelerating structure used, the travelling wave accelerator, operating in the 3000 MHz frequency band, is particularly appropriate for electrons, while the sequence of gaps, separated by drift tubes, is useful for non-relativistic particles such as protons or heavy ions, and is excited at a lower frequency. All linear accelerators have the advantage over orbital accelerators (Sect. **8.3**) of providing an external beam without difficulty; they have the disadvantage that their accelerating fields are also defocusing, and special measures are necessary to concentrate the ion beam in the non-relativistic region of velocities.

8.2.1 Drift tube accelerators

The earliest accelerators (Wideroe 1928, Sloan and Lawrence 1931, Beams and Snoddy 1934) were of this type. The principle is illustrated in Fig. 8.6. In the Sloan–Lawrence accelerator, operated at approximately 30 MHz, a number of field-free drift tubes of length $L_1, L_2, \ldots, L_n$, separated by small accelerating gaps, were connected alternately to the output terminals of an oscillator of free-space wavelength λ. The length of the drift tubes is such that the field in a gap just reverses in the time that a particle takes to pass from

† B. L. Donnally and G. Thoeming, *Phys. Rev.*, **159**, 87, 1967.

one gap to the next. If the voltage across each gap at the time of passage of the particles is V then the particle energy at entry of the drift tube numbered n (Fig. 8.6) is neV (for initial energy eV) and the particle velocity is (if $v \ll c$)

$$v_n = \sqrt{\frac{2neV}{M}} \tag{8.6}$$

where M is the mass of the particles being accelerated. The frequency of the oscillator is c/λ and, for a time of flight of half a cycle, the length of drift tube n must therefore be

$$L_n = \frac{1}{2} v_n \frac{\lambda}{c} = \frac{1}{2} \beta_n \lambda \tag{8.7}$$

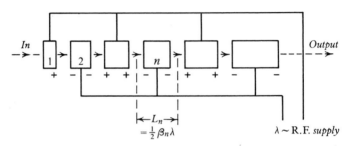

Fig. 8.6 Principle of the Sloan–Lawrence linear accelerator.

For non-relativistic energies it follows from eqs. (8.6) and (8.7) that

$$L_n \propto \sqrt{n} \tag{8.8}$$

It also follows from (8.7) that the size of a linear accelerator for a given output energy is determined jointly by the wavelength λ and the gap voltage V. If the energy gain per gap is held constant, the accelerator size is directly proportional to wavelength. The particles emerge in bunches corresponding to peak field at the gaps and resonance is only possible for particles passing at fields very close to this value.

In the method of excitation used by Beams and Snoddy the drift tube structure was connected to appropriate points of a loaded twin-wire transmission line. A voltage pulse was sent down the line and the loading was arranged so that the pulse travelled at the same speed as the particles, so that the two arrived at successive accelerating gaps in synchronism. The dimensions of the structure are given by (8.7).

The availability of high powers at short wavelengths as a result of radar transmitter development has enabled drift-tube accelerators to offer higher energies in a given length. More important, however, has been the realization that the accelerator need not be designed for exact resonance between the maximum accelerating field and the particles, since the motion can be *phase stable.* This discovery led, soon after its enunciation by McMillan and Veksler in 1945, to the design of an electron synchrotron, a synchrocyclotron and a linear accelerator at Berkeley. In its application to the linear accelerator (Alvarez, 1948) the principle of phase stability may be discussed as indicated in Fig. 8.7(a) which shows the amplitude of the electric field across two successive accelerating gaps as a function of time. The drift tube structure is

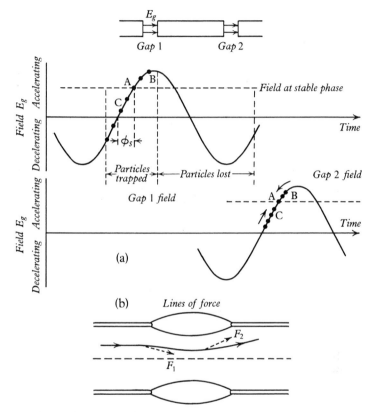

Fig. 8.7 (a) Phase stability in a drift-tube ion accelerator. The dots show the phase angles with respect to the gap field of a bunch of particles of uniform velocity arriving at gap 1. At gap 2, there is increased bunching about the stable phase ϕ_s (45° in this particular example). (b) Radial defocusing of particles passing through a cylindrical gap in a field increasing with time.

arranged so that a particle which crosses a gap with a phase angle ϕ_s (point A) with respect to the alternating field maintains this phase angle unchanged and is in exactly the same phase at the next gap transit. Particles with phase angles greater than ϕ_s (points B) will receive a larger acceleration in the gap, will traverse the drift tube more quickly and will move towards A in phase at the next gap; particles with phase angles less than ϕ_s (points C) will be less accelerated and will also move towards A in phase. Particles corresponding to point A thus have *stable phase*, and if ions of random phase with respect to the accelerating field are injected into the drift tube structure all particles

211

within a certain phase range (Fig. 8.7(a)) will be trapped and will oscillate about the point of stable phase. If the point of stable phase is moved to the peak of the accelerating field, the possibility of oscillation about this point disappears and trapping is much less efficient.

This desirable feature of phase (or axial) stability leads to radial instability because the stable phase point is on the rising part of the voltage wave. Figure 8.7(b) illustrates this; the shape of the lines of force is such that a non-axial particle experiences a focusing force F_1 on entering the gap and a defocusing force F_2 on leaving it. Since the overall field is increasing with time at the stable phase point the defocusing force predominates. In drift tube accelerators radial stability has been restored either by the use of grids across drift tube entrances to eliminate the unwanted curvature of the lines of force or by compensating the defocusing by means of quadrupole magnets within the tubes themselves.

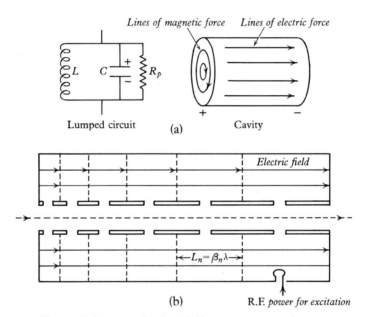

Fig. 8.8 (a) Resonant circuits. (b) Alvarez resonant accelerator.

The principle of the 32 MeV proton accelerator of Alvarez, which has been the pattern for all subsequent heavy particle linear accelerators, is illustrated in Fig. 8.8. Basically the problem in design is the transfer of energy from an electromagnetic field to a charged particle, that is to say just the reverse of the problem of the design of a transmitter valve. At rather low frequencies, as in the Sloan–Lawrence accelerator, the particles pass essentially from one plate of a condenser forming part of an oscillatory circuit to another. At higher frequencies, the LC circuit becomes a cavity resonator, excited by a magnetic loop. This is very similar to the resonance tube in sound, except that it is tuned by radius, rather than length variation. It is also very similar in performance to the LC circuit, and in particular may be represented at the resonant frequency by a shunt resistance R_p. The particles are accelerated by the electric field in the cavity and the energy gain in passage through the cavity, operated at power input W, is proportional to V_{max} where

$$\frac{V_{max}^2}{2R_p} = W \tag{8.9}$$

A single cavity could in principle be used for any specified final energy but it follows from (8.9) that it is more economical to divide the available power W between n similar cavities. The final energy is then proportional to

$$n\sqrt{\frac{W}{n}} = \sqrt{nW}$$

The arrangement of a succession of re-entrant cavities resonant at the same frequency to form an Alvarez accelerator is shown in Fig. 8.8(b). The successive re-entrant tubes become drift tubes and the end walls, which carry no net current, may be removed so that the drift tubes are then arranged in just one long cylindrical cavity. This is excited in its lowest resonant (standing wave) mode in which the lines of electric force are to a good approximation parallel to the axis and the field is uniform along the length. The particles travel through the drift tubes while the field is in the decelerating phase, and traverse one complete section in each cycle. The *section* length thus increases with particle velocity according to the equation

$$L_n = v_n \frac{\lambda}{c} = \beta_n \lambda \tag{8.10}$$

Radial focusing is provided by quadrupole magnets (Sect. **8.4**). The high radiofrequency power required to excite the cavity cannot be supplied continuously and the accelerator is operated from a pulsed transmitter with a duty cycle (on–off ratio) of about 1 per cent. A d.c. injector supplies ions of energy 500–4000 keV to the main accelerator, and an improvement in intensity is sometimes obtained by incorporating a special cavity to 'bunch' the injected beam at approximately the selected stable phase angle of the main radiofrequency field.

Linear accelerators have been built both for protons and for heavier ions; the principle is similar in each case. For a given structure and wavelength (8.10) requires that the velocity increments at each accelerating gap should be the same for each particle. A range of values of Ze/M, corresponding to different heavy ions, in different charge states, can therefore be accelerated by adjusting the radiofrequency voltage so that the gap field E is proportional to M/Z. In practice different structures are usually used for protons and heavy ions. The performance of the Alvarez proton accelerator at Berkeley is shown in Table 8.1.

The main advantages of the linear accelerator as a source of nuclear projectiles are the good collimation, the high homogeneity, the relatively high intensity of the beam (Table 8.1) and the possibility of extension of the machine to extremely high energies. A serious disadvantage for experiments requiring coincidence counting is the sharply bunched nature of the output, which increases the ratio of random to real coincidences in the counter systems. It is also difficult to vary the output energy except by the insertion of absorbing foils and in this and the preceding respect the linear accelerator is much inferior to the electrostatic generator. The major technical limitation in the extension of linear accelerators towards higher energies and higher intensities is in the development (and maintenance) of the necessary high-power oscillators. It is possible however that the situation will be completely transformed by the development of *superconducting* (cryogenic) accelerators in which the power dissipation will be very small.

TABLE 8.1 Performance of linear accelerators

Machine	Berkeley proton accelerator[a]	Stanford electron accelerator (SLAC)[b]
Energy	31·5 MeV	20 GeV
Sectionalization	47 drift tubes	$960 \times 3 \cdot 05$ m coupled sections
Length	12 m	3 km
Frequency	202·5 MHz	2856 MHz
Pulse length	400 μs	2·5 μs
Pulse repetition rate	15 Hz	1–360 Hz
Peak power input	2·3 MW	245×16 MW
Shunt impedance	280 MΩ	53 MΩ m^{-1}
Mean current	1 μA	30 μA
Energy spread	0·5%	1·3%

[a] L. W. Alvarez *et al.*, *Rev. sci. Instrum.*, **26**, 111, 1955.
[b] R. B. Neal, *Physics Today*, **20**, April 1967, p. 27.

8.2.2 *Wave-guide accelerators*

A standing wave pattern in a cavity, such as that developed in the ion accelerator (Sect. 8.2.1) may also be regarded as a superposition of two progressive waves moving in opposite directions. One of these waves travels with the particles and accelerates them. This suggests the feasibility of an equivalent form of accelerator in which bunches of particles are continuously accelerated by a progressive wave in a metal guide. This type of accelerator becomes particularly simple when the particles are moving with relativistic velocity, since the wavelength of the accelerating field is then constant. Wave-guide accelerators are therefore especially suitable for electrons, since these particles have a velocity of $0 \cdot 98c$ for an energy of only 2 MeV, which may easily be provided at injection by an electrostatic accelerator.

In familiar types of wave guide the phase velocity of the travelling wave is always greater than the velocity of light, but it may be reduced by loading the guide with a series of diaphragms, at a spacing giving 3 to 5 per free space wavelength (Fig. 8.9). Electrons are usually injected so that they travel in bunches near, but slightly earlier than the peak field of the travelling wave, as shown in Fig. 8.7. The electrons gain energy (i.e. increase in effective mass) continuously from the wave, rather than discretely at the gaps as in the drift tube accelerator. If the electron velocity is less than the velocity of light, there will be phase stability, as discussed in Sect. 8.2.1, and a wave guide section with variable disc loading ('tapered guide') can therefore be used to accelerate electrons over a certain range of non-relativistic velocities for feeding into a

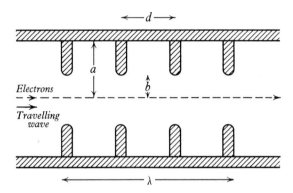

Fig. 8.9 Disc-loaded circular waveguide. The dimensions a, b, can be chosen to reduce the phase velocity of a travelling wave to the velocity of light (or lower).

uniformly loaded guide. The radial defocusing which accompanies axial stability can be simply counteracted for electrons by applying a small axial magnetic field from an external solenoid. When the particle velocity becomes approximately c, the defocusing force vanishes (Ref. 8.10) and no focusing fields are necessary. There is also no definite axial stability and the mechanical construction of the loaded guide must be of high quality in order to ensure that the phase velocity of the wave does not deviate seriously from c.

Travelling-wave electron accelerators usually operate from pulsed magnetrons or klystron amplifiers with a wavelength of about $0 \cdot 1$ m. In the successful and efficient 4 MeV accelerator constructed by Fry, Harvie, Mullett and Walkinshaw and developed commercially as an X-ray generator, the microwave power emerging at the end of the guide is returned to the beginning and recirculated. Electrons from a filament are injected at a voltage of about 50 kV and are accelerated in a guide section of variable loading so that the phase velocity increases to match the particle velocity. An axial magnetic field is provided throughout the 2 metre length. The electron beam itself absorbs about 30 per cent of the power provided by the magnetron oscillator; the remainder is used to build up the fields in the guide. The figure-of-merit of such an accelerator is its 'shunt impedance' (see eq. 8.9) per unit length which determines the ratio of the energy gain by an electron per unit length to the power dissipated in the guide in the same length. The shunt impedance is large when the energy losses in the guide walls are small, since the input power required to build up the accelerating fields is then correspondingly reduced.

Outstanding accelerators of the travelling wave type have been built at Stanford University. A 600 MeV machine,† designed under the general direction of W. W. Hansen, has been responsible for much of our knowledge of nuclear charge distributions (Sect. 11.2.1). In the remarkable 3 km Stanford Linear Accelerator (SLAC) whose performance is shown in Table 8.1, power is fed into a disc-loaded wave guide of high mechanical precision from 245 klystron amplifiers driven in synchronism from a master oscillator. Electrons are injected into a short tapered section in which the phase velocity and longitudinal accelerating field both increase, and are bunched at a phase near the peak field of the travelling wave. Phase oscillations are damped by the effect of the rapidly increasing mass and the bunch moves through the main length of uniformly loaded guide with the velocity of light. No auxiliary focusing is used but despite this a beam is obtained, once perturbing magnetic fields have been neutralized, at the end of the guide. This is possible because of the relativistic shortening of the guide in the electron frame of reference; from the point of view of an observer at rest with respect to the electron the guide appears to be less than 1 m long and there is not much time for lateral divergence. The machine is supplied with microwave power from the klystrons in pulses of 2·5 µs length. About half this time is required to build up the fields in the guide and electrons are injected and accelerated during the remaining part of the pulse.

Electron linear accelerators may also be used to produce and to accelerate positrons. A heavy metal radiator is placed in the electron beam at an intermediate point along the accelerator and the positrons emerging (as a result of direct production or photoproduction via bremsstrahlung) can be accelerated if the phase of the travelling wave is adjusted. The positrons, and electrons which are accelerated simultaneously 180° out of phase, can be used to fill storage rings (Sect. **8.5**) for the study of (e^+e^-) collisions. Two-quantum annihilation of the positrons in flight can be used to furnish a beam of homogeneous photons (cf. Ex. 5.28, p. 658).

The electron linear accelerator, as shown by the figures in Table 8.1, is a machine of poor duty cycle and it is expensive in radiofrequency power although superconducting techniques will improve performance substantially. Upward extension of energy is not limited by the radiative losses inherent in the operation of orbital electron accelerators, which do however offer a greatly improved duty cycle.

8.3 Orbital accelerators 1930–53

If a particle of mass M and charge e moves in a plane perpendicular to the lines of force of a uniform magnetic field of flux density B the radius r of its path is

† M. Chodorow *et al.*, *Rev. sci. Instrum.*, **26**, 134, 1955.

related to its velocity v by the equation

$$Bev = \frac{Mv^2}{r} \tag{8.11}$$

The angular velocity of the particle is

$$\omega = \frac{v}{r} = \frac{eB}{m} \text{ rad s}^{-1} \tag{8.12}$$

and its momentum is

$$p = Mv = Ber \tag{8.13}$$

from (8.11). These equations are true for relativistic velocities, providing that the mass M is not the rest mass M_0 but is related to it by the equation

$$M = \frac{M_0}{\sqrt{1 - v^2/c^2}} \tag{8.14}$$

The *total* energy of the particle moving in the orbit of radius r is, by the formulae of the special theory of relativity,

$$\begin{aligned} E &= Mc^2 \\ &= \sqrt{p^2c^2 + M_0^2 c^4} \\ &= \sqrt{(Ber)^2 c^2 + M_0^2 c^4} \end{aligned} \tag{8.15}$$

from (8.13). The *kinetic* energy T is given by

$$T + M_0 c^2 = E$$

or

$$T(T + 2M_0 c^2) = p^2 c^2 = (Ber)^2 c^2 \tag{8.16}$$

In the non-relativistic approximation $T \ll M_0 c^2$ and then

$$T = \frac{p^2}{2M_0} \tag{8.17}$$

For extreme relativistic velocities $T \gg M_0 c^2$ and

$$E \approx T = Berc \tag{8.18}$$

Equation (8.15) gives the total energy of a particle moving at radius r in an orbital accelerator in which the magnetic flux density is B. If the speed of the particle is uniform, the orbit is a circle. The way in which the final energy is attained will now be described for the different types of accelerator.

8.3.1 The standard cyclotron (fixed field, fixed frequency)

The first accelerator to produce high velocity protons without the use of correspondingly high voltages was the cyclotron described by Lawrence and Edlefsen in 1930 and developed by Lawrence and his collaborators, notably

Livingston† in the succeeding years. The cyclotron was based on the principle of magnetic resonance, which is fundamental to the majority of present-day orbital accelerators and which subsequently also became important in the measurement of nuclear magnetic moments (ch. 4) and of fundamental constants. The principle is illustrated in Fig. 8.10(a) for a particle of mass M and charge e in a magnetic flux density B_0. In the non-relativistic approximation

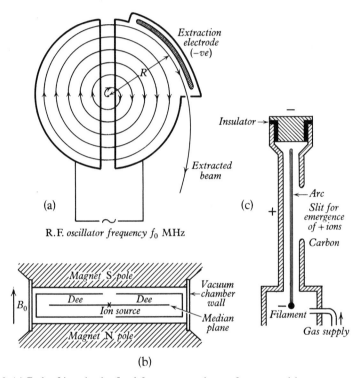

(a)

R.F. oscillator frequency f_0 MHz

(b)

(c)

Fig. 8.10 (a) Path of ions in the fixed frequency cyclotron from central ion source to extracted beam. (b) Vertical section showing dees and walls of vacuum chamber in which they are supported. (c) Ion source construction. The arc is constrained to the vertical direction by the main magnetic field.

($M = M_0 =$ constant) it is clear from (8.12) that the angular velocity is independent of the radius of the orbit. The 'cyclotron frequency' or number of revolutions per second is

$$f_0 = \frac{\omega}{2\pi} = \frac{eB_0}{2\pi M_0} \tag{8.19}$$

$$= 15 \cdot 25 \text{ MHz per tesla for protons.}$$

It follows from (8.19) that if the particle is accelerated while it is moving in the field B_0, then so long as the charge and mass remain constant, the cyclo-

† E. O. Lawrence and M. S. Livingston, *Phys. Rev.*, **40**, 19, 1932.

tron frequency will also be constant; it is this fact that makes the fixed-frequency cyclotron possible.

In the practical application, two dee-shaped electrodes are supported in a vacuum tank (Fig. 8.10(b)) leaving a fairly narrow gap between their opposing edges. The electrodes are excited from an oscillatory circuit at the cyclotron frequency f_0, so that an alternating electric field of this frequency appears across the dee-gap. The field B_0 is applied at right angles to the plane of the dee-electrodes and a source of ions (frequently a hot cathode arc source, Fig. 8.10(c)) is placed centrally in the dee-gap. A positive ion of low velocity emerging from the source will be accelerated towards the negative electrode and will enter the field free space within the electrode, in which it describes a circular arc. This returns it to the dee-gap, where it receives a further acceleration because of the synchronism between the applied voltage and the orbital frequency f_0. In the ideal case this synchronism is maintained and the particle describes a path consisting of semicircles of increasing radius until it reaches the maximum radius R permitted by the dimensions of the electrodes. The kinetic energy is then, according to (8.17) and (8.13),

$$T = \frac{p^2}{2M_0} = \frac{e^2 B_0^2 R^2}{2M_0} \qquad (8.20)$$

For normal operation of the cyclotron this means simply that the particle velocity at the final radius is equal to the circumference of the final orbit divided by the period of the radiofrequency voltage, i.e.

$$v = \frac{2\pi R}{1/f_0} \quad \text{or} \quad f_0 = \frac{v}{2\pi R} \qquad (8.21)$$

In fact, the resonance condition (8.19) is fulfilled exactly at one particular radius only, because of the relativistic increase of mass of the ion as it is accelerated. For a *radially uniform field* of flux density $B = B_0$ the resonant frequency when the particle velocity is v is

$$f = \frac{eB_0}{2\pi M} = f_0\sqrt{1 - v^2/c^2} = f_0\frac{M_0 c^2}{E} \qquad (8.22)$$

from (8.14) and (8.15). It follows that f should fall off as the particle energy increases, i.e. with increasing radius. For a *fixed frequency* $f = f_0$ the resonance condition may be preserved by letting B increase with radius from its low energy value B_0. From (8.12)

$$B = \frac{2\pi M}{e} f_0 = \frac{B_0}{\sqrt{1 - v^2/c^2}} = B_0\frac{E}{M_0 c^2} \qquad (8.23)$$

where B_0 is given by (8.19). For a 10 MeV proton ($M_0 c^2 = 938$ MeV) the necessary increase in B is thus about 1 per cent.

Unfortunately, if the cyclotron magnetic field is azimuthally uniform, it is not possible to allow it to increase radially without introducing a defocusing force. Figure 8.11 shows the lines of force for radially varying fields and it is apparent that an ion beam circulating in the median plane of the field will only experience a force confining it to this plane in case (b) in which there is a radial *decrease* of field. A stabilizing force of this type is necessary to prevent

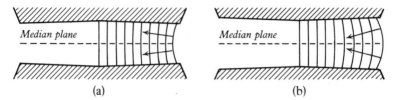

Fig. 8.11 Axial forces in radially varying field in a cyclotron.

the beam diverging and the resonance condition is therefore not met. This cannot be avoided in the simple type of cyclotron so far described and it is necessary to design the magnetic field† so that focusing is provided and to accept the fact that the ions will not be in synchronism with the accelerating voltage wave. In practice the radiofrequency is chosen to be slightly less than the resonant value at the centre so that the ions first of all revolve too rapidly and move ahead of the peak voltage. As they move out, however, they begin to lose phase in the radially decreasing magnetic field until they are about 90° behind (Fig. 8.12). This excursion over the accelerating phase range sets

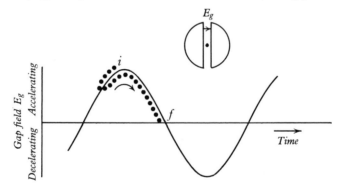

Fig. 8.12 Phase excursion in fixed-frequency cyclotron. The curve shows the time variation of the gap field on one side of the cyclotron. The dots represent the phase excursion of a bunched beam starting (i) at the centre at peak electric field with the radiofrequency lower than the resonant frequency. Subsequent motion takes place in a radially decreasing magnetic field. Acceleration ceases at a phase displacement of 90° (f) but an earlier phase is chosen for extraction.

the practical limit to the energy obtainable from a fixed frequency cyclotron with a given radiofrequency accelerating voltage. Higher energies may be obtained by increasing the radiofrequency voltage and thereby reducing the number of turns necessary for a given energy, but voltage breakdown soon

† The required radial variation is obtained by a process of *shimming*, in which either thin iron rings or pole-face conductors are used to adjust the field.

becomes a limitation. The highest velocity obtainable is about $0.2c$, corresponding to deuterons of 35 MeV, but such energies are now more easily obtainable with azimuthally varying field machines (Sect. 8.3.3).

The particles in a cyclotron perform vertical and horizontal oscillations about the mean orbits with frequencies given by (App. 3)

$$\left.\begin{array}{l} f_v = f\sqrt{n} \\ f_r = f\sqrt{1-n} \end{array}\right\} \tag{8.24}$$

where f is the orbital frequency and n is the field index (Sect. 7.1.3) which is given by

$$n = -\frac{r}{B_z}\left(\frac{\partial B_z}{\partial r}\right)$$

i.e. positive n indicates a radially *decreasing* field and negative n the opposite type of variation. In fixed frequency cyclotrons n is positive and small and the vertical oscillation period is much longer than the period of revolution, while the radial period is nearly equal to the revolution time. The vertical oscillation amplitude decreases as the beam moves outwards into regions of larger n, and the radial amplitude increases; this aids the extraction of the beam.

There is no phase stability in the cyclotron, since although the particles traverse a considerable phase range in the progress of acceleration to extreme radius there is no effect tending to restore them to a stable phase. However, a marked bunching of particles occurs during the first half-turn and this bunching is maintained throughout the acceleration process. Investigations of cyclotron beam waveform show that the final high-velocity beam occupies about 10 per cent of the radiofrequency cycle. This provides ready-made pulsing of the beam and time-of-flight measurements of particle energies (Sect. 7.2.1) can be based on this waveform.

The circulating beam of a cyclotron may be extracted at exteme radius by applying a negative voltage to an insulated deflecting electrode (Fig. 8.10). Extraction takes place in a region of rapidly falling magnetic field and the beam is well focused vertically but divergent horizontally.

Extraction efficiencies of about 50 per cent can be obtained in this way and the beam may then be refined by analysing magnets and concentrated on distant targets by quadrupole magnetic lenses (Sect. **8.4**).

If higher extraction efficiency is required it may be advantageous to accelerate *negative ions*, assuming that a suitable ion source is available. When these ions reach the extraction radius, they may be converted into positive ions by passage through a thin foil[†] and the resulting change of curvature in the magnetic field enables the extracted beam to bend away from the internal circulating beam.

† J. H. Fremlin and V. M. Spiers, *Proc. phys. Soc.*, **68**, 398, 1955.

The fixed-frequency cyclotron has proved a valuable and versatile instrument. It can be made into a variable energy machine by changing the radio-frequency circuit, although it is not as flexible as a d.c. accelerator in this respect. The large beams available are useful for isotope production, while the quality of the external beams can be rendered high enough for precise nuclear reaction experiments. Many different types of ion can be accelerated by appropriate adjustment of magnetic field and if saturation effects result in defocusing fields at some particular radius, these can be corrected by pole-face conductors. The characteristics of a typical cyclotron are given in Table 8.2.

8.3.2 The synchrocyclotron (fixed field, variable frequency)

The limit on the energy obtainable in a fixed frequency cyclotron imposed by the relativistic increase of mass is removed in the synchrocyclotron by the introduction of frequency modulation. This is only possible in practice because of the existence of phase-stable orbits. Consider a bunch of ions crossing the dee-gap of a cyclotron (Fig. 8.13) with exactly the resonant velocity, but with a range of phase angles with respect to the radiofrequency voltage. Resonant ions (A) which cross the gap at zero field circulate indefinitely in this phase (if magnetic field and radiofrequency are maintained constant) and if the field index n is less than 1 the radial and vertical oscilla-

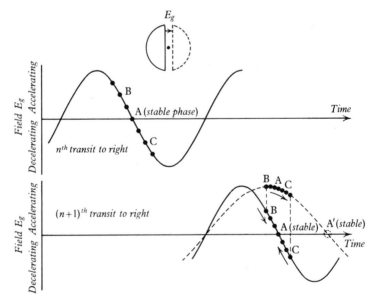

Fig. 8.13 Phase stability in a synchrocyclotron. The dots show the phase angles with respect to the gap field of a bunch of particles of uniform (resonant) velocity at a particular gap transit. At the next transit there is a further bunching towards the phase of zero field if the radiofrequency is constant. If the radiofrequency decreases (dotted curve) the bunch finds itself in an accelerating field.

tions will be stable and the orbit will be in equilibrium. Ions B crossing earlier than A will be accelerated and according to (8.22) their revolution frequency will decrease, so that at the next gap transit they are delayed and move towards A in phase. Similarly ions crossing later than A are decelerated and, from (8.22), acquire an increased revolution frequency so that these also move towards the stable phase (A) at the next transit. It may be shown that a stable *phase oscillation* covering a large range of phase angles about the zero-field point A, is generated. The phenomenon resembles that already discussed for the linear accelerator (Sect. 8.2.1) but it will be noticed that in orbital accelerators the phase stable point on the voltage wave is located on the side where the amplitude is *decreasing* with time. This is a direct consequence of the basic cyclotron equation (8.19) for variable mass; in a linear accelerator, increase of velocity causes a particle to arrive earlier at the gap, but in a cyclotron the opposite is true.

Once the phase stability of equilibrium orbits was realized it became clear that the energy of the particles could be increased indefinitely by decreasing the frequency of the dee-voltage. The dotted curve in the lower part of Fig. 8.13 shows the effect of this; the bunch now experiences an accelerating field and moves towards a new phase stable position A′, in which it would have a larger energy corresponding to the reduced circulation frequency. In practice the frequency variation cycle is designed for the particles to remain in a synchronous accelerating phase, so that (apart from the phase oscillation) there is continuous increase of energy. Such a procedure would have been possible ideally without phase stability but synchronization would have been extremely difficult. With phase stable orbits it is extremely simple, since the frequency change need not, and indeed must not, be rapid; the orbit expands until it reaches a limit set by the maximum radius available or by the onset of vertical amplitude increase. This latter limit occurs when n has increased to 0.2 from the value of about 0.05 which is held for vertical focusing over the main region of acceleration. At this value $f_v = 0.2^{1/2}f$ and $f_r = 0.8^{1/2}f$ so that $f_r = 2f_v$ and energy can be fed from radial into vertical oscillations. This coupling also occurs in cyclotrons but is less serious because the high energy gain per turn results in rapid traversal of the $n = 0.2$ 'resonance'. The energy attained for a given field and fixed radius in a synchrocyclotron may be calculated from (8.15).

The existence of phase stability has the important practical consequence that many more turns may be described in a synchrocyclotron than in a conventional machine. It is therefore possible to use low dee voltages and to dispense with one of the dees, making the radiofrequency power requirement less, and access to the machine much better. The remaining dee may be supported and fed with power symmetrically and the vacant space opposite may be used for targets and for extraction channels. The frequency modulation necessary for operation is imposed by including a variable condenser, driven mechanically, in the resonant circuit. The frequency range necessarily depends on the final energy and on the magnetic field drop and may be calculated from (8.19); from about 20 MHz to about 15 MHz is typical for proton synchrocyclotrons. A linear fall of magnetic field is generally used, with extra iron near the edge to reinforce the field when leakage occurs. The property of phase stability makes 'shimming' for the correct field law much less critical than in the conventional cyclotron. Despite this, large variations in the field law are not tolerable and because of its size the synchrocyclotron cannot readily be converted into a variable energy machine in which the different values of mean field would lead to different degrees of radial fall-off.

Ions are supplied from a source of the type (Fig. 8.10(c)) used in the conventional cyclotron. In typical operation, bunches of ions are accepted at the beginning of each modulation cycle over a period of 100 μs and with a modulation frequency of 100 Hz, so that both the duty cycle of 10^{-2} and the mean

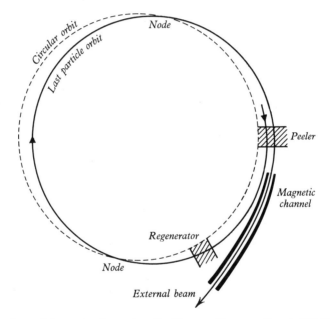

Fig. 8.14 Synchrocyclotron extraction system (Le Couteur, *Proc. roy. Soc.*, A, **232**, 236, 1955).

circulating current of about 1 μA are worse than those of the conventional cyclotron. They may be improved if the mechanical problems of high modulation frequency can be overcome. It is also possible† to improve the duty cycle (at expense of peak current) by using an auxiliary electrode (Cee) with an independent radiofrequency supply to slow down the rate of frequency decrease near extreme radius. Beam extraction by deflection methods tends to be more difficult than in the cyclotron because the orbit spacing is much less (≈0·1 mm instead of perhaps 5 mm) but this difficulty has been overcome by the ingenious peeler-regenerator technique developed by Le Couteur,

† G. Huxtable, P. S. Rogers and F. M. Russell, *Nucl. instrum. Meth.*, **23**, 357, 1963.

Crewe and Gregory, working on the 410 MeV synchrocyclotron of the University of Liverpool. The principle, suggested by Tuck and Teng, is shown in Fig. 8.14 and is essentially to increase the amplitude of radial oscillation of the circulating particles until they enter the mouth of a magnetically screened channel through which they can be conducted away from the main magnetic field. This is mechanically impossible without special measures because of the small spacing of successive orbits. If, however, the particles are allowed to pass first through a region in which the field is made to decrease linearly with radius (peeler) and then through a second region in which the field increases linearly with radius (regenerator) the free radial oscillation can be augmented. Extraction by this process is arranged to start at a radius at which the field index for the main synchrocyclotron field is small; the frequency of radial oscillation $f_r = f\sqrt{1-n}$ is approximately equal to the revolution frequency. A particle whose orbit has an antinode of radial oscillation (outward) between the peeler and regenerator receives an outward impulse in the former and an inward impulse in the latter and the oscillation amplitude builds up into radial instability. It was found that ultimately an increase in radius of 25 mm could be obtained after one turn and this is amply sufficient to bring the beam into a magnetic channel.

In the Liverpool experiments, 10 per cent of the circulating beam of about 1 µA entered the magnetic channel and 2 per cent of the beam was subsequently focused by an auxiliary magnet in a spot of area 1·5 cm^2 nearly 12 m from the cyclotron. Similar results have since been obtained with other machines; the performance in terms of particle flux is at least 1000 times better than that of extraction methods relying on scattering from foils, and has the great virtue of depending only on a static field system. It has been found that the extractor functions also as an energy selector and the homogeneity of the external beam is better than 0·5 per cent.

The successful operation of the proton synchrocyclotron has been of the greatest importance for the development of high energy physics. The principle was tried successfully on the 37-in machine at Berkeley in 1946 and almost immediately the large 184-in machine at Berkeley was also converted to frequency modulation. An astonishing number of important experiments, including artificial meson production, were carried out with the beams of 380 MeV alpha particles and 350 MeV protons which then became available. The only limit to the energy which may be reached by synchrocyclotrons is imposed by the cost of the magnet, and appears to have been reached at a proton energy of about 700 MeV. The performance figures for a well-known machine of this type are given in Table 8.2.

8.3.3 *The isochronous cyclotron (fixed field, fixed frequency)*

The disadvantages of the synchrocyclotron are its reduced beam intensity in comparison with a fixed frequency machine and the complexity of the radiofrequency system. In order to operate a high energy cyclotron at fixed frequency, i.e. as an isochronous machine, it is necessary to let the mean magnetic field increase with radius according to equation (8.23), which together with equation (8.12) gives

$$B = \frac{B_0}{\sqrt{1 - v^2/c^2}} \doteq B_0 \left(1 + \frac{1}{2}\frac{v^2}{c^2}\right) = B_0 \left(1 + \frac{r^2\omega^2}{2c^2}\right)$$

To provide focusing under these conditions the azimuthally varying magnetic field (A.V.F.) originally suggested by L. H. Thomas is used. In such a field, also known as sector-focusing, there are alternate high and low field regions. Figure 8.15 shows a simple example of this type of field. A closed

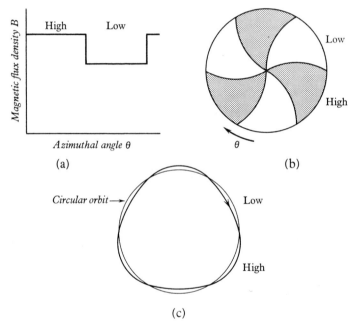

Fig. 8.15 The isochronous cyclotron: (a) Idealized field variation with azimuth at fixed radius. (b) Plan of sectors of spiral ridge cyclotron; simple radial ridges may also be used. (c) Closed orbit in azimuthally varying field.

orbit in such a field is non-circular, as shown in Fig. 8.15(c) and as a particle in such an orbit crosses a sector boundary (either high → low or low → high) *radial* components of velocity arise. The field variation at the boundaries gives rise to *azimuthal* components and the new $\mathbf{v} \times \mathbf{B}$ force is axially focusing in both types of transition region. This new focusing force can be made strong enough to compensate the defocusing (Fig. 8.11(a)) due to the radially increasing field necessary to preserve isochronism. The sector focusing may be improved by using spiral instead of radial sector boundaries. Several such cyclotrons with design energies of the order of 100 MeV for protons are now operating; they avoid the phase excursion shown in Fig. 8.12 and offer undiminished intensity to full radius.

Extraction of a positive ion beam may be achieved by the regenerative method described for the synchrocyclotron; alternatively a negative ion beam and charge reversal extraction (Sect. 8.3.1) may be employed. The efficiency of the latter method offers attractive features for high energy (≈ 500 MeV) cyclotrons to be used for pion production.†

8.3.4 The electron synchrotron (variable field, fixed frequency)

For an electron accelerator the basic cyclotron equation (8.19) may be written, using (8.15)

$$f = \frac{ec^2 B}{2\pi E} = \frac{ec^2 B}{2\pi \sqrt{m_e^2 c^4 + (BeRc)^2}} \tag{8.25}$$

where R is the orbit radius and B the magnetic flux density. In the relativistic region of velocities (say above an energy of 1 MeV), we have $BeRc \gg m_e c^2$ and

$$f \longrightarrow \frac{c}{2\pi R} \tag{8.26}$$

which means simply that the electrons move round a given orbit with constant frequency. If the electron energy E is increased by a radiofrequency electric field the orbit will expand, but if at the same time the magnetic field increases proportionately to E the orbit radius remains unchanged. The electron synchrotron is based on this principle.

Figure 8.16(a), (b) shows the layout of a machine of this type. The magnetic field is annular and in small synchrotrons this leads to the construction shown in the figure in which the magnetic circuit has to be completed outside the vacuum chamber for reasons of space. In larger machines this arrangement, which results in inaccessibility to the vacuum chamber, can be avoided. The machine is operated with an a.c. field of a frequency of the order of 100 Hz and thin magnet laminations are therefore necessary. The vacuum chamber in the smaller synchrotrons is a glass or ceramic 'doughnut' supported in the magnet gap. High resistance metallized surfaces are necessary in order to avoid storage of charge without excessive eddy currents.

Electrons are injected into the synchrotron from an electron gun just inside the chamber wall under the influence of a voltage of 50–100 kV. A surprisingly large fraction of these electrons is trapped into orbits in which acceleration can take place and the initial acceleration to about 2 MeV takes place as in the betatron (Sect. 8.3.6). When the magnetic field has risen to the value corresponding to this energy, saturation in the flux-bars (Fig. 8.16(a)) is arranged to occur and the conditions necessary for betatron acceleration are no longer satisfied. The electron velocity however is then

† Such a machine is under construction in British Columbia, Canada (TRIUMF project; see Ref. 8.13).

0·98*c* and the conditions necessary for synchrotron action are fulfilled. A radiofrequency voltage is applied to a resonator forming part of, or mounted in, the vacuum system, and the particles receive an energy increment per turn just sufficient to keep them near the equilibrium orbit in the rising magnetic field. A stable oscillation about the synchronous phase A develops

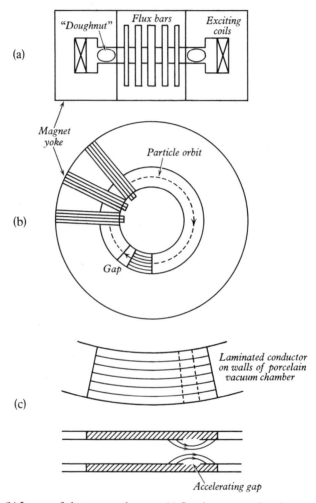

Fig. 8.16 (a), (b) Layout of electron synchrotron. (c) Synchrotron accelerating cavity (Ref. 8.11).

as discussed in the case of the synchrocyclotron (Fig. 8.13); if an electron crosses the resonator too early (B) it receives extra energy and therefore moves to an orbit of slightly larger radius according to (8.15); since it is already moving with velocity *c*, its velocity does not alter but its orbit takes longer to describe and in the next transit of the resonator it will have moved towards the stable phase. Similar arguments apply to the electrons crossing too late (C). Both phase and free oscillations are damped as the magnetic field rises and the final cross-section of the beam can be very small. The final energy is given by (8.18) as

$$T \approx E = eB_{max}Rc \qquad (8.27)$$

where B_{max} is the peak flux density.

The magnet inductance forms part of an oscillatory circuit with a large condenser bank chosen to give the required repetition frequency. The energy oscillates between the inductive and capacitative form and only ohmic losses have to be supplied by the driving circuit. The frequency of the accelerating voltage is constant and the accelerating resonator is effectively a quarter-wave coaxial line, shorted at one end and developing a maximum voltage difference at the open end which forms the resonator gap (Fig. 8.16(c)). The resonator is bent into an arc of radius equal to that of the equilibrium orbit, and would in free space occupy an angle of 90° since the electrons are travelling with the same velocity as electromagnetic waves. The angular extent of the resonator may be reduced by filling the line with dielectric, e.g. the wall of the vacuum chamber.

The main emphasis in electron synchrotron experiments has been on photodisintegration. For this purpose it is unnecessary to extract the electron beam and it is normally allowed to expand outwards at the peak of the magnet cycle and strike a heavy target, from which the main bremsstrahlung spectrum originates. In high energy machines the electrons themselves are interesting for nucleon scattering experiments and fast pulsing magnetic extractors have yielded an efficiency of 50 per cent. This makes the electron synchrotron, with its duty cycle of up to 10 per cent, a serious competitor for counter experiments with the linear accelerator, for which there is no problem of extraction.

The major limitation of electron synchrotrons is one which is common and fundamental to all orbital electron accelerators. The electrons moving in circular orbits of radius R are under radial acceleration and must therefore radiate energy. The total loss per turn increases as $(E/mc^2)^4$ for constant radius where E is the electron energy and this may amount to many hundred eV per turn in the larger synchrotrons. Some compensation for radiative loss occurs automatically as a result of phase stability, but ultimately it becomes impossible to supply the losses and no further acceleration can take place. The synchrotron radiation has a spectrum with maximum intensity in the far ultra-violet and a bluish glow originating from the orbit (and emitted at each point in a cone of angle $\approx mc^2/E$) is easily visible using suitable optical arrangements. This radiation is useful for spectroscopic experiments.

The successful operation of an electron synchrotron was first demonstrated by Goward and Barnes (1947) who converted a 4 MeV betatron into an 8 MeV synchrotron by the addition of an accelerating cavity. Modern machines use the principle of alternating gradient focusing and several now operate in the 5–10 GeV energy range; the performance of one of these is shown in Table 8.2.

8.3.5 *The proton synchrotron (variable field, variable frequency)*

In the electron synchrotron, resonance between the orbital frequency of a charged particle in a rising magnetic field and a radiofrequency accelerating voltage permits acceleration to take place in an orbit of constant radius. This principle may also be applied to protons, and an annular magnet then suffices so that a proton synchrotron is a much more economical machine than the

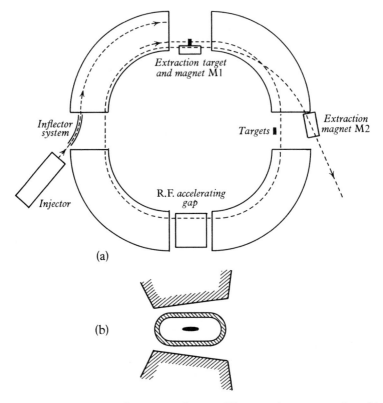

(a)

(b)

Fig. 8.17 (a) Main components of proton synchrotron. The extraction target carries a thin 'lip' at its outer edge. (b) Cross-section of synchrotron magnet gap. In the initial stages of acceleration the beam fills the whole cross-section of the vacuum chamber. The magnetic field decreases with increasing radius.

synchrocyclotron for the same final energy. Until the protons reach relativistic velocities, however, the orbital frequency in the proton synchrotron is energy dependent and provision for variation of accelerating frequency must be made.

Figure 8.17(a) shows the main components of a proton synchrotron. The annular magnet may be a continuous ring or, more conveniently, an arrangement of quadrants with intervening straight sections. The cross-section of the magnet gap is shown in Fig. 8.17(b); the pole tips are shaped to give a field index n of the order of $0 \cdot 7$, for which the radial and vertical oscillations are stable. The magnet is laminated to reduce eddy current losses. The vacuum ring, made of stainless steel, porcelain or plastic, is supported in the magnet gap.

The magnet current is supplied from a d.c. generator which develops an open circuit voltage V. When the contactor in the magnet circuit closes, the initial rate of growth of current i (and flux density B in the magnet gap) is determined by the equation

$$V = L\frac{\mathrm{d}i}{\mathrm{d}t} \tag{8.28}$$

where L is the inductance of the magnet. The growth to full current takes about 1 s and in the subsequent decay of current it may be arranged to return the inductive energy stored in the magnet to the generator, so that only ohmic losses have to be supplied. The cycle is repeated every 2–5 s.

For a given radius, each field corresponds to a definite energy. The synchrotron must be so designed that, accepting the magnetic field laws as a basic datum, the particle energy is increased by a radiofrequency electric field applied along the orbit in just the way required to match the magnetic field. The particles will then stay near the same orbit throughout the acceleration period. The minimum radiofrequency voltage amplitude necessary can be worked out from the magnetic field variation, and the frequency necessary at a given total energy E in order that the particles shall remain in an accelerating phase is given by the equation already used in discussion of the electron synchrotron,

$$f = \frac{ec^2B}{2\pi E} = \frac{ec^2B}{2\pi\sqrt{M_0^2c^4 + (BeRc)^2}} \tag{8.29}$$

In this case $BeRc \approx M_0c^2$ and the frequency increases with E (i.e. with B) in contrast with the case of the synchrocyclotron. The final energy E is given by (8.15).

The acceleration is in principle phase stable although this stability may be impaired if the frequency variation is imprecise. A bunch of particles moves round near the equilibrium orbit with an azimuthal spread determined by the limits of phase stability and particles move from end to end of this bunch as it revolves in consequence of the phase oscillation. The phase range contracts from about 180° to about 90° during acceleration and the superimposed free oscillations are also damped so that the beam shrinks in cross-section.

The accelerating voltage is supplied from a tubular electrode in the vacuum system (Fig. 8.16(b)) or from a magnetic toroid through which the vacuum box passes (Fig. 8.17(a)). In the former case the electrode forms part of an oscillatory circuit which must be kept in tune by inductance variation over the considerable range of frequency necessary for the acceleration. In the latter method the toroid forms effectively the core of a transformer of which the exciting circuit is the primary winding and the equilibrium orbit the secondary.

at Brookhaven, Livingston considered the possibility of building a synchrotron with successive magnetic sectors facing inwards to and outwards from
the centre in order to compensate for the effects of magnetic leakage. This
led to the occurrence of a reversed n value in alternate sectors and Courant
therefore studied the effect of this on orbital stability. It was soon found that
there was net focusing after a pair of positive and negative sectors, and more
important still both n values could be made extremely large with great
advantage. The general theory of this type of focusing was given by Courant,
Livingston and Snyder in 1953; it was subsequently found that it had been
enunciated two years earlier by Christofilos in Athens.

The principle of alternating gradient (AG) focusing is simply that with a
large value of n, alternately positive and negative, there will be strong vertical
focusing and radial defocusing in the sector with positive n and the converse
in the sector with negative n; while overall, for a large range of n values and

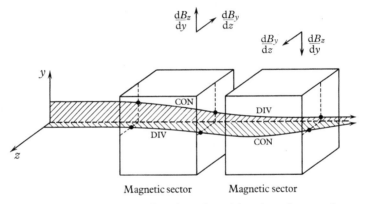

Fig. 8.18 Passage of two rays, representing charged particles, through magnetic sectors with
reversed field gradients. There is net focusing after the two sectors (Ref. 8.2).

for suitable separation of the sectors, there will be net convergence (Fig.
8.18). This result has now become familiar as the basis of magnetic quadrupole lenses (Ref. 8.5); it is also the principle of the converging-diverging
achromatic lens combination in optics. The advantage of AG focusing for
accelerators is that because the free oscillation periods become much shorter,
the amplitudes are also reduced and a considerable spread of momentum can
be accommodated in a small radial space.

The original proposal envisaged the use of n values of about 3600. Unfortunately this was found later to demand an impossible accuracy in magnet
construction in order that perturbations due to misalignment should not set
up instabilities. In practice therefore n values have been reduced to about
300, but there is still an enormous saving in magnet costs for a given energy
and the large proton accelerators now operating at energies above 20 GeV
are AG machines. The main properties of the CERN proton synchrotron,
which came into operation at Geneva on 24 November 1959, are included in

Table 8.2. A Russian machine of 70 GeV is operating at Serpukhov and a 200 GeV accelerator, to be extended to 500 GeV, is under construction at Batavia, Illinois, U.S.A; a 150–300 GeV machine is being built at the CERN laboratory.

TABLE 8.2 Performance of orbital accelerators

Machine	Particle energy (MeV)	Maximum stable orbit diameter (m)	Orbital frequency (MHz)	Pulse repetition rate (Hz)	Output current or particles per pulse	Magnet weight (10^3 kg)
Fixed Frequency cyclotron (Birmingham)[a]	20(d)	1·5	10·3	C.W.	500 μA	254
Synchrocyclotron (Liverpool)[b]	410(p)	3·9	29·2 – 18·9	110	1 μA	1670
AG electron synchrotron[c]	4000(e)	70·19	1·36	50	10 μA	366
CG proton synchrotron[d]	7000(p)	23·63	355 kHz – 2 MHz	0·5	3×10^{12}	6600
AG proton synchrotron (CERN)[e]	27000(p)	200	—	0·3	2×10^{12}	3450

All accelerators operating with radiofrequency fields exhibit a fine time structure of the beam on a scale determined by the radiofrequency.

NOTES p = protons, d = deuterons, e = electrons

[a] *Nature*, **169**, 476, 1952
[b] M. J. Moore, *Nature*, **175**, 1012, 1955
[c] NINA, Daresbury Nuclear Physics Laboratory Report DNPL 1 1967
[d] NIMROD, Rutherford High Energy Laboratory Report NIRL/R/44 1965
[e] J. B. Adams, *Nature*, **185**, 568, 1960.

The characteristics of strong focusing accelerators have been intensively studied theoretically. One of the interesting features of the stability in phase of the circulating particles is that there is a critical energy E_c at which the position of stable phase moves from the rising voltage part of the radiofrequency cycle to the falling part. The former mode of acceleration is similar to that in the linear accelerator (Fig. 8.7) but the associated electric defocusing is a small effect in comparison with the strong magnetic focusing; the latter mode is as in the constant gradient (CG) synchrotron or synchrocyclotron (Fig. 8.13).

8.5 Survey and future prospects

From 1930 onwards it seems that at each point as a particular accelerator appeared to have reached its limiting performance, either fundamentally or for economic reasons, a new principle has emerged permitting further progress. An excellent account of the sequence of advances is given in Ref. 8.6.

Electrostatic accelerators are numerous and popular and their maximum energy will be extended to 30–40 MeV by the tandem principle. They are pre-eminently suitable for the study of individual nuclear energy levels and the general problem of the structure of complex nuclei. The apparent limitation of output energy associated with d.c. machines was removed by the evolution of resonance acceleration, which was already being tried at the time of Cockcroft and Walton's original experiment. The fixed frequency cyclotron rapidly advanced to an energy of 25 MeV and a current of the order of a milliampere. Such machines are used extensively for studying the mechanisms of nuclear reactions, for which the highest resolution is not necessary. The cyclotron energy was limited by the relativistic mass increase, but the means of avoiding this was soon forthcoming in the azimuthal variations of field proposed by Thomas (1938); the idea of alternating gradient focusing was contained in this proposal but was not explicitly developed. The Thomas-type field was not introduced into post-war accelerator construction because of the discovery of phase stability. This basic advance enabled the focusing problem to be separated from the problem of acceleration and led to the development of the synchrocyclotron as the first proton accelerator of the 100 MeV region, as well as the evolution of the present generation of linear accelerators and synchrotrons. The limit to synchrocyclotron size is set by magnet cost, which is proportional to $E^{3/2}$ in the non-relativistic region, and in practice it seems unlikely that energies exceeding 750 MeV with currents of the order of 1 μA will be possible for economic reasons. These machines have had as their especial province the study of nucleon scattering and meson production.

The main disadvantage of the synchrocyclotron, particularly for accurate counter experiments, is the relatively low mean intensity of the beam (Table 8.2) and its sharply pulsed nature. The synchrocyclotron is being replaced by the isochronous cyclotron, which has assumed an important role not only in nuclear structure work but as a high intensity meson source. The energy of these machines will probably be limited to about 750 MeV for economic reasons, as for synchrocyclotrons. A possible competitive machine is the superconducting linear accelerator which could essentially have a duty cycle similar to that of the cyclotron. Even in non-superconducting form Alvarez-type linear accelerators have contributed significantly to nuclear reaction studies.

After the evolution of the synchrocyclotrons the next step in progress towards higher energies was the construction of the constant gradient proton synchrotron, with a probable limit of 15 GeV and a current which might reach 1 μA. The discovery of the antinucleon and the exploitation of K-meson and hyperon properties are triumphs for these machines. Hardly had the proton synchrotron limit been brought in view when it was very considerably extended by the discovery of AG focusing on which the 20–70 GeV accelerators of the present and higher energy machines of the future are based. Electron synchrotrons of this type are probably limited by radiation losses, for reasonable radial dimensions, to about 20 GeV, but should up to this limit be strongly competitive with non-superconducting linear accelerators of the Stanford type for which the consumption of radiofrequency power is very large.

The limit to the energies obtainable by accelerators is hard to define, but some there must be, whether physical or economic, and it is probable that 1000 GeV is further than accelerator physics can advance under the principles known at present, which may be taken to include the use of superconducting magnets, and of separated function magnetic structures. In these, independent magnets are used for bending and for focusing, with some advantages in flexibility of machine design. There is in fact one new principle of some promise, due to Veksler, which is under development. This is to accelerate protons entrapped in a ring of electrons which provides an attractive potential well. In this *electron ring accelerator*, sometimes known as the collective ion accelerator, the protons move at the same velocity as the electron bunch and very high proton energy gains are in principle possible because of the proton/electron mass ratio. The main problem is to stabilise the electron bunch, containing the protons, so that it can be accelerated in a linear accelerator. This is achieved by using a longitudinal magnetic field as a guide field and by making the electrons relativistic so that the repulsive Coulomb forces are compensated by attractive magnetic forces due to the electron motion. This principle, although attractive, does not yet furnish a practical accelerator and a demand for higher energies will for many years to come probably have to be met by AG synchrotrons. An unfavourable return in centre-of-mass energy in a moving proton/stationary proton collision for increasing accelerator expenditure has not escaped attention,† and for some purposes it is possible to envisage colliding beam experiments. Magnetic storage rings‡ with synchrotron type structures and *rf* cavities in which beams of electrons and positrons injected from a synchrotron or linear accelerator circulate in opposite directions have already been tried successfully and offer an enormous advantage in available energy, though with low intensity. The CERN Intersecting Storage Ring (ISR) system when operating with two oppositely circulating beams of 28 GeV protons in concentric but interlaced orbits yields a c.m. energy equivalent to that provided by a synchrotron of energy 1600 GeV. Such technical achievements offer exciting prospects but whatever success is achieved, the effective energy obtained is still likely to be low from the point of view of the cosmic-ray physicist. Ultra-high energy particles, as far as can be foreseen, will always have to be accelerated in the cosmos, and studied perhaps at the top of the earth's atmosphere.

---

† The total energy in the c.m. system is $\sqrt{2m_p c^2 (E + m_p c^2)}$ where E is the total laboratory energy of the incident proton.

‡ G. K. O'Neill, *Scientific American*, **215**, No. 5, 54, 1966.

References

1. FLUEGGE, S. (ed.), 'Nuclear Instrumentation I', *Encyclopedia of Physics*, Vol. **44**, Springer, 1959.
2. LIVINGSTON, M. S. (ed.), *The Development of High Energy Accelerators*, Dover Publications, 1966.
3. ROSENBLATT, J. *Particle Acceleration*, Methuen, 1968.
4. PERSICO, E., FERRARI, E. and SEGRÈ, S. E. *Principles of Particle Accelerators*, Benjamin, 1968.
5. NEWTH, J. A. 'Devices for the detection of energetic particles', *Rep. progr. Phys.*, **27**, 93, 1964.
6. JUDD, D. L. 'Conceptual Advances in Accelerators', *Ann. rev. nucl. Sci.*, **8**, 181, 1958.
7. VAN DE GRAAFF, R. J. 'Tandem Electrostatic Generators', *Nucl. instrum. Meth.*, **8**, 195, 1960.
8. THONEMANN, P. C. 'The Production of Intense Ion Beams', *Progr. nucl. Phys.*, **3**, 219, 1953.
9. ROSE, P. H. and GALEJS, A. 'Production and Acceleration of Ion Beams in Tandem Accelerators', *Progr. nucl. tech. Instrum.*, **2**, 1, 1967.
10. FRY, D. W. and WALKINSHAW, W. 'Linear Accelerators', *Rep. progr. Phys.*, **12**, 102, 1949.
11. FREMLIN, J. H. and GOODEN, J. S. 'Cyclic Accelerators', *Rep. progr. Phys.*, **13**, 295, 1950.
12. PICKAVANCE, T. G. 'Cyclotrons', *Progr. nucl. Phys.*, **1**, 1, 1950.
13. RICHARDSON, J. R. 'Sector focused cyclotrons', *Progr. nucl. tech. Instrum.*, **1**, 1, 1965.
14. PICKAVANCE, T. G. 'Focusing in High Energy Accelerators', *Progr. nucl. Phys.*, **4**, 142, 1955.
15. REICH, K. H. 'Beam Extraction Techniques for Synchrotrons', *Progr. nucl. tech. Instrum.*, **2**, 161, 1967.
16. ADAMS, J. B. 'Rethinking the 300 GeV machine', *Science Jnl.*, **6**, No. 9, page 58, Sept. 1970.

C Static properties of nuclei

9 Nuclear models (I), the nuclear ground state and the nuclear level spectrum

A nucleus, like other quantum mechanical systems, possesses a set of characteristic energies, or excited states. Of these the most stable state, or *ground state* is that in which nuclei are normally found, and it is for the ground state that the most extensive survey of nuclear properties has been made.

The purpose of a nuclear model is to provide a practical means of predicting nuclear properties. Ideally the calculation of nuclear structures should be based on a knowledge of the law of force between nuclear constituents and on techniques for applying this law in a complex many-body system. Progress has indeed been made in this direction and realistic forces based on two-body data (ch. 18) are increasingly being used in nuclear structure calculations. The nuclear many-body problem is distinct because the number of particles is limited, and the addition of another particle can profoundly affect the properties of the system. A comprehensive solution of this problem, with a complex and partially known interaction, will clearly be very difficult and the simplifications offered by a suitable basic model are still vital.

In the present chapter we shall mainly consider the nuclear shell model, which has been outstandingly successful both in predicting the static properties of the ground state (chs. 10, 11) and in indicating the broad structure of the nuclear level spectrum. In Chapter 12 we shall see how structure models of this type can be extended to give quite detailed predictions of the level spectrum, and how other features of nuclear behaviour such as collective motion can be introduced. We shall postpone, except for a brief preliminary mention, all consideration of models primarily designed to describe nuclear reactions. These are important for the determination of the characteristics of nuclear levels, and some unification with the structure type of model has in fact been possible.

9.1 The first nuclear models

One early nuclear model postulated an interacting assembly of electrons and protons as reasonable to describe a nucleus capable of spontaneous emission of both α- and β-particles. Such a structure is consistent with the observation (ch. 10) that nuclear masses are approximately integral multiples of the mass of the proton. This model fails for many reasons, chief among which are that:

(a) the observed spins and statistics of many nuclei disagree with the model. Even-mass nuclei of odd charge, e.g. ^{2}H, ^{14}N obey Bose statistics and have integral spin, which cannot be explained on the proton–electron model,

(b) the observed nuclear magnetic moments are of the order of the nuclear magneton $eh/4\pi m_p$ rather than the Bohr (electron) magneton $eh/4\pi m_e$,

(c) an electron confined within nuclear dimensions $\approx R$ would, in accordance with the uncertainty principle, have an uncertainty of momentum of about $\hbar/R$. The corresponding kinetic energy $\hbar c/R$ is ≈ 20 MeV for $R \approx 10^{-14}$ m and a potential energy considerably greater than this would be necessary to contain electrons within a nucleus. There is no evidence for such a strong attraction between protons and electrons.

(d) it is no longer necessary, in view of the success of Pauli's neutrino hypothesis and Fermi's theory of beta decay, to postulate the emission of pre-existing electrons in β-radioactivity.

The discovery of the neutron in 1932 (Sect. 14.1.3) made it possible to avoid the difficulties of the electron–proton model and a neutron-proton model is the basis of modern theories of nuclear structure. The first task of the neutron–proton model was to give some account of nuclear binding energies and for this purpose several specific versions of the model have been widely discussed. These are:

(a) *The Fermi gas model*, in which neutrons and protons, existing independently but attracting in pairs are confined within a cube of volume equal to the nuclear volume and are described by plane waves. This is exactly as in the electron theory of solids, and as in that theory, the total energy of such a system may easily be calculated. For a nuclear dimension R and for A particles of mass M the de Broglie wavelength must be $\lambda \approx R/A^{1/3}$ and the corresponding momentum p is then given by

$$p = \frac{hA^{1/3}}{R}$$

The kinetic energy per particle will be proportional to $p^2/2M$, i.e. to $A^{2/3}/MR^2$ and for the whole nucleus to $A^{5/3}/MR^2$. The potential energy on the other hand is proportional to the number of interacting pairs, i.e. to $\frac{1}{2}A(A-1)$ and for A large enough the potential energy is the main term. The nucleus would therefore collapse according to this model and properties of the neutron–proton interaction (ch. 18) must be postulated to prevent this. One such property usually assumed is *short range*, which limits the number of interacting pairs. The Fermi gas model is particularly useful in describing collision phenomena in high-energy nuclear processes because the idea of occupied states of low momentum explains the long mean free path that is observed, and the maximum allowed momentum determines the energy spread of particles after collision. The model obviously applies best to heavy nuclei; in such nuclei the neutron number always exceeds the proton number because of the need for extra neutrons to compensate the disruptive effect of the Coulomb force between protons.

(*b*) *The liquid droplet model*, (N. Bohr and F. Kalckar) which essentially starts from the idea of continuous nuclear matter. This model concentrates on the strong interaction between the neutrons and protons of a many-body system and predicts the existence of the many closely spaced levels actually found by experiment. Bohr's compound-nucleus theory of nuclear reactions (ch. 15) receives a natural interpretation by this type of model. The outstanding success of this concept shows that strong interaction is a feature which must appear in some way in more general theories of wider application. The liquid drop model may also be adapted to predict nuclear binding energies and forms the basis of the semi-empirical mass formula; like the Fermi gas model it is best applicable to heavy nuclei, and predicts no discontinuities of nuclear binding energy with N or Z.

(*c*) *The quasi-atomic or shell model*, which antedated the Bohr droplet model and offered, particularly for the lighter nuclei, the possibility of simple prediction of many nuclear properties in addition to binding energy. It differs from the Fermi gas model in using instead of plane waves the wave functions of a particle moving in a spherically symmetrical potential, and because of the significance of angular momentum in this problem, the model offers the possibility of a spectroscopic classification of nuclear properties and interpretation of observed periodicities. The existence of such periodicities is considered, together with the present development of the single-particle shell model, in Sections **9.2–4**.

9.2 Empirical evidence for the regularity of nuclear properties

Periodicity in atomic properties such as valency and ionization potential has been known for well over a century and forms the basis of the familiar periodic classification of the elements due to Mendeléev. It receives a natural interpretation in terms of the filling of the successive levels of a screened Coulomb

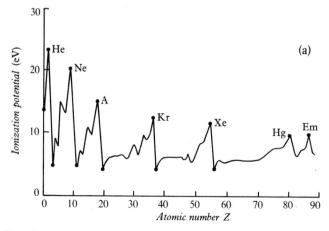

Fig. 9.1 (a) Periodicity of an atomic property—the first ionization potential—as a function of atomic number Z.

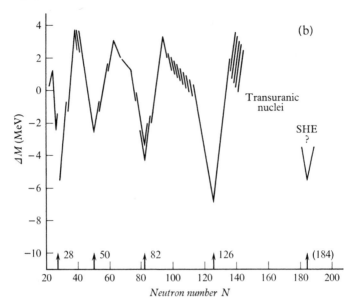

Fig. 9.1 (b) Periodicity of a nuclear property: ΔM is the observed nuclidic mass less the mass predicted by a smoothly varying mass formula (Chapter 10). In the detailed figure from which this is derived (Kummel *et al.*, *Nuclear Physics*, **81**, 129, 1966) ΔM is plotted against neutron number N for the complete range of atomic numbers; here only the general trend is shown. The region of superheavy elements, not discussed by Kummel, is speculative.

potential by electrons whose number in a given sub-level is limited to two by the Pauli exclusion principle. A typical graph of an atomic property (ionization potential) as a function of atomic number (Z) is given in Fig. 9.1(a).

It was early proposed from a consideration of nuclear binding energies and abundances (Elsasser, Guggenheimer and others, 1934) that a similar periodicity should exist in nuclear properties. Speaking before the Chemical Society on April 19, 1934, the centenary of the birth of Mendeléev, Rutherford concluded 'It may be that a Mendeléev of the future may address the Fellows of this Society on the "Natural Order of Atomic Nuclei" and history may repeat itself.' but it was not until the late 1940's, when sufficient nuclear data had accumulated, that the detailed nature of the nuclear periodicity began to emerge. Figure 9.1(b) shows a basic nuclear property (actual nuclear mass minus a smoothly varying predicted mass for a number of isotopic sequences) as a function of neutron number N; evidence for regularity is obvious.

Historically the development of the shell model in its modern form was retarded because it seemed difficult to envisage any nuclear structure of strongly interacting particles which could provide a strong central potential of the sort known in the atom. The feature of strong interaction was fundamental to the successful liquid droplet model (Sect. **9.1**) and this model seemed to exclude the possibility of the long mean free path for a nucleon in nuclear matter required by the shell model. Recently it has been realized that the Pauli principle operates to lengthen mean free paths because collisions in nuclear matter cannot take place if they lead to states of motion which are already occupied by other nucleons. It has also been found experimentally that nuclear reactions at energies higher than those considered by Niels Bohr do exhibit resonance phenomena of exactly the sort predicted by the simple one body type of interaction with a potential well originally suggested by Bethe as a reaction mechanism and embodied in the shell model. Another objection, that nuclear binding energies were badly predicted by shell model calculations, has also been removed by recent advances in our understanding of the nature of the shell-model potential. This potential is in fact now regarded as an important intrinsic property of nuclear matter.

We now list for convenience the main nuclear properties on which the idea of the shell model rests; most of these will be discussed in more detail in subsequent chapters. They are:

(a) Discontinuities of nuclear mass (Fig. 9.1(b)).
(b) Discontinuities of nucleon binding energy, especially the neutron binding energy as measured by the (n, γ) or the (d, p) reaction (Sect. **10.3**).
(c) Anomalies in both total abundance and relative abundances of isotopes and isotones as a function of nucleon number, N or Z (Sect. **10.5**).
(d) Excitation energy of the first excited state of nuclei, particularly even N–even Z nuclei (Fig. 9.9).
(e) Energies of α- and β-decay (ch. 16).
(f) Nuclear reaction cross-sections and level densities.

Properties (b)–(f) are consequences of (a), i.e. of the existence of irregularities due to shell closures in the mass formula based on the liquid drop model (Sect. **10.3**). In addition there are other properties connected with the regular sequence of available orbits for nucleons, which is also a consequence of shell filling. These are:

(g) the ground state spins of both stable and unstable nuclei,
(h) the parity of nuclear ground states,
(i) the magnetic dipole and to some extent electric quadrupole moments of nuclear states,
(j) the comparative half-lives of β-emitters,
(k) nuclear isomerism.

The properties of (a)–(f) suggest that nuclei containing 2, 8, 20, 50, 82 or 126 neutrons or protons are particularly stable. Some phenomena also suggest the addition of the number 28. Before considering how the shell model suggested by Mayer and by Haxel, Jensen and Suess is able to predict these 'magic numbers', we examine some direct evidence for the model that has become available in recent years.

9.3 Direct evidence for shell structure

The ideas underlying the shell model have received striking confirmation by the direct observation of proton shells through the (p, 2p) reaction. In this process protons of energy 50–400 MeV are used to bombard a nucleus, e.g. ^{12}C, producing a *knock-out* direct reaction

$$^{12}\text{C} + \text{p} \longrightarrow \ ^{11}\text{B} + 2\text{p}$$

The outgoing proton energies are measured in coincidence and geometrical conditions can be imposed which minimize energy loss to the recoil nucleus ^{11}B. The overall loss of energy in the reaction then measures the binding of

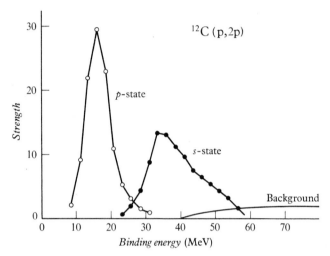

Fig. 9.2 Strength of the *p*-state and *s*-state contributions to the ^{12}C(p, 2p) reaction (A. N. James *et al.*, *Nucl. Phys.* **A133**, 89, 1969).

the knocked-out proton in the ^{12}C nucleus. Figure 9.2 shows the result obtained for this nucleus, which is expected to have protons in both *s* and *p* states according to any quasi-atomic model. Similar results have been obtained for nuclei as heavy as ^{120}Sn, in which a *g*-state may be seen. By altering the angle with respect to the incident beam at which the two protons are observed, the momentum distribution of the struck proton may be explored, and the angular momentum of the corresponding bound state may be inferred. Similar experiments have been made with other particles, including electrons.

9.4 The single-particle shell model

The starting point of all shell models is the solution of the Schrödinger equation for a particle moving in a spherically symmetrical *central* field of force, i.e. in a field of force in which the potential energy V of the particle with respect to the centre is a function $V(r)$ of its radial distance r only. The origin of the field of force in a nucleus is not essential to the argument; it is sufficient that it exists. This problem has been discussed in general terms in Sect. **3.1**; the solutions to the three-dimensional Schrödinger equation may be written

$$\Psi = R_{nl}(r)Y_l^m(\theta,\phi) \tag{9.1}$$

where $Y_l^m(\theta, \phi)$ are the spherical harmonic functions discussed in Appendix 1. The principal, azimuthal and magnetic quantum numbers n, l, m have also been discussed in Sect. **3.1**. The azimuthal quantum number l of a single particle state determines the *parity* of the wave function as $(-1)^l$. The quantum numbers n, l are related to the number of nodes in Ψ as a function of r, θ and ϕ; there are $(n-1)$ altogether of which l are in the angular co-ordinates and $(n-l-1)$ $(=n_r)$ in the radial variation. The number $n_r(\geqslant 0)$ is known as the *radial quantum number*. We will occasionally denote the principal quantum number as n_{sp} to emphasize its role in atomic spectroscopy.

The spherical harmonic functions are of general application, but the radial functions $R_{nl}(r)$ can only be obtained if $V(r)$ is specified. For $V(r) = -Ze^2/4\pi\varepsilon_0 r$ we have the case of the hydrogen atom and, allowing for electron spin, atomic shells containing $2\sum_0^{n-1}(2l+1) = 2n^2$ electrons arise. These consecutive shells contain

$$2, 8, 18, 32, 50, \dots$$

electrons if they fill regularly. In fact the higher shells begin filling before the lower ones are complete and the atomic 'magic' numbers correspond in some cases to the completion of subshells. From Fig. 9.1(a), the main atomic shell effects occur at the atomic numbers

$$2, 10, 18, 36, 54, 86, \dots$$

These are of course quite different from the nuclear magic numbers.

The hydrogen-like degeneracy of the atomic levels with respect to l is removed when the central field becomes non-Coulomb. In the alkali atoms for instance the potential has the 'screened' Coulomb form $-(Ze^2/4\pi\varepsilon_0 r)e^{-Kr}$ and since this falls off with r more rapidly than in the unscreened case, the wave functions with high l-values, for which the electrons are further away from the nucleus, are less strongly bound.

In the case of a nucleon of mass M moving in a static, spherically symmetric potential field $V(r)$ with angular momentum $l\hbar$, the radial part of the Schrödinger equation may be written

$$\frac{d^2}{dr^2}(rR_{nl}) + \frac{2M}{\hbar^2}\left\{E_{nl} - V(r) - \frac{l(l+1)\hbar^2}{2Mr^2}\right\}(rR_{nl}) = 0 \tag{9.2}$$

where the suffix nl envisages a dependence of both the eigenfunction R and the energy eigenvalue E on principal quantum number and angular momentum. The radial shape of the nuclear field is now well known from scattering experiments (ch. 11) but the form of $V(r)$ is not suitable for simple solution of (9.2). Since, however, the shell model is not primarily concerned with total binding energies, but only with levels corresponding to the states of motion of nucleons, it is sufficient to consider only simple forms of potential function.

The range of the potential is taken to be short so that the energy levels are well spaced, as shown in Fig. 3.1(a). We then find the following results:

(a) *The square well potential* (Fig. 9.4(a)). Qualitatively the effect of an attractive force of this sort is just opposite to that of a screened Coulomb potential. The square well has no singularity at the origin, and the *s*-states, which have wave functions finite at the origin, are less strongly disturbed. If a given *principal* quantum number, as used in the case of the hydrogen atom, is considered, then states of high *l*-value are found to be more strongly bound than those of lower *l*.

The radial wave functions in this case are analytically of the form

$$R_{nl}(r) = \frac{A}{\sqrt{Kr}} \mathcal{J}_{l+1/2}(KR) \tag{9.3}$$

where A is a constant, $\mathcal{J}_{l+1/2}$ is a Bessel function, and the wave number K is defined by the equation

$$K^2 = \frac{2M}{\hbar^2}(E_{nl} - V) \tag{9.4}$$

where E_{nl} is the total (negative) energy and $V(=-U)$ is the well depth. It will be convenient to measure energies from the bottom of the well and then

$$K^2 = \frac{2M}{\hbar^2} E'_{nl} \tag{9.4a}$$

where E'_{nl} is positive. It will also be convenient at this point to depart from conventional spectroscopic notation in which $n(=n_{sp})$ is related to the number

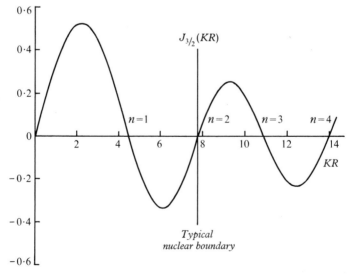

Fig. 9.3 Graph of the Bessel function $\mathcal{J}_{3/2}(x)$ against x, where $x = KR$.

of nodes in the complete wave function; instead we shall use $n = n_{sp} - l$, which is the number of radial nodes n_r (not counting the origin or potential boundary) plus one $(n_r + 1)$. This will convert a sequence of states such as

$$1s \ 2p \ 2s \ 3d \ 3p \ 3s \ ...(n_{sp}l)$$

into $\qquad\qquad 1s \ 1p \ 2s \ 1d \ 2p \ 3s \ ...(n_{sp} - l, l)$

in which n now indicates the serial order of appearance of a given angular momentum $l\hbar$.

For the square well potential permitted values of K are selected by a boundary condition. In the simple case of a well of infinite depth the wave function has to vanish at the nuclear boundary $r = R$, i.e.

$$R_{nl}(R) = 0 \tag{9.5}$$

with n as defined above. Figure 9.3 is a graph of a typical Bessel function $\mathscr{J}_{l+1/2}(KR)$ for $l = 1$. The zeros of the Bessel function occur at definite numerical values X_{nl} of the argument and with the present definition of $n \ (= n_r + 1)$ this number may be used to enumerate the zeros as shown in the figure. For a given nuclear radius R it will be possible to satisfy the boundary condition (9.5) by choosing K so that 1, 2, ... oscillations of the Bessel function take place within the distance R. The more oscillations that are included, the larger must be K and the larger is the energy according to (9.4). The level energies are given by putting

$$KR = X_{nl} \tag{9.6}$$

i.e.

$$E'_{nl} = \frac{K^2 \hbar^2}{2M} = \frac{X_{nl}^2 \hbar^2}{2MR^2} \tag{9.6a}$$

For $R = 8 \times 10^{-15}$ m the quantity $\hbar^2/2MR^2$ is 0·34 MeV.

The X_{nl} values defining the first few states of the square-well potential are shown in Table 9.1. It will be seen that for a given n the energies increase (i.e. binding decreases) with l. If however we consider the variation for a given principal quantum number $n_{sp} = (n + l)$, then the binding increases with l. The effect of l-variation is also seen to be comparable with that of n. This is a

TABLE 9.1 Single particle states of the infinite square well

	X_{nl}		
	---	---	---
	1st zero	2nd zero	3rd zero
l	$n = 1$	$n = 2$	$n = 3$
0	3·14	6·28	9·42
1	4·49	7·72	10·90
2	5·76	9·09	12·32
3	6·98	10·41	

general feature of short range potentials, as contrasted with the long range Coulomb potential of the hydrogen atom, and is the reason why it is useful to specify states by radial and azimuthal quantum numbers independently.

The levels (9.6a) are shown in Fig. 9.4(a); it will be seen that the order from the bottom of the well is

$$1s \ 1p \ 1d \ 2s \ 1f \ 2p \ 1g \ 2d \ 1h \ 3s \ \dots$$

where $1f$, for instance, means the first level of orbital momentum 3, i.e. the first zero in the Bessel function $\mathcal{J}_{7/2}$. If 2 nucleons occupy each of these states (i.e. with opposite spins) the occupation numbers $2(2l+1)$ predict shell closures at total particle numbers (neutrons *or* protons):

$$2, 8, 18, 20, 34, 40, 58, \dots$$

These are not the nuclear magic numbers. The situation is not altered by calculating the levels for a finite rather than infinite well since the level *order* is found to be the same although the excitations alter.

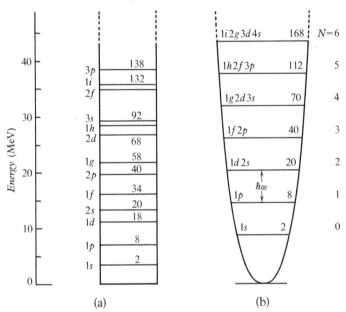

Fig. 9.4 Level sequence for nucleons in a potential well, showing spectroscopic classification of levels and total number of nucleons which may be accommodated up to the indicated excitation. (a) Infinite square well potential, radius 8×10^{-15} m. (b) Oscillator potential, showing uniform spacing of levels. The levels are of even (odd) parity when the oscillator number N is even (odd).

(b) *The harmonic oscillator potential* (Fig. 9.4(b)). For this well the potential energy may be written

$$V(r) = -U + \tfrac{1}{2}M\omega^2 r^2 \tag{9.7}$$

where ω is the frequency of the corresponding simple harmonic oscillations of the particle. The solution of (9.2) for an infinite well may be expressed in terms of Hermite polynomials (for Cartesian coordinates) or of an associated Laguerre polynomial (for polar coordinates). In the one dimensional case it is well known that the energy levels (again measured from the bottom of the well) are given by

$$E'_n = (n+\tfrac{1}{2})\hbar\omega \tag{9.8}$$

and in the general three-dimensional case by

$$E'_{n_1 n_2 n_3} = (n_1 + n_2 + n_3 + \tfrac{3}{2})\hbar\omega \tag{9.9}$$

or

$$E'_N = (N + \tfrac{3}{2})\hbar\omega$$

where n_1, n_2, n_3 are integers specifying the wave functions and $N = n_1 + n_2 + n_3$ (≥ 0) is the oscillator quantum number. When the angular dependence of the wave function is examined, it is found that for each N value there is a degenerate group of levels with different l-values such that $l \leq N$ and even (odd) N corresponds to even (odd) l. Specifically

$$N = 2n_r + l = 2n + l - 2$$

so that for $N = 2$ both s and d states occur with the same energy. The number of neutrons or protons which may be accommodated in the levels described by the oscillator number N is found to be $(N+1)(N+2)$, allowing for two spin directions in each substate. The levels are shown in Fig. 9.4(b), and in Table 9.2. It will be seen that the order is

$$1s; \; 1p; \; 1d, \; 2s; \; 1f, \; 2p; \; \dots$$

and shell closures occur at particle numbers

$$2, 8, 20, 40, 70, 112, \dots$$

which again is not the series of nuclear magic numbers.

TABLE 9.2 Single-particle states of infinite oscillator well

N	E'_N	l-values	$(N+1)(N+2)$
0	$\tfrac{3}{2} \times \hbar\omega$	0	2
1	$\tfrac{5}{2}$	1	6
2	$\tfrac{7}{2}$	0, 2	12
3	$\tfrac{9}{2}$	1, 3	20
4	$\tfrac{11}{2}$	0, 2, 4	30

(c) *The spin-orbit coupling model.* Several attempts have been made to modify the potential to yield the observed magic numbers. The most successful is that proposed by Mayer and by Haxel, Jensen and Suess in 1949 according to which a *non-central component* should be included in the force acting on a nucleon in a nucleus. If this non-central force is an interaction depending on the relative orientation of the orbital angular momentum and the spin momentum of the nucleon (so far neglected except for its role in prescribing only two particles per orbital state) then a different periodicity can easily arise. It is assumed in this model that the *spin-orbit force* separates the motion of a nucleon with orbital momentum l into sub-states with total angular momentum quantum number $j = l \pm \frac{1}{2}$ and that the level with the higher spin is the more stable. With this assumption the levels of the oscillator potential are enumerated as follows

$$1s_{1/2}; \quad 1p_{3/2} \; 1p_{1/2}; \quad 1d_{5/2} \; 2s_{1/2} \; 1d_{3/2}; \quad 1f_{7/2}; \quad 2p_{3/2} \; 1f_{5/2} \; 2p_{1/2} \; 1g_{9/2}; \quad \ldots$$

allowing that some anharmonicity of the well or tendency to square well shape already lowers the states of high angular momentum for a given oscillator number. Each state of given j may accommodate $2j + 1$ neutrons and $2j + 1$ protons. If now the $g_{9/2}$ level (for $N = 4$) is lowered so much by the spin-orbit potential that it merges with the levels of oscillator number $N = 3$, and similarly for higher N-values, the shell closures occur at particle numbers

$$2, 8, 20, 28, 50, 82, 126$$

exactly as required by experiment. The sequence of levels is shown in detail in Fig. 9.5. The order of sub-levels such as $s_{1/2}$, $d_{3/2}$ may be altered, without affecting the magic numbers, by adjusting the strength of the assumed spin-orbit force.

The origin of the spin-orbit coupling employed in this version of the shell model is not yet clear. The attractive interaction potential is usually assumed proportional to a term

$$V_S(r)\mathbf{s}.\mathbf{l} \qquad (9.10)$$

where $\mathbf{s}$ and $\mathbf{l}$ are the spin and orbital vectors for the nucleon. This form ensures that the orbits are split by an energy which increases with l-value, as required by the model. The potential resembles that which would arise from a simple magnetic effect but such effects are much too weak to give the necessary splitting. There is evidence for the existence of a spin-orbit force in the scattering of protons and neutrons by ^{4}He, which displays a split p-doublet very clearly and gives observable polarizations of the scattered particles. The polarization of high-energy nucleons scattered by nuclei also confirms the presence of a spin-orbit potential.

To summarize, the single-particle shell model or quasi-atomic model with spin-orbit coupling, provides a sequence of nuclear energy levels classified according to orbital momentum quantum number l (i.e. parity), total angular momentum j, and radial quantum number n. There are $2(2l + 1)$ nucleon states for given l, and $(2j + 1)$ states for given j. If there are x states of given nlj filled with nucleons we speak of a *configuration* $(nlj)^x$. The final total wave function of the nucleus will of course depend on the coupling of the individual angular momenta of the nucleons in the configurations and we neither expect nor find that the single particle model provides good wave functions in all cases. The important fact is that it sets up a plausible set of single particle levels; the vital part played by these levels in coordinating nuclear properties will be examined in subsequent chapters.

Continuation of the shell structure shown in Fig. 9.5 to higher nucleon numbers suggests a double closed shell at $Z = 126$, $N = 184$ (cf. Fig. 9.1(b)).

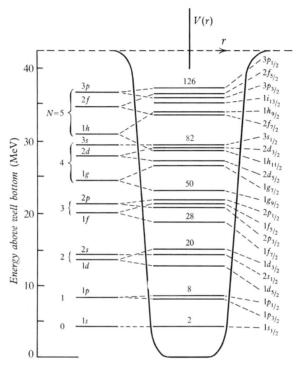

Fig. 9.5 Single particle levels in a potential well, allowing for spin-orbit coupling. The basic states, shown on the left, correspond to a potential $V(r)$ of the form shown, which is intermediate in shape between an oscillator and square well potential. The oscillator numbers are shown and it is evident that if the spin-orbit effect is sufficient to overcome the oscillator spacing then the observed magic numbers result.

However, more detailed considerations (Ref. 9.5) indicate that *superheavy elements* of long life against all forms of decay may be found in an "island of stability' near $Z = 114$, $N = 184$. The nucleus of maximum stability $\binom{294}{110}X$ might have a lifetime comparable with the age of the earth, and its presence in existing minerals is therefore not excluded.

9.5 The nuclear level spectrum

The structure of the succession of nuclei may be envisaged in terms of a single-particle potential of constant depth and of a radius increasing with mass

The level spectrum of the nucleus ^{61}Ni, determined by a method of high resolution, is shown in Fig. 9.7(a). The characteristic features of this and similar level systems are (*a*) a level spacing of a few hundred keV near the ground state, and (*b*) an increase of level density with increasing excitation energy. The levels shown are discrete, i.e. the width Γ is very much less than the level spacing D, although as the excitation energy increases beyond the value shown in the figure this is no longer true because of an increase of Γ due to increasing probability of particle emission and a decrease of D in accordance with the general trend already shown in the figure. Level spectra of the general type shown in Fig. 9.7(a) are found for all nuclei although the precise level density at a given excitation depends on the mass number. Such spectra are essentially those of a many-body nuclear model, as was first pointed out by Niels Bohr; they stand in sharp contrast with expectation from the single-particle shell model, according to which only a few widely spaced levels are expected in the range of energy shown in Fig. 9.7(a) (cf. Fig. 9.5). If, however, the nuclear level spectrum is examined by methods of low resolution, which average over the narrow levels, and if the *probability* of exciting levels in a reaction is plotted rather than the actual level positions, the result shown in Fig. 9.7(b) is obtained. This shows exactly the features expected from the single-particle model, namely well-separated peaks, whose behaviour in many cases agrees with the angular momentum values suggested by the model. The interpretation of the *gross structure* indicated by Fig. 9.7(b) in terms of the *fine structure* of Fig. 9.7(a) is one of the current problems of nuclear theory. A complete theory would of course predict these states, but the single-particle potential cannot be realistic. If residual two-body interactions are considered (Sect. **12.2**) many new states, based on mixtures of configurations of the same spin and parity, appear.

In the (d, p) reaction, a neutron is added to a target nucleus (e.g. ^{60}Ni in Fig. 9.7) to form states of a final nucleus (^{61}Ni). If the final state is a nearly pure neutron single-particle state, the cross section is large; the extent to which the cross section departs from the value expected for a single-particle state is measured by the *spectroscopic factor* S (Sect. 15.6.4). By determining cross sections for stripping to individual states it is possible to locate the main single particle states and to trace their excitation over a wide range of nuclei.

It is not possible to give a detailed theoretical account of individual levels in regions of high level density. Level densities can be predicted under particular assumptions, such as that of the Fermi gas model (Sect. **9.1**), and compared with experimental measurements, but this is more useful to the theory of nuclear reactions than to models of nuclear structure. It is however possible (ch. 15) to deduce from experiment certain averaged properties of levels such as total widths, by application of statistical methods. The ground state and the well-separated low-lying levels up to perhaps a few MeV in excitation, are likely to be described by the simpler types of nuclear motion; they are well known experimentally and have been surveyed for an extensive range of nuclei, both stable and unstable, throughout the periodic system. As a result of these surveys the following important regularities in the low-lying levels have been observed:

(*a*) *Isomeric states* (Sect. **13.5**) are found in groups of nuclei with neutron or proton numbers just below the magic numbers. These are bound states of long life, for which the expected radiative transition is slow because of a large spin change. Figure 9.8 shows the distribution of known isomeric states for nuclei of odd mass number. The single-particle shell model is able to account for this pattern if it is supposed that successive single-particle levels lie rather close together towards the end of a shell so that the first excited state of a nucleus may well have a spin considerably different from that of the ground state. Usually the isomeric state is one

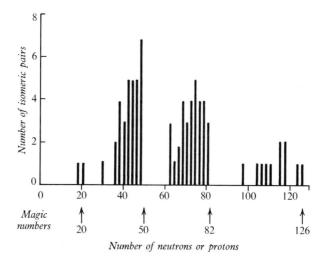

Fig. 9.8 Frequency distribution of odd-*A* isomeric nuclei (Ref. 10.4).

which has moved down from a higher oscillator level because of the spin-orbit effect, and this leads to a parity change in addition to the large spin change (e.g. $g_{9/2} \rightarrow p_{1/2}$ in Fig. 9.5). Isomeric transitions of $E3$ and $M4$ type (Sect. 3.5.2) are therefore frequently found.

(*b*) *The first excited states of even–even nuclei* show a systematic behaviour. These excited states usually have spin 2 and even parity and their excitation energy reaches a series of maxima at closed shells (Fig. 9.9). Between closed shells the energy varies in a systematic way with mass number. Even mass nuclei also frequently possess a low-lying state of spin 3 and odd parity while at higher excitation states of spin 1 and odd parity form the giant dipole resonance (Sect. 15.4.1).

(*c*) *Rotational bands of levels* with energies proportional to $I(I+1)$ where I is the spin of the level, are found in many nuclei with mass numbers lying between closed shells. These states have large quadrupole moments, and are interpreted in terms of a collective rotation (Sect. 12.3.3).

Features such as (*b*) and (*c*), derived from observations of the dynamical behaviour of nuclei, are not obvious consequences of the single-particle shell model. They must be accounted for, together with spin and static moment anomalies, in any wholly satisfactory extension of this model. In making this extension it will be necessary to discard the concept that a closed shell is a tightly bound, rigid system.

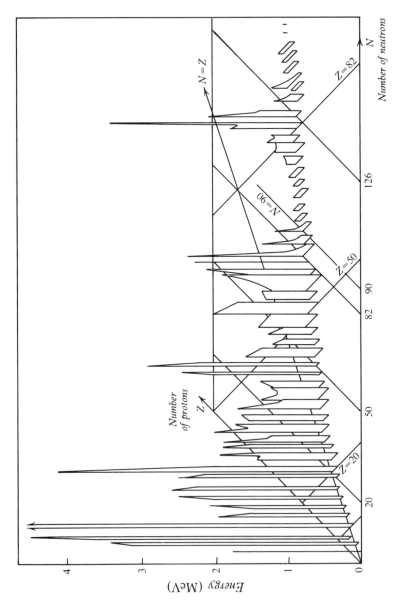

Fig. 9.9 Energy of first excited 2^+ state in even-even nuclei as a function of Z and N (Ref. 12.8).

References

1. BLIN-STOYLE, R. J. 'Structure of the Nucleus', *Contemp. Phys.*, **1**, 17, 1959.
2. FLOWERS, B. H. 'The Nuclear Shell Model', *Progr. nucl. Phys.*, **2**, 235, 1952.
3. EDEN, R. J. 'Nuclear Models', *Progr. nucl. Phys.*, **6**, 26, 1957.
4. MAYER, M. G. and JENSEN, J. H. D. *The Elementary Theory of Nuclear Shell Structure*, Wiley, 1955.
5. SEABORG, G. T. 'Elements beyond 100', *Ann. rev. nucl. Sci.*, **18**, 53, 1968; SEABORG, G. T. and BLOOM, J. L. 'The Synthetic Elements', *Scientific American*, **220**, April 1969, page 57.

IO The mass and isotopic abundance of nuclei

Mass and charge were the first properties to be determined for a wide range of atomic nuclei. Nuclear abundance, already known in part from chemical surveys, could be expressed as isotopic abundance as soon as mass spectrographic analyses were completed. Mass values are basic in any theory of nuclear structure but like many other nuclear phenomena can only be predicted accurately from the properties of nucleons if a complicated many-body problem is solved. In this chapter we first of all consider the inverse problem, namely the extent to which the existence of a range of stable nuclei, with known binding energy, can yield information on the nucleon–nucleon force. We then incorporate these general properties in a semi-empirical approach to the mass data, based on the idea of continuous nuclear matter, as embodied in the liquid-drop model (Sect. **9.1**). The effects of shell structure are noted as relatively small but important corrections to the underlying mass sequence.

10.1 The mass tables; binding energy

Two scales for mass measurement are now recognized.

(*a*) The *absolute scale*, related to the kilogramme
(*b*) The *atomic mass scale*, defined by setting the mass of one atom of the nuclide ^{12}C equal to 12·000... atomic mass units (a.m.u., symbol m_u or u.

The absolute value of the atomic mass unit is obtained by noting that for 1 mole of ^{12}C (0·012 kg)

$$0·012 \text{ kg} = N_A \times 12m_u$$

where N_A is Avogadro's number. This gives

$$1 \, m_u \equiv 1·661 \times 10^{-27} \text{ kg}$$

$$= 931·48 \text{ MeV}$$

The absolute mass scale is rarely used because of its numerical inconvenience and because it conceals the physical content of the results of mass measurements. For the purpose of nuclear physics all masses are based on the atomic scale; they are given for *neutral atoms* rather than for stripped nuclei mainly because in effect these are the masses that are determined directly by mass spectrometry. The mass $M(A, Z)$ of a nuclide of mass number A and atomic number Z is related to the nuclear mass M_N by the equation

$$M(A, Z) = M_N + ZN_A m - B(Z) \tag{10.1}$$

where N_0 is Avogadro's number, m is the mass of an electron and $B(Z)$ is the total electron binding energy, expressed in atomic mass units. In this equation $B(Z)$ is only about 10^{-4} per cent of the mass $M(A, Z)$ and its effect is therefore usually negligible. Although mass tables give the atomic masses of neutral atoms, the term 'nuclear mass' is often used. This will normally mean the neutral atom mass unless especially qualified.

The atomic mass scale is by definition relative rather than absolute. A relative scale has the advantage that atomic masses may be expressed more accurately in relative units (a.m.u.) than in absolute units (kg) because mass spectrometry is normally based on comparative rather than absolute measurements.

TABLE 10.1 Atomic mass table (from Ref. 10.3)

	Atomic mass a.m.u.	$B(A, Z)$ MeV	B/A MeV/nucleon	$S_n(A, Z)$ MeV	$P_n(A, Z)$ MeV
$^{131}_{54}\mathrm{Xe}$	130·90509 ±0·5	1103·5	8·42	6·6	—
$^{132}_{54}\mathrm{Xe}$	131·90416 ±0·5	1112·5	8·43	9·0	1·2
$^{133}_{54}\mathrm{Xe}$	132·90582 ±0·4	1119·0	8·41	6·5	1·1
$^{134}_{54}\mathrm{Xe}$	133·90540 ±0·5	1127·4	8·41	8·4	0·92
$^{135}_{54}\mathrm{Xe}$	134·90702 ±11	1134·0	8·40	6·6	0·78
$^{136}_{54}\mathrm{Xe}$	135·90722 ±11	1141·9	8·40	7·9	1·2
$^{137}_{54}\mathrm{Xe}$	136·91110	1146·3	8·3	4·4	—

The errors in the mass values (a.m.u.) refer to the last significant figure.

The masses of all stable isotopes have now been determined accurately by the refined techniques of mass spectrometry. For unstable isotopes the determination of energy releases in nuclear reactions and in decay processes provides information of comparable accuracy, and also connects together many mass values through consecutive processes. The combination of this body of data in such a way as to minimize errors (Ref. 10.3) leads to the present mass tables, of which a small section is presented in Table 10.1. The accuracy claimed in masses of the order of 30 a.m.u. is better than 1 part in 10^6, corresponding to a determination of Q-values to a few keV. Nuclear mass changes are linked to energy releases in nuclear reactions through Einstein's equation

$$\Delta E = c^2 \Delta M \tag{10.2}$$

and it is one of the functions of the mass tables to predict such energy changes where they have not been determined experimentally.

The first obvious conclusion to be drawn from an inspection of the mass tables is that the atomic masses of the isotopes are nearly whole numbers when expressed in terms of $^{12}C = 12 \cdot 000...$; the more obvious choice of taking $^1H = 1 \cdot 000...$ does not yield atomic masses so near to whole numbers for other isotopes. Despite this, the isotopic masses are sufficiently near to multiples of the mass of the hydrogen atom to suggest that nuclei are built up of particles of mass comparable with that of the proton. The neutron mass had to be determined by reaction methods; its similarity to the proton mass renders the neutron–proton type of nuclear structure an obvious hypothesis. The masses are not in fact exact multiples of any combination of the neutron and proton mass, and they are not whole numbers. The difference between the exact atomic mass of an isotope $M(A, Z)$ and its mass number A is known as the *mass defect*, i.e.

$$\Delta = M(A, Z) - A \tag{10.3}$$

and the mass defect per unit mass number or *packing fraction* was defined by Aston, long before the neutron–proton model of the nucleus was known, as

$$P = \frac{\Delta}{A} = \frac{M(A, Z) - A}{A} \tag{10.4}$$

If the neutron–proton model of the nucleus is assumed, it is possible to calculate the *total binding energy B* of the nucleus. This is the work required to break down the nucleus into its constituent nucleons, or alternatively, the energy released in the building up of the nucleus from these constituents. Thus

$$B(A, Z) = ZM_H + (A - Z)M_n - M(A, Z) \tag{10.5}$$

where M_H, M_n and $M(A, Z)$ are the masses of the hydrogen atom, the neutron and the nucleus in a.m.u., i.e. including sufficient electron masses to ensure neutral systems in each case. The binding energy B obviously measures nearly the same quantity as Aston's mass defect Δ and the *average binding energy per nucleon B/A* conveys for the neutron–proton model the same information as Aston's packing fraction. Table 10.1 shows the total binding energy $B(A, Z)$ and the binding energy per nucleon for some of the xenon isotopes.

It is often useful experimentally to refer to the binding energy of a specific nuclear particle, such as a proton, neutron or α-particle, which may be detached or absorbed in a nuclear reaction. If a neutron is added to a nucleus $(A-1, Z)$ in a (n, γ) capture reaction for example, an energy equal to the binding energy of the neutron in the nucleus (A, Z) is evolved. This energy is also known as the neutron *separation energy* $S_n(A, Z)$. The separation energy S_n or neutron binding energy B_n (strictly the binding energy of the last neutron) is related to the total binding energies of the nuclei concerned by the equation

$$
\begin{aligned}
S_n(A, Z) &= B_n(A, Z) \\
&= B(A, Z) - B(A-1, Z) \\
&= M(A-1, Z) - M(A, Z) + M_n
\end{aligned}
\Bigg\} \qquad (10.6)
$$

The separation energies for protons and neutrons are greater for even Z or N than for odd Z or N, as may be seen for neutrons in Table 10.1. This is due to attractive forces between pairs of nucleons, with opposite spin, occupying the same level. The *pairing energy* which thus arises may be defined for neutrons as

$$
P_n(A, Z) = (-1)^{N\frac{1}{4}}\{2S_n(A, Z) - S_n(A-1, Z) - S_n(A+1, Z)\}, \; Z \text{ even}
$$
$$(10.7)$$

This formula compares a given nuclide with nuclides containing one more or one less neutron. Pairing energies for the xenon isotopes are shown in Table 10.1.

10.2 Survey of nuclear binding energies

The results of the mass spectrometry of the stable nuclei were expressed by Aston (1927) not only in a table of masses, but also in a graph of the packing fraction as a function of mass number. It is now more usual to present this

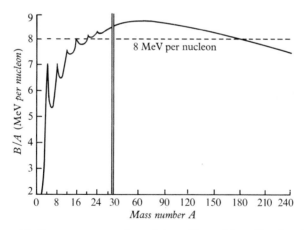

Fig. 10.1 Average binding energy per nucleon of the stable nuclei as a function of mass number (Ref. 10.4).

information as a graph of the average binding energy per nucleon as a function of mass number (Fig. 10.1). The main features of this diagram are:

(*a*) A positive binding energy for all nuclei, which means that any nucleus is more stable than an unconnected assembly of its constituent neutrons and protons. This will be so if there are attractive (nuclear) forces between these constituents within the nuclear volume, as indicated also by scattering experiments (ch. 5). Since nuclei do not collapse these forces must in effect become repulsive for very close distances of approach of the nuclear particles.

(*b*) A rapid increase of binding energy per nucleon for the light nuclei with a notable peak at $A = 4$ (^{4}He) and further peaks at $A = 4n$ (^{8}Be, ^{12}C, ^{16}O, ^{20}Ne, ^{25}Mg). This reflects the peculiar stability of the α-particle structure. Such stability arises naturally in any theory of nuclei in which a given orbital state may contain just two protons and two neutrons, with opposite spins in each case.

(*c*) Approximately the same binding energy per nucleon (7.3 to 8.7 MeV) for all nuclei with A greater than 16, so that for all but the lightest nuclei it is a good approximation to suppose that B is proportional to A. For a nucleus in which each particle interacted with every other particle the total binding energy B would be proportional (Sect. **9.1**) to A^2 approximately and the experimental facts therefore show that the nuclear constituents interact with only a limited number of their neighbours. This phenomenon is known as the *saturation of nuclear forces* (ch. 18). The peak in binding energy at $A = 4$ suggests that a neutron interacts strongly with one other neutron and two other protons; the fifth, sixth and seventh particles, in ^{5}Li, ^{6}Li, ^{7}Li for example, are less strongly bound. The maximum binding energy per nucleon occurs for A about equal to 60 (Fe, Ni, Co).

(*d*) A gradual decrease in B/A from a maximum of 8.7 MeV per nucleon at $A = 60$ to 7.3 MeV per nucleon at $A = 238$. This is associated with the disruptive effect of the nuclear charge, which ultimately sets a limit to the number of elements which can be formed.

(*e*) The absence of large magic number effects (except for the four-structure noted in (*b*)) in the *average* nucleon binding energy. This is in agreement with the ideas of the shell model because the large binding energies of the inner shells will submerge the discontinuities of the binding energy of the last nucleon. These can however be seen by comparison with the predictions of a smoothly varying function (Sects. **9.2** and **10.3**). For the heavy elements it is now known that shell corrections to the nuclear binding energy lead to the possibility of more than one quasi-stable configuration and hence to the phenomenon of fission isomerism (App. A4.2).

Mass values also show the onset of nuclear deformation (Sect. **12.3**) at a neutron number $N = 88$.

Points (*a*), (*b*) and (*c*), which imply saturating attractive forces between nucleons, are of basic importance and form a prime requirement to be imposed on any theory of the nucleon–nucleon interaction (ch. 18). They suggest strongly that in respect of binding energy at least a heavy nucleus behaves in analogy with a drop of liquid, in which as a result of the short range of the intermolecular forces each molecule experiences only attractions from its nearest neighbours and thus acquires a fixed potential energy.

10.3 The semi-empirical mass formula

Many attempts have been made to express the main features of Fig. 10.1 analytically. The mass formulae which have been set up have proved useful in predicting the binding energy of unstable nuclei and in evaluating the energy release to be expected in particular processes of nuclear division such as fission or α-particle emission. The earliest formulae, developed by Weis-zäcker and by Fermi, included no terms to represent shell structure, and were constructed as follows:

(i) A neutron–proton liquid drop model of constant density is assumed. The volume is then proportional to A and the nuclear radius to $A^{1/3}$, i.e.

$$R = r_0 A^{1/3} \tag{10.8}$$

where r_0 is a constant now usually taken to be about $1 \cdot 2 \times 10^{-15}$ m (ch. 11).

(ii) The first term in the mass formula is the sum of the atomic masses of Z hydrogen atoms and $(A - Z)$ neutrons.

(iii) The neutrons and protons are held together by short-range attractive forces. These forces contribute the main binding term and reduce the mass of the nucleus below that of its constituents by an amount proportional to the number of nucleons, since the binding energy per nucleon is constant. The second term is known as the *volume energy* and is written $a_v A$ where a_v is about 14 MeV per nucleon in unlimited nuclear matter. The volume energy is an important constant in the theory of nuclear matter. It is the difference between the attractive effect of the nuclear forces and the kinetic energy arising from the confinement of a nucleon to the volume corresponding to the uniform density assumed in (10.8). Calculation of the binding energy and equilibrium density of nuclear matter from two-body data (ch. 18) is highly complex, but significant progress has been made by Brueckner, Bethe and others (Ref. 10.6).

(iv) As in the theory of liquids, this simple expression neglects the surface in which nucleons are less effective for binding than in the interior of the nucleus. The main binding term must therefore be corrected by a disruptive term proportional to surface area, i.e. $+a_s A^{2/3}$ (*surface energy*).

(v) There is, from Fig. 10.1, clearly a tendency to maximum stability when the number of neutrons and protons in a light nucleus is equal. This means that if $Z \neq A/2$ the nucleus is less strongly bound than it might be. Since the neutrons and protons move in different potential wells, because of the Coulomb effect, the total kinetic energy is proportional (Sect. **9.1**) to $(Z^{5/3} + N^{5/3})/MR^2$. This exceeds the value for a nucleus with an equal number of neutrons and protons by the so called *asym-*

metry energy

$$\left[Z^{5/3} + N^{5/3} - 2\left(\frac{A}{2}\right)^{5/3} \right] / MR^2$$

which may be expanded in terms of the small quantity $(A - 2Z)$ to give (to second order)

$$\frac{a_a}{A}(A - 2Z)^2 \qquad (10.9)$$

where a_a is a constant.

(vi) The nuclear charge Ze is confined within a volume $\frac{4}{3}\pi R^3$. A simple calculation of the potential energy of such a charge distribution gives a disruptive potential energy of approximately $\frac{3}{5}(Ze)^2/4\pi\varepsilon_0 R$ or, alternatively, $+a_c(Z^2/A^{1/3})$ (*Coulomb energy*).

(vii) For nuclei of even A, Z and N may be both even or both odd. If these numbers are even, the nucleons may be grouped into stable pairs, with spins opposed, and the nucleus will be correspondingly more stable than in the case Z, N odd. A term $\delta(A, Z)$ approximately equal to the pairing energy $P(A, Z)$ (10.7), corrects for this effect.

If all these terms are collected together the atomic mass may be written

$$M(A, Z) = ZM_H + (A - Z)M_n - a_v A + a_s A^{2/3}$$

$$+ \frac{a_a}{A}(A - 2Z)^2 + \frac{a_c Z^2}{A^{1/3}} + \delta(A, Z) \text{ a.m.u.} \qquad (10.10)$$

with $\delta(A, Z)$ negative for A even, Z even; positive for A even, Z odd; and zero for A odd. The average binding energy per nucleon may be obtained directly from this formula using (10.5), and a graph of this quantity against A gives a curve of the general shape of that given in Fig. 10.1. The coefficients in (10.10) may be calculated theoretically, as in the case of a_c, under certain simplifying assumptions, but with such values the equation would have little practical value. It is better to determine values for the coefficients empirically by fitting the equation to known masses, and by insisting that it should predict the correct proton to neutron ratio (or Z/A value) for the medium weight and heavy elements. Fermi proposed the constants $a_v = 14.0$ MeV, $a_s = 13.0$ MeV, $a_a = 19.3$ MeV, $a_c = 0.58$ MeV, $|\delta| = 33.5A^{-3/4}$ MeV for A even. Other sets of values are also used, all agreeing that the main binding term in the mass formula is about 15 MeV per nucleon.

A comparison between measured mass values and the predictions of the semi-empirical formula has already been given in Fig. 9.1(b), which very clearly shows the shell effects. This figure is adapted from the work of Kummel *et al.*,† who have proposed a more complex mass formula in which terms to represent the shell discontinuities are included.

† *Nuclear Physics*, **81**, 129, 1966.

From eq. (10.6) it is clear that shell effects in the sequence of nuclear masses will also be apparent in neutron separation energies. These have been displayed by comparing experimental results with the predictions of formula (10.10). The radiative capture of a slow (thermal) neutron by a nucleus A

$$A + n \longrightarrow (A+1) + \gamma + S_n(A+1, Z) \tag{10.11}$$

releases a total energy per capture equal to the neutron separation energy for the nucleus $(A+1)$. The same quantity may be measured by determining the energy release Q in the (d, p) reaction with nucleus A as target.

$$A + d \longrightarrow (A+1) + p + Q \tag{10.12}$$

Since the deuteron is equivalent to a neutron and proton bound together with an energy $\varepsilon(= -2 \cdot 22 \text{ MeV})$, comparison of (10.11) and (10.12) shows that

$$S_n = Q + |\varepsilon| = Q + 2 \cdot 23 \text{ MeV} \tag{10.13}$$

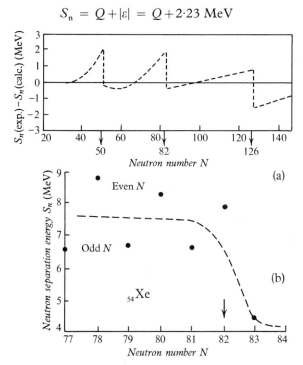

Fig. 10.2 (a) Difference between observed and calculated neutron separation energies as a function of neutron number (Ref. 10.4). (b) Variation of neutron separation energy in Xe isotopes with neutron number, showing drop in S_n for the 83rd neutron.

Many measurements of energy releases in (n, γ) and (d, p) reactions have been made. Figure 10.2(a) shows the results obtained for nuclides with a neutron number N between 40 and 140; the quantity plotted is the difference between the observed separation energy and the smoothly varying separation energy predicted from the semi-empirical mass formula. There is clear evidence for breaks at the magic numbers $N = 50$, 82 and 126. It should be noted that the separation energy for the nucleon which completes a closed shell is not markedly greater than that of the two or three nucleons which precede it. The shell closure is seen particularly as a sharp drop in the separation energy for the (magic + 1) and (magic + 2) nucleons (Fig. 10.2(b)).

Binding energy discontinuities have been found for other particles. In the region of the rare earth elements neutron-deficient isotopes formed in nuclear reactions are found to be α-active. If the energy of the α-particles is examined

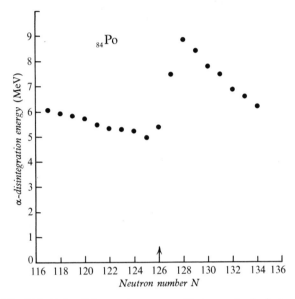

Fig. 10.3 Alpha-disintegration energy of isotopes of polonium.

for a particular series of isotopes of a given element it is found to show a maximum for $N = 83$ or 84, suggesting that after $N = 82$ the next two neutrons are less tightly bound than for $N \leqslant 81$. At neutron number $N = 126$ the same effect appears clearly, as is seen in Fig. 10.3 which shows the α-decay energy for the isotopes of polonium ($Z = 84$) plotted as a function of neutron number. The most recent data of mass spectroscopy for the heavy stable nuclei show anomalies in pairing energy near the neutron numbers $N = 90$ and 116. Other nuclear properties, such as electric quadrupole moments and the pattern of excited states also change at these numbers. These discontinuities differ from those at the magic numbers of the shell model and represent a transition from a nuclear structure in which collective vibration is the dominant mode of motion to one characterized by collective rotation (ch. 12).

10.4 Nuclear stability

If the known nuclei are arranged in order of increasing N and Z on a rectangular grid, all nuclides of the same N (isotones) appearing in horizontal lines and all nuclides of the same Z (isotopes) in vertical lines, the diagram so obtained is known as a Segrè chart, or neutron–proton diagram (Fig. 10.4). On this diagram lines of constant A intersect the axes at $45°$. The stable nuclei cluster about the line $N = Z = A/2$ for small mass numbers, but the medium weight and heavy nuclei have many excess neutrons. The unstable nuclei form a fringe to the band of stability and also intermingle with the stable isotopes for many elements.

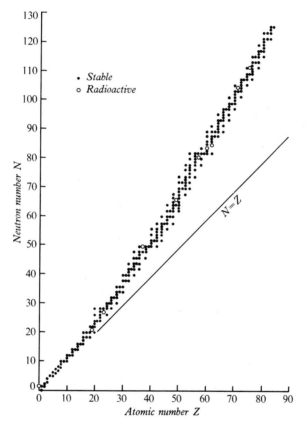

Fig. 10.4 Neutron–proton diagram (Segrè chart) of the naturally occurring nuclides with Z less than 84 (Ref. 10.4). The present diagram may not be found accurate in all details.

These features are readily interpreted in terms of the semi-empirical mass formula (10.10). In the case of odd-A nuclei, for which δ may be put equal to zero, the nucleus of maximum stability for a given A is obtained by setting dM/dZ equal to zero. This gives

$$Z = \frac{A}{1 \cdot 98 + 0 \cdot 015 A^{2/3}} \tag{10.14}$$

if Fermi's constants are used, and shows that Z falls short of $A/2$ as A increases. Physically this is due to the fact that Coulomb repulsion between protons is a long-range force and becomes increasingly important with respect to the short-range nuclear forces as the nuclear charge increases. The repulsive effect has to be balanced in a stable nucleus by the presence of extra neutrons to provide extra attractive interactions. Isobars of odd A, with Z different from the integer nearest to the value predicted by (10.14) are expected to be unstable.

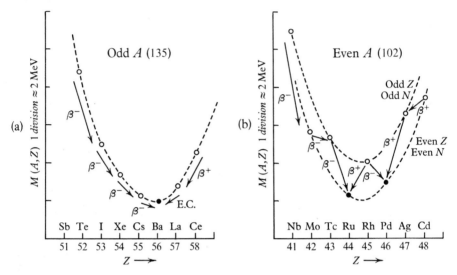

Fig. 10.5 Stability of isobars, showing atomic mass M plotted against atomic number Z. Open circles represent unstable, and full circles stable nuclei. (a) Odd A (135), for which there is one stable nucleus, $^{135}_{56}$Ba. (b) Even A (102), for which there are two stable isobars, $^{102}_{44}$Ru and $^{102}_{46}$Pd, both of even–even type (Ref. 10.4).

If a third coordinate representing the atomic weight $M(A, Z)$ is added to the neutron–proton diagram a three-dimensional region known as the mass surface is obtained. Stable nuclei then group round the bottom of a valley or trough in this surface. Intersections of the mass surface by planes of constant A define groups of isobars comprising both stable and unstable nuclei. Formula (10.10) shows that for A constant $M(A, Z)$ is a quadratic function of Z so that the curves of intersection are parabolas. For odd A one parabola only is obtained (Fig. 10.5(a)) but for even A the term $\delta(A, Z)$ in 10.10 must be included and the masses lie alternately on two distinct parabolas (Fig. 10.5(b)) separated by an energy $2\delta(A, Z)$. (A closer examination shows that there are two parabolas for odd A, corresponding to Z even and Z odd. The difference is smaller than in the case of even A and will be disregarded.)

It is clear that for odd A there will be one stable nucleus, although exceptionally (^{113}In, ^{113}Cd; ^{123}Sb, ^{123}Te) two isobars of adjacent Z may have so closely similar a mass that transitions between them are unobservable because of their low probability. For even A there may be two or even three stable isobars. Transitions between isobaric nuclei take place in the direction of

increasing stability by positron or negatron (β) emission or by electron capture
(Sect. **16.6**) as shown in Fig. 10.5(a),(b). The atomic mass $M(A, Z)$ of a
neutral atom undergoing a decay process of this type is related to the mass
$M(A, Z \pm 1)$ of the product by the inequalities:

$$M(A, Z) > M(A, Z+1) \qquad \text{for electron decay to} \atop (A, Z+1)$$

$$M(A, Z) > M(A, Z-1)+2N_A m \quad \text{for positron decay to} \atop (A, Z-1)$$

$$M(A, Z) > M(A, Z-1)+E_e \qquad \text{for electron capture} \atop \text{decay to } (A, Z-1)$$

$$(10.15)$$

where N_0 is Avogadro's number, m is the electron mass and E_e(a.m.u.) is the
ionization energy for the electron captured in the neutral atom of charge

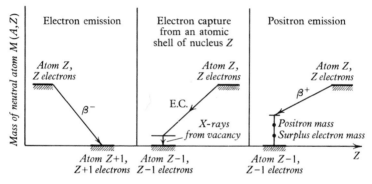

Fig. 10.6 Beta-decay processes for an atom (A, Z). In electron emission and electron capture (from
an atomic shell), the total number of electrons always matches the nuclear charge. In electron
capture the nuclear transition must provide the energy of ionization for the absorbed electron.

$Z-1$. In positron decay the formation of a neutral atom implies the release
of an atomic electron as well as the emission of a positron and the atomic mass
change must supply the mass of both of these particles. The relationships
(10.15) are illustrated in Fig. 10.6, in which energies associated with the rearrangement of outer atomic electrons are neglected.

It is also assumed that no particle of finite rest mass other than an electron
is emitted in β-decay, i.e. that the neutrino rest mass is zero. When the transition energy becomes very small it may be that only one of the lightly-bound
outer electrons can be captured. If there is insufficient energy even for this,
then both odd isobars (A, Z) and $(A, Z-1)$ may be stable; the rarity of this
occurrence is good evidence for the zero rest mass of the neutrino since a
finite rest mass for this particle would obviously widen the β-stability limits.
The lifetime associated with very low β-transition energies may be extremely
long even on a geological time scale and if so the corresponding nuclei are
found in nature. Examples are given in Table 2.3.

For even-A nuclei it is possible for a nucleus (e.g. ^{40}K, ^{64}Cu) to exhibit both positron and negatron decay. It is also possible in principle for many even-A nuclei to decay by double β-emission but the existence of a large number of isobaric pairs for even A is evidence for the slowness of this type of transition. Double β-decay appears to have been established† for ^{130}Te with a lifetime of about 10^{21} yr. Figure 10.5(b) also shows that even-A, odd-Z nuclei should always be unstable and this is true except for four light nuclei (^{2}H, ^{6}Li, ^{10}B, ^{14}N) to which it is wrong to expect the mass formula to apply because of the rapid variations of nuclear binding energy for small A (Fig. 10.1).

The width of the valley in the mass surface in which stable nuclei are found is determined by the criterion of β-stability. Other types of spontaneous

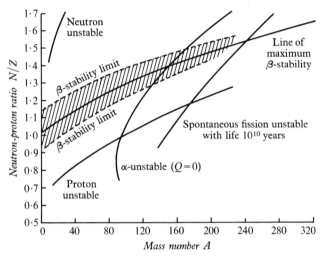

Fig. 10.7 Nuclear stability limits predicted by the semi-empirical mass formula (Segrè, *Experimental Nuclear Physics*, Vol. III).

decay, such as α-particle emission or fission may be energetically possible, but are impeded for the majority of nuclei by the Coulomb barrier. This barrier also affects positron decay but when its effect becomes serious, e.g. for heavy nuclei, electron capture provides an alternative and rapid means of decay. The stability limits predicted by the semi-empirical mass formula for various types of decay are shown in Fig. 10.7 in which the neutron–proton ratio N/Z is plotted against the mass number A. The line of maximum stability predicted by (10.14) is surrounded by the region of stable isobars determined by diagrams of the type (10.5(a),(b)) or eq. (10.15).

The β-stable nuclei are stable against proton or neutron emission but for considerably larger or smaller values of N/Z emission of one or other of these particles becomes energetically possible. Neither neutron nor proton radio-

† T. Kirsten *et al.*, *Phys. Rev. Lett.*, **20**, 1300, 1968.

activity has yet been observed as a competitive mode of ground state decay, although the emission of nucleons from excited (virtual) states of nuclei is a familiar phenomenon of nuclear reactions.† Proton emission has also been seen as a competitive process in the decay of isomeric states of high spin.‡

The β-stable nuclei with $A > 150$ are unstable with respect to the emission of α-particles, but for low mass numbers the ratio of disintegration energy to potential barrier is unfavourable for α-emission. The main importance of α-decay as a stability limit occurs for $Z > 83$, where the nuclei expected to be β-stable are α-active. The only elements in this region found in nature are those such as uranium and thorium which have half-lives comparable with the age of the earth.

The fact that there is a maximum in the binding energy curve at $A \approx 60$ suggests that the spontaneous splitting or 'fission' of heavy nuclei into two fragments of nearly equal size may be energetically possible. It may be shown from the semi-empirical mass formula that the critical parameter for such a process in a nucleus (A, Z) is the parameter Z^2/A; spontaneous fission may be expected to occur with a lifetime of about 10^{10} years if the numerical value of Z^2/A exceeds 37. This value is already exceeded for some nuclei with $Z < 92$ and for the heaviest nuclei spontaneous fission competes successfully with α-emission as a mode of decay. The probability of fission increases rapidly with Z and is usually taken to be the factor that limits the number of trans-uranic elements that may be prepared to those with $Z < 106$, although shell structure calculations suggest the possibility of extra stability in the region of $Z = 114$.

Shell closure effects are seen in the energy release in β-decay just as they are in particle separation energies, since discontinuities in the mass surface are involved in a similar way. These effects are shown clearly if the energy of β-disintegration is plotted against neutron number for nuclei with a given value of the *isotopic number* $I = N - Z = A - 2Z$. Such nuclei have the same excess of neutrons over protons and might be expected to show a regular dependence of decay energy on mass number. In fact there are irregularities near the magic numbers. The yields of fission products, which are mainly electron emitters, are also enhanced at $N = 82$ and 83, perhaps because the energy available for the formation of these products in the fission process is slightly larger than for adjacent neutron numbers.

10.5 Isotopic abundance

Naturally occurring nuclei are either stable or unstable with a half-life of the order of 10^9 years or longer. The naturally occurring radioactive elements with shorter periods derive from long-lived parents. The relative abundance

---

† When the virtual state is formed by a preceding beta transition, the observed proton or neutron emission has the period of the beta decay. Both delayed neutrons (Sect. A.4.1) and protons originating in this way have been observed.

‡ K. P. Jackson *et al.*, *Phys. Lett.*, **33B**, 281, 1970.

of the observed stable species depends on the process of creation, which may have singled out particular nuclear types for preferential formation and on the nuclear stability limits discussed in the last section which may have caused the disappearance of some of the preferentially created species.

The elements of highest atomic abundance in the universe are hydrogen and helium, which are not retained gravitationally in the earth's atmosphere. The elements of the earth's crust (Table 10.2) appear to represent very roughly the general cosmic abundance if this may be assessed by analysis of meteoritic material. The naturally occurring nuclides with $Z \leqslant 83$ are shown in Fig. 10.4 and the relative abundance of the lighter even–even nuclides in Fig. 10.8. These bodies form more than 85 per cent by weight of the crust.

TABLE 10.2 Abundant nuclides of the earth's crust (Ref. 10.4)

Nuclide	$^{16}_{8}O$	$^{28}_{14}Si$	$^{56}_{26}Fe$	$^{40}_{20}Ca$	$^{24}_{12}Mg$	$^{27}_{13}Al$	$^{23}_{11}Na$	$^{39}_{19}K$
% abundance by weight	48	26	5	3·5	2·0	8·5	2·8	2·5

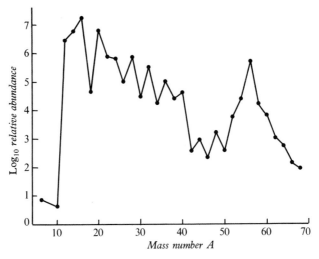

Fig. 10.8 Relative isotopic abundance in the universe of light even-even nuclei, taken from a larger diagram given by H. E. Suess and H. C. Urey, *Rev. mod. Phys.*, **28**, 53, 1956. The 'iron peak' near $A = 56$ is obvious.

In detail, isotopic abundances show anaomalies at $A = 56$, 90, 135 and 200 and these mass values may be correlated either directly, or before decay of active parent nuclei, with the magic numbers $N = 28$, 50, 82 and 126. It is probable that nuclei containing these neutron shells were produced preferentially in the element-building reactions. Relative abundance of the isotopes of a given element (Z constant) and the isotones of a given neutron number (N constant) show the following main anomalies:

(a) the number of stable and long-lived isotopes is greater for $Z = 20, 28, 50$ and 82 than for near-by elements,

(b) the number of stable and long-lived isotones is greater for $N = 20, 28, 50,$ 82 and 126 than for near-by N values,

(c) the abundance of even Z, even N isotopes relative to the abundance of the other isotopes of a given element is greater for $N = 50$ (^{88}Sr) and $N = 82$ (^{138}Ba, ^{140}Ce).

These anomalies and a number of similar ones are clearly correlated with the exceptional stability of nuclei containing magic numbers of neutrons or protons.

TABLE 10.3 Frequency distribution of stable nuclides (Ref. 10.4)

A	Z	N	Number of cases
odd	odd	even	50
odd	even	odd	55
even	odd	odd	4
even	even	even	165

The distribution of stable nuclides according to neutron and proton number is given in Table 10.3, taken from Ref. 10.4. The individual numbers depend on whether some of the very long-lived nuclei found in nature (Table 2.3) are included or not. The approximate equality of number of odd Z–even N and even Z–odd N nuclides means that neutrons and protons behave in an equivalent way in nuclear structure and there is no reason for a nucleus with an odd proton to be more stable than one with an odd neutron. The preponderance of even Z–even N nuclides is explained by the strong tendency of nucleons to form stable pairs. The presence of two unpaired odd nucleons, as in nuclides of odd Z and odd N, results in instability in all but the lightest nuclei ^{2}H, ^{6}Li, ^{10}B and ^{14}N, for which space-exchange forces (Sect. 18.1.3) are sufficiently strong to produce binding.

References

1. ASTON, F. W. *Mass Spectra and Isotopes*, Arnold, 1933.
2. MAYNE, K. I. 'Mass Spectrometry', *Rep. progr. Phys.*, **15**, 24, 1952.
3. MATTAUCH, J. H., THIELE, W. and WAPSTRA, A. H. *Nuclear Physics*, **67**, 1, 1965.
4. EVANS, R. D. *The Atomic Nucleus*, McGraw-Hill, 1955.
5. FEATHER, N. *Nuclear Stability Rules*, Cambridge University Press, 1952.
6. For discussions of nuclear matter see BROWN, G. E. *Nuclear Models*, North Holland, 1967; MOSZKOWSKI, S. A. 'Nuclear Structure and the Nucleon–Nucleon interaction', *Rev. mod. Phys.*, **39**, 657, 1967.

11 Nuclear charge, nuclear radius and nuclear moments

A nuclear charge value was first determined in the α-particle scattering experiments of Geiger and Marsden (ch. 2). The later and more accurate work of Chadwick showed that the nuclear charge was closely equal to Ze, and all subsequent experiments have confirmed this conclusion. The electrical neutrality of atoms and the neutron–proton nuclear structure are well tested hypotheses and together they imply that the charge of the proton and the electron are equal and opposite to a high degree of approximation. The exact equality of these charges has not remained unquestioned, particularly by the astrophysicist, but any difference appears to be beyond the present sensitivity of experimental investigation, and we shall assume that the nuclear charge is an integral multiple of the absolute charge of the proton.

As far as the early α-particle scattering experiments were concerned the nuclear charge could have been distributed through any volume of radius less than 3×10^{-14} m; it could have been located at a structureless point, as appears now to be a good approximation for Dirac-type particles such as the electron or muon. Extension of the α-particle experiments to light target nuclei (Sect. 5.3.1) first revealed deviations from inverse square law scattering which implied the existence of non-Coulomb forces at small distances or a finite nuclear size or both effects. Finite size is now an accepted and characteristic nuclear property and methods of determining it will be surveyed in the present chapter. Some evidence for shell effects in nuclear size has been found, but these are considerably less obvious than the effects in separation energy, since nuclear radii vary only as the cube root of the mass number.

The electromagnetic moments which arise as a result of the distribution of nucleons within a finite nucleus are not integral multiples of a fundamental unit, as is the case for angular momentum. These moments (ch. 4 and App. 2) are determined in a very direct way from the nuclear wave function, and if this wave function is essentially that of a single nucleon moving in a spherical field, then a simple and accurate prediction of moments may be expected. The single-particle shell model also offers precise information on parity and angular momentum.† From Fig. 9.5 there should be a change of ground state spin of odd mass nuclei at the completion of each sub-shell with given (l, j) and there may also be a change of parity. The spin and parity of the sequence

† We shall use the term spin for nuclear angular momentum for convenience, remembering that for a given nucleus it is a combination of the angular momenta of orbital motion and of intrinsic spin.

of single-particle excited states in an odd mass nucleus is also evident. If these predictions fail, it follows that the nuclear wave function must involve configurations other than that of a spherical core plus a single particle. If more than one particle is involved then the resultant *spin* may change as a result of a particular type of vector coupling, but so long as a sequence of shell model states is preserved the parity should remain predictable.

Nuclear statistics is invariably associated with the integral or half-integral nature of the nuclear spin. It depends only on the number of nucleons in the nucleus and not on their mode of motion, so that it offers no information on nuclear models beyond the simple and basic confirmation of the neutron–proton structure.

In this chapter we shall be primarily, though not exclusively, concerned with the properties of the nuclear ground state. Nuclear moments, nuclear radii and quantum numbers exist for unstable nuclei and for the excited states of all nuclei. In principle these are accessible to spectroscopic measurement and work of this sort is rapidly increasing. Most of the existing information on nuclear levels and unstable nuclei has come from a study of nuclear reactions including β-decay (ch. 16) and isomeric transitions (ch. 13).

11.1 The Coulomb barrier and the nuclear potential well

The nuclear potential wells considered in Chapter 9 (Fig. 9.4) were invoked mainly to provide a set of single-particle levels. The radius R is an important parameter of the square well, but for the purpose of defining levels the 'external' features of the well are less important. The finite well shown in Fig. 9.4 should, however, be taken to refer specifically to neutrons, since it implies that a charged particle approaching a nucleus from infinity feels no force until it comes within the range of the assumed potential. This cannot be true, and although neglect of the Coulomb potential may have no large effect on the order of single-particle levels, this potential is extremely important for radioactive decay and for low energy nuclear reactions. From the point of view of an incident proton or α-particle, the nuclear potential must be represented as in Fig. 11.1, which shows (a) the finite attractive nuclear well appropriate for neutrons, (b) the repulsive Coulomb potential for a point charge, and for a finite charge distribution, (c) the combination of (a) and (b) for a fairly light nucleus ($Z \approx 20$) and (d) the combination of (a) and (b) for uranium ($Z = 92$). The combination of a long-range electric force with a short-range nuclear force of opposite sign is consistent with the α-particle scattering experiments, according to which the scattering first of all falls off as the incident energy increases.

In a real nucleus the neutron well (Fig. 11.1(a)) and the proton well (Fig. 11.1(c)) must be filled with particles to about the same level, since otherwise one type of nucleon would decay into the other. Superposition of these two diagrams then shows that there will be a neutron-rich surface layer.† Evidence for and against this effect is discussed in Sect. **11.4**.

† M. H. Johnson and E. Teller, *Phys. Rev.*, **93**, 357, 1954.

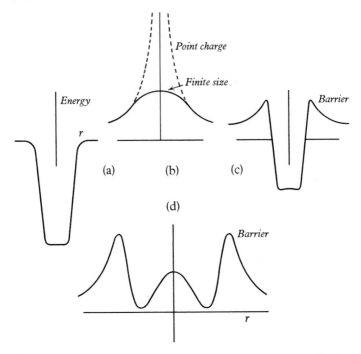

Fig. 11.1 Nuclear and Coulomb potentials. (a) Nuclear well, for neutrons. (b) Coulomb potential. (c) Addition of (a) and (b) for $Z=20$. (d) Addition of (a) and (b) for $Z=92$.

The potential well for charged particles is often idealized for the purposes of calculation as shown in Fig. 11.2(a). The radius R is then the distance from the centre at which the spherically symmetrical nuclear force becomes essentially zero and the height of the potential (Coulomb) barrier is given by

$$B = \frac{zZe^2}{4\pi\varepsilon_0 R} \tag{11.1}$$

for an incident particle of charge ze. For a uranium nucleus R is about 8×10^{-15} m and the barrier height is about 17 MeV per incident charge.† The α-particles from all naturally occurring radioactive bodies would therefore be scattered in accordance with the Rutherford law from this nucleus. Uranium itself, however, is an α-particle emitter and the observed particles of energy 4·2 MeV (^{238}U) must traverse the barrier. This is impossible classically but wave mechanically such penetration is reasonable and can be calculated (Fig. 11.2(b) and Sects. **14.2** and **16.2**). Classically, the α-particle ·

† A crude but useful estimate of the barrier height is

$$B = \frac{Z}{A^{1/3}} \text{ MeV}$$

which corresponds to a radius

$$R = 1\cdot4 \times 10^{-15} A^{1/3} \text{ m} = 1\cdot4 A^{1/3} \text{ fm}$$

has negative kinetic energy inside the barrier, but quantum mechanically the energy may be regarded as indefinite to an extent specified by the uncertainty principle during the time of passage through the barrier. At the point of emergence from the barrier (X in Fig. 11.2) the particle has zero kinetic energy but it is then accelerated by Coulomb repulsion from the nucleus until it attains its final energy at a large distance. The nuclear radius R in (11.1) is

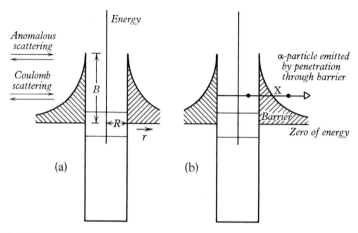

Fig. 11.2 Idealized potential well of radius R with barrier of height B showing on the energy scale: (a) Energies for which charged particles are scattered according to the Coulomb law and anomalously. (b) Emission of α-particles in a radioactive decay by tunnelling through the potential barrier.

that already used in Chapter 10 in discussion of the semi-empirical mass formula based on the liquid drop model of constant density and is written

$$R = r_0 A^{1/3} \tag{11.2}$$

The implied assumptions of Fig. 11.2 and eq. (11.2) that the nucleus has a sharp edge and that it is spherical have each had to be modified.

The radius R indicated in Fig. 11.2 is known as a *potential radius* since it represents the distance at which nuclear forces are first felt by a proton or neutron. Because of the finite range of nuclear forces this radius is slightly larger than (*a*) *the radius of the matter distribution and* (*b*) *the radius of the charge distribution*. The matter distribution can be inferred, under certain assumptions (Sect. **11.4**) from the potential radius. The charge radius (Sect. **11.2**) can be obtained directly by exploring the electromagnetic field of a nucleus with probes that are not sensitive to nuclear forces, such as high energy electrons or muons.

In this chapter we shall regard the spatial distribution of nuclear charge, i.e. the density of the nuclear protons, as the most important measure of nuclear size. This is not only because it is measurable with the greatest precision but because, owing to the strong attraction between neutrons and protons, it must be closely correlated with the total nucleon distribution. If the charge density is $\rho(r)$ (Fig. 11.5(b)) then the mean square radius of the distribution is given by

$$\overline{r^2} = \frac{\int_0^\infty r^2 4\pi r^2 \rho(r)\,dr}{\int_0^\infty 4\pi r^2 \rho(r)\,dr} \tag{11.3}$$

Some experiments determine primarily this quantity rather than the detailed form of $\rho(r)$. If $\overline{r^2}$ is known then the radius of the equivalent uniform spherical distribution of charge ('square well') is given by

$$R = \sqrt{\tfrac{5}{3}} (\overline{r^2})^{1/2} \tag{11.4}$$

We shall assume for the purposes of comparison of experimental data for a range of mass numbers A that R is related to A by the constant density formula (11.2).

11.2 The radius of the nuclear charge distribution

11.2.1 *Elastic scattering of fast electrons*

The reduced de Broglie wavelength $\lambda/2\pi$ of a 200 MeV electron is 10^{-15} m and such particles, which interact strongly only with electric charges, are suitable for studying the nuclear proton density. Low-energy electrons have too long a wavelength to respond to structural details and higher energy particles will perhaps interact chiefly with individual nucleons.

The principle of the Stanford linear electron accelerator, which has been used by Hofstadter in a series of electron scattering experiments of outstanding importance, was described in Sect. 8.2.2. Beams of energy up to 550 MeV were used for work on the lightest nuclei while for most of the heavier nuclei energies between 100 and 190 MeV were used in the first instance. The experimental arrangement is shown in Fig. 11.3. Scattered electrons are

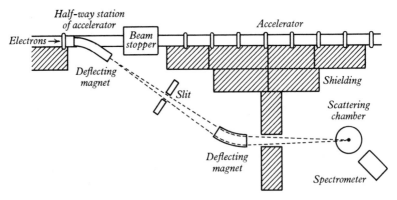

Fig. 11.3 Arrangement for observing scattering of 190 MeV electrons with the Stanford linear accelerator (Ref. 11.2).

detected, after momentum analysis by a double-focusing spectrometer, in a Cherenkov counter suitably screened from the intense background radiations produced by the accelerator. Figure 11.4 shows the energy distribution of 185 MeV electrons scattered from carbon. For the purpose of charge density determination attention is concentrated on the absolute cross-section and

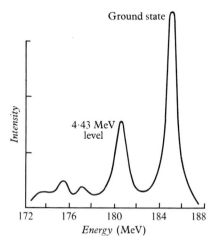

Fig. 11.4 Energy distribution of 185 MeV electrons scattered from carbon, showing elastic and inelastic peaks (Ref. 11.2).

angular distribution for the elastic peak; the inelastic scattering is important in the different context of the properties of excited levels (Sect. 13.6.3). The angular distribution of elastically scattered electrons for a carbon and a gold target are shown in Fig. 11.5(a), and the corresponding charge density functions in Fig. 11.5(b). The density for gold is typical of the results for heavy nuclei and may be expressed by the formula

$$\rho(r) = \frac{\rho_0}{1 + \exp\dfrac{r-c}{z}} \tag{11.5}$$

where c and z are adjustable parameters. This two-parameter fit, which characterizes the nuclear surface, is best obtained from the higher energy experiments. At low energies, say < 100 MeV, only the mean square radius is obtained.

The analysis of the observed angular distributions to give density functions was first approached in a way analogous to X-ray structure calculations. This is based on the Born approximation (Ref. 11.2) and gives a differential cross-section with

$$\sigma(\theta) = \left(\frac{Ze^2}{8\pi\varepsilon_0 E}\right)^2 \frac{\cos^2 \theta/2}{\sin^4 \theta/2} F^2 \tag{11.6}$$

where θ is the scattering angle (usually not corrected to the centre-of-mass system owing to the small mass of the electron, even at an energy E of 200 MeV, compared with that of the target) and F is a form factor. The quantity multiplying F^2 is the Mott expression for the scattering of a relativistic electron by a point charge, valid when $Ze^2/2\pi\varepsilon_0\hbar c \ll 1$. If the charge is spread out into a finite volume the scattering is less because of interference of waves from different parts of the volume. The form factor $F (\ll 1)$ is an integral, in accordance with Huygens' principle, of the charge density $\rho(r)$ times a phase factor over the nuclear volume. It may be written explicitly in terms of the momentum q transferred to the nucleus in the scattering:

$$F = F(q) = \int_0^\infty e\ i\mathbf{q}.\mathbf{r}/\hbar\ \rho(r)\ \mathrm{d}r \qquad (11.7)$$

where $q = 2E_i/c \sin \theta/2$ for incident electron energy E_i. The scattering falls to a small value compared with the point charge value at a scattering angle θ for which the phase factor reaches the interference minimum and this gives

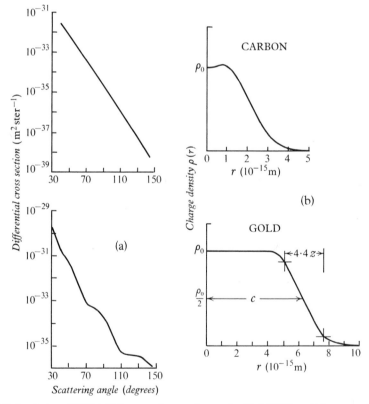

Fig. 11.5 Scattering of 183 MeV electrons from carbon and gold. (a) Differential cross-section per unit solid angle, showing diffraction features in the case of gold. (b) Density distribution $\rho(r)$ derived from scattering data (Ref. 11.2).

an estimate of nuclear size in a familiar way $(\lambda/R \approx \theta)$. Experimental values of F obtained from observations such as those of Fig. 11.5(a) do not show sharp minima but can be compared with calculations for different charge distributions. Unfortunately, this relatively simple procedure is only suitable for light nuclei since for heavier targets the Born approximation using plane waves is invalid owing to distortion of the incident and scattered waves in the Coulomb fields. A full phase-shift analysis (ch. 14) then becomes necessary; this does not give the sharp minimum of the plane wave approximation and is in better agreement with experiment.

The results of electron scattering experiments for a wide range of nuclei indicate that the half-way radius c (Fig. 11.5(b)) is proportional to $A^{1/3}$, that the surface thickness parameter z is approximately constant at about 0·5 fm, and that the radius R of the equivalent uniform distribution may be written in the form (11.2) with

$$
\left.
\begin{aligned}
r_0 &= 1\cdot32 \text{ fm} \quad \text{for } A < 50 \\
r_0 &= 1\cdot21 \text{ fm} \quad \text{for } A > 50
\end{aligned}
\right\} \tag{11.8}
$$

The hypothesis of constant nuclear density $[\propto A/(4\pi R^3/3)]$ is verified.

An important application of high-energy electron scattering is the exploration of the electric and magnetic structure of the proton and the neutron. These particles have anomalous magnetic moments and this is explained following Pauli by assuming that each dissociates into a core and a meson cloud, which contributes the anomalous part of the moment. The nucleons have a radius of about 10^{-15} m and exploration of the meson cloud demands the use of the highest possible electron energy (6 GeV has been used). The outcome of such work is discussed in Sect. 21.4.3.

For low energy electrons, the neutron–electron interaction may be represented by a potential well of 3,700 eV.[†] This is insufficient to permit the formation of a bound neutron–electron system.

11.2.2 Bound electrons

The *optical isotope shift* has already been mentioned (Sect. 4.2.2) as an example of the effect of nuclear properties on atomic spectra. The penetration of the wave functions of optical *s*-electrons into the nuclear volume was shown in 1932 by Breit to result in a decrease in the electron binding energy by an amount depending on the spatial extent of the charge. The modified Coulomb potential due to the finite charge distribution is shown in Fig. 11.1(b). This is a small effect and cannot be obtained from observations with a single nucleus because the point charge energy values cannot be calculated sufficiently accurately. The difference in the energy values for two isotopes such as ^{196}Hg–^{198}Hg can, however, be calculated in terms of the change in nuclear radius due to the different mass (providing that the correction for the change

[†] V. E. Krohn and G. R. Ringo, *Phys. Rev.*, **148**, 1303, 1966.

of mass itself can be made). This is an even smaller effect, amounting to perhaps 1 part in 10^6 at the most, but the techniques of optical spectroscopy are sufficiently accurate to yield useful results, and have indeed been successful with sources of radioactive ^{197}Hg containing as few as 5×10^{12} atoms. The isotope shift essentially determines the change in $\overline{r^2}$ between the two nuclei compared; the shifts observed are consistent with $r_0 = 1 \cdot 2$ fm.

The uncertainties in the calculation of electronic energy level shifts are considerably reduced for the inner levels, which may be studied by X-ray methods. The $K\alpha$ X-ray shifts have been measured accurately for several stable pairs of isotopes using a curved crystal spectrometer. The results are also consistent with $r_0 = 1 \cdot 2$ fm. Both optical and X-ray isotope shifts are enhanced for a neutron number $N = 90$, presumably owing to the incidence of nuclear deformation.

A change in nuclear radius also occurs when a nucleus is excited. For long lived isomeric states the corresponding change of frequency in optical lines (*isomer shift*) may be observed (e.g. in ^{197}Hg).

11.2.3 Bound muons (muonic atoms)

The muon, unlike its parent particle, the π-meson, is known experimentally to interact with nuclei mainly through the Coulomb force. Recent accurate measurements of the magnetic moment of the muon yield a value which is close to that expected for a Dirac particle of spin $\frac{1}{2}$, such as the electron (Sect. 19.3.1). The electron and muon are therefore both suitable particles for studying the nuclear charge distribution. The accuracy of high energy muon scattering experiments is limited by source strength, but the intensities of stopped muons are amply sufficient to permit the observation of muonic atoms. A positive muon approaching a nucleus is repelled and disappears by decay into electron and neutrinos with a mean life of $2 \cdot 2$ μs. A negative muon however can enter a bound state, losing energy by radiative transitions, in the same way that an electron is captured by a positive ion. The Bohr theory of hydrogen-like atoms shows that muonic orbits are smaller in radius than electronic orbits of the same quantum number by a factor $m/m_\mu = 1/207$, so that the muon can exist in bound states whose wave functions lie mainly within those of the electrons of an ordinary atom. It may be shown that such a meson captured into an atom will pass rapidly, in perhaps 10^{-13}–10^{-14} s, from loosely bound 'optical' levels to the tightly bound mesonic X-ray levels; the energy appears as radiation or Auger electrons from the ordinary outer shells of the atom. The binding energies of the muonic levels are a factor $m_\mu/m = 207$ greater than those of the similar electron levels and the $K\alpha$ X-ray line of a muonic atom may have an energy of several MeV. The nuclear size effect, which is difficult to observe in the ordinary electronic transitions, is greatly magnified in the muonic atom because of the proximity of the muonic orbits to the nucleus; the nuclear radius for $Z = 45$ is about equal to the Bohr radius for a muon in the K-shell. In heavy atoms the K-orbit may be entirely within the nucleus and such penetration greatly reduces the binding energy of the muon in the corresponding state. The reduction of energy in comparison with that expected for a point nucleus may be calculated for postulated charge distributions and compared with that deduced from the observed energy for the $L \longrightarrow K$ X-ray. The lifetime of the muonic $2p$ state is probably about 10^{-18} s, corresponding to a natural width of 1 keV. The muon terminates its career either by normal decay in the K-shell or, for penetrating orbits, occasionally by nuclear capture with star production.

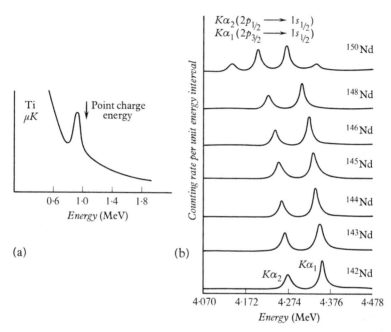

Fig. 11.6 Spectra of muonic K X-rays. (a) From Ti, using NaI scintillation detector (Fitch and Rainwater, 1953). (b) From Nd isotopes, using Ge(Li) counter (E. R. Macagno *et al.*, *Phys. Rev.*, **C1**, 1202, 1970).

In the work of Fitch and Rainwater† which established the value of the muons as a nuclear probe, π-mesons from a cyclotron decayed into μ-particles which were then allowed to enter a block of the material under investigation. The π-mesons were removed from the beam of incident particles by a suitably chosen copper absorber. The μK X-rays were detected in a sodium iodide crystal; Fig. 11.6(a) shows the type of result obtained. The availability of Ge(Li) detectors has very largely improved the accuracy of such measurements, as shown in Fig. 11.6(b), and has made it easily possible to resolve the $2p_{3/2, 1/2}$ spin-orbit fine structure, giving separated $K\alpha_1$ and $K\alpha_2$ lines ($2p_{3/2, 1/2} \longrightarrow 1s_{1/2}$ muonic transitions).

The muonic X-ray observations determine the root mean square radius of

† V. L. Fitch and J. Rainwater, *Phys. Rev.*, **92**, 789, 1953.

the nuclear charge distribution for spherical nuclei, and results are in general agreement with those from electron scattering. For the heavier elements, two-parameter fits to a Fermi distribution may be made, giving the half-density radius and skin thickness. An equivalent uniform charge distribution with $r_0 = 1 \cdot 15 \pm 0 \cdot 03$ fm is consistent with the data.

Figure 11.6(b) also shows the isotope shift in muonic atoms (of neodymium). Like the electronic X-ray isotope shift, the muonic X-ray shift can be calculated accurately and comparison with the optical shifts can be used to remove some of the uncertainties of the latter. The isotope shifts also give information on the nuclear charge distribution and there is general agreement between the results of optical, electronic X-ray and muonic X-ray measurements (Ref. 11.5).

11.2.4 Coulomb energy of nuclei

The semi-empirical mass formula (Sect. **10.3**) contains a term

$$a_c \frac{Z^2}{A^{1/3}} = \frac{3}{5} \frac{e^2}{4\pi\varepsilon_0 r_0} \frac{Z^2}{A^{1/3}} = E_c \qquad (11.9)$$

representing the Coulomb potential energy due to the nuclear charge. Surveys of nuclear binding energies can therefore in principle lead to a value for r_0, but in practice ambiguity arises because of the presence of other adjustable constants in the formula. This difficulty can be avoided if binding energies of isobars are compared.

Figure 11.8 shows schematically the structure of three nuclei; nucleus (a) contains equal numbers of neutrons and protons and nuclei (b) and (c) differ from (a) by the addition of a proton and neutron respectively. If the main binding energy of these nuclei is assumed to arise from (nn), (pp) and (np) interactions it is clear that nuclei (b) and (c) differ in this respect by the replacement of a number of (pp) bonds by the same number of (nn) bonds. If nuclear forces are *charge symmetric* the potential energies associated with (pp) and (nn) interactions with the nucleon pairs in the same spatial states are the same. The difference in binding energy between the odd mass isobaric nuclei (b) and (c), is then to a good approximation due to the difference in Coulomb energy arising from the nuclear charge. This can be calculated with the assumption of a particular charge distribution; for a uniform spherical distribution of radius R the expression (11.9) may be used as a first approximation. The difference in binding energy between the two isobars shown in Fig. 11.7, which are known as *odd mirror nuclei* since they differ merely by exchange of neutron and proton numbers, is

$$\Delta E_c = \frac{3}{5} \frac{e^2}{4\pi\varepsilon_0 r_0 A^{1/3}} [Z^2 - (Z-1)^2]$$

$$= \frac{3}{5} \frac{(2Z-1)e^2}{4\pi\varepsilon_0 r_0 A^{1/3}} \qquad (11.10)$$

For $A > 3$ the nucleus of greater charge Z is unstable with respect to its mirror isobar and decays into it by positron emission or electron capture, e.g.

$$^{13}_{7}N \longrightarrow {}^{13}_{6}C + \beta^{+}(+\text{neutrino}) \tag{11.11}$$

and the maximum kinetic energy T_0 of the positron spectrum is determined by the Coulomb energy change ΔE_c and by the fact that a proton changes into a neutron and a positron. In terms of the *nuclear* masses of these particles

$$T_0 = \Delta E_c - (M_n - M_p) - m_e c^2$$
$$= \Delta E_c - 1 \cdot 80 \text{ MeV} \tag{11.12}$$

and since $A = 2Z - 1$ for the mirror nucleus of greater charge this becomes, substituting from 11.10,

$$T_0 = \frac{3}{5} \frac{e^2}{4\pi\varepsilon_0 r_0} A^{2/3} - 1 \cdot 80 \text{ MeV} \tag{11.13}$$

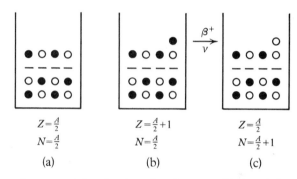

$$Z = \frac{A}{2} \qquad Z = \frac{A}{2} + 1 \qquad Z = \frac{A}{2}$$
$$N = \frac{A}{2} \qquad N = \frac{A}{2} \qquad N = \frac{A}{2} + 1$$

(a)   (b)   (c)

Fig. 11.7 Schematic representation of positron decay between odd mirror isobars. If nuclear forces are charge symmetric, the addition of a proton (●) to the nucleus (a) results in the same increase of binding energy as does that of a neutron (○).

Experimental measurements of the decay energy T_0 of the odd mirror nuclei can therefore be used to obtain r_0. An equivalent method is the determination of the energy balance in a (p, n) reaction between isobars,[†] e.g. $^{13}C(p, n)^{13}N$.

Figure 11.8 shows the existing data, which verify the relation (11.13) in fair detail for nuclei up to $A = 40$. Close examination of the most accurate experimental information, including evidence obtained from even mass nuclei assuming charge independence (Sect. **12.1**), reveals discontinuities in the Coulomb energies at major closed shells, e.g. at $N = Z = 2$, 8 and 20. Differences also appear between nuclei of mass number $4n + 1$ and $4n + 3$ which may be attributed to pairing energy effects. The method of this section has not so far been applied to nuclei with $A > 41$. In this region of mass numbers the odd mirror nuclei are both unstable and are found at an

† J. D. Anderson, C. Wong and J. W. McClure, *Phys. Rev.*, **138**, B615, 1965.

increasing distance from the line of β-stability so that their half-lives tend to be short and their decay schemes are not fully known. In addition the increasing Coulomb forces seriously disturb the neutron–proton symmetry of structure which is such a marked feature of lighter nuclei.

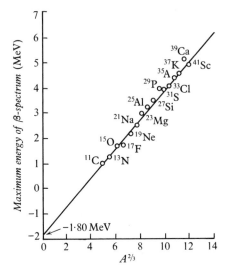

Fig. 11.8 Maximum positron energy for the light nuclei with $A = 2Z - 1$ (Ref. 11.6).

The first application of this method to the determination of nuclear radii gave $r_0 = 1 \cdot 45$ fm which is in disagreement with electron scattering results. Corrections however must be introduced for the effect of the Pauli exclusion principle in keeping the protons apart and for $A < 50$ it now appears that

$$r_0 = 1 \cdot 28 \pm 0 \cdot 05 \text{ fm} \tag{11.14}$$

11.3 The potential radius

11.3.1 Alpha particle scattering

With the advent of strong beams of α-particles accelerated in cyclotrons to an energy of about 40 MeV, it has been possible to extend the classical experiments on α-particle scattering to the heaviest nuclei of the periodic table. The measurements usually made are (a) the energy variation of the scattering at a given angle, and (b) the angular distribution of scattering at a given energy. Figure 11.9(a, b) shows the type of results obtained for heavy nuclei. Both experiments show that the scattering falls below the Coulomb value at an angle or energy corresponding to a certain critical distance of approach, at which it is assumed that strong absorption sets in. This distance is related to the nuclear radius.

For light nuclei, high energy α-particles are affected by the nuclear force even in glancing collisions and the angular distribution in this case gives a very convincing demonstration of the finite nuclear size. Figure 11.9(c) shows that marked diffraction-like peaks appear and these may be simply (and indeed quantitatively) interpreted as in the case of electron scattering by multiplying the point charge scattering formula by a form factor which takes account of interference between waves scattered from different points of the nuclear volume. In the case of α-particle scattering the strong nuclear interaction confines interference effects to the rim of the nucleus, since α-particles which penetrate into the nucleus will not usually emerge without loss of energy, and it is therefore a good approximation to treat the scattering as

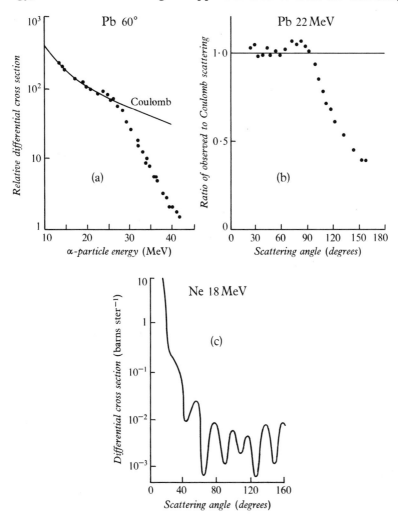

Fig. 11.9 Scattering of α-particles by nuclei. (a) Heavy nucleus, variation of scattering with energy. (b) Heavy nucleus, variation of scattering with angle. (c) Light nucleus, variation of scattering with angle. (Eisberg and Porter, *Rev. mod. Phys.*, **33**, 190, 1961.)

diffraction from an absorbing disc of radius $R_{abs} = R + R_\alpha$ where R_α is the radius of the α-particle. The familiar small-angle optical formula

$$I(\theta) \propto \left[\frac{\mathcal{J}_1(2kR_{abs}\sin\theta/2)}{2kR_{abs}\sin\theta/2} \right]^2 \qquad (11.15)$$

where k is the wave number ($= 2\pi/\lambda$) then gives the intensity observed as a function of angle and permits a value for R_{abs} and thence R to be obtained. Alternative methods of analysis, including that of the optical model (ch. 15) have been applied to the same problem. The values for r_0 obtained depend on the mass number; a recent study† for nuclei with A between 12 and 124 indicates $r_0 = 1.52$ fm.

11.3.2 Alpha radioactivity

The wave mechanical theory of α-decay is discussed in Sect. **16.2**. Essentially it expresses the α-emission probability as a product of an internal disintegration probability and an exponential factor measuring the probability of a particle leaking out through a barrier as shown schematically in Fig. 11.2. The penetrability factor is a rapidly varying function of the nuclear radius and could be used to determine this quantity if the internal disintegration probability were known. Unfortunately no reliable theoretical predictions for the quantity exist and usually it is regarded as the unknown in this type of calculation, to be deduced from penetrability values based on assumed interaction radii. Wilkinson‡ has shown that for even–even nuclei the intrinsic emission probabilities group together most closely for an effective barrier of the type shown in Fig. 11.2 with $r_0 = 1.57 \pm 0.06$ fm. The nuclear charge distribution, as determined by electron scattering, lies well within this barrier.

11.3.3 Reaction and scattering cross-sections in general

Alpha-particle scattering and α-radioactivity have a special significance for nuclear size determination mainly for historical reasons. Practically all nuclear processes in fact similarly involve a radius of interaction and can provide values for this parameter which are comparable in accuracy with those obtained from the α-particle data. Analysis usually requires, in addition to penetrability calculations, some particular theory of nuclear reactions. These theories (ch. 15) depend very much on the energies involved.

The experiments which have been used to evaluate r_0 in (11.2) are:

(i) *The total cross-section for the interactions of neutrons with nuclei.*

For neutrons of energy 10–20 MeV a simple classical argument (Sect. 14.2.2) suggests that the total cross-section should be:

$$\sigma_t = 2\pi R^2$$

† J. C. Faivre, H. Krivine and A. M. Papiau, *Nuclear Physics*, **A108**, 508, 1968.
‡ *Phys. Rev.*, **126**, 648, 1962.

or more accurately

$$\sigma_t = 2\pi(R + \lambda)^2$$

where λ is the reduced de Broglie wavelength of the incident particle. The absorption cross-section in the high-energy limit is:

$$\sigma_{abs} = \pi R^2$$

In these formulae R is a potential radius and includes a contribution due to the range of nuclear forces. Neglecting this correction it is found that

$$r_0 = 1 \cdot 25 \pm 0 \cdot 02 + (0 \cdot 6 \pm 0 \cdot 15)A^{-1/3} \text{ fm} \tag{11.16}$$

(ii) *Angular distributions for the elastic scattering of protons in the energy range 5–300 MeV.*

These give characteristic diffraction patterns on a scale determined by the ratio λ/R as in α-particle scattering. Very precise measurements in the region of 10–50 MeV show that the concept of a sharp nuclear boundary is unsuitable, since it leads to too much backward scattering. A diffuse nuclear edge, specified as a potential with a radial dependence

$$f(r) = \frac{1}{1 + \exp\dfrac{r - R_1}{a}} \tag{11.17}$$

where R_1 is the 'half-way' radius, gives good agreement with experiment. This radial dependence of the potential is similar to the electron scattering results for the charge density of the heavier nuclei (eq. 11.5), and the surface thickness parameter a agrees closely with the quantity z found from electron scattering. The radius parameter for the equivalent square well is given by optical model analysis as

$$r_0 = 1 \cdot 33 \text{ fm} \tag{11.18}$$

These determinations of potential radius usually involve some ambiguity since the quantity yielded by the analysis is Ur_0^n where U is the depth of the potential well assumed in the analysis and $n \approx 2$. The well depth and r_0 may be found separately when extra data, such as the energy variation of a cross-section, are available.

11.4 The matter radius

The potential radius exceeds the matter radius because of the finite range of nuclear forces. In a reformulated version of the scattering analysis, or optical model (Sect. **15.5**), Greenlees, Pyle and Tang[†] introduce an effective nucleon–nucleon force and fold this in with an assumed matter density $\rho_m(r)$ to obtain the scattering potential. Comparison with observed proton

[†] *Phys. Rev.*, **171**, 1115, 1968.

scattering then yields $\rho_m(r)$ which is related to individual neutron and proton densities by the expression

$$\rho_m(r) = \rho_p(r) + \rho_n(r) \tag{11.19}$$

The charge distribution $\rho_p(r)$ is known (Sect. **11.2**), so the experimental results can be used to predict the neutron distribution $\rho_n(r)$. For $A \gtrsim 100$ the latest analysis of this type gives clear evidence that the neutron distribution extends about 0·15 fm beyond the proton distribution, as suggested by Johnson and Teller (Sect. **11.1**).

The conclusion that the heavier nuclei have a neutron-rich skin is supported by observations[†] that negative kaons, which should be absorbed in the outer regions of the nuclear density, appear to interact more frequently with neutrons than with protons (Sect. 20.5.3). On the other hand, pion absorption experiments are consistent with equal neutron and proton density distributions and this is the expectation on the general ground that neutrons are strongly attracted to protons by a charge–exchange force. Theoretical guidance on this question is given by shell model calculations for ^{208}Pb (Rost, Elton[‡]) in which nuclear potentials capable of predicting observed single particle states are used to generate single-particle wave functions consistent with electron scattering results. For this nucleus at least there is some support for a neutron rich surface. Other calculations disagree on the actual magnitude and form of the excess distribution, and the question is still open both experimentally and theoretically.

A high energy experiment which directly measures matter radii by use of a neutral probe has recently been reported by Alvensleben *et al.*[§] Using bremsstrahlung from a 7·5 GeV electron synchrotron these authors studied the photoproduction of ρ^0-mesons from nuclei

$$\gamma + A \longrightarrow A + \rho^0 \tag{11.20}$$

and concluded that the observed angular distributions were consistent with a matter distribution of the form 11.17 with $r_0 = 1\cdot12 \pm 0\cdot02$ fm which is close to the electron scattering value for the charge distribution.

11.5 Parity of nuclear levels

The predictions of the single-particle shell model for the parity of many nuclear levels and in particular for that of the ground state, are unambiguous. Nucleons have even intrinsic parity by convention and the parity of a nucleon state in a potential well is determined immediately by the orbital motion. The angular part of the wave function (App. 1) is proportional to $P_l (\cos \theta)$ and the known properties of the Legendre polynomials show that the parity of the nucleon state is then $(-1)^l$. The parity of the Z protons of the nucleus is

[†] E. H. S. Burhop, *Nuclear Physics*, **B1**, 438, 1967.
[‡] *Phys. Lett.*, **26B**, 689, 1968.
[§] *Phys. Rev. Lett.*, **24**, 792, 1970.

therefore even if only even-l states, or an even number of odd-l states, are occupied and is odd if an odd number of odd-l states are filled. The parity of the N neutrons is similarly determined. The parity of the nucleus as a whole is the product of the parities of the neutrons and protons, and since their motions are independent this parity is simply determined by the distribution of the A nucleons between even and odd orbital states. Thus we have

for $A = 6$, configuration $1s^4 1p^2$, parity even $(+)$,
for $A = 7$, configuration $1s^4 1p^3$, parity odd $(-)$,
for $A = 19$, configuration closed shell $+1d^2 2s^1$, parity even $(+)$.

This classification is independent of the intrinsic spin of the nucleons and of the total angular momentum quantum number I of the nuclear level. The parity of the nuclear ground state is thus predicted directly from the occupation numbers of the shell model states. For a given nucleus, excited states exist corresponding to the elevation of a nucleon from one singe-particle orbit to another, and the parity of this sequence of single-particle states is immediately obvious. Thus for $A = 13 (1s^4 1p^9)$ the ground state is of odd parity, but the configurations $(1s^4 1p^8 2s)$ and $(1s^4 1p^8 1d)$ with even parity should also be found at an excitation of a few MeV. The observed spectrum for ^{13}C and ^{13}N exhibits these states.

Methods of parity investigation strictly determine only change of parity between two states. The specific methods which have been much used are:

1. *Direct nuclear interactions*, and in particular the (d, p) stripping reaction (Sect. 15.6.4) of which a typical example is

$$^{16}\text{O} + \text{d} \longrightarrow \, ^{17}\text{O} + \text{p} + 1 \cdot 92 \text{ MeV} \qquad (11.21)$$

It has been found that a large number of nuclear reactions may be well described on the simple assumption that a nucleon taking part in the process enters or leaves a definite shell-model state of the target nucleus. Since such states have a definite orbital momentum $\mathbf{l}$ the process imposes a particular condition on the wave-function of the emitted particle, and the angular distribution of this particle has a characteristic form. From the position of the strong forward maximum in the diffraction-like pattern the angular momentum $\mathbf{l}$ taken into the target nucleus by the absorbed particle can be deduced. A great many examples of this type of reaction have been studied; the angular distribution analysis is above all a determination of *parity* change, although useful limits may also be set on nuclear angular momentum (I-value).

2. *Resonance reactions, decay-scheme analysis and angular correlations.* Although direct interactions probably give the least ambiguous information on parity changes, practically every nuclear process is sensitive to this quantity. In heavy particle reactions the orbital quantum number of a particle may determine the ease with which it can penetrate a potential barrier and the yield is then determined by the parity change. Decay schemes involving

electromagnetic radiation are a prolific source of information, since spin and parity changes determine multipolarity and the electric or magnetic character of the radiation (Sect. 3.5.2) which in turn determine observed conversion coefficients (ch. 13). Angular correlations between successive γ-quanta (Sect. 17.4.2) do not yield parities directly, but this information can be obtained from simultaneous measurements of polarization.

3. *Electromagnetic moments.* As will be shown in Section **11.7**, the static magnetic moment of an odd-*A* nucleus often gives a clear indication of the *l*-value of the odd nucleon. Similar but more involved deductions can be made for even-*A* nuclei. Electric quadrupole moments can be used in the same way.

The results of a large number of investigations of this type confirm in detail the parities of the sequence of single-particle states predicted by the shell model. Many states besides single-particle states exist, and parity is also a good quantum number for these, but the corresponding nuclear motion is more complicated and involves the behaviour of more than one nucleon.

11.6 Angular momentum (spin) and statistics

The methods by which the spin and statistics of stable nuclei have been determined are listed in Table 4.1. The spins of unstable nuclei and of the excited states of all nuclei are mainly inferred from nuclear reaction and decay scheme analyses, including angular correlation studies (ch. 17) although atomic beam methods are increasing in ability to deal with small quantities of radioactive material.

Nuclear spins are small integral multiples of the fundamental spin $\frac{1}{2}\hbar$, which is known experimentally to be the spin of both neutron and proton. This fact suggests that in the neutron–proton model the individual neutrons and protons in the same spatial state tend to pair off with opposite total angular momenta *j*.† In the single particle shell model it is specifically assumed that this takes place not only within the closed shells constituting the nuclear core, but also for the equivalent nucleons outside such shells. The model then predicts that odd-*A* nuclei *have a ground state spin given by the j-value of the last unpaired nucleon.* The spin sequence is then directly obtainable from Fig. 9.5 and the experimental values largely confirm this; spin values of odd *mass* nuclei are all odd multiples of $\frac{1}{2}\hbar$ and their values agree in detail with what is expected from the parities of the corresponding orbits and the spin-orbit coupling hypothesis. Allowance must be made in certain cases for the effect of the pairing energy in states of high *j* which, for example, makes the proton configuration $p_{3/2}^3 f_{5/2}^2$ (^{75}As, spin $=\frac{3}{2}$) more stable than the configuration $p_{3/2}^4 f_{5/2}$ (spin $=\frac{5}{2}$). Other exceptions to the simple sequence

---

† This may be compared with the atomic case, in which according to Hund's Rule, the ground state of a configuration of equivalent electrons is the state of highest multiplicity. The difference arises because electrons mutually repel, so that they tend to stay apart spatially and the spins can then be parallel.

(e.g. ^{23}Na, $d^3_{5/2}$ spin $=\frac{3}{2}$) cannot be explained in this way, and make it clear that the simple shell model stands in need of extension. The statistics of odd mass nuclei has been determined in a number of cases and is always of the Fermi–Dirac type. This is expected for an odd number of nucleons obeying these statistics, since the complete exchange of nucleons between a pair of odd mass nuclei just multiplies the wave function of the system by $(-1)^A$.

Even-A nuclei have either Z, N both even, or both odd. The single-particle shell model (and all other nuclear models) predicts that:

even N–even Z nuclei have zero ground state spin

owing to the pairing effect. Experimentally, no exception is known to this rule. The even–even nuclides are the largest group of even-A nuclei; for the four stable odd–odd nuclei and for the much larger number of unstable nuclei of this class the single-particle model makes no prediction because the angular moments of two unpaired nucleons must be combined in a way that is not specified by the model. Experimentally most odd–odd nuclei have $I > 0$, reflecting a tendency for the odd proton and neutron to align with parallel spins as in the deuteron. More detailed theoretical prediction requires an extension of the simple shell model. Even-A nuclei obey Einstein–Bose statistics; it was the experimental determination of this for ^{14}N which, among other reasons, rendered the proton–electron model of the nucleus untenable.

11.7 Nuclear magnetic dipole moments

The experimental determination of nuclear magnetic moments, like that of spins, has so far been mainly possible for the ground states of stable nuclei; the methods used, which generally give the nuclear g-factor g_I or the hyperfine interaction constant a in the first place are listed in Table 4.1. It has recently become possible in certain special cases to deduce magnetic moments of excited states of nuclei by observing the perturbation of angular correlations involving these states by a superimposed magnetic field (ch. 17). The most direct method of magnetic moment determination, the nuclear Zeeman effect, has in essence been used in a few favourable cases in which γ-ray lines of natural width are observed as a result of the Mössbauer effect (Sect. 13.6.4).

The single-particle shell model predictions for the magnetic moments of odd mass nuclei are not as simple as for spins because of the different g-factors g_s and g_l for intrinsic spin and orbital motion. For a single nucleon moving in a potential well with orbital angular momentum $\mathbf{l}$ and total angular momentum $\mathbf{j}$, where

$$\mathbf{j} = \mathbf{l} + \mathbf{s} \qquad (11.22)$$

the calculation of the nuclear g-factor g_I $(I = j)$† follows that of the atomic

† For a single-particle nucleus the spin I is just the total angular momentum j of the odd particle.

g-factor for a single electron atom. By using vector diagrams, it may be shown that

$$\mu_I = g_I \mu_N I \tag{11.23}$$

where

$$g_I = \frac{1}{2} \left[(g_l + g_s) + (g_l - g_s) \frac{l(l+1) - \frac{3}{4}}{j(j+1)} \right] \tag{11.24}$$

This expression can now be evaluated for a single proton or neutron using known *g*-factors. Two cases arise, corresponding to the two possible orientations of the nucleon spin with respect to the orbital vector $j = l \pm \frac{1}{2}$.

For an odd *neutron* ($g_l = 0$, $g_s = -3\cdot826 = 2\mu_n/\mu_N$, where μ_n is the neutron moment) we find

$$\left. \begin{array}{ll} \mu_I = \frac{1}{2} g_s \mu_N = \mu_n & \text{for } I = j = l + \frac{1}{2} \\[2mm] \mu_I = \frac{-j}{j+1} \frac{1}{2} g_s \mu_N = \frac{-j}{j+1} \mu_n & \text{for } I = j = l - \frac{1}{2} \end{array} \right\} \tag{11.25}$$

and for an odd *proton* ($g_l = 1$, $g_s = 5\cdot585 = 2\mu_p/\mu_N$)

$$\left. \begin{array}{ll} \mu_I = [(j - \frac{1}{2}) + \frac{1}{2} g_s] \mu_N = (j - \frac{1}{2})\mu_N + \mu_p & \text{for } I = j = l + \frac{1}{2} \\[2mm] \mu_I = \frac{j}{j+1} [(j + \frac{3}{2}) - \frac{1}{2} g_s] \mu_N = \frac{j}{j+1} [(j + \frac{3}{2})\mu_N - \mu_p] & \text{for } I = j = l - \frac{1}{2} \end{array} \right\} \tag{11.26}$$

If a figure showing μ_I as a function of $I = j$ is drawn we obtain the *Schmidt diagrams*, as in Fig. 11.10. The lines drawn in this figure have a meaning only at the half-integral spin values corresponding to observed nuclear spins. Figure 11.10 also gives observed values of μ_I for a number of nuclei with known spin; if these are truly single-particle nuclei, the magnetic moments (which are not quantized) should fall on the Schmidt lines.

Inspection of Fig. 11.10 shows that there is some correlation between observed moments and the Schmidt lines. In particular the magnetic moment follows the spin change at shell closure. With the exception, however, of a few single particle nuclei, such as ^{3}H, ^{3}He, ^{17}O, ^{39}K, observed moments tend to fall between the Schmidt lines, though rather nearer to one than to the other. It is this feature which enables magnetic moments to be used, with some confidence, in deducing the orbital quantum number, and hence the parity of the nuclear ground state. Thus we have, as an example of the method for odd proton nuclei:

$$^7_3\text{Li} \quad I = \tfrac{3}{2} \quad \mu_{\text{exp}} = 3\cdot26 \quad \mu_{\text{calc}} = 3\cdot79 \ (p\text{-orbit})$$
$$= 0\cdot12 \ (d\text{-orbit})$$

in units of μ_N, so that a *p*-orbit is indicated, and

$$^{23}\text{Na} \quad I = \tfrac{3}{2} \quad \mu_{\text{exp}} = 2\cdot22 \quad \mu_{\text{calc}} = 3\cdot79 \ (p\text{-orbit})$$
$$= 0\cdot12 \ (d\text{-orbit})$$

so that some mixing of states is indicated, as is already expected since $I = \frac{3}{2}$

is a spin anomaly, and

$$^{39}_{19}\text{K} \quad I = \tfrac{3}{2} \quad \mu_{\text{exp}} = 0\cdot22 \quad \mu_{\text{calc}} = 0\cdot12 \ (d\text{-orbit})$$

$$^{45}_{21}\text{Sc} \quad I = \tfrac{7}{2} \quad \mu_{\text{exp}} = 4\cdot76 \quad \mu_{\text{calc}} = 5\cdot79 \ (f\text{-orbit})$$

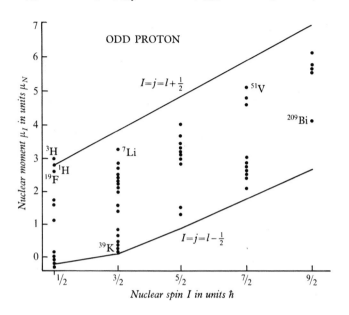

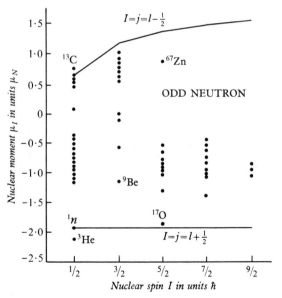

Fig. 11.10 Schmidt diagrams showing magnetic moment of odd-mass nuclei as a function of nuclear spin. The lines show the magnetic moments expected for a single nucleon at the spin values $j = l \pm \tfrac{1}{2}$, where l is the orbital momentum quantum number (Ref. 11.10).

which shows the transition between shells at magic number 20. For odd-neutron nuclei the evidence is not quite so clear but we note

$$^{129}_{75}\text{Xe} \quad I = \tfrac{1}{2} \quad \mu_{\text{exp}} = -0\cdot78 \quad \mu_{\text{calc}} = -1\cdot91 \text{ (s-orbit)}$$

$$^{137}_{81}\text{Ba} \quad I = \tfrac{3}{2} \quad \mu_{\text{exp}} = 0\cdot94 \quad \mu_{\text{calc}} = 1\cdot1 \quad \text{(d-orbit)}$$

Attempts have been made to modify the single particle model to account for the deviations from the Schmidt lines, but the only successful approach is the generalization of the model into the individual particle model or the collective model (ch. 12). The magnetic moments of even–even nuclei are zero as expected since the nuclear spin is zero and the charge distribution is spherically symmetrical.

The nucleons are mirror nuclei and have moments $\mu_n = -1\cdot91$ μ_N and $\mu_p = 2\cdot79$ μ_N. The mirror nuclei (^{3}H, ^{3}He) might also in a first approximation be expected to have just these moments. This expectation will however be modified if there are mesonic currents in the nucleus. Such currents may well be of opposite effect in the two members of a mirror pair and if so, the deviations of the moments from the nucleon value should be equal and opposite. There is some experimental evidence for this.†

11.8 Nuclear electric quadrupole moments

The existence of nuclear electric quadrupole moments (App. 2) was first demonstrated spectroscopically and spectroscopic methods of determining this quantity, which are mainly observations of the quadrupole coupling constant b, are listed in Table. 4.1. Electric quadrupole effects are observable in the spectra of muonic atoms where they are enhanced in comparison with the electronic case (Ref. 11.5). Nuclear methods of determining this moment, which is the lowest order electric moment observable as a static nuclear property, will be discussed in Sect. 12.3.5. For the purpose of comparing observed quadrupole moments with predictions of the single-particle shell model we require to know the quadrupole moment for a single proton moving in a shell model orbit; this has the same sort of significance for the electric moment as has the nuclear magneton for the magnetic dipole moment.

From the general definition of quadrupole moment (App. 3) we need to evaluate the average value of the quantity

$$(3z^2 - r^2) \tag{11.27}$$

for a proton of total angular momentum j. The calculation gives

$$Q_{1p} = -\frac{2j-1}{2j+2}\overline{r^2} \tag{11.28}$$

where $\overline{r^2}$ is the mean square radius of the charge distribution as defined in Sect. **11.1**. The single proton in orbital motion thus has a negative quadrupole

† E. D. Commins and H. R. Feldman, *Phys. Rev.*, **131**, 700, 1963.

moment, which corresponds pictorially to a disc-shaped charge distribution surrounding the axis Oz.

An odd neutron moving in a potential well has no electric quadrupole moment since it is uncharged. There is, however, a small quadrupole moment in an actual nucleus (A, Z) because of a recoil effect. If the neutron is considered to describe an orbit of radius δ about the centre, the Z protons of the nucleus describe an orbit of radius δ/A and the quadrupole moment expected is

$$Q_{In} = \frac{Z}{A^2} Q_{Ip} \tag{11.29}$$

which is of the same sign as the single proton moment but much less in magnitude.

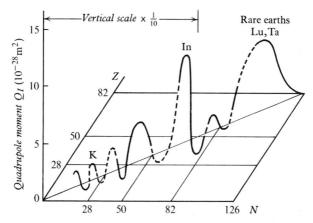

Fig. 11.11 Values of nuclear electric quadrupole moment Q_I as a function of N and Z. The values up to [115]In are multiplied by 10. Dotted lines show regions in which Q_I is not known. The moments are very large in the rare earth region (Ref. 11.11).

Since $\overline{r^2} = \frac{3}{5}R^2$ where R is the radius of the equivalent uniform charge distribution, the quadrupole moment of odd proton nuclei should be of the order of the nuclear radius squared, i.e. 10^{-28}–10^{-29} m². The ratio Q_I/R^2 gives a useful estimate of the extent to which Q_I differs from the single particle value. The results of measurements of electric quadrupole moment are shown plotted against Z and N in Fig. 11.11; the moments change in sign at the magic numbers. Closed shell nuclei have zero quadrupole moment as would be expected from the symmetry of their nucleon orbits. Even–even nuclei in general also have zero quadrupole moment, but this must be so because they have spin $I=0$ and therefore a spherically symmetrical charge distribution. These general features are in agreement with the predictions of the single particle model. In detail, however, important deviations are found; these are:

(*a*) The majority of quadrupole moments are of positive sign, whereas the simple shell model with a spherical potential would predict equal numbers of positive and negative moments (Ref. 11.9).

(*b*) The quadrupole moments of odd-neutron nuclei are similar in magnitude to those of odd-proton nuclei instead of being much smaller, e.g.

$$^{35}_{16}\text{S} \quad Q_I = -0\cdot055 \times 10^{-28} \text{ m}^2;$$

$$^{35}_{17}\text{Cl} \quad Q_I = -0\cdot078 \times 10^{-28} \text{ m}^2$$

(*c*) The magnitudes of quadrupole moments for nuclei with nucleon numbers between closed shells in the region $150 < A < 190$ are 10–20 times the single particle value, i.e. $Q_I/R^2 \approx 10$–20.

11.9 Summary

Table 11.1 is a small section of a typical table of data of nuclear moments, covering the whole range of stable nuclei. Such tables show that the single-particle shell model provides an excellent framework for a general description of the properties of the nuclear ground state. Nuclear radii are mainly properties of continuous nuclear matter and do not reveal inadequacies of the shell model. The deficiencies that have been noted in spin values, magnetic moments and electric quadrupole moments however are found to require extensions of the model to cover interactions between equivalent nucleons outside a closed core. We consider these extensions in the next chapter.

TABLE 11.1 Data on nuclear moments

(From Ref. 11.11.) R is taken to be $1\cdot3A^{1/3} \times 10^{-15}$ m. The first three nuclei have an odd proton, the final two an odd neutron.

Element	A	Z	N	I	Assumed state of last particle	$\dfrac{\mu_I}{\mu_N}$	Schmidt value	Q_I (10^{-28} m^2)	$\dfrac{Q_I}{R^2}$
B	11	5	6	$\frac{3}{2}$	$p_{3/2}$	2·69	3·79	0·036	0·55
Cl	35	17	18	$\frac{3}{2}$	$d_{3/2}$	0·82	0·12	−0·079	−0·55
Ta	181	73	108	$\frac{7}{2}$	$g_{7/2}$	2·1	4·79	4·0	9·3
O	17	8	9	$\frac{5}{2}$	$d_{5/2}$	−1·89	−1·91	−0·004	−0·05
S	35	16	19	$\frac{3}{2}$	$d_{3/2}$	1·0	1·15	0·038	0·27

References

1. SCOTT, J. M. C. 'The Radius of a Nucleus', *Progr. nucl. Phys.*, **5**, 157, 1956.
2. HOFSTADTER, R. 'Electron Scattering and Nuclear Structure', *Rev. mod. Phys.*, **28**, 214, 1956; 'Methods of Measuring Nuclear Size', in *Methods of Experimental Physics*, Vol. 5, Part A, ed. L. C. L. Yuan and C. S. Wu, Academic Press, 1961.
3. FORD, K. W and HILL, D. L. 'The Distribution of Charge in the Nucleus', *Ann. rev. nucl. Sci.*, **5**, 25, 1955.

4. BURHOP, E. H. S. 'Exotic Atoms', *Contemp. Phys.*, **11**, 335, 1970.
5. DEVONS, S. and DUERDOTH, I. 'Mesonic Atoms', *Adv. nucl. Phys.*, **2**, 295, 1969; WU, C. S. and WILETS, L. 'Muonic Atoms and Nuclear Structure', *Ann. rev. nucl. Sci.*, **19**, 527, 1969.
6. EVANS, R. D. *The Atomic Nucleus*, McGraw-Hill, 1955.
7. ELTON, L. R. B. *Nuclear Sizes*, Oxford University Press, 1961.
8. WILETS, L. 'Shape of the Nucleus', *Science*, **129**, 361, 1959.
9. MAYER, M. G. and JENSEN, J. H. D. *Elementary Theory of Nuclear Shell Structure*, Wiley, 1955.
10. BLIN-STOYLE, R. J. *Theories of Nuclear Moments*, Oxford University Press, 1957.
11. KOPFERMAN, H. *Nuclear Moments*, Academic Press, 1958.
12. SMITH, K. F. 'Nuclear Moments and Spins', *Progr. nucl. Phys.*, **6**, 52, 1957.

I2 Nuclear models (II)

The earliest nuclear models (ch. 9) were of two types, those in which discrete nucleon states could be distinguished and those in which the emphasis was on the continuous properties of nuclear matter. Modern developments of models have proceeded on both lines and we now recognize:

(a) models of the *individual particle type*, in which shell structure is a basic feature, and

(b) models of the *strong interaction or collective type*, such as liquid-drop models, in which there is much correlation between the motion of nucleons.

These two types of model are not mutually exclusive because nuclear matter is far from being a classical liquid. In fact the Pauli exclusion principle operates to permit a nucleon in a nucleus to have the long mean free path required for the development of the observable shell structure. The interpretation of collective modes in terms of particle motion is the basis of the *unified model*.

Each sort of model gives a more complete account of the nuclear level spectrum than does the single-particle shell model and this is achieved by allowing some degree of coupling between nucleons. In this chapter we examine the effects of introducing these interactions. The approach will be a phenomenological one, although there is of course a deeper level of understanding at which lie the *microscopic theories*. In these the nuclear many-body problem is treated in terms of realistic interactions between nucleons. Basic to this description is the Hartree–Fock self-consistent field as used in atomic physics. This replaces the strong internucleon interactions of the many body system by a properly constructed field for non-interacting particles, plus residual interactions of the sort necessary in the phenomenological approach (Ref. 12.11). In all such treatments the presence of a hard core in the internucleon force (ch. 18) is a source of difficulty.

Since internucleon forces are explicitly involved in nuclear structure calculations it will be convenient at the outset to introduce the formal concept of isobaric spin which makes possible a concise account of the effect of charge independence of forces and provides a new quantum number for nuclear states.

12.1 Isobaric (isotopic) spin†

The similarity of structure of the odd mirror nuclei due to charge symmetry of nuclear forces has been noted in Sect. 11.2.4. If we compare the level systems of even-A nuclei further similarities are apparent which suggest that disregarding Coulomb effects the neutron–proton force is the same as the

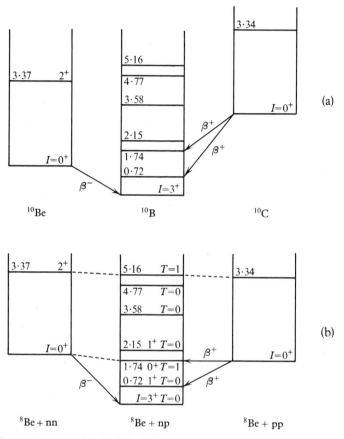

Fig. 12.1 Low-lying energy levels of the isobars of mass 10. Energies are marked in MeV and spins and parities are indicated by the symbols 0^+, 1^+, etc. (a) Masses of bare nuclei. (b) Nuclear masses corrected for Coulomb energy and for neutron–proton mass difference, showing isobaric triplet levels.

neutron–neutron and proton–proton force for nucleons in the same state of motion. In Fig. 12.1(a), the low-lying levels of the nuclei of mass 10 (^{10}Be–^{10}B–^{10}C) are shown, and the ground states are placed at levels corresponding to the respective *nuclear* masses. Beta-decay takes place between the isobars as

† The original term isotopic spin is widely used but has little to commend it.

shown. If now the nuclear masses are corrected by subtracting the Coulomb energy (Sect. 11.2.4) which is greater for ^{10}C than for ^{10}B and greater for ^{10}B than for ^{10}Be, and if in addition a correction is made for the neutron–proton mass difference, the relative level positions become as shown in Fig. 12.1(b). It is then found that there is a well-marked correspondence of binding energy and other properties between the low levels of the even–even isobars ^{10}Be, ^{10}C and that these levels have counterparts in the odd–odd isobar ^{10}B, which has in addition many other levels.

This can be understood in terms of the *charge independence* of nuclear forces. If an isobar of mass 10 is considered as a core (^{8}Be) plus two nucleons in the $p_{3/2}$ shell, then identical levels will be expected in each isobar if the (nn), (pp) and (np) interactions are the same. Because of the Pauli principle, not all states of motion accessible to a neutron–proton system will be permitted for pairs of identical nucleons and fewer states are therefore found in the even–even nuclei. The states that are common between the isobars are those in which the motion is such that a proton may be changed into a neutron, or vice versa, without offending the Pauli principle. The ^1S state for instance, in which two nucleons with spins opposed move with zero relative orbital motion is permitted to the three pairs (nn), (pp) and (np) but the ^3S state is permitted only to the system (np).

The isobaric spin formalism describes the three pairs as a *charge triplet* and the (np) system associated with states such as ^3S as a *charge singlet*. An isobaric spin quantum number $T = 1$ is ascribed to the states of the triplet with components m_T (often written T_z) $= 1, 0 - 1$ corresponding to the systems (pp), (np) and (nn), while for $T = 0$ only the (np) system exists, with $m_T = 0$. This concept, which will not be developed fully here (see Refs. 12.1 and 12.3), is only of value so long as the states of $T = 1$ have essentially the same binding energy, i.e. so long as nuclear forces are charge independent. There is then a degeneracy of the triplet states (Fig. 12.1(b)) which is removed when electrostatic forces (Fig. 12.1(a)) are allowed to act. This is in analogy with the removal of degeneracy connected with ordinary spin or orbital motion by the application of a magnetic field. We obtain a physical picture of the isobaric spin quantum number T by noting that $2T + 1$ is the number of isobaric nuclei in which the level of given T occurs. In Fig. 12.1, the states of ^{10}Be and ^{10}C are all of $T = 1$ and have $T = 1$ analogues in ^{10}B. States of $T = 0$ in ^{10}B cannot be found in the even–even nuclei.

The isobaric spin quantum number was first used in order to describe the neutron and proton as alternative states of the nucleon, distinguished only by charge. If we ascribe an isobaric spin $t = \frac{1}{2}$ to the nucleon then the components $t_z = -\frac{1}{2}$ and $t_z + \frac{1}{2}$ may be taken† to denote the neutron and proton respectively. In formal terms we speak of an operator $\hat{t}^2$ for the total isobaric spin of the nucleon with eigenvalue $t(t + 1)$ and an operator $\hat{t}_z$ for the third component with eigenvalues $\pm\frac{1}{2}$ indicating the actual charge state. For a state of

---

† The sign convention here is opposite to that used by Heisenberg in his discussion of exchange forces (Sect. 18.1.3) but is the more convenient in meson physics.

a complex nucleus there will be a total isobaric spin T with a component T_z ($\leqslant T$) given by

$$T_z = \frac{Z-N}{2} = \frac{A-2N}{2} = -\frac{A-2Z}{2}$$

where A is the mass number and N and Z the neutron and proton numbers of the nucleus in which the state occurs. For two nucleons, we obtain the results already quoted; in general T is integral for even A and half integral for odd A. For the $2T+1$ members of an isobaric multiplet, the atomic masses differ mainly because of the Coulomb interaction and are given by

$$M(A, T, T_z) = a + bT_z + cT_z^2$$

where a, b and c are functions of the mass number A and total isobaric spin T.

The concept of isobaric spin as a useful quantum variable under conditions of charge independence leads to selection rules for nuclear reactions which provide that this quantity shall be conserved. In the case of a scattering process such as

$$^{10}_{5}B + {}^{2}H \longrightarrow {}^{10}_{5}B^* + {}^{2}H$$

the colliding nuclei have $T=0$ and the isobaric spin selection rule therefore only permits the excitation of states of $T=0$ in the ^{10}B nucleus. For inelastic proton scattering on the other hand, e.g.

$$^{10}_{5}B + {}^{1}H \longrightarrow {}^{10}_{5}B^* + {}^{1}H$$

this restriction does not apply, since the nucleon has $t=\frac{1}{2}$ and by combination of the corresponding vectors states of both $T=0$ and $T=1$ in ^{10}B can be excited without violation of the conservation principle. Experimental results for these two reactions clearly indicate discrimination against the 1·74 MeV $T=1$ state of ^{10}B in inelastic deuteron scattering.

Neither the charge symmetry nor the charge independence of specifically nuclear forces is necessarily exact, but deviations are so small that the isobaric spin quantum number remains very generally useful. It is of vital significance in the classification of strongly interacting particles (ch. 21).

12.2 The individual (independent) particle model

The energy of a nucleus containing A nucleons may be written as a sum of kinetic and potential energy terms, i.e.

$$\text{Total energy} = \sum_{i=1}^{A} T_i + \frac{1}{2} \sum_{i=1}^{A} \sum_{j=1}^{A} V_{ij} \quad (i \neq j)$$

where V_{ij} is the potential energy for nucleons i, j. The *single-particle shell model* approximates this to the form

$$\sum_{i=1}^{A} (T_i + V_i) + \frac{1}{2} \sum_{i=1}^{A} \sum_{j=1}^{A} v_{ij} \quad (i \neq j)$$

(a) a configuration containing an even number of neutrons or protons has $I = 0$ in the ground state,

(b) a configuration containing an odd number of particles of one kind has $I = j$ in the ground state,

(c) the states of a nucleus with a j^2 configuration of two like particles (e.g. the $T = 1$ states of p^2 shown in Fig. 12.2(a) have spins 0, 2, 4, ...$(2j - 1)$ in order of ascending energy.

Predictions (a) and (b) are also those of the single-particle shell model and are independent of the type of coupling assumed. In the general case, when the classification in terms of quantum numbers (I, T) or (L, S, T) has been established, the order of the states must be found by solution of the Schrödinger equation. This type of calculation is familiar for the hydrogen atom but must be generalized in the nuclear case because of the several particles outside closed shells.

This type of treatment (Ref. 12.3) can be extended to the levels of a many particle configuration but then may involve a large number of basic states. Moreover the residual interaction assumed is only an approximation to the truth. In an alternative approach for the p-shell (Ref. 12.4) the *effective interactions* between pairs of particles in states with $j = 3/2, 1/2$, but coupled to different spins or isobaric spins, are taken as free parameters. Single-particle energies for the $p_{3/2}$ and $p_{1/2}$ states are taken from a nucleus in which these states are clearly identified. A number of well-established data on nuclear levels are then fitted to determine the free parameters. The effective interaction method has been applied in the (sd) and f shells and even with a limitation of the number of configurations gives a good account of many level positions. There are however states for which the shell model is unable to account and these are now associated with collective excitations (Sect. **12.3**). In a somewhat more basic approach (Ref. 12.5) the effective interactions themselves are calculated using realistic forces which fit the two-body scattering data.

The wave function of a nuclear state obtained by IPM calculations may be quite complicated especially when different configurations are involved. For ^{19}F, for instance, in which there are three nucleons outside the ^{16}O core and the $1d$ and $2s$ shells are filling, the wave function of a typical low-lying state is 12% d^3, 59% d^2s, 0% ds^2 and 29% s^3. With this detailed knowledge it is possible to calculate the main nuclear properties and it is found that magnetic moments and electric quadrupole moments are both better predicted. Magnetic moments between the Schmidt values are obtained because the single-particle moments are 'diluted' as a result of joint motion in the configurations of equivalent nucleons. The many-body description also explains the fact that odd neutron nuclei have appreciable quadrupole moments; the neutron can be regarded as having an effective charge of the order of $\frac{1}{2}e$. There are however still deviations from experiment, of which the most obvious is the existence of the extremely large electric quadrupole moments for the rare-earth elements. Some improvement can be obtained by taking account of the mixing of effects due to different configurations, which adds extra coherence to the motion. A more appropriate description of these extreme moments appears however to be offered by the collective model.

12.3 The collective model

12.3.1 General

If mixing of configurations in the shell-model sense is pushed to its extreme, the motions of all particles in the nucleus must be taken into account. This leads to a new type of model, in which the shape and angular momentum of the closed shells forming the nuclear 'core' play an important part. The significance of the shape of the nuclear core was first pointed out by Rainwater† and the quantitative development of the resulting 'collective' model was carried out by A. Bohr and Mottelson‡ using the analogy of the liquid drop, which had been discussed much earlier by N. Bohr and Kalckar. Many other authors have contributed to the details of the model, and striking successes in prediction have been obtained.

The collective model starts essentially with the idea that in order to explain the extremely large static electric quadrupole moments of nuclei lying between closed shells some co-operative motion of nuclear matter, resulting in a 'permanent' nuclear deformation, is necessary. It is supposed that this deformation (which vanishes for closed-shell nuclei), is produced by a polarizing effect of the individual or intrinsic motion of the nucleons outside closed shells. It is vital to the detailed development of the collective model that it should be possible to envisage a clear separation between the individual motion of the 'loose' nucleons and the collective motion of the core. This means that the single-particle energies associated with the shell-model states of the nucleons should be large compared with the rotational and vibrational energies of the core. It is then possible, by allowing some interaction between the two types of motion, to present a '*unified*' picture of nuclear motion in which both shell-model and collective features appear. This approach is a combination of the individual particle type of model with the strong interaction type; it not only explains the large quadrupole moments but also predicts a fine structure of the nuclear level spectrum owing to energies associated with the vibrational and rotational motion of the core.

The origin of a stable deformation may be discussed as follows: suppose that a nucleus has an ellipsoidal type of deformation (Fig. 12.3(a)), symmetrical about an axis Oz′ fixed in the body and specified by a deformation parameter β such that

$$\beta = \frac{\Delta R}{R_0} \tag{12.2}$$

† J. Rainwater, *Phys. Rev.*, **79**, 432, 1950.
‡ A. Bohr and B. R. Mottelson, *Dan. Mat. fys. Medd.*, **27**, No. 16, 1953.

where R_0 is the average nuclear radius and ΔR is the difference between the major and minor semi-axis of the ellipse.† The potential energy of the nucleus as a function of deformation may be represented as in Fig. 12.3(b), in which the different curves are plotted for different numbers of nucleons outside

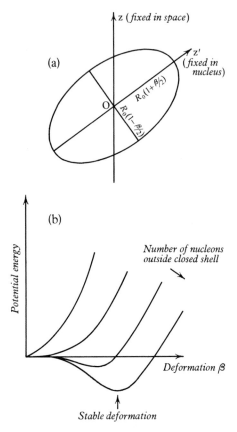

Fig. 12.3 Deformation of a nucleus. (a) Axially symmetrical deformed shape. (b) Potential energy of nucleus as a function of deformation (Ref. 12.8).

closed shells. For nuclei near closed shells the pairing forces (Sect. **10.1**) favour the grouping of these nucleons into pairs with zero angular momentum. The equilibrium shape is then spherical, and the collective motion is a *vibration* about this shape. As the number of loose nucleons increases, the effect of longer range forces is felt more, the frequency of the collective vibration decreases, and finally the spherical shape becomes unstable and the nucleus acquires a permanent deformation. This deformation appears both in the loose nucleons and in the nuclear core with which they interact, but for sim-

† Frequently β is defined by expressing the nuclear radius for an angle θ with the axis of symmetry as $R(\theta) = R_0[1 + \beta Y_2^0(\theta)]$. This is equivalent to (12.2) with the addition of a numerical factor of 1·06.

plicity we shall consider it to be a property of the core alone. The collective motion now becomes a vibration about the equilibrium shape, and (more important for many nuclear spectra) a *rotation* of the nuclear orientation which maintains the deformed shape.

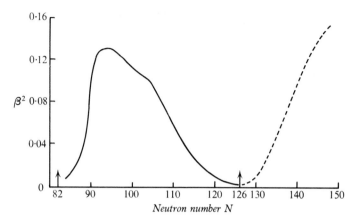

Fig. 12.4 Variation of the nuclear deformation parameter β (plotted as β^2) with neutron number in the range between the closed shells at $N = 82$ and 126. This includes the rare earth nuclei with their large quadrupole moments. The deformation disappears for the closed shell nuclei; at $N = 90$ a stable deformation suddenly appears (Ref. 12.9).

The deformation parameter β can be obtained from measurements of electromagnetic transition rates (Coulomb excitation and lifetime measurements), from the various isotope shifts and from the pattern of rotational energy levels under certain assumptions as to moments of inertia. The general variation of β with neutron number between the magic numbers 82 and 126 is shown in Fig. 12.4. The jump in β between $N = 88$ and $N = 90$ is entirely unconnected with any features of the single-particle shell model and represents a transition between the vibrational and rotational type of collective motion. The main region of stable deformation is that of the rare earth nuclei, but it also occurs for transuranic elements with $N \approx 150$, and in limited regions of the light nuclei (cf. Sect. **12.4**).

The coupling of angular momenta in a deformed nucleus resembles the coupling scheme for molecules and is shown in Fig. 12.5. As in the molecular case there is no rotation about the symmetry axis in the lowest states, but the intrinsic motion of the loose particles takes place about this axis, with a component angular momentum $K\hbar = \sum K_p \hbar$ where K_p refers to a single nucleon. The collective angular momentum perpendicular to the symmetry axis is **R** and the total angular momentum is **I**. The component of **I** along the axis fixed in space is $M\hbar$ and the component of **I** along the symmetry axis is just $K\hbar$. The absolute value of **R** is given by

$$|\mathbf{R}|^2 = [I(I+1) - K^2]\hbar^2 \tag{12.3}$$

It is only possible to make a simple investigation of this complex situation if it is assumed that the motion can be separated into:

(*a*) intrinsic nucleonic motion in the non-spherical potential,
(*b*) collective rotation,
(*c*) vibration about the static equilibrium shape.

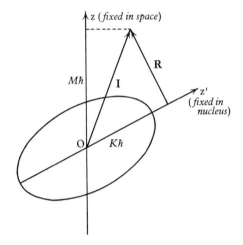

Fig. 12.5 Coupling scheme for angular momentum of deformed nuclei.

The excitation energies corresponding to these motions increase in the order (*b*), (*c*), (*a*); we now consider each of these separately.

12.3.2 Intrinsic states in a spheroidal field

The stable deformation β is produced by the polarizing action of nucleons moving in a spheroidal field, as distinct from the spherically symmetrical field assumed in the single-particle shell model. The sequence of states of a single particle under these circumstances has been calculated by Nilsson[†] as a function of deformation for a stationary field of oscillator type. Results of the type shown in Fig. 12.6 are obtained; for zero distortion the normal shell model ordering appears but for large deformation this is drastically altered.

In a non-spherical field, total angular momentum is no longer a constant of the motion, but for an axially symmetric field the component of angular momentum along the axis is conserved. A particle which occupies a state l_j in a spherical field ($=0$) can occupy $(2j+1)$ states with component angular momenta $K_p=j, (j-1), \ldots, -(j-1), -j$ along the symmetry axis in the non-spherical field. There is no energy difference between the states K_p and $-K_p$ so that each shell-model level splits into only $\frac{1}{2}(2j+1)$ levels in the

† S. G. Nilsson, *Dan. Mat. fys. Medd.*, **29**, No. 16, 1955.

spheroidal potential. These new single particle states are labelled by K_p and by the parity. In an actual nuclear spectrum, bands of rotational levels are built on the intrinsic levels defined by the number of particles in the nucleus and the deformation. In analysis of a spectrum the deformation may be estimated from the electromagnetic moments of the nucleus, and the intrinsic level order is then predictable.

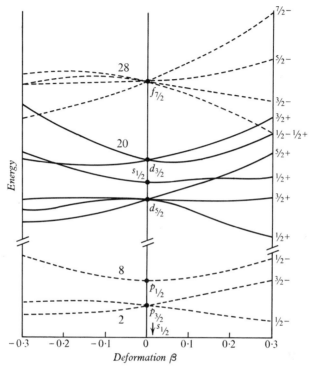

Fig. 12.6 Single particle states in a spheroidal potential as a function of the deformation parameter. The states are labelled by the quantum number K_p ($\leqslant j$) and the parity (Ref. 12.10).

The intrinsic states shown in Fig. 12.6 are populated by protons and neutrons independently in the building up of a stable nucleus, as in the single particle shell model. Two nucleons of each kind with oppositely directed orbital angular momenta ($\pm K_p$) can be associated with each intrinsic state and the nth state is completely occupied in a nucleus of mass $A = 4n$.

12.3.3 Rotational states

For nuclei far from closed shells, the rotational motion of a deformed nucleus does not affect the internal structure of the system. The energy of the rotation may be written classically

$$E_{\text{rot}} = \tfrac{1}{2}\mathscr{I}\omega^2 \tag{12.4}$$

where ω is the angular velocity and $\mathscr{I}$ is an effective moment of inertia. The rotation is not simply that of a rigid body; it was at first regarded rather as due to a hydrodynamical wave moving round the nucleus considered as a liquid drop. Fig. 12.7, taken from an article by Bohr, shows how constituent particles of a nucleus in particular modes of oscillation can lead to a rotation

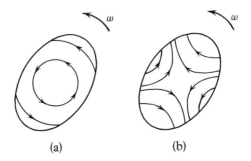

(a)   (b)

Fig. 12.7 Rigid and wave-like rotations of nucleus. In (a) the component particles move in circles round the axis of rotation. In (b) the particles oscillate along the paths shown and only the geometrical shape rotates (irrotational flow).

of the geometrical shape. For the irrotational motion of the particles (Fig. 12.7(b)) the effective moment of inertia about an axis perpendicular to the symmetry axis is determined by the deformation and may be written

$$\mathscr{I} \approx \mathscr{I}_0 \beta^2 = \mathscr{I}_0 \left(\frac{\Delta R}{R_0}\right)^2 \tag{12.5}$$

where $\mathscr{I}_0$ is the 'rigid' moment given by

$$\mathscr{I}_0 = \tfrac{2}{5} MAR_0^2 \tag{12.6}$$

for rotation of a mass MA. If the rotational angular momentum $\mathscr{I}\omega = |\mathbf{R}|$ is inserted in (12.4) we find, using (12.3)

$$E_I = \frac{h^2}{2\mathscr{I}} [I(I+1) - K^2] \tag{12.7}$$

For the value of K given by the intrinsic motion, this equation determines a rotational band of levels superimposed on the energies of the intrinsic motion. For $K = \tfrac{1}{2}$ a special formula is necessary (Ref. 12.8). The first term is familiar as the rotational energy of a diatomic molecule.

For *even–even nuclei* in their ground state, the particles fall alternately into states of opposite K_p and there is no contribution to the total angular momentum from the intrinsic motion ($K = 0$). There is symmetry about a plane perpendicular to the nuclear axis and hence, as in the case of the homonuclear diatomic molecule obeying Bose statistics,

$$I = 0, 2, 4, 6, \ldots \text{(parity even)} \tag{12.8}$$

in order that the wave function shall be invariant for a rotation of 180°. This type of rotational band is found in many nuclei in the known regions of deformation, $A \approx 24$, $150 < A < 190$ and $A > 200$. By putting $I = 0, 2, 4, \ldots$ in (12.7) it may be seen that the ratio of excitation of the successive states is $E_4/E_2 = 10/3$, $E_6/E_2 = 7$, $E_8/E_2 = 12$ and these characteristic values are verified for many even–even nuclei. Figure 12.8(a) shows the predicted spacings of the levels of a rotational band, compared with observation for the nucleus ^{238}U (Fig. 12.8(b)).

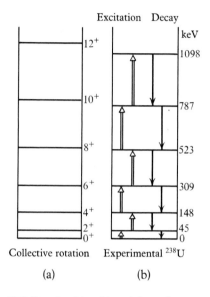

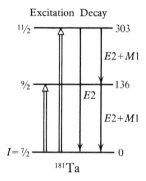

Fig. 12.8 Rotational band in a deformed even–even nucleus. (a) Theoretical series of levels. (b) Experimental results for ^{238}U obtained by Coulomb excitation (F. S. Stephens et al., Phys. Rev. Letters, **3**, 435, 1959).

Fig. 12.9 Rotational states in odd-A nucleus showing radiative transitions upward (excitation) and downward (decay).

The ground state of an even–even nucleus always has $I = 0$, and the first excited state usually has $I = 2^+$ in agreement with prediction. The systematic variation in energy of the observed 2^+ states in even–even nuclei is shown in Fig. 9.9; these states are usually strongly excited in $E2$ Coulomb transitions (Sect. 13.6.2), and have low energies between closed shells. The moments of inertia calculated from the level spacing can be used to obtain the deformation parameter β, from (12.5), but it is more instructive to use independent determinations of β (Sect. 12.3.5) to check the value $\mathcal{J}_0$. It is found that the collective rotation requires a moment of inertia which is less than the rigid body value (12.6) but greater than the hydrodynamical value.

In *odd-A nuclei* the last odd particle contributes an angular momentum K (which may be deduced from the Nilsson orbits, Sect. 12.3.2) along the nuclear axis and the allowed values of spin are

$$I = K, K+1, K+2, \ldots \text{ (half integral)} \qquad (12.9)$$

where K is then the spin of the ground (rotationless) state. The states of the rotational band have all the same parity, which is that of the intrinsic motion, and $E2$ Coulomb excitation can now populate two states $(K \to K+1, K+2)$ instead of one only as in even–even nuclei. Energy values again agree with prediction from (12.7) and moments of inertia are similar to (though not identical with) those derived from nearby even–even nuclei. A typical spectrum is shown in Fig. 12.9. Rotational states in odd-A nuclei are known with spins up to $I = 21/2$.

12.3.4 Vibrational states

Nuclei with relatively few particles outside closed shells have a spherical equilibrium form, and the collective motion takes the form of an oscillation of the loose particles about the spherical surface. In this type of motion the nucleus possesses a certain number of vibrational quanta or 'phonons' of energy $\hbar\omega_l$ and angular momentum $l\hbar$ in accordance with the quantum mechanical picture of the harmonic oscillator. Since there is no stable deformation for these nuclei the static moments are not enhanced, as in the case of nuclei far from closed shells.

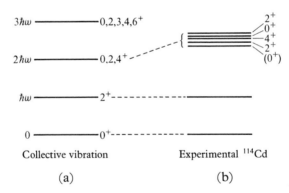

Fig. 12.10 Vibrational levels in even–even nuclei. (a) Theoretical series of levels. (b) Experimental results for ^{114}Cd (*Nuclear Data Sheets*).

The simplest vibrational spectra are found for even–even nuclei in which there is no contribution to the nuclear spin from the intrinsic motion. The basic vibrational spectrum is due to quadrupole 'phonons' and is given in Fig. 12.10 together with permitted spin values; in practice the degeneracy between the different levels is resolved and the expectation is a 0^+ ground state, a 2^+ first excited state (single phonon) and then a triplet of states $0^+2^+4^+$ formed by coupling two phonons. Classically an octupole phonon has about the same energy as two quadrupole phonons and this may produce a 3^- state near to the triplet. The energies of these states vary regularly according to the distance of the nucleus concerned from closed shells.

The vibrations of permanently deformed nuclei include oscillations of the parameter β and of a further parameter γ which determines the relative deformation of the three axes of the ellipsoid. The β-vibrations preserve an axis of symmetry, but γ-vibrations do not. Rotational bands may be built upon each vibrational state.

12.3.5 *Static and transition moments*

The collective model gives a good account of static magnetic dipole moments in the regions of strong deformation. Gyromagnetic ratios for the core and for the intrinsic motion are necessary and with reasonable values for these quantities the Schmidt lines (Fig. 11.10) are shifted in the right direction. The large static electric quadrupole moments shown in Fig. 11.11 are naturally explained by the collective model as due to the equilibrium quadrupole deformation of the core. The moment corresponding to the deformation β for a nuclear charge number Z, is

$$Q_0 = \frac{4}{5} ZR_0^2 \beta \qquad (12.10)$$

but the spectroscopically observed ground state moment (App. 2) is only

$$Q_I = \frac{I(2I-1)}{(I+1)(2I+3)} Q_0 \qquad (12.11)$$

Radiative transitions between the levels of rotational bands of strongly deformed nuclei may be simply related to the static quadrupole moment Q_0. General expressions for these transition probabilities, which measure the lifetime of an excited state, are given in Chapter 13. They contain, in addition to known energy dependent factors, *reduced transition probabilities B* characteristic of the nuclear states concerned. Successive levels in a rotational spectrum have spins differing by one unit (odd A) or two units (even A) and since these levels have the same parity, $M1$ and $E2$ transitions are expected. For even–even nuclei, electric quadrupole excitations from the 0^+ ground state to the 2^+ first excited state have a reduced transition probability

$$B(E2) = \frac{5}{16\pi} e^2 Q_0^2 \qquad (12.12)$$

Observation of these probabilities, mainly by the method of Coulomb excitation (Sect. 13.6.2) yield Q_0 values; it is not possible to obtain these moments spectroscopically for even–even nuclei since the moment Q_I cannot be observed in a state of $I = 0$. For odd-A nuclei similar reduced transition probabilities $B(E2)$ and $B(M1)$ may be obtained and these exceed the single particle values (ch. 13) by a large factor. They are, however, consistent with the values of Q_0 deduced spectroscopically. Similar measurements show the decrease of Q_0 as closed shells of neutrons at $N=82$ and $N=126$ are approached (Fig. 11.11). It may be concluded that the collective model gives a good account of both static and transition moments of deformed nuclei.

12.4 Prediction of nuclear level spectra

From the results outlined in Sections **12.2** and **12.3** it will now be clear that there are two general models available for the prediction of level spectra and other nuclear properties. One of the most interesting results is the particular application of the two models to the level spectra of ^{19}F. This nucleus has been studied in detail by Elliott and Flowers† using the methods of the individual particle model, and the even-parity level spectrum predicted is shown in comparison with the experimental levels in Fig. 12.11(a), (b). By suitable choice of the two-body potential a very close correspondence can be reached.

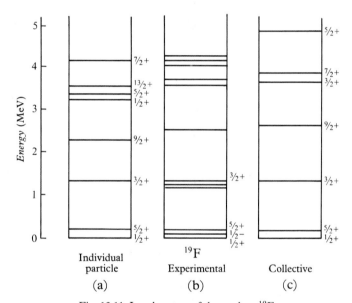

Fig. 12.11 Level system of the nucleus ^{19}F.

In Fig. 12.11(c), however, are shown the results of an analysis by Paul‡ of the level system of the same nucleus in terms of mixed $K=\frac{1}{2}$ and $K=\frac{3}{2}$ bands of the collective model. The agreement is striking, and suggests that the two methods of approach are indeed equivalent. Similar rotational bands have been identified in other light nuclei such as ^{25}Al and ^{29}Si. In the case of mass 19 the two models also give equally good descriptions of the radiative transition probabilities, magnetic moments, β-decay probabilities and the detailed

† J. P. Elliott and B. H. Flowers, *Proc. roy. Soc.*, A, **229**, 536, 1955.
‡ E. B. Paul, *Phil. Mag.*, **2**, 311, 1957.

configuration of the last three particles, i.e. the percentage of d and s states in the corresponding wave function.

The most important parameter for an understanding of the general nature of nuclear spectra is the number (n) of particles outside a closed shell, or the number needed to complete a shell. Fig. 12.12 illustrates the change of level spectra as n increases from the doubly magic nucleus $^{208}_{82}$Pb. In ^{207}Pb, single particle levels of the spherical potential are found, in ^{206}Pb there is a two-particle (hole) spectrum which can be treated by the individual particle model. From ^{190}Os to ^{152}Sm, the even–even nuclei show the rotational spectra of the deformed shape, but at ^{152}Gd the deformation is small and the vibrational spectrum of the spherical shape is seen.

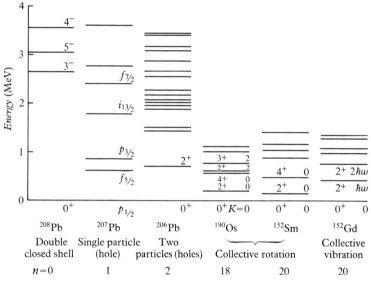

Fig. 12.12 Sketch of level spectra in nuclei with increasing number (n) of particles outside, or holes in, a closed core.

The most striking states of the nuclear spectrum for a wide range of even-mass nuclei, are low-lying levels of spin 2^+ and 3^- and a broad 1^- state (or superposition of states) at 15–20 MeV excitation. The 1^- states are concerned in the *giant resonance* of photodisintegration (ch. 15). These states, and those which derive from them by coupling to single particles in odd-mass nuclei, have each been discussed in terms of both independent particle and collective theories.

12.5 Study of the nuclear level spectrum

Frequent reference has been made in preceding sections to methods of studying the excited states of nuclei. These are collected together here for convenience, and to serve as an introduction to subsequent chapters. The

D

Dynamical properties of nuclei

I3 Radiative processes in nuclear physics

The emission and absorption of radiation by nuclei is of especial importance among the many types of nuclear reaction because the electromagnetic field is well understood. Corrections arising from field theory are negligible and observations of radiative processes can lead to reliable values for nuclear properties. The general nature of the electromagnetic interaction, and especially its connection with invariance principles, is discussed in Chapter 21.

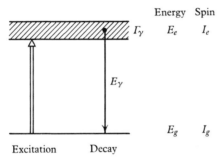

Fig. 13.1 Excitation and decay of a bound excited state at mean energy E_γ, above a nuclear ground state. The width Γ_γ is shown greatly exaggerated.

Radiation itself is a property of two states as shown in Fig. 3.3; in order to avoid ambiguity in discussion we redraw this figure with an altered nomenclature (Fig. 13.1). The probability of both upward and downward transitions between the two states depends on (a) the energy difference, (b) the multipolarities of the transitions allowed by the selection rules (Sect. 3.5.2) and (c) the wave functions of the states. Experimentally the multipolarity is determined if possible by conversion phenomena (Sects. **13.2, 13.3**) and the change of parity and the maximum change of spin in the transition are then known. The wave functions determine the absolute transition probability $T(L)$ for a given multipolarity L and this may be calculated from detailed nuclear models for comparison with experiment. It gives information on nuclear structure similar to that obtained from the static moments discussed in Chapter 12. The general selection rule

$$I_e + I_g \geqslant L \geqslant |I_e - I_g|$$

holds for all radiative transitions.

The radiative transition probability is the reciprocal of the mean life τ_γ of the level concerned for emission of radiation and is therefore connected with the radiative width Γ_γ by

$$T = \frac{1}{\tau_\gamma} = \frac{\Gamma_\gamma}{\hbar}$$

where

$$\Gamma_\gamma \tau_\gamma = \hbar = 6 \cdot 6 \times 10^{-16} \text{ eV s} \qquad (13.1)$$

It is of course just equal to the 'radioactive' decay constant λ of the exponential decay of the excited state (Sect. 2.3.4). If neither I_e nor I_μ is zero, transitions of mixed multipolarity, e.g. $M1 + E2$ may occur between the states. The mixing ratio δ, given by

$$\delta^2 = \frac{\Gamma_\gamma(E2)}{\Gamma_\gamma(M1)}$$

must be obtained, e.g. from conversion coefficients or angular correlation experiments, before the individual radiative widths can be extracted.

The radiative width of bound levels is usually about $0 \cdot 1$ eV or less, corresponding to a lifetime τ_γ of about 10^{-14} s or more from (13.1). The decay of virtual levels depends on particle emission probabilities as well as radiative processes and may take place rapidly, in a time comparable with the time for a proton or neutron to cross the nucleus. This is typically $\tau \approx R/v \approx 10^{-21}$ s and from (13.1) it follows that the total width (Sect. **9.5**) of such levels may be as much as $0 \cdot 5$ MeV.

We shall be concerned in this chapter with transitions between discrete nuclear states to which quantum numbers may be assigned, and with electronic phenomena in the radiating atom. Processes such as bremsstrahlung and external pair production will not be discussed; their relevance to our knowledge of the electromagnetic interaction is outlined in Sect. **21.2**.

13.1 The lifetime–energy relation

In Sect. 3.5.1 a semi-classical expression for the decay rate of an excited state emitting radiation was given. It was also stated that in the *long wavelength approximation* $(a \ll \lambda)$ it is possible to consider the total radiation from an oscillating charge distribution as a series of terms of increasing multipolarity and decreasing intensity. In most nuclear problems, since $a\ (\approx R) < 10^{-14}$ m for all nuclei, this approximation will be valid.

A more detailed classical calculation for the transition probability for multipolarity L and energy E_γ (Fig. 13.1) gives (Ref. 13.1)

$$T(L) = \frac{\mu_0 c^2}{4\pi} \times \frac{8\pi(L+1)}{L[(2L+1)!!]^2} \frac{1}{\hbar} \left(\frac{E_\gamma}{\hbar c}\right)^{2L+1} B_{eg}(L) \text{ s}^{-1} \qquad (13.2)$$

where

$$(2L+1)!! = 1.3.5 \cdots (2L+1) \qquad (13.3)$$

and $B_{eg}(L)$ is known as the *reduced transition probability*; it must be evaluated for a specified nuclear model. This probability contains all specifically nuclear quantities and is obtained from an experimentally determined $T(L)$ by dividing out the calculable energy-dependent factors shown in (13.2). The suffix indicates that the downward transition from the excited state to the ground state is considered. The reduced transition probability B_{ge} for the upward excitation process shown in Fig. 13.1 is related to the downward probability B_{eg} by introducing the statistical weight $2I + 1$ of the states. This gives

$$B_{ge}(L) = B_{eg}(L)\frac{2I_e + 1}{2I_g + 1} \qquad (13.4)$$

In the *single-particle shell model* radiation may be considered to arise as a result of the transition of a single proton from one orbital state to another. This approach was adopted by Weisskopf† who shows that for this model a

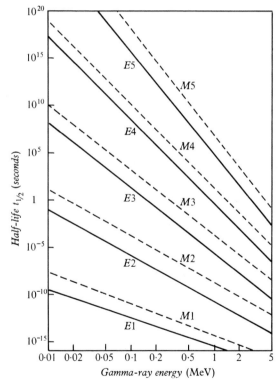

Fig. 13.2 Lifetime–energy relations for γ-radiation according to the single particle formula of Weisskopf.

† V. F. Weisskopf, *Phys. Rev.*, **83**, 1073, 1951, and Ref. 13.1.

rough estimate for the case of zero orbital momentum in the final state is

$$B(EL) = \frac{e^2}{4\pi} \left(\frac{3R^L}{L+3} \right)^2 \quad \text{for electric radiation} \quad (13.5a)$$

and

$$B(ML) = 10 \left(\frac{\hbar}{m_p cR} \right)^2 B(EL) \quad \text{for magnetic radiation} \quad (13.5b)$$

where R is of the order of magnitude of the nuclear radius. The factor 10 in $B(ML)$ is introduced to allow for magnetic radiation originating from re-orientation of intrinsic spins. The lifetime–energy relations based on these estimates have been very widely used; they are shown in Fig. 13.2, from which it is clearly seen that lifetimes become long as energy decreases and as multipolarity increases. The formulae also give *single-particle radiative widths* directly, e.g.

$$\left. \begin{aligned} \Gamma_\gamma(E1) &= 0{\cdot}07 E_\gamma^3 A^{2/3} \\ \Gamma_\gamma(M1) &= 0{\cdot}021 E_\gamma^3 \\ \Gamma_\gamma(E2) &= 4{\cdot}9 \times 10^{-8} A^{4/3} E_\gamma^5 \end{aligned} \right\} \quad (13.6)$$

where Γ_γ is the radiative width in eV, E_γ the transition energy in MeV and A the mass number of the nucleus. In the individual particle model radiative widths would be expected to be smaller, because of the sharing of the radiative moment among the several particles of the configuration. The $M1$ transition probability would vanish (Ref. 13.1) in the absence of spin–orbit coupling.

If the nuclear core participates in the radiative process, transitions may sometimes be described by using an *effective charge* of the order of $1{\cdot}5e$ for protons and $0{\cdot}5e$ for neutrons. In the fully collective model, in-phase motion of particles gives radiative widths which are much larger than the single particle value.

13.2 Internal conversion

The lifetime formula of Weisskopf for radiative transitions, and similar formulae, give the probability of decay of a bare point nucleus, completely stripped of the atomic electrons. Usually this is not the case, and since many electronic wave functions (in particular all wave functions of s-states) have finite amplitudes at or near the nucleus it is possible for nuclear excitation energy to be removed by the ejection of an atomic electron. The total probability per unit time of decay of the excited nucleus is then given by $\Gamma/\hbar$ where we write for a bound state

$$\Gamma = \Gamma_\gamma + \Gamma_e \quad (13.7)$$

in which Γ_e is the width for emission of electrons. In both cases the total energy of nuclear excitation is removed, but the electron emission process is generally described for historical reasons as internal conversion. This does not imply that the process follows the emission of radiation and the two processes must be regarded as competing alternatives. That this is so has been demonstrated conclusively† in the case of ^{90}Nb whose 24 s lifetime for isomeric decay (Sect. **13.5**) has been altered (by $3 \cdot 6 \pm 0 \cdot 4 \%$) by changing the chemical nature of the Nb compound used and hence affecting the electron distribution surrounding the active nucleus. If the number of electrons observed per excited nucleus is N_e and the number of γ-rays is N_γ we define the internal conversion coefficient α as

$$\alpha = \frac{N_e}{N_\gamma} = \frac{\Gamma_e}{\Gamma_\gamma} \tag{13.8}$$

and α may have any value between 0 and ∞. From (13.7) it follows that if the total width Γ is determined then the radiative width Γ_γ is given by

$$\Gamma_\gamma = \frac{\Gamma}{1 + \alpha} \tag{13.9}$$

or, in terms of mean lives,

$$\tau = \frac{\tau_\gamma}{1 + \alpha} \tag{13.10}$$

The internal conversion process is responsible for the sharp homogeneous lines found in β-ray spectra. Figure 7.4(b) shows a typical spectrum, obtained by magnetic analysis. From such results the energy E_γ of the nuclear transition may be obtained very accurately because the kinetic energy of an electron ejected from the K-shell , for instance is

$$E_\gamma - E_K \tag{13.11}$$

where E_K is the binding energy of an electron in the K-shell, which is known precisely.‡

The magnetic spectra show that for each γ-ray there are in fact several conversion lines corresponding to the ejection of electrons from different atomic shells, e.g. K, L_I, L_{II}, L_{III}, M_I, M_{II}, ... and the total conversion coefficient is therefore

$$\alpha = \frac{N_K + N_L + N_M + \cdots}{N_\gamma} \tag{13.12}$$

$$= \alpha_K + \alpha_L + \alpha_M + \cdots$$

The differences in energy between the various groups of internal conversion electrons corresponding to a single nuclear transition agree exactly with the

---

† J. A. Cooper, J. M. Hollander and J. O. Rasmussen, *Phys. Rev. Lett.*, **15**, 680, 1965.
‡ A table of $K, L, \ldots$ binding energies is given in Ref. 13.2, App. 2.

energies of the lines of the X-ray spectrum of the atom containing the excited nucleus. The internal conversion electrons must be distinguished carefully from externally produced electrons arising from the photoelectric interaction of the alternative γ-radiation with matter through which it passes.

The emission from an atom of an internal conversion electron leaves a vacancy in one of the atomic shells. As in the case of the photoelectric effect with γ-radiation (Sect. 5.4.3) either *K* X-rays or Auger electrons may then be emitted. The fluorescence yield, which gives the probability of emission of *K* X-rays when a *K*-shell vacancy is created is shown in Fig. (5.22).

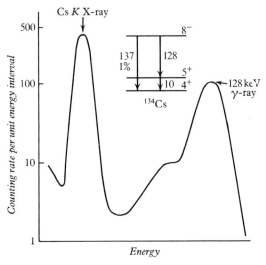

Fig. 13.3 Determination of *K*-conversion coefficient for the 128 keV transition in ^{134m}Cs by observation of relative intensity of *K* X-rays and nuclear γ-rays. The superscript *m* denotes an isomeric state (Sect. **13.5**). (Sunyar *et al., Phys. Rev.*, **95**, 570, 1954.)

Internal conversion coefficients are determined by making absolute measurements of the intensity of γ-radiation and of the associated conversion electrons, X-rays or Auger electrons produced in radioactive decay or in nuclear reactions. Alternatively the intensity of any one of these radiations may be compared with that of a preceding or following nuclear transition, or with the yield of photoelectrons produced externally from a converter foil. Magnetic spectroscopy may be used or the radiations may be detected in scintillation or semiconductor spectrometers. A typical example of the latter method is given in Fig. 13.3 which shows the pulse height spectrum of the radiations from ^{134m}Cs. The peaks are due to the nuclear γ-ray and to the *K* X-rays associated with the internal conversion electrons. Among the corrections to be applied in obtaining absolute yields is one for fluorescence yield.

The importance of the experimental study of internal conversion lies in the information which it may give about the multipolarity and parity change of the nuclear transition. The total conversion coefficient α is a ratio of intensities and if a nucleus of infinitesimal size is assumed, it does not depend on the actual value of the nuclear transition probability. The dependence on the multipolarity L arises because the amplitude of multipole fields near the nucleus is determined by L and is relatively higher for the higher multipoles.

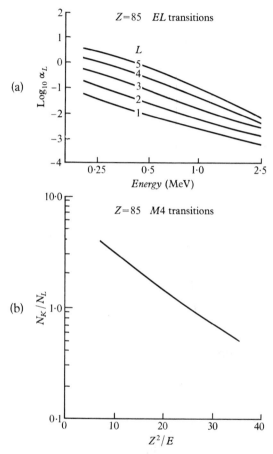

Fig. 13.4 Internal conversion. (a) Variation of K-conversion coefficient for electric multipoles (L) with energy for $Z=85$. (b) N_K/N_L for $M4$ transitions as a function of Z^2/E for $Z=85$ and E in keV (Ref. 16.4, pp. 342 and 345).

Extensive tables of K- and L-shell internal conversion coefficients, based on calculations with relativistic wave functions, now exist (Refs. 13.3, 13.4); Fig. 13.4(a) shows how the coefficient α_K varies with transition energy for a number of multipolarities in a typical nucleus.† The rapid decrease of α with transition energy and the relatively poor discrimination between different multipolarities for high transition energies are made clear in this figure.

† It is conventional to use α for electric transitions and β for magnetic transitions.

Internal conversion coefficients are small for light nuclei and the determination of these coefficients is therefore mainly useful for heavy elements and low transition energies.

Although the emission of electrons from the K-shell is the most probable process of internal conversion when sufficient energy is available, because of the relatively greater electron density near the nucleus, the intensity of L and M electrons is usually easily observable. The ratio N_K/N_L of total emission of K-electrons to total emission of L-electrons also depends on multipolarity and nuclear transition energy and may be easier to measure accurately than conversion coefficients since it can be obtained directly from a high-resolution magnetic spectrum of the type shown in Fig. 7.4(b). The variation of N_K/N_L with transition energy in a particular case is shown in Fig. 13.4(b).

The internal conversion coefficient gives no information about nuclear structure beyond that which is contained in the radiative lifetime if a point nucleus is assumed. This is also true if corrections are made for the effect of finite nuclear size on the electronic wave functions. However, further corrections which are structure-dependent arise for a finite nucleus because the atomic electrons penetrate within the nuclear charge and current distribution.† These latter corrections, which can only be calculated if a model is assumed, are important for the $E0$ monopole transitions between states of zero spin (Sect. **13.4**) for which internal conversion is the main mode of decay. These transitions may also compete with alternative transitions such as $M1$ or $E2$ when the spins of the initial and final states are the same and may become significant if the alternative transitions are retarded.

13.3 Pair internal conversion

If the transition energy E_γ exceeds $2mc^2$, i.e. if $E_\gamma > 1$ MeV an excited nucleus may emit a positron–electron pair as an alternative to γ-ray emission and electron internal conversion. The theory of this process, which is electromagnetic in that it takes place in the Coulomb field of the excited nucleus, shows that the probability of pair internal conversion increases with transition energy, is greatest for small multipolarities and is almost independent of Z. These differences from ordinary internal conversion arise because the pair creation process requires only the elevation of an electron in a negative energy state to a positive energy and the supply of such electrons is unlimited, in contrast with the situation for ordinary electrons in an atom. Pair conversion thus becomes important under exactly the conditions under which ordinary internal conversion is becoming small and the phenomenon provides a powerful method of studying the radiative transitions of light nuclei. In such experiments care must be taken to distinguish between internal pairs and ordinary pair production in matter by the competing γ-ray.

Figure 13.5(a) gives the yield of internal pairs as a function of transition energy and multipolarity for electric transitions. It is clear that discrimination

† E. L. Church and J. Weneser, *Phys. Rev.*, **104**, 1382, 1956.

between the different multipoles is only good when the yield is poor (i.e. at low energy) and accurate experiments are necessary to distinguish between the multipoles by measurement of pair conversion coefficients. A more sensitive method is to measure the angular correlation between the positron and electron. This method has been extensively applied by Devons and his

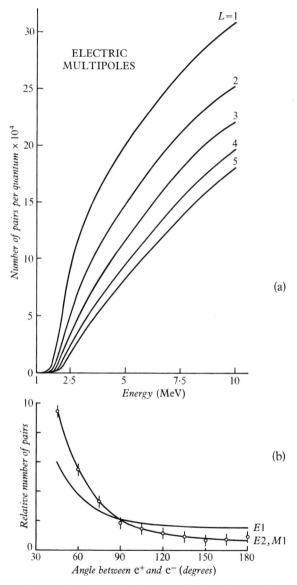

Fig. 13.5 Pair internal conversion. (a) Yield of pairs as function of energy for electric multipole radiation (Rose, *Phys. Rev.*, **76**, 678, 1949). (b) Angular correlation between positron and electron for the high energy transition in the ^{7}Li(p, γ)^{8}Be reaction. The full lines are theoretical curves (Devons and Goldring, *Proc. phys. Soc. Lond.*, **67**, 413, 1954).

collaborators to determine multipolarities in the light nuclei; Fig. 13.5(b) shows results for the transition $Be^8* \longrightarrow Be^8 + \gamma$ which determine the radiation to be of *M1* or *E2* type. In a development by Warburton *et al.*† a lens-type pair spectrometer, with a special selection of angles, has been used to increase the sensitivity of the method to multipolarity; it has been applied for *E0*, *E1*, *M1*, *E2*, *M2* and *E3* transitions in light nuclei.

13.4 Zero–zero transitions

In general internal conversion proceeds because of the multipole field associated with the nuclear transition and if this field falls off with distance fairly slowly, contributions to the yield of internal conversion electrons come from all over the atom. In the particular case of $I=0$ to $I=0$ transitions, single quantum radiation is strictly forbidden (Sect. 3.5.2) and multipole fields do not exist outside the nucleus. The only contribution to internal conversion effects then involves electrons (such as the *K*-electrons) whose wave functions penetrate the nucleus, within which electromagnetic fields may still exist. *Total internal conversion* $(\alpha = \infty)$ arising in this way is observed as a single homogeneous electron line, without associated γ-radiation, in the spectrum of RaC′ (1·414 MeV level) and of ^{72}Ge (0·7 MeV level). The lifetimes of these levels are longer than would be expected for radiative transitions of the particular energy and both transitions are assumed to be of the monopole type $0^+ \longrightarrow 0^+$. Pair internal conversion can also occur in $0^+ \longrightarrow 0^+$ transitions and it is known for the first excited states of ^{16}O and ^{40}Ca.

Transitions of the form $0^+ \longrightarrow 0^-$ cannot occur by any of the internal conversion processes just discussed. Two radiations are required, passing through a virtual state of suitable properties, e.g. $0^+(E1)1^-(M1)0^-$, with the emission of two quanta, would be a possible mode of de-excitation. Alternatively two conversion electrons or one conversion electron and one photon might be emitted. No clear example of this process of decay has yet been detected.

13.5 Nuclear isomerism

Figure 13.2 shows that electromagnetic transitions of high multipolarity and low energy are relatively slow processes. Excited states of nuclei which can only decay by such transitions may therefore have a long life. Nuclei excited to these states will differ from unexcited nuclei only in their radioactive properties, and not in charge or mass number. Such excited nuclei are said to be *isomeric* with respect to their ground state.

Nuclear isomerism was first established for UX$_2$ and UZ (Sect. 2.4.3). Since this discovery, and since the suggestion of von Weizsäcker that the phenomenon of slow radiative decay might be associated with a high spin difference, many examples of nuclear isomerism have been found, both among

† *Phys. Rev.*, **133**, B42, 1964.

the stable and radioactive nuclei. In the case of UX_2 the radiative transition is so slow that the energetically possible β-decay is able to compete with it; in β-stable nuclei an isomeric transition is usually accompanied by internal conversion electrons, since the conditions for isomerism are just those for high internal conversion coefficients. In this latter case the experimentally observed radiation is often an electron line of low energy decaying with a half-life short compared with that expected for a nuclear β-decay of the corresponding energy. The occurrence of internal conversion shortens the experimentally observed lifetime of a nuclear isomer, and the partial lifetime for the radiative transition is obtained by multiplying the observed lifetime by the factor $(1 + \alpha)$ where α is the internal conversion coefficient. In the case of nuclear isomeric transitions the X-rays following conversion electrons are

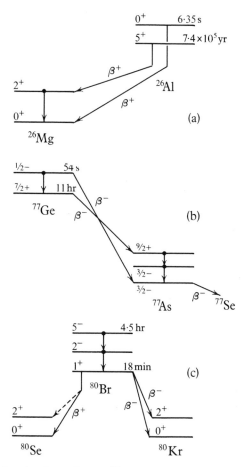

Fig. 13.6 Examples of nuclear isomerism. (a) ^{26}Al, in which the isomeric state decays by positron emission and the radiative transition is not seen. (b) ^{77}Ge, in which electron emission competes with γ-ray emission from the $\frac{1}{2}^-$ state. (c) ^{80}Br in which the 5^- state decays wholly by radiative transition.

characteristic of the radiating atom, not of its daughter as in the case of β-decay of *K*-capture. There is no clear de-limitation of the range of isomeric lifetimes but it has been customary to regard them as measurable by the simple direct timing techniques of radioactivity. There is no essential difference between states with such lifetimes, of perhaps $t_{1/2} > 10^{-5}$ s, and the much shorter-lived states which may be studied in special ways (Sect. **13.6**). A nucleus excited to an isomeric level which may be regarded as metastable is indicated by a superscript *m*, e.g. ^{80m}Br.

Typical examples of isomeric transitions are shown in Fig. 13.6. Usually, but not always, the high spin level is the higher of the isomeric pair. Isomers may be produced in all types of nuclear reaction including β-decay but the yield may be low if the transference of a large amount of angular momentum is necessary. In such cases it is possible to reach the isomeric level by exciting a higher level of lower spin, from which cascade transitions can take place. The frequency of occurrence of nuclear isomers is plotted as a function of the odd nucleon number for nuclei of odd *A* in Fig. 9.8. There is a characteristic grouping just below the major closed shells at *Z* or *N* = 50 and 82 and a less marked distribution near *N* = 126. These so-called 'islands of isomerism' receive an immediate interpretation in terms of the single-particle shell model (Sect. **9.4**) and form an important part of the evidence for the validity of this model.

The long life of isomeric states, and the possibility of producing strong sources of isomeric activities by neutron irradiation, have stimulated many attempts at their separation. Among the successful methods employed are:

(*a*) *the Szilard-Chalmers reaction*, in which the radiative transition releases the radiating atom from its molecule so that it can be separated chemically;
(*b*) *electric collection*, in which use is made of the fact that atoms containing isomeric nuclei are generally produced in an ionized state, and may be collected on a charged plate.

The long life may also permit the direct measurement of the spins of both states concerned in a radiative transition by atomic beam methods (Sect. 4.4.4). This has been done for ^{134m}Cs $(I = 8)$ and ^{134}Cs$(I = 4)$; conflict of these results with the assignment of the transition on the basis of conversion coefficients as *E*3 led to the discovery of an extra low energy transition in the decay scheme (Fig. 13.3).

13.6 Determination of transition probabilities (Ref. 13.5)

13.6.1 Lifetimes of bound states $(\tau \leqslant 10^{-6}$ s$)$

(*a*) *Comparison with alternative transitions.* The earliest estimate of the lifetime of radiative transitions was made by comparing the intensity of γ-radiation with the intensity of long-range α-particle emission in the decay of certain naturally occurring radioactive elements. Thus in the case of ThC′

the main group of α-particles observed has an energy of 8·78 MeV but for every million such particles 170 α-particles of energy 10·54 MeV are emitted. These arise because the ThC′ nucleus is formed in excited states, as well as its ground state, by β-decay from ThC and these excited states can either decay directly by γ-ray emission to the ground state of ThC′ or by long-range α-particle emission to ThD. The competition between the long-range α-particle emission and the radiative process is determined theoretically by their respective lifetimes and the α-particle lifetime may be assumed to be given by the probability of penetration of the nuclear potential barrier (Sect. **14.2**). This is of course only a rough estimate since it neglects internal structure effects, but it was sufficiently good to indicate γ-ray lives of the order of 10^{-13} s.

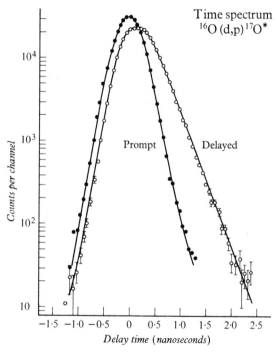

Fig. 13.7 Yield of coincidences between timing signal from a pulsed accelerator and γ-radiation from a target as a function of delay of the accelerator signal. The curve marked 'Prompt' is for radiation from a state of effectively zero life-time. (J. Lowe and C. L. McClelland, *Phys. Rev.*, **132**, 367, 1963).

(b) *Delayed coincidence method* $(\tau \approx 10^{-6}$ s-10^{-11} s) In this method, originally applied by Dunworth† in a study of ThC′, a signal is produced in a detector by a radiation emitted by a parent body. The decay of the daughter body is observed in a second detector with a certain delay due to the finite lifetime of the daughter state. If the first signal is delayed electronically,

† J. V. Dunworth, *Nature*, **144**, 152, 1939.

coincidences between the two signals may be observed and from their dependence on delay time the mean lifetime τ may be deduced.

In more recent applications of this method to the successive transitions of a decay scheme, scintillation counters have often been used. The main requirement of the counters and circuits is fast response, and by use of wideband amplifiers, and by a choice of photomultipliers, Bell, Graham and Petch† were able to measure lifetimes down to 10^{-11} s. In fast timing experiments of this type, energy selection is usually achieved using 'slow' channels independently coupled to the detectors. The fast and slow outputs are finally combined in a triple coincidence unit.

The delayed coincidence method is also applicable to work with pulsed or modulated accelerators. In an experiment at Brookhaven a quartz target was bombarded by a 2·75 MeV deuteron beam modulated at a frequency of 7·5 MHz. A special time-compression system was used to produce pulses of duration 10^{-10} s on the target. Radiation from the first excited state of ^{17}O formed by the reaction ^{16}O(d, p)^{17}O* was detected by a plastic scintillator in coincidence with a signal derived from the radiofrequency pulse circuit and delayed by a time T from the beam pulse; Fig. 13.7 shows the results, from which a mean lifetime of $2·63 \pm 0·08$ ns was obtained. In an alternative scheme for shorter lifetimes microwave modulation of a bombarding beam and of conversion electrons is employed.‡

(c) Recoil distance (plunger) method ($\tau \approx 10^{-7}$ s–10^{-11} s) In the original version of this method, used by Jacobsen for RaC' ($t_{1/2} = 1·6 \times 10^{-4}$ s), ions of the active element recoiled from a parent body (RaC) and moved down an evacuated tube with a velocity $v = 5 \times 10^3$ m s^{-1}. Alpha particle decay was observed as a function of distance x of travel down the tube and from the fall-off of activity with distance the lifetime could be deduced using the formula

$$A_x = \text{constant} \times e^{-x/v\tau} \tag{13.13}$$

The method was adapted to cases of much shorter half-life by Devons and his collaborators, whose apparatus is shown in principle in Fig. 13.8(a). In a typical experiment excited ^{16}O nuclei were produced in the reaction

$$^{19}\text{F} + \text{p} \longrightarrow {}^{16}\text{O}* + \alpha + 2·1 \text{ MeV} \tag{13.14}$$

from which the recoiling ^{16}O nucleus leaves a thin target with an initial velocity of about 10^6 m s^{-1}. The collimator and counter define a narrow region of path from which radiative decay of excited nuclei which have travelled a certain distance from the target may be observed. This distance is adjusted, conveniently by mounting the target on a micrometer head or *plunger*, and the number of radiative decays of ^{16}O* as a function of distance

† *Canad. jnl. Phys.*, **30**, 35, 1952; Ref. 13.2, page 905.
‡ A. E. Blaugrund, Y. Dar and G. Goldring, *Phys. Rev.*, **120**, 1328, 1960.
§ *Proc. phys. Soc. Lond.*, **67**, 134, 1954; **68**, 18, 1955.

($\approx 10^{-5}$ m) is observed. In this work a mean life of $7 \cdot 2 \pm 0 \cdot 7 \times 10^{-11}$ s was measured for the 6·06 MeV (pair emitting) state of ^{16}O.

The lifetimes of *atomic states* can be determined in a similar way using the technique of *beam foil spectroscopy*. Excited ions are produced by passing a beam of heavy ions through a thin foil and the intensity of radiation at right angles to the beam is determined as a function of distance from the foil.

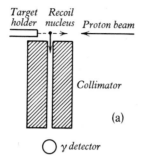

(a)

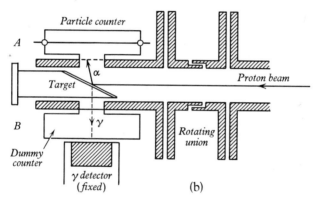

(b)

Fig. 13.8 Lifetimes by recoil techniques. (a) Recoil-distance method. (b) Recoil-Doppler method (Devons *et al.*, *Proc. phys. Soc. Lond.*, **A68**, 18, 1955).

The availability of Ge(Li) gamma radiation detectors of high resolution has permitted an elegant development of the plunger method.† If an excited nucleus emits a gamma ray while it is moving, the observed energy shows a Doppler shift and is $E_\gamma = E_0 (1 \pm v/c \cos \theta)$; if the nucleus emits the radiation after it is brought to rest, the energy is unshifted, $E_\gamma = E_0$. If therefore the target is fixed and the plunger is used to receive the recoil ions the spectrum observed at a certain angle with the beam direction shows two peaks whose relative height gives the fraction of nuclei decaying before and after reaching the plunger. This varies with plunger displacement and permits a direct calculation of lifetime; the recoil velocity is known from the observed Doppler

† T. K. Alexander and K. W. Allen, *Can. jnl. Phys.*, **43**, 1563, 1965.

shift which can be an order of magnitude greater than the resolution width of the detector.

(d) *Doppler Shift Attenuation Method* (DSAM) $(\tau \approx 10^{-11}$ s-10^{-15} s) For lifetimes less than 10^{-11} s the recoil distance method becomes difficult because the plunger displacements that have to be measured accurately are $< 10^{-5}$ m. It is however still possible, following Devons, Manning and Bunbury, to use the Doppler shift to measure these lifetimes. The principle of the method is indicated in Fig. 13.8(b), which again relates to the thin target reaction ^{19}F(p, α)^{16}O*. Gamma ray energy spectra were recorded in coincidence with the reaction α-particle and there were two geometrical positions, *A* as shown and *B* with the target and particle counter rotated through 180°. These two observations determined the gamma ray Doppler shift for recoil into vacuum (*B*) and into a known target backing (*A*). The former gives the full Doppler shift and the latter an attenuated shift determined jointly by the lifetime under investigation and the range-velocity relation for the recoil ions in the target backing. A value $\tau \geqslant 2 \times 10^{-12}$ s was found for the 6·13 MeV level of ^{16}O and this is consistent with the upper limit found for the same level by the recoil distance method.

The Doppler shift attenuation method has been developed by the use of heavy ions as bombarding particles to increase recoil velocities, and by the application of Ge(Li) counters to give precision determinations of line shape. Much work has been devoted to investigation and understanding of the energy-loss processes for heavy ions in solids, which determine the time scales for the method, and the accuracy ($\approx 20\%$) of its results. These are derived from measurements of the line centroid shift as a fraction of the maximum possible shift or on detailed line shapes. The DSAM is most effective when the lifetime is comparable with the slowing down time of the recoiling ion; for a given material the line moves from a wholly shifted position to a wholly unshifted position as the lifetime increases.

(e) *Lifetimes less than* 10^{-16} s *(particle decays)* The recoil distance method has been applied[†] in the lifetime range below 10^{-14} s by use of the phenomenon of blocking in single crystals. If an excited nucleus is produced at a lattice site then for the reasons outlined in Sect. 5.3.7, a charged particle emitted from this nucleus at the lattice point will be excluded from the directions of channelling in the crystal. If however the lifetime is sufficiently long, the recoil velocity of the nucleus due to the production reaction may remove it sufficiently far from the atomic rows to permit observation of the so-called 'blocking angle'. For the reaction ^{70}Ge(p, p')^{70}Ge* in a germanium crystal a lifetime of $36 \pm 23 \times 10^{-18}$ s was obtained.

Even shorter lives can be estimated in certain special cases by the method of *proximity scattering*[‡] which is applicable to sequential reactions of the type ^{12}C$+$d$\longrightarrow$ ^{13}N*$+$n, ^{13}N*$\longrightarrow$ ^{12}C$+$p. The energy spectrum of the proton

† M. Maruyama *et al., Nucl. Phys.,* **A145**, 581, 1970.
‡ J. Lang *et al., Nucl. Phys.,* **88**, 576, 1966.

is influenced by interaction with the neutron to an extent determined by the $^{13}N^*$ decay lifetime. For this reaction a result of 0.7×10^{-20} s was obtained.

13.6.2 Coulomb excitation

The radiative lifetime measurements described in the preceding sections cover the region from the 'isomeric' lives of 10^{-6} s down to about 10^{-15} s, which corresponds to a width of 0.6 eV. It is, however, not always convenient to produce the recoil velocities necessary for measurement of the shorter lives. Fortunately, phenomena which are determined by the radiative *width* become easier to observe as the mean life diminishes and may be used to measure transition probabilities when direct lifetime measurements are impracticable. The most generally applicable method is that of Coulomb excitation.

A charged particle passing near to a nucleus (Fig. 5.3(a)), with an energy considerably less than the height of the mutual potential barrier so that nuclear effects may be disregarded, follows a classical trajectory prescribed by the Coulomb law of force $zZe^2/4\pi\varepsilon_0 r^2$. The deflection of the particle (Rutherford scattering) is associated with an electric field at the nucleus which varies with time. Quantum mechanically this field may be represented as a process of emission and absorption of equivalent photons whose momentum changes create the Coulomb force between the two particles; the photon spectrum in this representation of the field is continuous so that photons of exactly the right energy to excite levels of the nucleus are always available. The cross-section for this process (Coulomb excitation) depends on the trajectory of the incident particle, and also on the width of the nuclear level concerned, which determines the absorption of energy from the electromagnetic field. Wide levels are easily excited and measurement of the cross-section for Coulomb excitation (by determining the yield of radiation from the excited level) is equivalent to a determination of lifetime.

One of the first demonstrations of Coulomb excitation was given by Huus and Zupančič who observed the yield of 136 keV radiation when tantalum was bombarded by protons of energy between 1.0 and 2.2 MeV. The radiation was detected in a sodium iodide crystal and the energy spectrum and yield curve are shown in Fig. 13.9, together with a theoretical prediction for the latter. Care was taken to discriminate against X-radiation and proton bremsstrahlung.

The cross-section for the production of radiation by Coulomb excitation using protons may be written

$$\sigma_L = f(E, Z, \Delta E)B_{eg}(L) \qquad (13.15)$$

where E is the energy of the incident particle, Ze the charge of the bombarded nucleus, $\Delta E \, (= E_\gamma$ in Fig. 13.1) the energy of the transition concerned and f is a calculable function. The yield curves for different multipole transitions $E1$, $M1, E2, \ldots$ differ and may be used to yield information on nuclear spins and parities. The reduced transition probabilities $B(L)$ are obtained by direct application of (13.15). In some cases direct comparison with measurements made by other methods is possible and agreement is satisfactory.

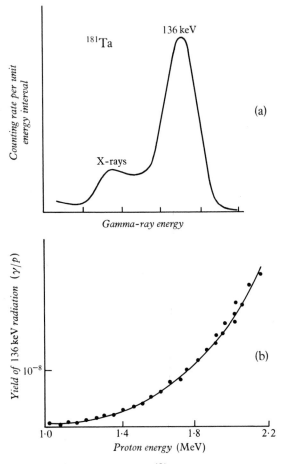

Fig. 13.9 Coulomb excitation of 136 keV level in ^{181}Ta. (a) Spectrum of radiation from target taken in scintillation spectrometer. An absorber is used to reduce the yield of X-rays. (b) Yield of 136 keV radiation as a function of proton energy, compared with theory for $E2$ excitation (full curve) (Huus and Zupančič, *Det. Kgl. Dansk. Vidensk. Selskab*, **28**, 1, 1953).

Coulomb excitation has been observed throught the periodic system and with many types of bombarding particle, including heavy ions such as ^{14}N and ^{40}A which are particularly effective because of their high charge and mass. A particularly important application of Coulomb excitation is in the study of rotational levels of the collective type excited by electric quadrupole ($E2$) transitions and in fact, practically all Coulomb excitation processes are of $E2$ type (cf. Fig. 12.8). In cases when $|\Delta I| = 1$ the subsequent radiative transition may be a mixture of $M1$ and $E2$ components and analysis of the mixture may be made by the methods normally applied to radioactive decay schemes.

The use of heavy ions has permitted the observation of higher order processes, e.g. the excitation of a 4^+ level in an even–even nucleus by the double $E2$ sequence $0^+ \longrightarrow 2^+ \longrightarrow 4^+$. Another double $E2$ process of some importance is the *reorientation* effect in which a state of spin I which has been Coulomb-excited makes an $E2$ transition from one magnetic substate (I, m_1) to a second substate (I, m_2) as a result of interaction of the electric field of the bombarding particle with the electric quadrupole moment Q_0 of the excited state. The effect is now a useful method for determination of this moment.

Coulomb excitation of nuclear levels using *electrons* is in principle possible but is experimentally difficult to observe as an emission process because of the high yield of bremsstrahlung quanta. If the level is of the long-lived isomeric type, delayed activity may be observed, but the radiative width is then small and the yield is low. Excitation of an isomeric level of ^{115}In by electron bombardment has been reported but the process is essentially equivalent to photo-excitation, and practically all recent work of this type has been conducted with bremsstrahlung sources or monochromatic photons (Sect. 8.2.2). Electrons interact with nuclei mainly through the electromagnetic field, and the barrier limitation imposed on heavy particles does not apply so that states of higher excitation can be reached. The observation of the inelastically scattered electrons is in fact an important method of determining transition probabilities and is discussed in the following section.

13.6.3 Inelastic scattering of electrons

In Sect. 11.2.1 the technique and results of experiments on the *elastic* scattering of ≈ 200 MeV electrons were described; essentially this work yields the nuclear charge density. Figure 11.4 shows peaks corresponding to *inelastic* scattering with excitation of known levels of the target nucleus ^{12}C, and the differential cross-section for this process can also conveniently be expressed in the form of eq. (11.6). The charge density appearing in the form factor $F(q)$ (where q is the momentum transfer to the nucleus) is now a *transition* charge density, which may be predicted by nuclear models. It may be shown (Ref. 13.7) that for electric multipole transitions and for small q^2 the quantity F^2 is proportional to the radiative width of the state. Analysis of the cross-section measurements then yields Γ_γ. Furthermore the q dependence of the cross-section indicates the multipolarity of the transition. Electron scattering can excite $E0$ transitions and is generally more sensitive to the details of nuclear structure than lifetime measurements because of the possibility of varying the momentum transfer to the target nucleus.

13.6.4 Nuclear resonance processes

Suppose that the level E_e shown in Fig. 13.1 exists in a nucleus C, which may be represented as C* when excited. It may then be possible to produce the excitation by capture of a bombarding particle or photon (*a*) in a suitable initial nucleus X, i.e.

$$X + a \longrightarrow C^*$$ (13.16)

These processes are of great importance for the theory of nuclear reactions and are discussed in Chapter 15. For the present purpose we note only that the cross-section for the production of C* as the energy of *a* is varied may be expressed for well-defined states by the resonance formula

$$\sigma_a = \pi \lambda^2 g \frac{\Gamma_a \Gamma}{(E - E_0)^2 + \Gamma^2/4}$$ (13.17)

where Γ is the total width of the level E_e and Γ_a is the partial width of the level for re-emission of *a*. The energy *E* is the excitation energy of the compound nucleus created by absorption of the incident particle (Sect. 15.2.1) and $E_0 (= E_\gamma)$ is the resonance energy, at which there is exact excitation of the level. The quantity λ is the reduced channel wavelength (Sect. 14.2.1) and *g* is a statistical factor.

If now the emission of γ-radiation from C* is observed, the cross-section for this mode of decay is

$$\sigma_{a\gamma} = \sigma_a \frac{\Gamma_\gamma}{\Gamma} = \pi \lambda^2 g \frac{\Gamma_a \Gamma_\gamma}{(E - E_0)^2 + \Gamma^2/4}$$ (13.18)

Under suitable conditions this formula can be used to deduce Γ_γ from experimental cross-sections; it is of the general form of the cross-section (13.15), in which the radiative transition probability (here $\Gamma_\gamma/\hbar$) is multiplied by a factor representing the mechanism of excitation.

The statistical factor is

$$g = \frac{2I_e + 1}{(2s + 1)(2I_g + 1)}$$ (13.19)

where *s* is the intrinsic spin of the incident particle. For radiation we take $2s + 1 = 2$, corresponding to two independent directions of the polarization vector.

(*a*) *Proton capture.* For light elements and proton energies of 1–3 MeV the proton capture (p, γ) reaction shows sharp resonances. From (13.18) the cross-section at resonance ($E = E_0$) is

$$\sigma_0 = 4\pi \lambda^2 g \frac{\Gamma_p \Gamma_\gamma}{\Gamma^2}$$ (13.20)

and if γ-emission and proton re-emission are the only decay modes of the excited level, $\Gamma = \Gamma_p + \Gamma_\gamma$. Usually Γ_p (≈ 10 keV) is very much greater than Γ_γ (≈ 1 eV) and then

$$\sigma_0 = 4\pi \lambda^2 g \frac{\Gamma_\gamma}{\Gamma}$$ (13.21(*a*))

By observing the variation of σ with energy the total width Γ can be obtained, and then σ_0 gives $g\Gamma_\gamma$ for the virtual level excited.

(*b*) *Neutron capture.* For elements of medium weight and slow neutrons sharp capture resonances are also found. They are normally observed as dips in transmission curves for particular solid samples as a function of neutron energy. In this case $\Gamma = \Gamma_n + \Gamma_\gamma$ and Γ_n ($\approx 10^{-3}$ eV) is much less than Γ_γ so that the resonant cross-section is

$$\sigma_0 = 4\pi\lambdabar^2 g \frac{\Gamma_n}{\Gamma} \qquad (13.21(b))$$

The radiative width Γ_γ must thus be determined directly from the width Γ ($\approx \Gamma_\gamma \approx 0\cdot1$ eV) of the transmission curve. Unfortunately these widths are so small that they may be seriously affected by the Doppler effect. This arises because an incident thermal neutron finds the atoms of the absorber moving with random thermal velocities and the effective absorption cross-section must be obtained by averaging over a Maxwell distribution of relative velocities. The observed width of the transmission curve and the cross-section at resonance then both involve, in addition to Γ, the Doppler width

$$\Delta = p_n \sqrt{\frac{2kT}{M}} = 2\sqrt{m_n E_n \frac{kT}{M}} \qquad (13.22)$$

where $p_n = \sqrt{2m_n E_n}$ is the momentum of the incident neutron, M is the mass of the absorbing atom, and T is the absolute temperature. For the resonance level in ^{239}U at a neutron energy of 6·7 eV, $\Delta = 0\cdot06$ eV for $T = 373$ K; this is comparable with radiation widths and must be allowed for in calculation of Γ_γ for virtual levels excited by neutrons. In proton capture reactions Γ is usually $\gg \Delta$ and such corrections are unimportant.

(*c*) *Resonant scattering and absorption of γ-radiation.* A case of particular interest arises when nuclear levels are excited by radiation. From (13.17) the cross-section for absorption at resonance is

$$\sigma_0 = 4\pi\lambdabar^2 g \frac{\Gamma_\gamma}{\Gamma} \qquad (13.23)$$

which for radiation of energy 500 keV and $\Gamma_\gamma = \Gamma$ is of the order of 10^4 barns. The total electronic cross-section for attenuation of radiation of this energy in a light element such as oxygen or carbon is of the order of 10^2 barns. It might therefore be expected that nuclear attenuation could be easily observed by increasing the energy of the incident radiation until resonance absorption appeared. If a bremsstrahlung spectrum is used for the experiment it is necessary to average the nuclear cross-section over the energy interval ΔE accepted at one setting of the detector. When (13.17) is averaged in this way the effective cross-section becomes

$$\bar{\sigma}_0 = 2\pi^2 \lambdabar^2 g \frac{\Gamma_\gamma}{\Delta E} \qquad (13.24)$$

and since the factor $\Gamma_\gamma/\Delta E$ may be 10^{-4} or less the nuclear effect becomes small compared with electronic absorption. If large angle scattering for a specimen is observed rather than absorption the situation is more favourable since the non-resonant (electronic) differential scattering cross-sections are peaked in the forward direction. It has thus been possible to observe nuclear lines in the spectrum of bremsstrahlung at about 145° from light elements such as boron[†] and carbon, and to deduce the width of the levels concerned (≈ 0.5 eV). Monochromatic sources of radiation from nuclear reactions including slow neutron capture, and from positron annihilation, are now available and may be used in their appropriate energy ranges.

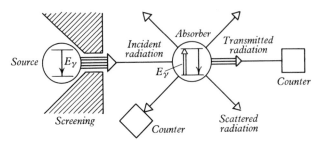

Fig. 13.10 Scattering and absorption experiment.

For the narrowest states, say $\ll 0.5$ eV, it is in general necessary to use resonance radiation, as in optics, i.e. to excite a level by means of the radiation emitted in the decay of similar levels in an active source. An experimental arrangement for the detection of resonant absorption and of the subsequent re-emission of the resonance line is shown in Fig. 13.10. The cross-section at exact resonance is given by (13.23) and in the case of bound levels with which this process is mainly concerned the only process that increases Γ above Γ_γ is internal conversion. Introducing the conversion coefficient α the resonant cross-section becomes

$$\sigma_0 = 4\pi\lambda^2 g\frac{1}{1+\alpha} = 2\pi\lambda^2\frac{2I_e+1}{2I_g+1}\frac{1}{1+\alpha} \tag{13.25}$$

The incident radiation may itself have a Lorentz line shape and the total absorption effect is then obtained by folding this with the absorption cross-section (13.17). For a thin absorber the result is a Lorentz absorption dip of width 2Γ and depth corresponding to a cross-section $\sigma_0/2$. For most values of α encountered in practice this is still a large cross-section ($\approx \lambda^2$) but measurement of it would be of little interest since it determines nothing new. In practice, however, the situation is drastically altered in a useful way by the existence of recoil and Doppler effects. In the emission of the resonance radiation of energy E_γ ($=E_0$) from a stationary source atom of mass M, a

[†] L. Cohen, R. A. Tobin and J. McElhinney, *Phys. Rev.*, **114**, 590, 1959.

momentum $p_\gamma = E_\gamma/c$ must be imparted to the emitting body, and the emitted radiation therefore only has energy

$$E_s \approx E_\gamma - \frac{p_\gamma^2}{2M} = E_\gamma - \frac{1}{2} \frac{E_\gamma^2}{Mc^2} \qquad (13.26)$$

Similarly in the absorption process, the energy required for resonance is

$$E_a \approx E_\gamma + \frac{1}{2} \frac{E_\gamma^2}{Mc^2} \qquad (13.27)$$

The emission and absorption lines are therefore shifted in energy by the recoil effect by an amount

$$2R = \frac{E_\gamma^2}{Mc^2} \qquad (13.28)$$

In addition, both source and absorber atoms are taking part in thermal motion and the lines are broadened by the Doppler effect. The Doppler width is now

$$\Delta = p_\gamma \sqrt{\frac{2kT}{M}} = \frac{E_\gamma}{c} \sqrt{\frac{2kT}{M}} \qquad (13.29)$$

The order of magnitude of these effects is shown for three cases of interest in Table 13.1. It can be seen that the recoil effect is negligible in *atomic*

TABLE 13.1 Recoil and Doppler effects for electromagnetic radiation

Radiating system	Wavelength (m)	Line energy E_γ	Γ eV	2Δ (300 K) eV	$2R$ eV
Sodium atom	$5 \cdot 89 \times 10^{-7}$	2·11 eV	$4 \cdot 5 \times 10^{-8}$	$6 \cdot 6 \times 10^{-6}$	$2 \cdot 1 \times 10^{-10}$
^{198}Hg nucleus	3×10^{-12}	412 keV	$2 \cdot 1 \times 10^{-5}$	0·4	0·9
^{57}Fe nucleus	$8 \cdot 6 \times 10^{-11}$	14·4 keV	$4 \cdot 6 \times 10^{-9}$	0·02	0·004

transitions and resonance absorption of Doppler broadened lines takes place as shown in Fig. 13.11(a). For a *nuclear* transition with free recoil on the other hand $2R$ is always $\gg \Gamma$ for low-lying bound levels, and is often $> 2\Delta$. Nuclear resonance is therefore a weak effect unless recoil losses are restored or eliminated; the situation is as shown in Fig. 13.11(b). The introduction of recoil and thermal terms into (13.25) creates a dependence of the resonant cross-sections on Γ_γ itself and permits measurement of this quantity if a sufficient effect can be achieved.

Recoil energy loss can be restored, and resonance between Doppler-broadened lines obtained, by the following means (Ref. 13.8):

(i) By mounting a radioactive source on the tip of a high-speed rotor and using the Doppler effect to increase the frequency of the emitted radiation. From (13.28) the speed required is given by

$$E_\gamma \frac{v}{c} = \frac{E_\gamma^2}{Mc^2} \tag{13.30}$$

and for a nucleus of mass 200 and $E_\gamma = 500$ keV the value $v = 820$ m s^{-1} is found.

(ii) By heating the radioactive source so that the Doppler broadening is largely increased. Temperatures of up to 1500 K have been used.

(iii) By utilizing a preceding transition, such as β-decay, γ-decay or production in a nuclear reaction to provide the necessary velocity for Doppler shift of frequency.

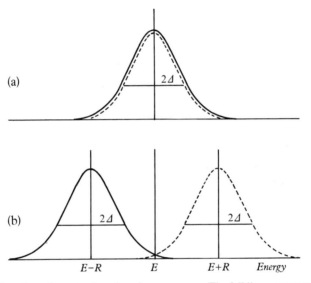

Fig. 13.11 Effect of recoil on atomic and nuclear resonance. The full line represents the intensity distribution in the incident radiation and the dotted line gives the absorption cross-section of the absorber as a function of the homogeneous incident energy. (a) Atomic resonance, $R \ll \Delta$. (b) Nuclear resonance, $R > \Delta$.

Nuclear γ-resonance was first demonstrated by Moon† using method (i). Figure 13.12(*a*) shows the principle of apparatus used by Davey and Moon for a study of the 412 keV radiation originating in the transition

$$^{198}\text{Au} \longrightarrow {}^{198}\text{Hg*} + \beta^- + \bar{\nu}; \quad {}^{198}\text{Hg*} \longrightarrow {}^{198}\text{Hg} + \gamma \tag{13.31}$$

The source was mounted on the tip of a high-speed rotor driven electromagnetically and resonance was sought in the back-scattered radiation from a sample of mercury as the rotor speed varied. In order to distinguish the

† P. B. Moon, *Proc. phys. Soc., Lond.*, **64**, 76, 1951.

resonant effect from non-resonant scattering comparison with the scattering from lead was made at each rotor speed. Figure 13.12(b) shows the observed ratio as a function of speed, and clearly exhibits the increase of yield expected as the emission peak of Fig. 13.11(b) moves towards the absorption line with increasing rotor speed. From the observed cross-section for the resonant scattering effect, with appropriate account taken of the thermal broadening, which is responsible for the line shape shown in Fig. 13.11, a total width of $(2 \cdot 1 \pm 0 \cdot 4) \times 10^{-5}$ eV for the 412 keV level of ^{198}Hg was found in agreement with, but more accurate than, observations by the delayed-coincidence technique.

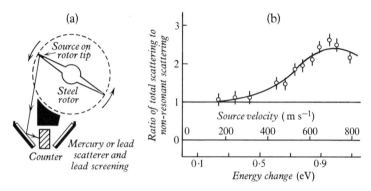

Fig. 13.12 Demonstration of nuclear resonant scattering with free recoil for the 412 keV transition in ^{198}Hg (Moon, 1951; Davey and Moon, *Proc. phys. Soc. Lond.*, **66**, 956, 1953). (a) Apparatus. (b) Ratio of scattering observed with mercury to scattering observed with lead as a function of rotor tip speed.

Like Coulomb excitation, nuclear resonance shows largest effects in cases of large width, i.e. of small lifetime. The method is rather more limited than Coulomb excitation since gramme quantities of the scatterer at least are required and the available velocities or preceding transitions may not always be adequate to supply the necessary recoil energy. The theory is however simple and direct and the detailed predictions such as the existence of a maximum scattering cross-section for the velocity $v = E_\gamma/Mc$ have been checked. The double transition $I_g \longrightarrow I_e \longrightarrow I_g$ is essentially the same as a cascade transition between levels of the same spins and the angular correlation of the incident and scattered radiation and polarization of the latter can therefore be predicted (Sect. 17.4.2).

(*d*) *The Mössbauer effect* (Ref. 13.9). Recoil energy loss in a resonant scattering experiment can be *eliminated* in certain special cases.

If a nucleus in a solid, emitting a γ-ray, is assumed to behave like a gaseous particle with thermal motion it can take up the recoil energy $E_\gamma^2/2Mc^2$ associated with γ-emission. If, however, the energy E_γ is small the recoil energy may be insufficient to loosen the emitting atom in its chemical binding. It then follows that not only the originating atom, but a large number of associated neighbour atoms amounting in the limit to the crystal as a whole, are available for absorbing the recoil energy. The mass M in (13.28) then becomes very large and the recoil shift is correspondingly very small. The spectrum of radiation emerging from a crystal under these conditions contains a sharp line of natural width Γ_γ and of unshifted frequency as shown in Fig. 13.13, corresponding to the fraction f of nuclear transitions in which the lattice energy does not change (zero thermal phonon process). In the remaining fraction $(1-f)$ of the transitions phonons are created or destroyed, changing

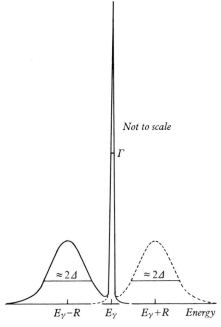

Fig. 13.13 Recoilless emission and absorption of γ-rays (not to scale). The full line represents the emission spectrum, and the dotted line the absorption cross-section; each contains a narrow line of natural width at the resonant energy and a Doppler distribution.

the lattice energy, and this populates a broad thermal spectrum, also shown in Fig. 13.13, of width corresponding to the mean thermal energy of the crystal. The same effect takes place on absorption, and if the sharp lines contain an appreciable fraction of the total spectral distribution, there is potentially available an extremely well-defined resonance phenomenon. In 1958 Mössbauer,† using radiation from ^{191}Ir, demonstrated this narrow resonance in absorption by cooling the source and absorber; this decreases the thermal motion and increases the fraction of emissions and absorptions in the sharp line. Mössbauer was also able to obtain a direct measurement of the line width Γ_γ by *slow* motion of the source; his results are shown in Fig. 13.14, in

† R. L. Mössbauer, *Zeits. für Physik*, **151**, 124, 1958.

which the velocity scale should be contrasted with that shown in Fig. 13.12(b) for recoil-shifted lines. The principle of Mössbauer experiments is as shown in Fig. 13.10 with the addition of a mechanism for moving the source or absorber.

The theory of the Mössbauer effect has been extensively investigated and it has been pointed out that the phenomenon of the absorption of *momentum* by a solid lattice without the absorption of *energy* is well known in Bragg reflection of X-rays from lattice planes. The Debye–Waller formula which gives the probability of this coherent X-ray process in comparison with incoherent scattering may be applied to the Mössbauer effect to give the

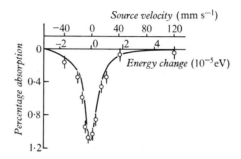

Fig. 13.14 Determination of natural line width in a resonant absorption experiment by movement of source, under conditions of recoilless emission and absorption.

fraction of emissions or absorptions occurring in the narrow line. Similar considerations were shown by Lamb, as early as 1939, to apply to the scattering of slow neutrons by crystals. The strongest Mössbauer effect known occurs in ^{57}Fe; the attenuation of recoilless 14·4 keV γ-radiation from this nucleus in an absorber of ^{57}Fe is primarily a nuclear effect in contrast with the usual predominance of non-resonant scattering processes.

The Mössbauer effect has not been much used for the determination of transition probabilities, but its overwhelming importance in general,† as well as in nuclear physics, is that it provides a method for measuring small frequency changes with a sensitivity of better than Γ_γ/E_γ ($\approx 3 \times 10^{-13}$ for ^{57}Fe). It has been used to resolve magnetic dipole and electric quadrupole hyperfine structure of nuclear states arising from interaction with the surrounding electron distribution or even with externally applied magnetic fields. The *isomer shift*, arising from the difference between the mean square radius of nuclear ground and excited states can be extracted from Mössbauer experiments using the given nucleus in different compounds as source and absorber. Such interrelations between nuclear and atomic properties are exploited extensively in solid state physics, metallurgy and chemistry.

† Application to testing the hypotheses of relativity theory are discussed in Ref. 13.9.

13.7 Comparison between experiment and theory

The methods described in this chapter have made available a large amount of data on transition probabilities, sometimes expressed directly in this form, sometimes given as lifetimes or radiative widths. In comparing these results with theoretical prediction it is necessary in the first place to group the transitions according to multipolarity. The observed radiative width is then expressed in terms of the single-particle, or Weisskopf width given by (13.1) and (13.2) for the particular multipolarity concerned. This procedure gives essentially the reduced transition probability as a fraction of the single particle value. The ratio

$$|M|^2 = \frac{\Gamma_\gamma \text{ (observed)}}{\Gamma_\gamma \text{ (Weisskopf)}}$$

then gives information on nuclear models and selection rules. The following conclusions emerge from extensive surveys (Ref. 13.10):

(a) For practically all transitions, except those of the $E2$ type, $|M|^2$ is less than one, so that observed transitions are slow compared with the single-particle processes.

(b) $E1$ transitions in the light nuclei have $|M|^2 \approx 0.03$. If a transition is found with a value of $|M|^2$, in relation to the Weisskopf dipole width, which is >0.02 it may almost certainly be identified as $E1$.

In the light nuclei with equal numbers of neutrons and protons (self-conjugate nuclei) a special selection rule for *isobaric spin* may reduce the probabilities for $E1$ emission. The strict rule is $\Delta T = \pm 1$ for $E1$ transitions in nuclei with $N = Z$ ($T_z = 0$) but the rule is frequently weakened by mixing of states and the retardation is typically a factor of about 10–50 beyond the average value of the transition probability.

(c) $M1$ transitions in the light nuclei have $|M|^2 \approx 0.15$. There is also some retardation due to isobaric spin rules in self-conjugate nuclei but this is less marked than for $E1$ transitions.

(d) $E1$ and $M1$ widths in medium weight and heavy nuclei have been surveyed by means of the (n, γ) capture reaction and are also in general less than the single-particle value, sometimes by a factor of as much as 10^5.

(e) $E2$ transitions in both light and heavy nuclei include many examples with $|M|^2 \approx 10$ to 100 times the single-particle value.

(f) The isomeric transitions, which are found in nuclei just below the magic numbers and are usually of $E3$ or $M4$ type, show relatively little spread of $|M|^2$ within each class.

(g) There is no apparent difference in strength between transitions associated with a single odd neutron or a single odd proton.

The interpretation of the observed radiative widths in terms of nuclear models is similar to the description given of static moments (ch. 12) and particle reduced widths (ch. 15). The single-particle strength is considered to be distributed among a number of possible transitions, so that any specific transition has a considerably reduced strength. Specific calculations, dependent on coupling schemes, have been based on the individual particle model for light nuclei near closed shells. As the number of loose particles increases, this type of calculation becomes difficult in detail, but still suggests that the single-particle transition probability should be multiplied by a factor (<1) depending on the number of available particles. Further away from closed shells the individual particle merges into the collective model as has already been discussed and an explanation of the large observed $E2$ strengths is provided in terms of coherent motion of nucleons. A similar explanation can account for the $E1$ giant resonance and the low-lying 3^- states corresponding to octupole phonons.

References

1. BLATT, J. B. and WEISSKOPF, V. F. *Theoretical Nuclear Physics*, p. 595, Wiley, 1952; MOSZKOWSKI, S. A. *Phys. Rev.*, **89**, 474, 1953.
2. SIEGBAHN, K. (ed.). *Alpha, Beta and Gamma Ray Spectroscopy*, North Holland, 1965.
3. ROSE, M. E. in Ref. 13.2, page 887.
4. SLIV, L. A. and BAND, I. M. in Ref. 13.2, page 1639.
5. DEVONS, S. 'The Measurement of Very Short Lifetimes', *Nuclear Spectroscopy*, Part A, ed. F. Ajzenberg-Selove, Academic Press, 1960; SCHWARZSCHILD, A. Z. and WARBURTON, E. K. 'The Measurement of Short Nuclear Lifetimes', *Ann. rev. nucl. sci.*, **18**, 265, 1968.
6. HUBY, R. 'Coulomb excitation', *Rep. progr. Phys.*, **21**, 59, 1958; D. Alburger in Ref. 13.2, page 745.
7. BARBER, W. C. 'Inelastic Electron Scattering', *Ann. rev. nucl. Sci.*, **12**, 1, 1962.
8. METZGER, F. R. 'Resonance Fluorescence in Nuclei', *Progr. nucl. Phys.*, **7**, 53, 1959; see also Ref. 13.2, page 1281.
9. O'CONNOR, D. A. 'The Mössbauer Effect', *Contemp. Phys.*, **9**, 521, 1968; see also Ref. 13.2, page 1293.
10. WILKINSON, D. H. 'Analysis of Gamma Decay Data', *Nuclear Spectroscopy*, Part B, ed. F. Ajzenberg-Selove, Academic Press, 1960.
11. BARTHOLOMEW, G. A. 'Neutron Capture γ-Rays', in *Nuclear Spectroscopy*, Part A, ed. F. Ajzenberg-Selove, Academic Press, 1960.

14 General features of nuclear reactions

By 1939 all the main types of nuclear reaction induced by particles with an energy up to about 10 MeV had been discovered. Post-war development of accelerators made higher energies available and processes involving multiple emission of particles were then more frequently observed, but no phenomena of an essentially novel character were revealed until meson-producing reactions were achieved in the laboratory in 1948 with energies of the order of 100 MeV per nucleon. Some years later (1955) the production of the anti-proton in proton–proton collisions, with a threshold energy of 5·6 GeV for target protons at rest, was a dramatic success for accelerator and detector technology. The post-war era is, however, also remarkable for the enormous increase in the amount of available information on reactions of familiar type, an increase which has led to confirmation, to extension and above all to a better understanding, of theories of reactions which were formulated earlier on rather meagre data.

Some of the important discoveries in the field of nuclear reactions are described in the following section, **14.1**. From the information presented by these early experiments, the theory of such processes began to take shape in the work of Bethe, Bohr, Breit, Peierls, Placzek, Weisskopf and Wigner. It was soon recognized that the observable features might be determined equally by the operation of 'external' factors such as the incident energy and barrier penetrability for this energy, and of 'internal' factors connected with nuclear structure. The evaluation of these internal factors from the data is based on the wave-mechanical theory of collision processes (Sect. **14.2**).

As a result of the development of theory and of the extension of experiment, several specific types of nuclear reaction mechanism can now be distinguished, each with some relevance to the nuclear models discussed in Chapter 9 and Chapter 12. The more important types (compound nucleus, direct reaction, optical model) will be discussed in Chapter 15. Some of the basic experiments of elementary particle physics are described in Chapters 19, 20 and 21.

14.1 The 'historical' reactions

14.1.1 The disintegration of nitrogen by α-particles

By 1919 many of the phenomena associated with the scattering of α-particles by light nuclei had been thoroughly investigated, mainly by Rutherford and his pupils at the University of Manchester. Among the unexplained 'anomalous' effects was the production of scintillations on a zinc sulphide screen placed near an α-particle source in air, but at a distance from it greater than the range of the α-particles. It was at first felt that these scintillations were due to projected H-atoms (or protons, as we now call them) from hydrogen present in the source assembly since the elastic collision of α-particles with hydrogen had already been observed and was known to give rise to just such

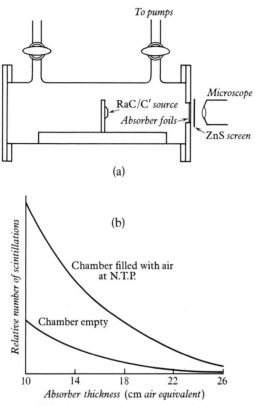

(a)

(b)

Fig. 14.1 The discovery of artificial transmutation. (a) Apparatus used by Rutherford to demonstrate the production of fast protons in the bombardment of air by α-particles from RaC'. The protons originating in the air-filled chamber passed through a thin window and through absorbing foils to the zinc sulphide screen. (b) Absorption curves obtained by variation of foil thickness (Rutherford, *Phil. Mag.*, **37**, 537 and 581, 1919).

a phenomenon. In order to study these particles Rutherford† used the apparatus shown in Fig. 14.1(a) and compared the number and the range of the particles observed on the screen when the RaC–C′ source assembly was evacuated with the corresponding values found when the box was filled with air. The absorption curves are shown in Fig. 14.1(b) and it will be noted that the introduction of air causes an *increase* in the rate of scintillation, although

†Sir E. Rutherford, *Phil. Mag.*, **37**, 581, 1919.

the range of the particles is unaltered. An extensive series of checks led to the conclusion that the extra particles were due to the introduction of *nitrogen* into the source chamber and the appearance of the light flashes, taken together with rough measurements of magnetic deflection, suggested that the particles produced were protons. Recoil nitrogen atoms would have had insufficient range to reach the screen. Rutherford concluded '. . . that the nitrogen atom is disintegrated under the intense forces developed in a close collision with a swift α-particle and that the hydrogen atom which is liberated formed a constituent part of the nitrogen nucleus'; both conclusions have been abundantly verified. The process observed is now written

$$^{14}\text{N} + \alpha \longrightarrow {}^{17}\text{O} + \text{p} - 1\cdot19 \text{ MeV} \tag{14.1}$$

and striking evidence for this interpretation was obtained by Blackett[†] using a Wilson cloud chamber. Plate 11 shows one of the pictures obtained in these experiments; it was found that disintegrations of the type (14.1) were occurring at a rate of about 1 for every 5×10^5 α-tracks.

Plate 11 Expansion chamber photograph showing ejection of a proton from a nitrogen nucleus by an α-particle (P. M. S. Blackett and D. S. Lees, *Proc. roy. Soc.* A, **136**, 325, 1932).

This was the discovery of artificial transmutation. The final sentence of Rutherford's paper was prophetic: '. . . if α-particles—or similar projectiles —of still greater energy were available for experiment, we might expect to break down the nucleus structure of many of the lighter atoms'.

[†] P. M. S. Blackett, *Proc. roy. Soc.*, A, **107**, 349, 1924; P. M. S. Blackett and D. S. Lees, *Proc. roy. Soc.*, A, **136**, 325, 1932.

14.1.2 The disintegration of lithium by protons (the Cockcroft–Walton† experiment)

The fundamental discovery made in 1932 by Cockcroft and Walton, working in the Cavendish Laboratory, Cambridge, under Rutherford, was that reactions of the type (14.1) could be reversed by the use of artificially accelerated protons. This was not the discovery of a new process since artificial disintegration had already been established; the importance of the Cockcroft–Walton experiment was that it gave enormous impetus to the development of accelerators of all types.

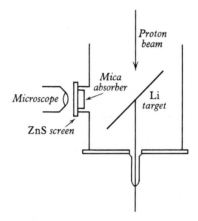

Fig. 14.2 Disintegration of elements by high velocity protons (Cockcroft and Walton, 1932).

The cascade generator used in this experiment has already been described (Sect. 8.1.1). Protons accelerated in the vacuum tube to an energy of 100 to 500 keV were allowed to strike a target of lithium (Fig. 14.2), opposite to which was placed a zinc sulphide screen which could be viewed through a microscope. Protons scattered from the target were intercepted and stopped by a mica screen before they could reach the zinc sulphide. With a proton current of about 1 µA at 125 keV bright scintillations were observed on the screen; the range of the particles producing these scintillations was shown by the use of mica absorbers to be homogeneous and equal to about 8 cm of air. The particles were also allowed to pass through a mica window into a shallow ionization chamber connected to an amplifier and oscillograph; the size of the pulses from the particles near the end of their range was the same as that from the α-particles of polonium.

From these and other observations Cockcroft and Walton concluded that the nuclear reaction

$$^{7}\text{Li} + \text{p} \longrightarrow {}^{4}\text{He} + {}^{4}\text{He} + Q \tag{14.2}$$

had been induced. The masses of the nuclei concerned were known in 1932

† J. D. Cockcroft and E. T. S. Walton, *Proc. roy. Soc.*, A, **137**, 229, 1932.

with rather low precision, but the energy release Q expected in accordance with the Einstein mass–energy relation $\Delta E = \Delta M . c^2$ seemed to agree well with that deduced from the observed range of the α-particles. The proposed mechanism was confirmed by simultaneous detection of the two α-particles both by scintillation screens and in the cloud chamber (Dee and Walton,† Plate 12).

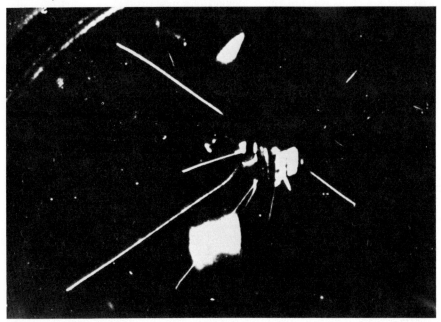

Plate 12 Expansion chamber photograph showing disintegration of lithium by protons with the emission of pairs of α-particles in opposite directions (P. I. Dee and E. T. S. Walton, *Proc. roy. Soc.* A, **141**, 733, 1933).

The variation of the yield of the α-particles from the lithium reaction as a function of proton energy (excitation function) was found by Cockcroft and Walton to have roughly the exponential shape predicted by the Gamow theory of barrier penetration (Sect. **14.2**). This theory, which had been published in 1928 and had already been applied to the problem of radioactive decay, encouraged the development of the fairly modest accelerator, since it appeared that the energies necessary to *surmount* barriers would not be required for penetration of an incident particle into a nucleus. The excitation function for the $^7\text{Li}(p, \alpha)^4\text{He}$ reaction was the first of a great many measurements of this type as a result of which the theory of barrier penetration has been exhaustively checked.

14.1.3 *The discovery of the neutron*

In the years immediately following the discovery of artificial transmutation, nearly all the light elements up to potassium were found by Rutherford and

† P. I. Dee and E. T. S. Walton, *Proc. roy. Soc.*, A, **141**, 733, 1933.

Chadwick to emit protons under bombardment by α-particles. It is typical of the breadth of Rutherford's vision that he was not content to assume that only protons could be emitted in, or as a result of, these transmutations; he asked students to look for induced radioactivity (and only failed to detect it because of inadequacy of experimental technique) and in 1920 he wrote 'Under some conditions, however, it may be possible for an electron to combine much more closely with the H nucleus forming a kind of neutral doublet. . . . The existence of such atoms seems almost necessary to explain the building up of the nuclei of heavy elements; . . .' He referred continually to this idea over the period 1920–30 and initiated experiments to detect such a particle in a hydrogen discharge; its possible existence was in his mind when in 1930 new results on the radiations emitted in the α-particle bombardment of light nuclei began to appear.

The first new result was the demonstration by Bothe and Becker of the production of what appeared to be electromagnetic radiation in the bombardment of elements such as Li, Be, B, Mg, Al with α-particles. The radiation, which was particularly intense in the case of a beryllium target, was detected by means of its ionizing effect in a point counter. It was suggested that for some of the target elements at least the radiation might be associated with excited states of residual nuclei produced in (α, p) reactions. Further work by Bothe, by Webster and by Mme Curie-Joliot showed that the beryllium radiation possessed a penetrating power considerably greater than that of any known γ-radiation[†] ($\mu_m = 2 \times 10^{-3}$ m^2 kg^{-1} for lead, cf. Fig. 5.17). Several other puzzling properties of this radiation emerged in the early experiments, including the facts that the attenuation coefficients measured with a point counter and an ionization chamber were different and that the radiation was of different hardness in different directions with respect to the incident α-particles. The most striking discovery of all however was made by Mme Curie-Joliot and M. Joliot who, thinking that the new radiation might have some affinity with cosmic radiation, sought for possible secondary effects and found that the radiation from beryllium was able to eject high speed *protons* from hydrogenous material. The protons had a range of about 26 cm of air and the only possible explanation for their production, if the incident radiation were electromagnetic, seemed to be the Compton effect. The quantum energy necessary was 35–50 MeV.

The ejection of protons as a result of a Compton process was not an attractive hypothesis. The cross-section for the process seemed clearly inconsistent with that expected from the Klein–Nishina formula (Sect. 5.4.4) and the energy of the radiation seemed too large to account for as a result of the α-beryllium interaction. Accordingly, Chadwick[‡] undertook further experiments and showed, using an ionization chamber and amplifier, that the radiation from beryllium was able to produce recoil atoms of many light elements

---

[†] The phenomenon of pair production, and the consequent minimum in the X-ray absorption coefficient, was not known in 1930.

[‡] J. Chadwick, *Proc. roy. Soc.*, A, **136**, 692, 1932.

as well as of hydrogen. This was very difficult to explain on the quantum hypothesis but was quite consistent with the supposition that the radiation was a stream of neutral particles (neutrons) of mass equal to that of the proton, which were able to project other atoms by the process of elastic collision. This proposal was able immediately to clarify all the properties of the new radiation.

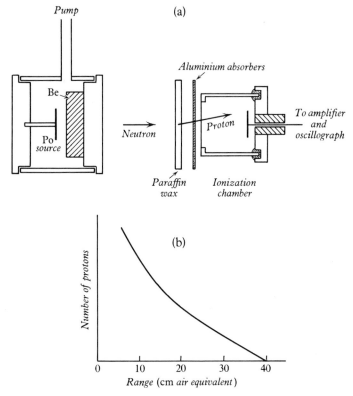

Fig. 14.3 Discovery of the neutron. (a) Apparatus, in which neutrons were produced by the (α, n) reaction in a block of beryllium and then ejected protons from a sheet of paraffin wax. (b) Number–range curve for recoil protons (Chadwick, 1932).

The apparatus of Chadwick was extremely simple; the ionization chamber shown in Fig. 14.3(a) was placed near a polonium–beryllium source assembly. When the source was placed near the ionization chamber the pulse rate shown by the oscillograph increased from 0·1 per minute to about 4 per minute; this rate was further considerably increased when a sheet of paraffin wax was placed in the path of the radiation. The former increase was interpreted as due to the production of recoil nitrogen atoms in the air of the ionization chamber; the latter increase as due to the ejection of recoil protons from the paraffin wax. The recoil protons were shown to have a range of about 40 cm of air (Fig. 14.3(b)) by the use of aluminium absorbing foils. Other elements placed near the ionization chamber and other gas fillings also furnished recoil particles. The range and the initial velocity $(4.7 \times 10^6 \text{ m s}^{-1})$ of the recoil particles in nitrogen was found from expansion chamber pictures taken by Feather.† Comparison of these figures with the initial velocity of the hydrogen recoils $(3.3 \times 10^7 \text{ m s}^{-1})$ showed that γ-rays of different quantum energy in each case would be required if the Compton hypothesis were accepted. If, however, a heavy particle is assumed (with no net charge so that the great penetration is explained) the two observed recoil velocities u are consistent providing that the mass M and velocity v of the primary particle satisfy the equation (Sect. 5.1.1)

$$u = \frac{2M}{M + M_s} v \qquad (14.3)$$

where M_s is the mass of the struck atom. For hydrogen and nitrogen

$$\frac{u_H}{u_N} = \frac{3.3 \times 10^7}{4.7 \times 10^6} = \frac{M + M_N}{M + M_H} \qquad (14.4)$$

and solving this equation $(M_N = 14, M_H = 1)$

$$M = 1.15 \text{ atomic mass units}$$

with an error of some 10 per cent.

The nuclear reaction proposed for the production of the neutrons in a beryllium target was

$$^9\text{Be} + \alpha \longrightarrow {}^{12}\text{C} + \text{n} \qquad (14.5)$$

and since the maximum velocity of the neutron was known from the recoil experiments, this reaction could in principle have been used to give a more accurate value of the neutron mass. In 1932, however, the mass of ^9Be was not known, and use had to be made of the comparable neutron-producing reaction in boron

$$^{11}\text{B} + \alpha \longrightarrow {}^{14}\text{N} + \text{n} \qquad (14.6)$$

From the observed maximum neutron energy and from the known masses, Chadwick found that $1.005 < M_n < 1.008$, a result very close to the present day value of 1.00867 a.m.u. This calculation completed an experiment which by the brilliance of its conception as much as by the force of its conclusion won immediate and widespread acceptance.

The early work of Chadwick included observations on the attenuation of neutrons by nuclear collisions which suggested reasonable values for nuclear radii (Sect. 11.3.3). In the expansion chamber experiments already mentioned Feather not only obtained evidence for the elastic collision

$$^{14}\text{N} + \text{n} \longrightarrow {}^{14}\text{N} + \text{n} \qquad (14.7)$$

† N. Feather, *Proc. roy. Soc.*, A, **136**, 709, 1932.

but also for the nuclear transmutation (Plate 13)

$$^{14}N + n \longrightarrow {}^{11}B + \alpha - 0 \cdot 16 \text{ MeV} \qquad (14.8)$$

Plate 13 Expansion chamber photograph of disintegration of nitrogen by neutrons (N. Feather, *Proc. roy. Soc.* A, **136**, 709, 1932).

and for other inelastic reactions which are now well-known. Such processes are the main cause of the attenuation of neutron beams in passing through matter; direct ionization by neutron–electron interaction was shown by Dee,† also using the expansion chamber, to be less than 1 ion pair in 3 metres of air.

It may be remarked that the overwhelming success of the neutron hypothesis did not divert Chadwick's attention from other processes possible in the α-beryllium interaction. There was good evidence from expansion chamber pictures of the presence of γ-radiation in addition to neutrons and he proposed that this could originate both as a result of direct capture

$$^{9}Be + \alpha \longrightarrow {}^{13}C^{*} \longrightarrow {}^{13}C + \gamma + 10 \cdot 6 \text{ MeV} \qquad (14.9)$$

and as a result of the formation of excited states of residual nuclei

$$\left. \begin{array}{c} ^{9}Be + \alpha \longrightarrow {}^{12}C^{*} + n \\ \\ ^{12}C^{*} \longrightarrow {}^{12}C + \gamma \end{array} \right\} \qquad (14.10)$$

The latter process, for the (α, p) rather than the (α, n) reaction, had already been envisaged by Bothe and Becker. That the foundations of so many fields of investigation in nuclear physics should have been laid in one short series of experiments remains one of the triumphs of human achievement.

14.1.4 Induced activity

One of the most thoroughly established characteristics of the phenomenon of natural radioactivity was that it appeared impossible to produce, modify or destroy an activity except by the normal processes of radioactive growth or decay. Despite this, Rutherford had always suspected that nuclear bombardment might induce activity, and had in fact sought for delayed emission of heavy particles following (α, p) or (α, n) reactions (Sect. 14.1.1). There is little doubt that induced activity would have been discovered much earlier than 1934 if Geiger–Müller counters had been more generally available,

† P. I. Dee, *Proc. roy. Soc.*, A, **136**, 727, 1932.

since the experiment of Curie and Joliot† which revealed this effect was essentially simple. These workers were studying the emission of positrons and electrons from light elements under α-particle bombardment, with the supposition that these might originate in pair internal conversion (Sect. **13.3**) when they noticed that the emission of positrons from an irradiated foil continued for some time after removal of the α-particle source. In the case of aluminium bombarded by α-particles from polonium, a positron activity decaying according to the normal radioactive law with a period of 3 min 15 s was observed; cloud chamber photographs established the sign of the emitted particles, and absorption measurements the energy. The production process proposed was

$$
\left.
\begin{array}{l}
{}^{27}\text{Al} + \alpha \longrightarrow {}^{30}\text{P} + \text{n} - 2\cdot69 \text{ MeV} \\[2mm]
{}^{30}\text{P} \longrightarrow {}^{30}\text{Si} + \beta^+ + \nu
\end{array}
\right\} \tag{14.11}
$$

In later experiments the aluminium was dissolved in acid and the induced activity was found to appear in the gaseous phase. This was to be expected if the active atoms were phosphorus (which forms phosphine, PH_3) but could not be explained if the active atoms were aluminium, since this element yields no sufficiently volatile compound under these conditions. This was the first application of radiochemical methods to the identification of induced activities.

Curie and Joliot suggested that activity might be induced by other agents. such as protons or deuterons. This was speedily verified; the active nitrogen isotope ${}^{13}\text{N}$ was produced by Curie and Joliot using α-particles,

$$
{}^{10}\text{B} + \alpha \longrightarrow {}^{13}\text{N} + \text{n} + 1\cdot06 \text{ MeV} \tag{14.12}
$$

by Crane, Lauritsen, Henderson, Livingston and Lawrence using deuterons,

$$
{}^{12}\text{C} + \text{d} \longrightarrow {}^{13}\text{N} + \text{n} - 0\cdot28 \text{ MeV} \tag{14.13}
$$

and by Cockcroft, Gilbert and Walton using protons,

$$
{}^{12}\text{C} + \text{p} \longrightarrow {}^{13}\text{N} + \gamma + 1\cdot94 \text{ MeV} \tag{14.14}
$$

These reactions lead to positron emitters because the primary process increases the proton to neutron ratio of the target nucleus. Shortly after the experiments of Curie and Joliot, Fermi‡ demonstrated the production of negatron emitters by the neutron capture process. In the case of aluminium the reaction

$$
\left.
\begin{array}{l}
{}^{27}\text{Al} + \text{n} \longrightarrow {}^{24}\text{Na} + \alpha - 3\cdot14 \text{ MeV} \\[2mm]
{}^{24}\text{Na} \longrightarrow {}^{24}\text{Mg} + \beta^- + \bar{\nu}
\end{array}
\right\} \tag{14.15}
$$

was established radiochemically using the neutrons from a radon–beryllium

† I. Curie and F. Joliot, *Comptes Rendus*, **198**, 254, 1934.
‡ E. Fermi, *Ricerca Sci.*, **1**, 283, 1934; *Nature*, **133**, 757, 1934.

source of strength 800 mCi. Fermi also pointed out that neutron reactions were not limited by energy loss of the incident particle in matter, or by Coulomb barrier effects, as in the case of charged particles. Induced activity was therefore expected, and found, for elements throughout the periodic system, including uranium. The possibility, first apparent in this work, of producing elements of atomic number greater than 92 led ultimately to the discovery of nuclear fission.

The process for inducing activity in heavy elements appeared to be the capture reaction, e.g.

$$\left. \begin{array}{l} ^{107}\text{Ag} + \text{n} \longrightarrow\ ^{108}\text{Ag} + \gamma + 7{\cdot}23\ \text{MeV} \\[2ex] ^{108}\text{Ag} \longrightarrow\ ^{108}\text{Cd} + \beta^- + \bar{\nu} \end{array} \right\} \tag{14.16}$$

since the active isotope could not usually be separated chemically from the target material, although the (n, 2n) reaction could not at first be excluded. One of the most important discoveries made in an extensive set of experiments by Amaldi, d'Agostino, Fermi, Pontecorvo, Rasetti and Segrè† was that neutron-induced activities could be increased largely by surrounding the source and target foil by hydrogenous material such as paraffin wax. This has the effect of slowing down the neutrons so that they spend longer near the target nuclei and have a greater chance of being captured ($1/v$-law of neutron absorption). The neutron slowing-down process is of the highest importance in the design of chain reacting systems.

14.1.5 Fission and the transuranic elements

The activities produced in the capture of slow neutrons by uranium presented an extremely complex picture during the years 1934–1938.‡ It was expected that they would be characteristic of uranium itself and of the transuranic elements now known as neptunium and plutonium which would be formed by successive β-decays following the capture of a neutron in ^{238}U

$$\left. \begin{array}{l} ^{238}_{92}\text{U} + \text{n} \longrightarrow\ ^{239}_{92}\text{U} + \gamma \\[2ex] ^{239}_{92}\text{U} \longrightarrow\ ^{239}_{93}\text{Np} + \beta^- + \bar{\nu} \\[2ex] ^{239}_{93}\text{Np} \longrightarrow\ ^{239}_{94}\text{Pu} + \beta^- + \bar{\nu} \end{array} \right\} \tag{14.17}$$

These processes are now known to take place, and the production of the 23-minute uranium isotope ^{239}U was recognized in the early experiments. It was thought that the other activities would be those of transuranic elements and radiochemical separations based on the expected behaviour of these new atoms were applied to separate the activities. These experiments must in fact have

† E. Fermi, E. Amaldi, B. Pontecorvo, F. Rasetti and E. Segrè, *Ricerca Sci.*, **2**, 280, 1934; E. Amaldi, O. d'Agostino, E. Fermi, B. Pontecorvo, F. Rasetti and E. Segrè, *Proc. roy. Soc.*, A, **149**, 522, 1935.
‡ E. Fermi, *Nature*, **133**, 898, 1934.

yielded minute quantities of neptunium and plutonium but the neutron sources used before 1939 were not sufficiently strong to make the identifications certain. Moreover it was quite clear that the main body of neutron-induced activity in uranium did not exhibit the expected chemical behaviour. Meitner, Hahn and Strassmann selected certain particular periods for detailed radiochemical study and concluded at first that several isotopes of radium must be formed, perhaps by two successive α-particle emissions from uranium. No specific evidence for these α-particle emissions was found, despite many attempts, and Hahn and Strassmann continued their experiments with the object of establishing the atomic number of the supposed radium isotope with greater certainty. In their first investigations, the activity had been extracted by use of a barium carrier, since this element is expected to behave radiochemically in the same way as radium. (The outer electron configurations are $5s^2 \, 5p^6 \, 6s^2$ for barium and $6s^2 \, 6p^6 \, 7s^2$ for radium.) The next step was to remove the radium itself from the barium salts by fractional recrystallization but at this point the surprising fact emerged that this separation could not be made. It was possible to recover known isotopes of radium which had been deliberately added as tracers but the neutron-induced activity of period 3·5 hr remained obstinately in the barium fraction, contrary to all expectation. It is a remarkable tribute to the power of the scientific method that Hahn and Strassmann should at this point have placed implicit trust in their results; the following words in the concluding paragraphs of their paper† have become a classic reference of scientific history '. . . als Chemiker müssten wir eigentlich sagen, bei den neuen Körpern handelt es sich nicht um Radium, sondern um Barium; denn andere Elemente als Radium oder Barium kommen nicht in Frage. . . . Als der Physik in gewisser Weise nahestehende "Kernchemiker" können wir uns zu diesem, allen bisherigen Erfahrungen der Kernphysik widersprechenden, Sprung noch nicht entschliessen'.

The conclusion of Hahn and Strassmann that alkaline earth metals were produced in the irradiation of uranium by neutrons was consistent with an earlier observation by Curie and Savitch that some of the activities were concentrated by extractions of lanthanum, which has an outer electronic structure $(5s^2 \, 5p^6 \, 5d^1 \, 6s^2)$ similar to that of actinium $(6s^2 \, 6p^6 \, 6d^1 \, 7s^2)$. It appeared incontrovertible that these bodies were connected by a genetic chain which should be written

$$\text{Ba} \xrightarrow{\beta-} \text{La} \xrightarrow{\beta-} \text{Ce} \qquad (14.18)$$

rather than

$$\text{Ra} \xrightarrow{\beta-} \text{Ac} \xrightarrow{\beta-} \text{Th} \qquad (14.19)$$

and it was not long before an explanation of this chemically based conclusion

† O. Hahn and F. Strassmann, *Die Naturwissenschaften*, **27**, 11, 1939.

was forthcoming from the physicists. In 1938 Meitner and Frisch,† accepting the conclusions of Hahn and Strassmann, proposed that the uranium nucleus, on absorption of a slow neutron, could assume a shape sufficiently deformed to promote its rapid division into two approximately equal masses; the name *fission* was suggested for this process. The energy released in such a transformation could be estimated simply from the semi-empirical mass formula (Sect. **10.3**) and the appearance of two heavy *fragments*, with kinetic energies of the order of 75 MeV, was predicted for each process. Ionization pulses corresponding to this energy were observed in a uranium-lined chamber irradiated by neutrons. News of the fission hypothesis was brought by Niels Bohr early in 1939 to a meeting of the American Physical Society, and in the outburst of experimental activity which immediately ensued, the main characteristics of nuclear fission were established in many laboratories

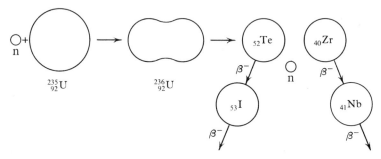

Fig. 14.4 Fission of the uranium-235 nucleus by a slow neutron. The excited compound nucleus develops a deformation, from which a transition to the energetically favoured state of two fragments occurs.

throughout the world. These experiments, mainly brief, simple and clear, bore much the same relation to the hypothesis of fission as did the work of Geiger and Marsden nearly thirty years before to that of the nucleus; they established the following facts, some of which are illustrated in Fig. 14.4 and further discussed in Appendix 4:

(*a*) The fission of uranium leads to a wide variety of nuclear fragments, which are produced in pairs such as the following

$$_{92}U + n \longrightarrow _{52}Te + _{40}Zr \left.\begin{array}{c} \\ \\ \end{array}\right\}$$

$$\text{or} \longrightarrow _{56}Ba + _{36}Kr$$

(14.20)

(*b*) The fission fragments recoil in opposite directions and have a range of about 3 cm of air (Plate 14) which is reasonable for particles of energy ≈ 75 MeV. Since this range is less than that of the normal radioactive α-particles from uranium, early attempts to observe 'long-range' particles from uranium under neutron irradiation were clearly misdirected;

† L. Meitner and O. R. Frisch, *Nature*, **143**, 239, 1939; O. R. Frisch, *Nature*, **143**, 276, 1939.

had the experiments been designed to reveal increased *ionization* fission would probably have been discovered by a physicist.

(*c*) The fission fragments are unstable against β-emission as they have excess neutrons in comparison with their charge. This is clear from the *N–Z* diagram, Fig. 10.4. Successive β-particles therefore appear in a fission chain, e.g.

$$^{40}_{54}\text{Xe} \xrightarrow{\beta-} \,^{140}_{55}\text{Cs} \xrightarrow{\beta-} \,^{140}_{56}\text{Ba} \xrightarrow{\beta-} \,^{140}_{57}\text{La} \xrightarrow{\beta-} \,^{140}_{58}\text{Ce (stable)}$$

leading to a stable nucleus.

(*d*) A few fast neutrons are produced simultaneously with the fission fragments shown in eq. 14.20. This circumstance led immediately to the conception of the possibility of a chain reaction.

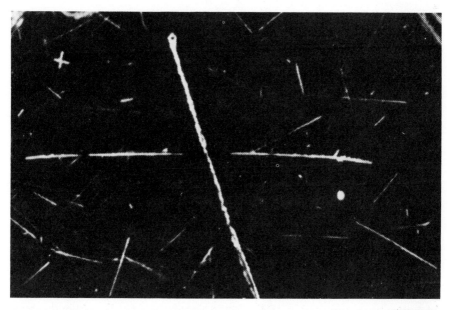

Plate 14 Expansion chamber tracks of fission fragments emerging from a uranium foil bombarded by slow neutrons (J. K. Bøggild *et al.*, *Phys. Rev.*, **71**, 281, 1947).

These and other properties of the fission process were co-ordinated early in 1939 in a paper by Bohr and Wheeler,† which gave a detailed theory of the process in terms of the liquid drop model. The outstanding point made in this paper (which was rapidly confirmed experimentally), was that the slow neutron fission process took place in the rare isotope ^{235}U of uranium.

The dramatic impact of the discovery of fission, despite what must be admitted to be a relatively small effect on the development of nuclear theory, diverted attention from the search for positive identification of transuranic

† N. Bohr and J. A. Wheeler, *Phys. Rev.*, **56**, 426, 1939.

elements. Success in this work, however, rapidly followed when the nature of the complicating fission process was understood; present knowledge of elements 93 to 105 is mainly due to a brilliant combination of physical and chemical techniques developed in the Radiation Laboratory and Department of Chemistry of the University of California at Berkeley and at the Dubna laboratory (JINR) in Russia. The following elements have been characterized:

$Z = 93$, Neptunium $Z = 99$, Einsteinium
$Z = 94$, Plutonium $Z = 100$, Fermium
$Z = 95$, Americium $Z = 101$, Mendelevium
$Z = 96$, Curium $Z = 102$ ⎫
$Z = 97$, Berkelium $Z = 103$ ⎬ names under discussion
$Z = 98$, Californium $Z = 104$ ⎪
     $Z = 105$ ⎭

The production processes for these elements involve increase of charge either by successive neutron capture followed by β^--decay, or by direct addition through α-particle or heavy ion bombardments, e.g.

$$^{249}_{98}\text{Cf} + ^{15}_{7}\text{N} \longrightarrow ^{260}_{105}\text{X} + 4\text{n} \tag{14.21}$$

Reactions between heavy ions of comparable mass seem necessary for the production of superheavy elements (Sect. **9.4**), and this is likely to be one of the exciting areas of future heavy ion physics (Ref. 14.8).

14.2 Wave-mechanical collision theory

14.2.1 Formal definitions

A simple type of induced nuclear reaction, which covers a great many cases encountered in practice, including most of the reactions discussed in Sect. **14.1**, may be represented by the equation already frequently used in this book

$$X + a \longrightarrow Y + b + Q \tag{14.22}$$

In this, X represents the target nucleus, a the bombarding particle or photon, Y the product nucleus, b an emitted particle or photon and Q denotes the kinetic energy released in the process (Sect. 5.1.1). The most useful broad classification of nuclear reactions in general distinguishes between

(a) *elastic scattering* ($Q=0$) represented by

$$X + a \longrightarrow X + a \tag{14.23}$$

in which the emergent particle is the same as the incident particle, and (*b*) *non-elastic or inelastic processes* ($Q \neq 0$) represented by equations of the type

$$
\begin{aligned}
X+a &\longrightarrow Y+b+Q \\
&\longrightarrow Y_1 + b_1 + Q_1 \\
&\longrightarrow Y_2 + b_2 + Q_2
\end{aligned}
\quad\Biggr\} \qquad (14.24)
$$

in which the kinetic energy of the products differs from that of the initial system. In *inelastic scattering*, *b* is the same type of particle as *a*, and the initial nucleus is left in an excited state. This process is written

$$ X+a \longrightarrow X^* + a' + Q \qquad (14.25) $$

The alternative processes $Y+b$, $Y_1 + b_1$, ... in (14.24) are known as *reaction channels* and the system $X+a$ defines the *incident channel*. The energy release Q is calculable from the masses of X, *a*, Y, and *b* is given by

$$
\begin{aligned}
Q &= [(M_X + M_a) - (M_Y + M_b)] \text{ a.m.u.} \\
&= [(M_X + M_a) - (M_Y + M_b)] \times 931 \text{ MeV}
\end{aligned}
\quad\Biggr\} \qquad (14.26)
$$

The observed energy of the product particles Y and *b* depends also on the energy E_a of the incident particle and the kinetic energy of the system $Y+b$ is then

$$ E_a + Q \qquad (14.27) $$

In the case of *exothermic reactions* Q is positive and the reaction may in principle proceed for zero bombarding energy. For *endothermic reactions*, Q is negative and there is a threshold bombarding energy below which the reaction is not observed; this is not $E_a = -Q$ because of the necessity to conserve linear momentum, which means that only the energy in the centre-of-mass system

$$ \varepsilon_a = E_a \times \frac{M_X}{M_X + M_a} \qquad (14.28) $$

is available for producing the reaction. The threshold energy in the laboratory system is thus

$$ E_T = -\frac{M_X + M_a}{M_X} \times Q \qquad (14.29) $$

The centre-of-mass energy ε_a is the *incident channel energy*; with this energy is associated a *channel wavelength* λ_a, corresponding to a particle of the reduced mass $M_a M_X / M_a + M_X$ possessing the channel energy.† The energy of particle *b* observed at a given angle θ_L with the direction of incidence of *a* may be calculated by a simple extension of the formulae given in Sect. 5.1.1.

† The suffix *a* of the channel wavelength will be omitted when this leads to no confusion.

In the most accurate work on nuclear reactions, and in all experiments involving β-particles and high-energy particles, a relativistic treatment must be given.

As in the case of Rutherford scattering (Sect. 5.1.4) the general laws of collision do not predict the probability of a nuclear reaction taking place. For such predictions, more detailed knowledge of the interaction mechanism is necessary and in general the classical approach found satisfactory to describe Rutherford scattering over a considerable range of energies is inadequate. This is because of the failure of the criterion (5.26) for classical orbits in the long-range Coulomb field as the velocity v increases; a wave mechanical treatment then becomes necessary. Classical calculations may again be used at high energies if only short-range potentials are important and if the reduced de Broglie wavelength of the incident particle is much less than the range of the forces, since the trajectory of the particle is then well-defined with respect to the size of the scattering centre.

The cross-section for a particular process, such as that represented by (14.22), will be written σ_{ab}. The *total cross-section* (Sect. 5.1.3) for the interaction of a with a specified target is then

$$\left. \begin{aligned} \sigma_t = \sigma_a &= \sigma_{aa} + \sigma_{ab} + \sigma_{ab_1} + \cdots \\ &= \sigma_{el} + \sigma_{inel} \end{aligned} \right\} \tag{14.30}$$

where $\sigma_{el}\ (=\sigma_{aa})$ is the *elastic cross-section* and $\sigma_{inel}\ (\sum_b \sigma_{ab})$ is the complete *non-elastic or inelastic cross-section*. This latter quantity is often written σ_r or σ_{abs} and referred to as the *reaction or absorption cross-section*. Such cross-sections may also be expressed as the integral of a differential cross-section as defined in Sect. 5.1.3.

The wave mechanical theory of collisions sets out to calculate these cross-sections in terms of assumed properties of an interaction region defined by the incident channel and bounded at a particular radius R. Several equivalent treatments have been developed and of these probably the most useful is the method of partial waves in which the effect of particles a with a definite angular momentum $l\hbar$ with respect to the target X is considered separately. A particular cross-section is then expressed as

$$\sigma_{ab} = \sum_0^{l_{max}} \sigma_{ab}^l \tag{14.31}$$

where $l_{max}\hbar$ is the maximum angular momentum contributing to the reaction. Since this is often specified uniquely by selection rules, the partial cross-sections σ_{ab}^l acquire a special significance.

14.2.2 *The method of partial waves*

(*a*) *Semi-classical approach.* The partial reaction cross-sections σ_r^l are limited in value by geometrical factors. This may be seen crudely from Fig. 14.5

which represents the interaction of a steady incident stream of projectiles with a fixed target nucleus at O (*a*) according to the particle picture, and (*b*) according to the wave picture. For simplicity the incident particles are supposed to be uncharged and effects connected with their intrinsic spin are neglected. The wave and particle pictures are not really distinct and it must

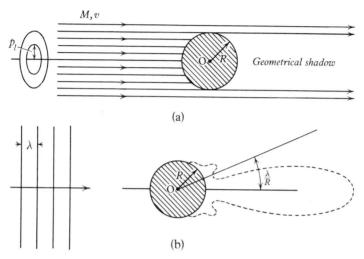

Fig. 14.5 Absorption cross-section ('black' nucleus). (a) Classical picture in which the cross-section πR^2 arises naturally as a target area. (b) Wave mechanical picture, in which an absorption cross-section πR^2 is also found for $\lambda \ll R$, but in order to create the shadow behind the target nucleus, elastic scattering occurs with a further cross-section of πR^2. The dotted line shows the angular distribution of this diffraction scattering; a minimum occurs at an angle $\approx \lambda/R$.

be assumed, in considering Fig. 14.5, that the incident particles have a wavelength $\lambda = h/Mv$ where Mv is the incident linear momentum. A particle with an impact parameter p_l (Sect. 5.1.4) then has an angular momentum

$$Mvp_l \qquad (14.32)$$

about the centre of the target nucleus. If the incident particle (or wave) and the target nucleus are to be regarded as one quantized system the angular momentum about O must be an integral multiple l of $\hbar$ and we have therefore

$$Mvp_l = l\hbar$$

or $\qquad (14.33)$

$$p_l = l\lambda$$

Since the impact parameter cannot be too closely defined, it is reasonable to assume that all particles with this angular momentum $l\hbar$ have impact parameters between p_l and p_{l+1}. The cross-section for removal of these particles from the incident beam then follows immediately from definition and cannot exceed

$$\pi(p_{l+1}^2 - p_l^2)$$

or (14.34)

$$\sigma_r^l \leqslant (2l+1)\pi\lambdabar^2$$

from (14.33). Particles with $p_1 > R$, the nuclear radius, will not interact and the critical l-value l_{max} is given by

$$p_l = l_{max}\lambdabar = R$$

or (14.35)

$$l_{max} = \frac{R}{\lambdabar} = kR$$

where $k = 2\pi/\lambda = 1/\lambdabar$ is the wave number of the incident particles.

The particle picture, Fig. 14.5(a), suggests a well-defined shadow and a total absorption cross-section πR^2. This is consistent with the wave picture if a large number of partial waves is effective, i.e. if $l_{max} \gg 1$, for then

$$\sigma_r = \sum_l \sigma_r^l \leqslant \pi\lambdabar^2 \sum_0^{R/\lambdabar} (2l+1)$$

$$\leqslant \pi(R + \lambdabar)^2$$ (14.36)

but the wave picture requires some elastic scattering in addition, as will be discussed later. For a nucleus such as ^{27}Al, (14.36) will be a good approximation for $l \approx 10$ or $E_a \approx 25$ MeV for protons.

(b) *Wave-mechanical treatment of the scattering of uncharged, spinless particles.* A particle approaching a target nucleus at O in a direction parallel to the axis of z with velocity v is represented (Fig. 14.5(b)) by the plane wave.

$$\psi_{inc} = e^{ikz} \times \text{time factor}$$ (14.37)

The wave number k refers to the centre-of-mass system, i.e. to a particle of the reduced mass possessing the channel energy (14.28). The wave amplitude is set equal to unity, so that there is one particle per unit volume and the incident beam is then v particles per unit area per second. The position of the particle with respect to the xy axes cannot be specified in this representation and the angular momentum $l\hbar$ with respect to the nucleus can have many values. It is therefore convenient to replace the plane wave by an equivalent series of spherical waves each of which represents a particle with a definite angular momentum about the nucleus and with no component of this angular momentum in the direction of incidence. This can be done using the asymptotic form of a standard mathematical transformation of the wave function (14.37). It will be assumed that the forces are of short range, so that Coulomb scattering is excluded.

First of all we postulate a continuous sequence of interactions with the assumed incident beam. The problem then becomes one of the steady state and the time factor in (14.37) may be eliminated. The incident wave is then expressed

$$\psi_{\text{inc}} = e^{lkz} = e^{ikr\cos\theta}$$

$$\approx \frac{1}{kr} \sum_{0}^{\infty} (2l+1) i^l P_l(\cos\theta) \sin\left(kr - \frac{l\pi}{2}\right) \tag{14.38}$$

$$= \frac{1}{kr} \sum_{0}^{\infty} (2l+1) i^l P_l(\cos\theta) \frac{e^{i(kr-l\pi/2)} - e^{-i(kr-l\pi/2)}}{2i} \tag{14.39}$$

The latter form has the physical meaning that the plane wave can be considered as a series of spherical waves $e^{-i(kr-l\pi/2)}$ converging on O, together with a coherent superposition of waves $e^{i(kr-l\pi/2)}$ diverging from O. The interaction potential may affect the outgoing waves in phase or in amplitude and phase; in the former case we have *elastic scattering* with a certain angular distribution and in the latter there are *inelastic processes* in addition. The steady state wave function when the interaction takes place may therefore be written

$$\psi = \frac{1}{kr} \sum_{0}^{\infty} (2l+1) i^l P_l(\cos\theta) \frac{\eta_l e^{i(kr-l\pi/2)} - e^{-i(kr-l\pi/2)}}{2i} \tag{14.40}$$

where η_l is a complex constant representing the effect of the scattering centre. The modulus of η gives the change in amplitude and the argument the change in phase. We may also regard this wave as arising from the superposition of a *scattered wave* ψ_{sc} on the incident wave, i.e.

$$\psi = \psi_{\text{inc}} + \psi_{\text{sc}} \tag{14.41}$$

A suitable form for the outgoing scattered wave, which is itself a superposition of partial waves, is

$$\psi_{\text{sc}} = \frac{f(\theta)e^{ikr}}{r} \tag{14.42}$$

and this defines an elastic *scattering amplitude* $f(\theta)$. From (5.10) the differential cross-section for elastic scattering can be obtained at once using the definition (5.9); the number of particles crossing unit surface in the incident plane wave per second is v and from (14.42) the number crossing unit area per second at a large distance r in the scattered beam is

$$v|\psi_{\text{sc}}|^2 = \frac{v|f(\theta)|^2}{r^2} = v|f(\theta)|^2 \, d\Omega$$

so that

$$d\sigma_{\text{el}} = \sigma_{\text{el}}(\theta) \, d\Omega = |f(\theta)|^2 \, d\Omega \tag{14.43}$$

where $d\Omega$ is the element of solid angle. The scattering amplitude may be connected with the quantity η_l representing the effect of the target nucleus by writing ψ, ψ_{inc} and ψ_{sc} in (14.41) explicitly. This gives, using the identity $i^l = e^{i(l\pi/2)}$

$$f(\theta) = \sum_0^\infty f_l(\theta) = \frac{1}{2ik} \sum_0^\infty (\eta_l - 1)(2l+1)P_l(\cos\theta) \tag{14.44}$$

If $d\sigma_{el}$ (eq. 14.43) is integrated over the angular range 0 to π, using this value of $f(\theta)$, it will be found that cross terms vanish and eq. (14.31), with $b = a$, follows.

We now consider the different types of interaction that may occur. For *elastic scattering* only there must be no loss of incident particles and therefore in (14.40)

$$|\eta_l|^2 = 1$$

so that

$$\eta_l = e^{2i\delta_l} \tag{14.45}$$

where δ_l is a real quantity. This gives

$$f(\theta) = \frac{1}{2ik} \sum_0^\infty (2l+1)(e^{2i\delta_l} - 1)P_l(\cos\theta) \tag{14.46}$$

and δ_l has the physical significance of a *phase shift* in the asymptotic form of the partial wave l. This can be seen by substituting for η_l in (14.40) and converting back to the sinusoidal form. The last factor of the expression for ψ becomes

$$e^{i\delta_l} \sin\left(kr - \frac{l\pi}{2} + \delta_l\right) \tag{14.47}$$

instead of $\sin(kr - l\pi/2)$.

If there are *inelastic processes* then $|\eta_l|^2 < 1$ so that if η_l is expressed as in (14.45)

$$\delta_l = \alpha_l + i\beta_l \tag{14.48}$$

with β_l a positive quantity.

Cross-sections for interaction in the partial wave l can be written down directly by computing the corresponding flux. For *elastic scattering*, from (14.43) and (14.44)

$$d\sigma_{el}^l = \frac{1}{4k^2} |1 - \eta_l|^2 (2l+1)^2 [P_l(\cos\theta)]^2 \, d\Omega \tag{14.49}$$

and by integration we find,

$$\sigma_{el}^l = \frac{\pi}{k^2} (2l+1)|1 - \eta_l|^2 = 4\pi\lambda^2(2l+1)|e^{i\delta_l} \sin\delta_l|^2$$

$$= 4\pi\lambda^2(2l+1) \sin^2\delta_l \tag{14.50}$$

using $\eta_l = e^{2i\delta_l}$ and the normalization of the Legendre function given in Appendix 1. The quantity $\lambdabar e^{i\delta_l} \sin \delta_l$ is sometimes known as the *partial wave amplitude*.

The *inelastic cross-section* is obtained by computing the *net* flow of particles associated with the wave ψ. This gives, using (14.40)

$$\sigma_{\text{inel}}^l = \frac{\pi}{k^2}(2l+1)(1-|\eta_l|^2) \tag{14.51}$$

The *total cross-section* is the sum of σ_{el}^l and σ_{inel}^l

$$\sigma_t^l = \frac{\pi}{k^2}(2l+1)2(1-\text{Re}\eta_l) \tag{14.52}$$

From the physical nature of the problem ($|\eta_l| \leqslant 1$) we may deduce the following useful geometrical limitations on cross-sections

$$\left.\begin{array}{l} \sigma_{\text{el}}^l \leqslant 4\pi\lambdabar^2(2l+1) \\[4pt] \sigma_{\text{inel}}^l \leqslant \pi\lambdabar^2(2l+1) \\[4pt] \sigma_t^l \leqslant 4\pi\lambdabar^2(2l+1) \end{array}\right\} \tag{14.53}$$

The elastic cross-section reaches its maximum value $4\pi\lambdabar^2(2l+1)$ when $\eta_l = -1$ and the inelastic cross-section then vanishes. This means, according to (14.40), that the outgoing wave has the same intensity as the incoming wave but is shifted in phase by 180°. The inelastic cross-section reaches its maximum value $\pi\lambdabar^2(2l+1)$, in agreement with (14.34), when $\eta_l = 0$, corresponding to complete absorption of the partial wave; the elastic cross-section then has the same value and the total cross-section is $2\pi\lambdabar^2(2l+1)$. It is also evident from (14.50) and (14.51) that although elastic scattering may take place without absorption ($|\eta_l|^2 = 1$) the converse is not true. An absorption process in which particles of orbital momentum $l\hbar$ are removed creates an elastic scattering, known as *shadow scattering*, with an angular distribution characteristic of l.

The angular distribution of elastic scattering, in the present case in which Coulomb forces are neglected, is given immediately by (14.49). From the formulae for $P_l(\cos\theta)$ given in Appendix 1 it can be seen that if the maximum l-value entering into the scattering is L, then the angular distribution contains powers of $\cos\theta$ up to $\cos^{2L}\theta$ only. The angular distributions of inelastic scattering and reactions are not given individually by the present considerations.

(c) *The optical theorem.* From (14.52) the total cross-section for all partial waves is

$$\sigma_t = \frac{2\pi}{k^2}\sum_0^\infty (2l+1)(1-\text{Re }\eta_l) \tag{14.54}$$

Comparing this with the expression (14.44) for the scattering amplitude it follows that

$$\sigma_t = \frac{4\pi}{k} \operatorname{Im} f(0) \tag{14.55}$$

since

$$P_l(\cos \theta) = 1 \quad \text{for } \theta = 0$$

This useful relation between total cross-section and the imaginary part of the forward scattering amplitude is a general consequence of wave theory.

(d) *The black disc; diffraction scattering* ($\lambda \ll R$). If the scattering centre is a perfect absorber for all particles with a classical impact parameter up to R (Fig. 14.5(a)) we have

$$\begin{aligned} \eta_l &= 0 \quad \text{for } l \leqslant kR \\ \eta_l &= 1 \quad \text{for } l > kR \end{aligned} \tag{14.56}$$

We then find, as in the semi-classical approach,

$$\sigma_{\text{inel}} = \sum_0^{kR} \pi \lambda^2 (2l+1) = \pi(R+\lambda)^2 \tag{14.57}$$

but in addition we also have in the same way

$$\sigma_{\text{el}} = \sum_0^{kR} \pi \lambda^2 (2l+1) = \pi(R+\lambda)^2 \tag{14.58}$$

and consequently

$$\sigma_t = 2\pi(R+\lambda)^2 \tag{14.59}$$

a result which is not predicted classically. The elastic cross-section is due to shadow scattering in all the waves up to l. The angular distribution of this scattering (Fig. 14.5(b)) can be found by summing (14.44) for the limited series of partial waves defined by (14.56), from which

$$f(\theta) = \frac{1}{2ik} \sum_0^{kR} (-)(2l+1)P_l(\cos \theta) \tag{14.60}$$

This gives an outgoing wave in antiphase with the incident wave and thus creates the 'geometrical' shadow. For small θ and large l we may use a Bessel function approximation:

$$P_l(\cos \theta) \approx \mathcal{J}_0(l \sin \theta)$$

and replace the sum over l in (14.60) by an integral so that the differential cross-section becomes

$$d\sigma_{el} \doteq \frac{1}{k^2} \left[\int_0^{kR} l \mathcal{J}_0(l \sin \theta) \, dl \right]^2 d\Omega$$

$$= R^2 \left[\frac{\mathcal{J}_1(kR \sin \theta)}{\sin \theta} \right]^2 d\Omega \tag{14.61}$$

(using the relation $\int x \mathcal{J}_0(x) \, dx = x \mathcal{J}_1(x)$ as in optical diffraction). This particular type of shadow scattering is known as *diffraction scattering*. It is forward peaked and falls to its first minimum at a c.m. angle $\theta \approx 3 \cdot 8/kR$. Since it deviates particles from the incident beam it contributes to the total cross-section.

(e) *The phase shift; hard sphere scattering* ($\lambda \gg R$). The properties of the scattering centre are contained in its radius R, which determines the number of partial waves mainly concerned in an interaction and in the asymptotic phase shifts δ_l. In the simple case in which $\lambda \gg R$, $l = 0$ only is possible and the elastic scattering amplitude is

$$f_0(\theta) = \lambda e^{i\delta_0} \sin \delta_0 \tag{14.62}$$

This is *s*-wave scattering and is *isotropic*, with a differential cross-section

$$d\sigma_{el}^0 = \lambda^2 \sin^2 \delta_0 \, d\Omega \tag{14.63}$$

If the scattering centre is an impenetrable sphere, then the wave amplitude must vanish at the surface $r = R$ and for the case of *s*-waves (14.40) gives

$$\eta_0 = e^{-2ikR} \tag{14.64}$$

so that

$$\delta_0 = -kR = -R/\lambda \tag{14.65}$$

and the *s*-wave cross-section is

$$\sigma_{el}^0 = 4\pi \frac{d\sigma_{el}^0}{d\Omega} \approx 4\pi R^2 \quad \text{if} \quad \lambda \gg R \tag{14.66}$$

This is the scattering cross-section expected for slow neutrons at energies between nuclear resonances (ch. 15) but in slow neutron physics (App. 5) it is customary to use a quantity a, defined so that

$$a = -f_0(\theta) \tag{14.67}$$

and known as the *scattering length*, instead of R.

As the incident energy increases more phase shifts are required to account for the angular distribution and cross-section, and as the limit $\lambda \ll R$ is approached the scattering cross-section tends to the value $2\pi R^2$, as for the black disc, or completely absorbing sphere (Ref. 14.9, pp. 193 and 324). The phase shifts may be predicted from particular models of the scattering centre, e.g. a square well potential. If the scattering is due to an attractive force the kinetic energy is increased and the de Broglie wavelength is decreased within the range of the force; this displaces the partial wave towards the origin, corresponding to a *positive* phase shift δ_l in the asymptotic wave (14.47) (Ref. 14.2, page 885). Conversely, for a repulsive force the de Broglie wavelength is increased and δ_l is *negative*.

(*f*) *Elastic resonance scattering.* In both high and low energy nuclear physics the dependence of partial wave amplitudes on energy may reveal resonant states. For this purpose we disregard angular dependence and write the amplitude in the form

$$f_l(k) = \frac{e^{i\delta_l} \sin \delta_l}{k} = \frac{1}{k \cot \delta_l - ik} \tag{14.68}$$

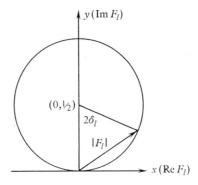

Fig. 14.6 Amplitude F_l for pure elastic scattering in the complex plane, showing dependence on the phase shift δ_l.

From this we define a quantity

$$F_l(k) = kf_l(k) = \frac{1}{\cot \delta_l - i} = \frac{1}{\varepsilon - i}$$

where $\varepsilon = \cot \delta_l$. The real and imaginary parts of F_l are then

$$x = \operatorname{Re} F_l = \frac{\varepsilon}{\varepsilon^2 + 1} \quad \text{and} \quad y = \operatorname{Im} F_l = \frac{1}{\varepsilon^2 + 1}$$

and the relation between x and y is

$$x^2 = (1 - y)y \tag{14.69}$$

This is the equation of a circle of radius $\frac{1}{2}$ centred at the point $(x, y) = (0, \frac{1}{2})$ in the x–y plane (Fig. 14.6). The quantity F_l is represented by a chord of this circle drawn from the origin to a point which moves round the circumference as δ_l varies; the chord subtends an angle $2\delta_l$ at the point $(0, \frac{1}{2})$. The maximum value of $|F_l|$ is unity, for $\cot \delta_l = 0$ or $\delta_l = \pi/2$, and this is known as *resonance scattering*. Phase shift analysis of differential cross-section measurements first of all extracts the various δ_l as a function of energy and then examines these to see if any predict a behaviour of the sort shown in Fig. 14.6. If so, resonance parameters may be deduced.

(g) *Scattering of identical particles.* In a scattering process in which the incident and struck particles are the same (e.g. α-He scattering, Sect. 5.3.1) the particles scattered through angle θ_L to the left, say, in the laboratory system are accompanied by identical particles recoiling from incident particles scattered through an angle $\pi/2 - \theta_L$ to the right. In the centre-of-mass system the corresponding angles are θ ($=2\theta_L$) and $\pi - \theta$. If each of these scattered beams is described by a wave function of the type (14.42), then the total scattered intensity at angle θ might be given by the sum of the individual intensities

$$|f(\theta)|^2 + |f(\pi - \theta)|^2 \qquad (14.70)$$

This classical expectation however disregards the requirements of quantum statistics (Sect. **3.3**) according to which the wave-function for the pair of particles must be either symmetrical or antisymmetrical under interchange of particles. This means that (14.42) must be replaced by

$$\psi_{sc} = [f(\theta) \pm f(\pi - \theta)]\frac{e^{ikr}}{r}$$

and the scattered intensity at angle θ is given by

$$|f(\theta) \pm f(\pi - \theta)|^2 \qquad (14.71)$$

which corresponds, as is reasonable, to an addition of amplitudes rather than of intensities.

In the case of α-He scattering, the spins are zero and the total wave function must be symmetrical, i.e. the positive sign must be taken. The scattered intensity at the centre-of-mass angle $\theta = \pi - \theta = 90°$ ($\theta_L = 45°$) is then just twice the classical result, as found by experiment (Sect. 5.3.1). In the case of the scattering of identical particles with spin, both symmetrical and antisymmetrical wave functions are required and the results are somewhat more complicated. They are described in Ref. 14.6.

14.2.3 *Barrier penetration*

The formulae developed in Sect. 14.2.2 take no account of possible distortions of the incident plane wave on its way to the scattering nucleus or of reflection of the incident waves at the nuclear surface (except in the case of the hard sphere). For charged incident particles, the nucleus is surrounded by a potential barrier as shown in Fig. 14.7, and the incident particles must traverse this barrier before they can initiate a nuclear reaction. For neutrons there is no Coulomb barrier, but the finite potential step at the nuclear surface $r = R$ causes reflection.

The process of barrier penetration was described qualitatively in Sect. **11.1**.

We now wish to make these considerations more quantitative and to write the limiting inelastic cross-section (14.53) in the form

$$\sigma^l_{\text{inel}} \leqslant \pi\lambdabar^2(2l+1)T_l \qquad (14.72)$$

where T_l is the *barrier transmission coefficient* for particles of orbital angular momentum $l\hbar$. This quantity is the ratio of the number of particles entering

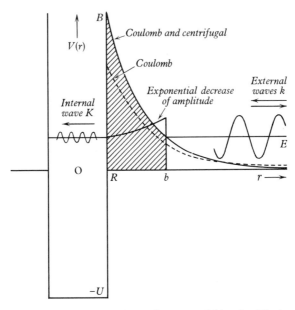

Fig. 14.7 Calculation of transmission coefficient for a potential barrier. The barrier penetration factor is the ratio of the particle densities at the nuclear surface and at $r = \infty$. The particle energy at infinity is E, and a factor $\approx 4k/K$ arises because of the change of wave number at the nuclear surface.

the nucleus per second ($= v|\psi|^2$ for a plane wave) to the number of particles incident on the barrier per second. It will be convenient to think of T_l as composed of two factors (Fig. 14.7):

(a) a *barrier penetration factor* P_l giving the probability of reaching the nuclear surface, and
(b) a *potential discontinuity factor*, familiar in all types of wave problem, due to the change of wave-number from k in free space to $K (\gg k)$ within the nucleus.

The potential discontinuity factor for *s*-wave neutrons is easily shown to be

$$\frac{4kK}{(k+K)^2} \approx \frac{4k}{K} \qquad (14.73)$$

in intensity and we therefore write in general

$$T_l \approx \frac{4k}{K} P_l \tag{14.74}$$

The barrier penetration factor P_l must be obtained by solving the Schrödinger equation for the incident particles in the potential field and finding the ratio between the beam intensity at a large distance and at the nuclear radius $r = R$. The general radial wave equation appropriate to this problem has already been given [(9.2) with $V(r)$ replaced by $zZe^2/4\pi\varepsilon_0 r$ for a particle of charge ze approaching a nucleus of charge Ze]. We are concerned in scattering problems with positive energy solutions of this equation and we note that the effective potential outside the range of nuclear forces is

$$\frac{zZe^2}{4\pi\varepsilon_0 r} + \frac{\hbar^2 l(l+1)}{2Mr^2} \tag{14.75}$$

where M is the mass of the incident particle. The second term is due to the centrifugal effect, which tends to keep particles of high angular momentum away from the nucleus.

For *neutrons with* $l = 0$, $P_0 = 1$ so that $T_0 = 4k/K$. For *charged particles*, the solutions of the Schrödinger equation are complicated and although there are several methods (one of which will be outlined in Section **16.2**) for obtaining penetration factors it is best to use numerical tabulations. For $l = 0$ and high barriers ($R \longrightarrow 0$) the dependence of P_0 on energy is shown by the expression (Ref. 14.3, p. 332)

$$P_0 = \frac{2\pi\eta}{e^{2\pi\eta} - 1} \tag{14.76}$$

where $\eta = zZe^2/4\pi\varepsilon_0 \hbar v$ and v is the velocity of the incident particle.

Formula (14.76) should only be used when $\eta \gg 1$ and then it becomes

$$P_0 \approx 2\pi\eta e^{-2\pi\eta} \propto \frac{1}{v} e^{-zZe^2/2\varepsilon_0 \hbar v} \tag{14.77}$$

This factor, first obtained by Gamow and by Gurney and Condon (although in somewhat different form) determines the yield of nuclear reactions at low energies. Thus from (14.72), (14.74) and (14.77) we expect the cross-section for a reaction such as the ^{7}Li (p, α) process (Sect. 14.1.2) to be of the form

$$\sigma^0_{inel} \approx \pi \lambda^2 \frac{4k}{K} P_0$$

$$\propto \frac{1}{v^2} e^{-zZe^2/2\varepsilon_0 \hbar v} \tag{14.78}$$

in which the exponential (Gamow) factor is by far the more important. A similar formula determines the lifetime for α-decay of radioactive nuclei (Sect. **16.2**).

Expressions for T_l for both charged particles and neutrons with $l \geqslant 0$ are given in Ref. 14.3 and are shown in Fig. 14.8 for a particular case; the suppression of proton-induced reactions owing to the barrier is obvious. It will be noted that the transmission coefficient is still less than unity at a proton energy equal to the barrier height.

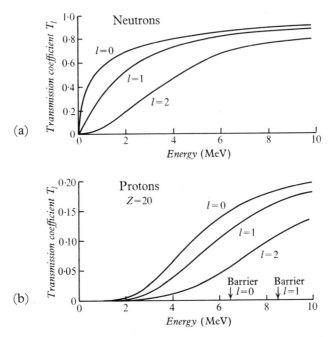

(a)

(b)

Fig. 14.8 Transmission coefficients for potential barriers. (a) Incident neutrons, $R = 5 \times 10^{-15}$ m. (b) Incident protons, $R = 4 \cdot 5 \times 10^{-15}$ m and $Z = 20$.

References

1. BEYER, R. T. *Foundations of Nuclear Physics*, Dover Publ., 1949.
2. EVANS, R. D. *The Atomic Nucleus*, Appendix C, McGraw-Hill, 1955.
3. BLATT, J. B. and WEISSKOPF, V. F. *Theoretical Nuclear Physics*, p. 358, Wiley, 1952.
4. ELTON, L. R. B. *Introductory Nuclear Theory*, 2nd ed., chapter 6, Pitman, 1965.
5. BOHM, D. *Quantum Theory*, chapter 21, Constable, 1951.
6. SCHIFF, L. I. *Quantum Mechanics*, 3rd ed., McGraw-Hill, 1968.
7. FRISCH, O. R. and WHEELER, J. A. 'The Discovery of Fission', *Physics Today*, **20**, No. 11, page 43, 1967.
8. FLEROV, G. N. and ZVARA, I. 'Synthesis of Transuranium Elements', *Science Jnl.*, July 1968, page 63; SEABORG, G. T. 'Elements beyond 100', *Ann. rev. nucl. Sci.*, **18**, 53, 1968; SEABORG, G. T. and BLOOM, J. L. 'The Synthetic Elements', *Scientific American*, **220**, April 1969, page 57.
9. DE BENEDETTI, S. *Nuclear Interactions*, Wiley, 1964.

I 5 Nuclear reactions (detailed mechanisms)

The wave mechanical theory outlined in Chapter 14 is not a theory of nuclear reactions; it provides only a framework for a formal description and sets certain limits on cross-sections. In this chapter we consider a number of specific models for nuclear reactions which permit calculation of the asymptotic phase shifts of the formal theory. We shall approach these models from the experimental point of view, beginning with the compound nucleus model of Niels Bohr, which has little connection with structure models of the individual particle type, continuing with the general optical model description of scattering which involves the basic single-particle structure, and then examining some of the reactions in which individual motion of nucleons is important. The special features of heavy ion reactions will be briefly mentioned.

All formulae presented in this chapter must be assumed to refer, when relevant, to the centre-of-mass system.

15.1 The origin of the compound nucleus hypothesis

The main nuclear phenomena known in 1935 suggested that reaction cross-sections for high-energy particles were of the order of nuclear dimensions, in agreement with the high energy limit πR^2 predicted by wave theory (ch. 14). The cross-sections for the capture of neutrons by nuclei were shown by Amaldi *et. al.* (Sect. 14.1.4) to increase beyond the nuclear area when the neutrons were slowed down by passage through hydrogenous material. This is also consistent with the results of wave theory since the low energy limit of the reaction cross-section is $\pi \lambdabar^2$ and $\lambdabar \approx 10^{-10}$ m for slow neutrons.

The first model which attempted to explain the variation of neutron cross-sections with energy was essentially the single-particle shell model, in which the incident neutron was supposed to move briefly in the potential well provided by the target nucleus. In this model the probability of scattering was always large, while the probability of capture was generally small but increased proportionately with the time spent by the incident particle near the nucleus, i.e. as $1/v$. For thermal neutrons the scattering and absorption cross-sections were expected to be about equal. It was also predicted that resonance anomalies would be seen both in absorption and scattering but these would be associated with the virtual single-particle levels and would be spaced in energy by perhaps 10 MeV. They would also be ≈ 1 MeV wide because of the short time ($\approx 10^{-21}$ s) spent by the neutron in the potential well. Variations of capture cross-section with energy in the thermal range were not likely to be affected by these resonances (Fig. 15.1(a)).

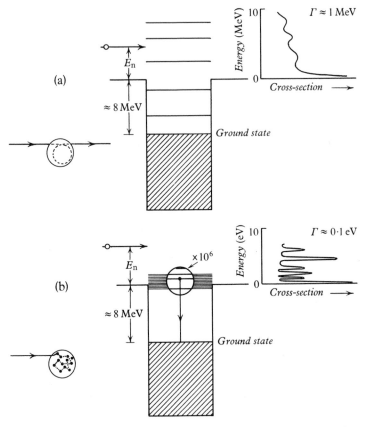

Fig. 15.1 Interaction of a neutron with a nucleus. The levels shown represent the states of excitation of the system (neutron + nucleus); a neutron of zero energy E_n produces an excitation of about 8 MeV. (a) Potential well model. (b) Compound nucleus model. Note the different energy scale of the cross-section curve.

This picture is in sharp disagreement with observation at several points. In the first place, many nuclei were found to have large absorption cross-sections for slow neutrons but to show very small scattering. Then the work of Moon and Tillman and of Amaldi and Fermi, using reactions of the type (cf. 14.16)

$$^{107}Ag + n \longrightarrow {}^{108}Ag + \gamma + 7.23 \text{ MeV}$$

$$^{108}Ag \longrightarrow {}^{108}Cd + \beta^- + \bar{\nu} \tag{15.1}$$

established that although slow neutron capture cross-sections varied as expected with the temperature of the surroundings, they also varied in an unexpected way when the incident neutrons passed through different types of absorber. The neutrons causing activation of silver were strongly absorbed by silver itself but much less strongly by other materials. This selective absorption is inconsistent with the prediction of a monotonic decrease of cross-section with increasing neutron velocity for all nuclei and gave the first indication that there are strong resonances in slow neutron capture cross-sections and that they are sharp and closely spaced. Later and more accurate work using neutron velocity selectors (Sect. **7.2**) has confirmed this conclusion and has shown that the level widths concerned are of the order of a few tenths of an electron volt and the level spacing perhaps a few electron volts (Fig. 15.1(b), and Fig. 15.5). The total cross-sections at resonance are mainly due to the capture process.

In order to explain these observations Niels Bohr† in 1936 introduced the concept of the compound nucleus which is a many-body system of strongly interacting particles formed by the amalgamation of an incident particle a with a target nucleus

$$X + a \longrightarrow C^* \tag{15.2}$$

In such a system, which has analogies with a liquid drop, the incident particle a has a short mean free path and shares its energy with the other particles of the system C^* so that it cannot be re-emitted until, as a result of further exchanges, sufficient energy is again concentrated on this or a similar particle. If the incident particle is a slow neutron this may be a very long time, sufficient to permit the emission of radiation. The capture process is then complete and the compound nucleus C is formed in its ground state either by emission of a single photon,

$$C^* \longrightarrow C + \gamma \tag{15.3}$$

or by the emission of a sequence of quanta in cascade. This explains the predominance of the capture reaction and the suppression of elastic scattering; the sharp, closely spaced resonances are also naturally accounted for as the characteristic modes of the many-body system.

The progress of many types of nuclear reaction can be described, in terms of a compound nucleus, as a *two-stage process*

$$X + a \longrightarrow C^* \longrightarrow Y + b \tag{15.4}$$

in contrast with the *single-stage process*

$$X + a \longrightarrow Y + b \tag{15.5}$$

envisaged by the potential well model and leading to the same products.

We shall write reactions in the form (15.4) when it is desired to draw attention to the compound nucleus, and in the form (15.5) when it is only necessary to indicate the initial and final system (cf. Sect. 14.2.1).

The second stage of the nuclear reaction according to Bohr's suggestion is

† N. Bohr, *Nature*, **137**, 344, 1936.

to be considered as independent of the first (*independence hypothesis*). In other words, the break-up of the compound nucleus C* into different reaction channels (Sect. 14.2.1) $Y+b$, Y_1+b_1, etc. should be determined only by the properties of the compound nucleus and not by its mode of formation. If this is true (and it is not always so) the cross-section for the process $X(a, b)Y$ may be written

$$\sigma_{ab} = \sigma_a \frac{\Gamma_b}{\Gamma} \tag{15.6}$$

where σ_a is the cross-section for the formation of a compound nucleus by particle a and Γ_b is proportional to the probability of breakup into channel b. The total 'width' Γ is equal to the sum of all partial widths Γ_b. The validity of the independence hypothesis depends on the relation between the level width Γ and the spacing D of energy levels of the compound nucleus. It has already been pointed out (Sect. **9.5**) that the nuclear level spectrum may be divided into ranges for which $\Gamma \ll D$ (resonance region) and $\Gamma \gg D$ (continuum).

15.2 Discrete levels of the compound nucleus ($\Gamma \ll D$)

15.2.1 *Cross-section formula for spinless particles*

In the resonance region of the nuclear spectrum, levels are discrete and the independence assumption is reasonable. Each level is characterized, as far as nuclear reactions are concerned, by the parameters listed in Sect. **9.5**, i.e. excitation energy E_0 above the ground state, angular momentum with quantum number I_e, parity ($\pm$) and partial widths for decay. We wish to find the cross-section for excitation of a well-defined nuclear level of this type by a particle or photon of energy E_a. As shown in Fig. 15.2 the excitation in the compound nucleus is

$$E = \varepsilon_a + S_a$$

where ε_a is the channel energy (eq. 14.28) and S_a is the separation energy for the particle a in the compound nucleus.

The excitation of a nuclear level by an incident particle is analogous to the excitation of the oscillations of an electrical circuit by an electromagnetic wave. We therefore expect the nuclear cross-section to vary with incident energy in the same way that the energy in a forced oscillation varies with incident frequency. The classical resonant circuit absorbs energy because of resistive losses; in the nuclear case, damping arises because of the possibility of decay, either via the incident channel, or through other open channels. Because of this possibility the nuclear state has a finite width Γ as already discussed in connection with the semi-classical theory of radiation (Sect. 3.5.1). The wave function of a decaying state of mean energy E_0 may be written

$$\psi_t = \psi_0 e^{-iE_0t/\hbar} e^{-\Gamma t/2\hbar} \tag{15.7}$$

which corresponds to an exponential decrease of intensity of excitation $|\psi_t|^2$ with a time constant $\tau = \hbar/\Gamma$. This wave function is not that of a stationary state but may be built up from a superposition of stationary states of slightly different energy as a Fourier integral:

$$\psi_t = \int_{-\infty}^{\infty} A(E)e^{-iEt/\hbar}\,\mathrm{d}E \tag{15.8}$$

where $A(E)$ is the amplitude of the state of energy E. From 15.7 and 15.8 we obtain, making a Fourier transform,

$$|A(E)|^2 = \frac{|\psi_0|^2}{4\pi^2} \frac{1}{(E-E_0)^2 + \Gamma^2/4} \tag{15.9}$$

and this gives the level shape (Fig. 15.2). It is exactly as for pure radiative decay except that particle emission is now included by using the total width

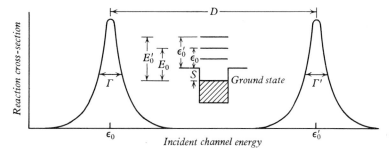

Fig. 15.2 Reaction cross-section as a function of incident channel energy ε in the resonance region of nuclear spectra.

Γ instead of the radiative width Γ_γ. The cross-section for excitation of the level by collision of particle a with nucleus X is therefore expected to have the form

$$\sigma_a = \frac{C}{(E-E_0)^2 + \Gamma^2/4} \tag{15.10}$$

where C is a constant.

To find C we use a simple statistical argument. Suppose (Fig. 15.3) that compound nucleus formation and decay take place in a box of volume Ω containing one nucleus X and one particle a. The number of states of motion of the particle, with momentum between p and $p+\mathrm{d}p$ is (Sect. 3.3)

$$\frac{4\pi p^2\,\mathrm{d}p}{h^3} \cdot \Omega \tag{15.11}$$

if the states are quantized. The probability of formation of the compound level per unit time is the probability that the nucleus X is contained within the small volume $\sigma_a v$ swept out by the effective collision area per second multiplied by the number of possible states of motion, i.e.

$$\frac{\sigma_a v}{\Omega} \cdot \frac{4\pi p^2 \, dp}{h^3} \cdot \Omega \tag{15.12}$$

integrated over the energy spectrum. This gives the probability,

$$\frac{4\pi}{h^3} \int_{-\infty}^{\infty} v\sigma_a p^2 \, dp = \frac{4\pi}{h^3} \int_{-\infty}^{\infty} \sigma_a p^2 \, d\varepsilon_a = \frac{4\pi}{h} \int_{-\infty}^{\infty} \frac{\sigma_a}{\lambda^2} \, d\varepsilon_a \tag{15.13}$$

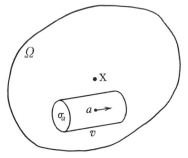

Fig. 15.3 Relation between formation and decay of a compound nucleus in a single channel.

We now assume that the variation of the channel wavelength λ of the particle over the level width Γ may be neglected and the probability of formation then becomes, using (15.10),

$$\frac{4\pi}{h\lambda^2} \frac{2\pi C}{\Gamma} = \frac{C}{\hbar \pi \lambda^2 \Gamma} \tag{15.14}$$

In a system containing a large number of particles and nuclei a, X in equilibrium this rate of formation would be balanced by the decay of the excited state back into the system $X + a$. This process has by definition the probability

$$\frac{\Gamma_a}{\hbar} \tag{15.15}$$

per unit time where Γ_a is the partial width of the compound level for emission of a. From (15.14) and (15.15)

$$C = \pi \lambda^2 \Gamma \Gamma_a \tag{15.16}$$

and the cross-section for the formation of the level becomes

$$\sigma_a(E) = \pi \lambda^2 \frac{\Gamma_a \Gamma}{(E - E_0)^2 + \Gamma^2/4} \tag{15.17}$$

This needs a slight modification if the level is formed by particles with orbital angular momentum $l > 0$. The spin of the level is then $I_e = l$ since intrinsic spins are assumed zero, and the level therefore has statistical weight $(2l + 1)$. Each of the substates can decay with equal probability and Γ_a should therefore be replaced by $g\Gamma_a$, where $g = 2l + 1$ for spinless particles. The cross-section formula is then

$$\sigma_a(E) = \pi\lambda^2 g \frac{\Gamma_a\Gamma}{(E - E_0)^2 + \Gamma^2/4}$$

(15.18)

For the process X(a, b) Y we then obtain, from (15.6)

$$\sigma_{ab}(E) = \pi\lambda^2 g \frac{\Gamma_a\Gamma_b}{(E - E_0)^2 + \Gamma^2/4}$$

(15.19)

This is the celebrated single-level, or *Breit–Wigner* formula for reaction cross-section. It is easily seen to be consistent with the limits derived in Sect. 14.2.2.

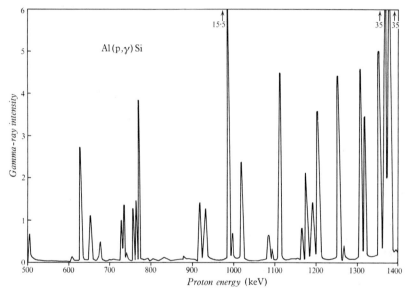

Fig. 15.4 Resonant yield of γ-radiation in the reaction ^{27}Al(p, γ)Si^{28}. The peaks indicate virtual levels at an excitation of about 12 MeV in the nucleus ^{28}Si (Brostrom *et al., Phys. Rev.,* **71**, 661, 1947).

Thus for *elastic scattering* through the compound state, with no other process possible, $\Gamma_a = \Gamma_b = \Gamma$ and at resonance ($E = E_0$)

$$\sigma_{el} = \sigma_{aa} = 4\pi\lambda^2(2l + 1)$$

(15.20)

Also for the total *inelastic cross-section* we write $\Gamma_b = \Gamma - \Gamma_a$ and then at resonance

$$\sigma_{inel} = \pi\lambda^2(2l + 1)\frac{\Gamma_a(\Gamma - \Gamma_a)}{\Gamma^2/4}$$

$$\leqslant \pi\lambda^2(2l + 1)$$

(15.21)

The *excitation function,* or variation of σ_{ab} with energy, is successfully described by (15.19) for many nuclear reactions. Important examples are the radiative capture of protons, and the capture and scattering of slow neutrons.

These processes have been examined in Sect. 13.6.4 from the point of view of the determination of radiative widths Γ_γ, but here we are concerned mainly with their relevance to the compound nucleus theory. For *proton capture* a typical excitation function is shown in Fig. 15.4, taken from the work of Brostrom, Huus and Tangen on the ^{27}Al (p, γ) ^{28}Si reaction. The sharp peaks in γ-ray yield as the energy of the proton beam is varied are due to levels in the compound nucleus ^{28}Si, which emits cascades of radiation in returning to the ground state

$$^{28}\text{Si}^* \longrightarrow {}^{28}\text{Si} + \gamma \qquad (15.22)$$

The widths shown in Fig. 15.4 are instrumental but by use of precision analysers true widths Γ of the order of 1 keV or less can be obtained from proton excitation curves.

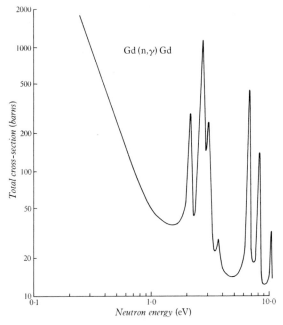

Fig. 15.5 Total cross-section for interaction of slow neutrons with gadolinium (Hughes and Schwartz, *Neutron Cross Sections*, BNL 325).

The total cross-section for *slow neutron interaction* in a heavy element is shown in Fig. 15.5, and the narrow, closely spaced levels of the Bohr theory are clearly evident. The widths in this case are, apart from Doppler broadening, mainly due to (n, γ) capture since slow neutron widths are small ($\approx 10^{-3}$ eV). These widths are however, directly proportional to velocity† ($\Gamma_n \propto v$)

† The width Γ_n is proportional to the density of states of motion of a free particle, i.e.

$$\Gamma_n \propto \frac{4\pi p^2}{h^3} \frac{\mathrm{d}p}{\mathrm{d}E} \propto v$$

Strictly, the energy dependence of the widths Γ should be included in (15.19).

and this fact can be used to predict the relative variation of scattering and absorption cross-section in an energy range such as that below 1 eV in Fig. 15.5. If no resonance lies near zero neutron energy then E may be neglected in comparison with E_0 in (15.19) and then

$$\sigma_{n\gamma} \propto \lambda^2 \Gamma_n \Gamma_\gamma \propto v \lambda^2 \propto \frac{1}{v} \qquad (15.23)$$

since the radiative widths Γ_γ for the 8 MeV capture radiation will vary little over the range involved. This result is the familiar $1/v$ law for neutron capture. For scattering on the other hand

$$\sigma_{nn} \propto \lambda^2 \Gamma_n^2 \propto \text{constant} \qquad (15.24)$$

In the vicinity of a resonance both cross-sections are much increased; analysis of their energy variation gives $\Gamma_n \approx 10^{-3}$ eV and $\Gamma_\gamma \approx 0.1$ eV in the neutron energy range of a few eV. As the neutron energy increases, the relative contribution of elastic scattering to the total cross-section also increases, and in the 100–1000 keV range, resonances in total cross-section are mainly due to scattering. The shape of the scattering resonances depends in detail on the presence of the interference effect discussed in the next section.

15.2.2 Elastic scattering

The cross-section for elastic scattering of spinless particles at an energy near a nuclear resonance is obtained by setting $\Gamma_a = \Gamma_b = \Gamma$ and the cross-section becomes

$$\sigma_{aa}(E) = \pi \lambda^2 g \frac{\Gamma^2}{(E-E_0)^2 + \Gamma^2/4} \qquad (15.25)$$

This formula can be obtained from the expression (14.50) by the substitution $\eta_l = e^{2i\delta_l}$ with

$$\tan \delta_l = \frac{\Gamma}{2(E_0 - E)} \qquad (15.26)$$

The final wave is shifted in phase by an angle which varies from approximately $0°$ (if $E_0 \gg \Gamma$) through $90°$ (at resonance) to $180°$ (for $E \gg E_0$) as the incident energy varies; a graphical representation of the scattering amplitude was given in Fig. 14.6.

Expression (15.25) represents only the compound nucleus part of the elastic scattering, i.e. that part which is associated with the adjacent level E_0. Experimentally it is observed that there is usually a background of elastic scattering for energies between resonances and that in the neighbourhood of a resonance there is an interference between this background scattering and the resonant scattering. The theoretical interpretation of the background, or average, scattering is best based on the optical model (Sect. **15.5**) which is constructed in such a way as to provide an average scattering phase angle;

the corresponding physical effect is known as *potential* or *shape-elastic scattering*. It is customary in single level treatments of nuclear scattering to represent the potential scattering for the partial wave l by a phase angle ϕ_l and to write

$$\delta_l = \beta_l - \phi_l \tag{15.27}$$

where

$$\tan \beta_l = \frac{\Gamma}{2(E_0 - E)}$$

gives the resonant phase angle. The elastic cross-section (14.50) becomes

$$\sigma_{el}^l = \sigma_{aa}^l = 4\pi\lambda^2(2l+1) \left| \frac{e^{2i(\beta_l - \phi_l)} - 1}{2i} \right|^2$$

$$= 4\pi\lambda^2(2l+1) \left| \frac{\frac{1}{2}\Gamma}{E_0 - E - (i\Gamma/2)} - e^{i\phi_l} \sin \phi_l \right|^2 \tag{15.28}$$

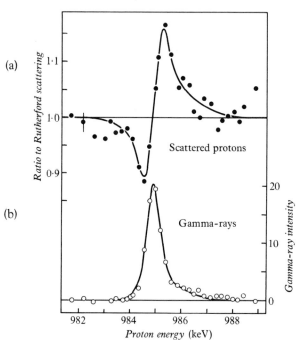

Fig. 15.6 Scattering and absorption of protons at 985 keV resonance in the $^{27}\text{Al} + \text{p}$ interaction. (a) Yield of scattered protons at 135°. (b) Yield of $^{27}\text{Al}(p, \gamma)\text{Si}^{28}$ reaction (Bender *et al.*, *Phys. Rev.*, **76**, 273, 1949).

and between resonances this tends for s-wave neutrons to the value $4\pi\lambda^2 \sin^2 \phi_0$. This is just the cross-section expected (Sect. 14.2.2) for the scattering from an impenetrable sphere of radius $R \approx \lambda\phi_0$ and the background effect is therefore sometimes known as hard sphere scattering. Although a hard sphere radius is convenient for formal analysis and may be given a specific form, such as $R = r_0 A^{1/3}$, it should not be identified with any physical dimension.

A formula such as (15.28) has been used to analyse the *elastic scattering of neutrons* by nuclei. For *charged particles* Coulomb scattering must be included and the formulae become more elaborate (Ref. 15.3). The precision of proton and α-particle scattering experiments using beams accelerated in electrostatic generators is very high, and the scattering process may be studied in detail.† Thus in the case of the scattering of protons by aluminium, interference between the nuclear and Coulomb scattering may be clearly seen. Figure 15.6(a) gives the yield of scattered protons near the 985 keV resonance of the compound nucleus ^{28}Si, and Fig. 15.6(b) shows the associated yield of the (p, γ) reaction. The reaction cross-section (15.19) contains no interference term. Analysis of these curves yields total and partial level widths, and also information on the angular momentum of the level (Ref. 15.6).

If Coulomb forces may be neglected, then (15.28) shows that the scattering cross-section far from resonances reduces to the nuclear potential scattering. In the hard sphere approximation (Sect. 14.2.2) the cross-section is then $4\pi R^2$ for the $l=0$ interaction, where R is the nuclear radius. Near the resonance level the cross-section increases to a value of approximately $4\pi \lambda^2$.

15.2.3 *Introduction of spin*

Suppose now that the nuclei a, X have spin quantum numbers s, I, and that I_e is the spin of the single excited state of C effective in the reaction. Then the incident orbital momentum vector **1** must satisfy the equation

$$\mathbf{s} + \mathbf{I} + \mathbf{1} = \mathbf{I}_e \tag{15.30}$$

The problem is simplified if the spins s, I of the initial system are combined into a number of equivalent *channel spins j* such that

$$\mathbf{s} + \mathbf{I} = \mathbf{j}$$

and

$$s + I \geqslant j \geqslant |s - I| \tag{15.30}$$

This equation gives either $(2s+1)$ or $(2I+1)$ distinct values of j according as $s <$ or $> I$. It may easily be seen that the total number of magnetic substates is the same, namely $(2s+1) \times (2I+1)$ for each description of the initial colliding system, i.e. for two spins s and I or for the appropriate number of channel spins j. In an unoriented system all these magnetic substates are equally populated, and for zero incident orbital angular momentum the probability of forming a given channel spin j from the combining spins s, I is

$$\frac{2j+1}{(2s+1)(2I+1)} \tag{15.31}$$

† See for instance, among many examples, the work of Pearson and Spear on the reaction ^{16}O(α, γ), *Nucl. Phys.*, **54**, 434, 1964.

To form the compound state of spin I_e from a given channel spin and orbital momentum l requires an l value given by

$$\mathbf{j} + \mathbf{l} = \mathbf{I}_e$$

and

$$j + I_e \geqslant l \geqslant |j - I_e|$$

$$(15.32)$$

with the limitation that only even or only odd values of l are permitted because of conservation of parity. The probability of forming I_e from l, j is, for unpolarized systems,

$$\frac{2I_e + 1}{(2l + 1)(2j + 1)} \qquad (15.33)$$

The two factors (15.31) and (15.33) therefore multiply the cross-section for the case of particles with spin and we obtain from (15.19)

$$\sigma_{ab}^l = \pi \lambda^2 \frac{2I_e + 1}{(2s + 1)(2I + 1)} \frac{\Gamma_a \Gamma_b}{(E - E_0)^2 + \Gamma^2/4} \qquad (15.34)$$

$$= \pi \lambda^2 g \frac{\Gamma_a \Gamma_b}{(E - E_0)^2 + \Gamma^2/4}$$

where

$$g = \frac{2I_e + 1}{(2s + 1)(2I + 1)} \qquad (15.35)$$

The maximum cross-section for elastic scattering is now

$$\sigma_{aa}^l = 4\pi \lambda^2 g \qquad (15.36)$$

and only reaches this value when no reactions are possible. The maximum inelastic cross-section is

$$\sigma_{ab}^l = \pi \lambda^2 g \qquad (15.37)$$

These relations are useful in finding values for the spin of compound states from observed cross-sections.

Equation (15.34) relates to the simple case of one channel spin and one orbital momentum in both initial and final states. In general, summations over permitted values of these quantities must be made. Since it is possible (by using polarized beams and targets) to study individual channel spins separately, contributions to a reaction yield from different channels are *incoherent* and may be added as intensities. This is in contrast with contributions from different l-values in the differential cross-section (Sect. 14.2.2) which are coherent, since they cannot be separated without destroying the character of the incident plane wave.

15.2.4 *Conservation theorems*

In any nuclear reaction I is a good quantum number since total angular momentum must be conserved. In special cases, total orbital and spin momenta, L and S are separately conserved. If nuclear forces are charge independent, the total isobaric spin T is also a good quantum number, and this is always true for the third component T_z of this quantity in strong interaction processes (Sects. **12.1** and 21.4.1). Parity seems to be conserved in nuclear reactions although in β-decay processes it is not. The conservation of the quantities I, T, and parity thus imposes selection rules on nuclear reactions; from observations of angular distributions and transition probabilities it may be possible to deduce values for these quantum numbers.

15.2.5 *Strength function; reduced widths*

The partial widths Γ_a, Γ_b used in preceding sections were introduced as parameters of the single-level formula. They may however be calculated if a particular model of nuclear structure is assumed. Suppose for instance that we postulate a very crude model in which one of the nuclear particles approaches the nuclear surface v times per second with an energy sufficient for it to be emitted. The mean life would then be $\tau_0 = 1/v$. This is increased because of the necessity for barrier penetration and the actual mean life is

$$\tau = \frac{\tau_0}{T_0}$$

where T_0 is the transmission coefficient, taken in this simple case for $l = 0$. The partial width for particle decay is then

$$\Gamma = \frac{\hbar}{\tau} = \frac{\hbar T_0}{\tau_0} \tag{15.38}$$

The mean life without barrier τ_0 may be related to the level spacing D of the nucleus at high excitation by the argument that if the nucleus is an oscillating system of frequency v then a set of energy levels of uniform spacing $D = hv = h/\tau_0$ might be expected. The Bohr liquid drop model, with its many closely spaced levels, is an approximation to such a system. Substituting in (15.38) we then have

$$\Gamma = \frac{D T_0}{2\pi} \tag{15.39}$$

and using the expression (14.74) for T_0 we obtain

$$\frac{\Gamma}{D} = \frac{2k}{\pi K} P_0 = \frac{2k}{\pi K} \quad \text{for } s\text{-wave neutrons} \tag{15.40}$$

in which K is the wave number of the particle in the nuclear potential well and k ($\ll K$) its wave number when emitted. This expression may also be obtained (Ref. 15.4) by averaging the single level formula (15.17) over an energy range containing many resonances and it then connects the average level width with the average level spacing. It is usual to convert this ratio to an energy independent form, defined as the *s*-wave *strength function*, by the reduction

$$\frac{\Gamma_0}{D} = \sqrt{\frac{E_0}{E}} \frac{\Gamma}{D} \qquad (15.41)$$

where E_0 is a standard energy. The strength function is a convenient parameter to characterize the structure of a nuclear level spectrum; it is proportional at a given energy to the average cross section for compound nucleus formation.

The strength function may also be defined in terms of a quantity known as the *reduced width* γ^2, and given by

$$\gamma^2 = \frac{\Gamma}{2kR} \qquad (15.42)$$

for *s*-wave neutron reactions where R is the hard sphere radius. From (15.40) we see that

$$\frac{\gamma^2}{D} = \frac{1}{\pi KR} \approx 1 \qquad (15.43)$$

Reduced widths are normally quoted as fractions of the single-particle reduced width for a square well; this may be shown to be $\hbar^2/MR^2$ MeV where M is the nucleon mass and R the radius of the well. Reduced widths predicted by the individual particle model are usually much less than this because the nuclear motion is usually more complicated than that of a single particle. Special cases are known however (e.g. in nuclear photodisintegration) in which reduced widths may be enhanced. In the limit of a highly coordinated nuclear motion, very large values of reduced widths may be found and nuclear levels are appropriately described by the collective model. In all cases this width is to be interpreted as the probability of a nucleus dissociating into a certain pair of particles at the nuclear surface. The single particle reduced width plays the same part in the interpretation of particle emission and absorption as does the 'Weisskopf' unit in the interpretation of radiative transitions.

15.3 Overlapping levels of the compound nucleus ($\Gamma \gg D$)

15.3.1 The statistical assumption

As the excitation energy of the compound nucleus increases, the *total* width of levels also increases because more channels become available and each channel may be more easily entered. Since the yield of a nuclear reaction follows the total width, sharp resonances are no longer observable when $\Gamma \gg D$ and contributions from many overlapping levels must be assumed at any given energy. In this *continuum region* it is no longer obvious that the assumptions of the Bohr theory of nuclear reactions will hold. It has been pointed out by Ericson† that the *statistical assumption* of random motion in the compound state implies that the excitation functions for reactions measured with good resolution should show marked *fluctuations*. Only if an energy average over a range large compared with a certain *coherence width* Γ is taken do such fluctuations disappear; the Bohr independence hypothesis may then be valid. Calculations relating to these average properties form the *statistical theory of nuclear reactions*.

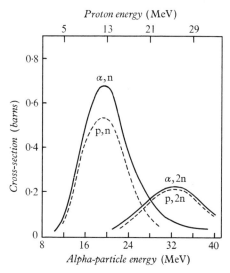

Fig. 15.7 Independence assumption for the continuum region of nuclear spectra. The full lines show the yield of ^{63}Zn and ^{62}Zn formed in α-particle bombardment of ^{60}Ni and the dotted lines the yield of these isotopes from the proton bombardment of ^{63}Cu. The energy scales are adjusted so that the excitation of the compound nucleus is the same for the two targets at the indicated bombarding energies (Ghoshal, *Phys. Rev.*, **80**, 939, 1950).

The fluctuations observed in high resolution experiments ($\Delta E \ll \Gamma$) in the continuum region are correlated over an energy range $\approx \Gamma$ because of the overlapping of states. Analysis of correlations between differential cross-sections for a given final state as a function of energy have been made with respect to energy interval and yield Γ, which in turn gives the lifetime of the compound state. For the reaction ^{27}Al(p, α)^{24}Mg the value $\Gamma = 45$ keV was found‡ for an excitation of 22 MeV in the compound nucleus. Similar correlations are found between differential cross-sections as a function of angle, and their analysis yields a quantity of the order of the nuclear radius. The analysis is the same as that used to determine the size of astronomical radio sources by correlating time-dependent radio fluctuations.

The independence of the energy-averaged processes of formation and decay for nuclei formed with excitations in the continuum region has been demonstrated directly. Figure 15.7 shows the results of Ghoshal for the yield

† *Phys. Rev. Lett.*, **5**, 430, 1960.
‡ B. W. Allardyce, W. R. Graham and I. Hall, *Nucl. Phys.*, **52**, 239, 1964.

of ^{62}Zn and ^{63}Zn formed by the reactions

$$
\left.
\begin{array}{l}
^{60}\text{Ni}+\alpha \longrightarrow {}^{64}\text{Zn} \longrightarrow {}^{62}\text{Zn}+2\text{n} \\
\qquad\qquad\qquad\quad \longrightarrow {}^{63}\text{Zn}+\text{n} \\
^{63}\text{Cu}+\text{p} \longrightarrow {}^{64}\text{Zn} \longrightarrow {}^{62}\text{Zn}+2\text{n} \\
\qquad\qquad\qquad\quad \longrightarrow {}^{63}\text{Zn}+\text{n}
\end{array}
\right\}
\qquad (15.44)
$$

and assessed by the resulting activity. It is clear that the different formation processes, if plotted on a scale of equal excitation energy for the compound nucleus ^{64}Zn, lead to nearly the same ratio of disintegration products.

In continuum theory it is assumed that at high energies when transmission coefficients are unity the cross-sections for scattering and absorption are given by the black nucleus values (14.57), (14.58) and (14.59).

15.3.2 Excitation functions

The statistical theory also expresses the cross-section for the nuclear reaction X(a, b) Y passing through a compound system of high excitation in the form (15.6). The cross-section σ_a for formation of the compound nucleus may be assumed to have the maximum value allowed for inelastic processes.† The decay width Γ_b is obtained by using relation (15.39) and integrating over the levels of the final nucleus Y that may be excited energetically. This brings in the level density $\omega_Y(E)=1/D_Y(E)$ of the nucleus Y at excitation E and this must be specified by a suitable model.

In its simplest form the Fermi gas model predicts that, omitting angular momentum dependence,

$$
\omega(E) = \text{constant} \times \exp 2\sqrt{aE} \qquad (15.45)
$$

where a is a parameter proportional to the single-particle level spacing near the Fermi energy. A modified treatment of the model gives the frequently-used form

$$
\omega(E) = \text{constant} \times E^{-2} \exp 2\sqrt{aE} \qquad (15.46)
$$

By use of such formulae the probability of decay as a function of the energy of the outgoing particle b can be calculated. The excitation energy of the residual nucleus is usually removed by a radiative transition, but if the excitation is high enough a second particle may be emitted.

From the expressions for σ_a and Γ_b the excitation function for the reaction X(a, b)Y may be calculated, including competition between alternative emission processes when multiple emission of particles is possible. The general forms of excitation function thus obtained are shown in Fig. 15.8. If a specific neutron energy is chosen and the energy-averaged (n, γ) capture cross-section is measured for a range of elements‡ the level density formula (15.46) may be tested as far as dependence on mass number A (through a) is concerned.

† M. M. Shapiro, *Phys. Rev.*, **90**, 171, 1953.
‡ D. J. Hughes, R. C. Garth and J. S. Levin, *Phys. Rev.*, **91**, 1423, 1953.

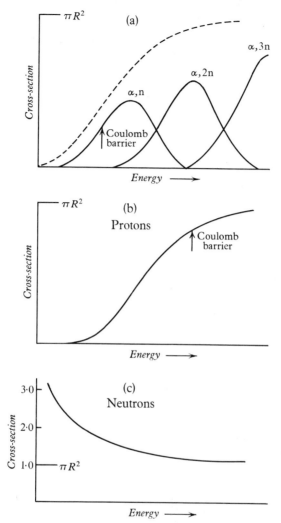

Fig. 15.8 Cross-sections in the continuum theory (schematic). (a) Competition in (α, xn) reactions. The dotted curve represents the theoretical cross-section for formation of the compound nucleus. (b) Reaction cross-section for protons. The rise is due to increasing barrier transmission. (c) Reaction cross-section for neutrons, showing transition from low energy value $\approx \pi \lambda^2$ to the high energy value $\approx \pi R^2$.

15.3.3 Evaporation spectra

The probability of the emission of a particle b with energy between ε and $\varepsilon + d\varepsilon$ is determined jointly by the energy ε and by the level density $\omega_Y(E)$ in the nucleus Y to which it corresponds. Specifically

$$I(\varepsilon) \, d\varepsilon \propto \varepsilon \omega_Y(E) \, d\varepsilon \qquad (15.47)$$

so that the particle spectrum falls off at both high and low energies (since

$E = \varepsilon_{max} - \varepsilon$). For nuclei with $A \approx 50$ and greater and for energies $\varepsilon_{max} \approx 10$ MeV, the spectrum has a continuous appearance, resembling that of the energy distribution of molecules in a gas. The analogy is often taken further to suggest that outgoing particles can be regarded as evaporation products of a highly excited nucleus. It is possible (Ref. 15.3) to define a *nuclear temperature T*, which is a parameter of the evaporation spectrum, and this is related to the level density of the residual nucleus, Y, at its maximum excitation. The level density as a function of E may be obtained directly from the spectrum of emitted particles using eq. (15.47). The density may also be obtained by direct counting of states observed in high resolution experiments.†

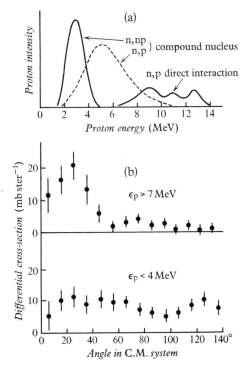

Fig. 15.9 Protons emitted in reactions of 14 MeV neutrons with medium weight nuclei such as Fe, Ni. (a) Component processes (Allan, *Proc. phys. Soc. Lond.*, **70**, 195, 1957). (b) Angular distributions of high energy (direct interaction) and low energy (compound nucleus) groups from ^{60}Ni(n, p)^{60}Co (March and Morton, *Phil. Mag.*, **3**, 577, 1958).

The angular distribution of evaporation particles is usually isotropic in the c.m. system. The spectrum however does contain a small number of high energy particles corresponding to excitation of low-lying states of Y. These are found to have an angular distribution with a marked forward peak and arise from non-compound nucleus processes, or *direct interactions*. Experimental data illustrating this point are shown in Fig. 15.9.

† A. Aspinall, G. Brown and J. G. B. Haigh, *Nucl. Phys.*, **52**, 630, 1964.

15.4 Special states

The underlying distribution of fine structure levels in the nuclear spectrum assumes a special character in two particular regions of excitation. These are discussed here, rather than in Chapter 12 (see however p. 319), in order to distinguish them from the fluctuations in cross-section mentioned in Sect. 15.3.1 (which are less relevant to nuclear structure) and because of their intimate connection with nuclear reaction theory.

15.4.1 *The photodisintegration resonance* (giant dipole resonance)

The ejection of neutrons or protons from nuclei by gamma radiation (γn or γp reaction) takes place preferentially at energies of about 15 MeV. Figure 15.10 shows the variation of yield with photon energy; the resonance width

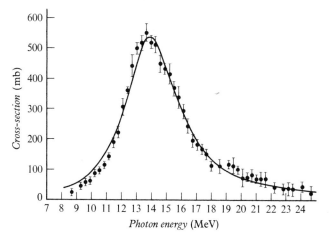

Fig. 15.10 Giant resonance of photodisintegration of ^{197}Au. The yield of neutrons is shown as a function of the energy of the monochromatic photons used to produce the reaction (S. C. Fultz *et al.*, *Phys. Rev.*, **127**, 1273, 1963).

is about 5 MeV. Closer examination of this resonance, especially by use of the inverse (pγ) reaction, reveals that it has some fine structure of states all of the same spin and parity. For even–even nuclei with a 0^+ ground state the so-called giant resonance is of 1^- character, corresponding to electric dipole excitations.

The giant resonance was described by Goldhaber and Teller (Ref. 15.12) as due to the collective oscillations of the neutron and proton distributions in a nucleus. As these move relatively to one another, a time-varying dipole moment is created. Wilkinson (Ref. 15.12) pointed out however that the shell model would predict such an effect as a result of the transitions of nucleons between major shells in the electromagnetic field. Quantitatively this gave rather too low a resonance energy. Later the addition of coupling between the excited nucleon and the hole in the shell that it had left (Ref. 15.12) showed that a state carrying most of the dipole strength would be located at an energy sufficiently high to explain the observed photoresonance.

15.4.2 Isobaric analogue levels

Figure 12.1 illustrates the fact that the level spectrum of a nucleus with $T = T_z = 0$ (equal number of neutrons and protons) contains states of $T = 1$ analogous to the levels of the isobaric $T = 1$, $T_z = \pm 1$ nuclei. These analogue states

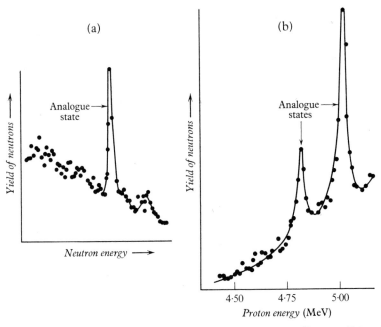

Fig. 15.11 Isobaric analogue states. (a) Neutron spectrum for the reaction ^{51}V (p, n) ^{51}Cr (adapted from J. D. Anderson and C. Wong, *Phys. Rev. Lett.*, **8**, 442, 1962). (b) Excitation function for the reaction ^{89}Y (p, n) ^{89}Zr (adapted from J. D. Fox, C. F. Moone and D. Robson, *Phys. Rev. Lett.*, **12**, 198, 1964).

may be connected by a (p, n) reaction e.g., (supposing that ^{10}Be were available as a target)

$$^{10}_{4}\text{Be} + \text{p} \longrightarrow {}^{10}_{5}\text{B}^* + \text{n} \qquad (15.48)$$

and this might be expected to have a high probability because of the similar structure of the initial and final states.

In heavier nuclei with a considerable neutron excess $T\ (\geqslant \frac{1}{2}[N-Z])$ was at first thought to have doubtful value as a quantum number because of the large Coulomb force, but it was found by Anderson and Wong that the (p, n) reaction still picked out analogue states for two adjacent isobars. Figure 15.11(a) shows the neutron spectrum for the reaction ^{51}V (p, n)^{51}Cr. Following on this work Fox, Moore and Robson showed that in many cases analogue states were located at such high excitation that they could be studied as *formation* resonances of the compound state rather than as final states in production reactions. Figure 15.11(b) shows their results for the reaction $^{89}Y(p, n)^{89}Zr$; the sharp states above the continuum are analogues, not of states of the target nucleus, but of the nucleus formed by adding a neutron to it, e.g., by the (d, p) reaction. Their excitation is mainly due to the Coulomb energy.

Isobaric analogue states have been found to have an underlying fine structure and they provide another example of the way in which correlated features of nuclear behaviour dominate certain regions of the background spectrum. They have proved to be of exceptional interest both to nuclear spectroscopy and to nuclear dynamics.

15.5 The optical model for nuclear scattering

The Bohr theory of the compound nucleus assumes essentially that as soon as an incident particle reaches the surface of a nucleus a compound state is formed, and that the subsequent decay is independent of the mode of formation. Moreover the probability of re-emission of the incident particle, although finite (so that narrow resonances at low energies result) is nevertheless small, so that the incident wave is heavily damped in the compound nucleus. In the particle picture this corresponds to a short mean free path in the compound nucleus. On the other hand the success of the nuclear shell model in describing the static properties of nuclei suggested that some type of permanence of single particle orbital motion, apparently inconsistent with the Bohr theory, must be possible. The theory of nuclear reactions is therefore faced with the problem already encountered in theories of nuclear structure (ch. 12), namely, how to ensure that models of an underlying individual particle type can yet show features of a more collective nature.

The optical model, which treats a nuclear reaction in analogy with the propagation of light through a partially absorbing medium, has proved useful in reconciling these apparently conflicting aspects of nuclear behaviour. A general formulation of such a model in terms of a complex scattering potential was proposed in 1940 by Bethe, but extensive development of the theory followed much later, after measurements of neutron cross-sections at both high and low energies had shown obvious conflict with the predictions of the Bohr theory. In the continuum region the total cross-section for neutrons (Sect. 15.3.1) should be given by the black nucleus formula

$$\sigma_t = 2\pi(R + \lambda)^2 \tag{15.49}$$

This predicts only a smooth dependence of σ_t on incident energy E (through λ) and on target mass number A (through R), and the nuclear radius at high energies as $\lambda \rightarrow 0$ should be given by

$$R = \sqrt{\frac{\sigma_t}{2\pi}} \qquad (15.50)$$

Nuclei were expected to become 'blacker' as the incident energy increased because of the increasing number of possible reaction processes. In contrast with this expectation the experiments of Cook, McMillan, Peterson and Sewell[†] in 1949 with 90 MeV neutrons from the 184-in synchrocyclotron, showed that nuclei appeared consistently *smaller* to 90 MeV particles than to lower energy (≈ 25 MeV) projectiles. This behaviour had already been predicted by Serber,[‡] who pointed out that at high energies incident particles would tend to interact with individual target nucleons rather than with the nucleus as a whole and that the probability of compound nucleus formation would accordingly be reduced. There would on the other hand be an enhanced probability for the bombarding particle to emerge without engaging in any interaction at all, and the target nuclei would therefore appear partially transparent, i.e. to have a smaller radius than expected. This idea was developed quantitatively by Fernbach, Serber and Taylor[§] into the high energy version of the optical model.

Low energy data also suggested an interaction which could be represented by a simple potential well. The scattering of slow neutrons by nuclei at energies not adjacent to resonance levels is expected in the compound nucleus theory to yield the 'hard sphere' cross-section of $4\pi a^2$ where $a \approx R$, the nuclear radius. The variation of cross-section with mass number should therefore follow that of R^2; i.e. should be proportional to $A^{2/3}$ and the scattering length should vary as $A^{1/3}$. When the available data for a wide range of elements were assembled by Ford and Bohm,[¶] (Fig. 15.12(a)), it became clear that discontinuities occurred at certain mass numbers and these were interpreted as due to resonance of the internal wave function in a nuclear potential well whose dimension depended on A. The spacing of these resonances in A corresponds to *single-particle* levels, not to the fine structure levels of the Bohr theory. Later, extensive studies of neutron total cross-sections in the energy range 1–3 MeV by workers at the University of Wisconsin and elsewhere, first collected together by Barschall[‖] revealed these single particle resonances very clearly. Figure 15.13 shows the results of this work; broad peaks are found for a given element as the neutron energy varies, and these peaks occur at different energies as A varies. This behaviour is again similar to that expected in the interaction of a particle with a potential well and is in disagreement with the monotonic energy dependence of averaged cross-sections indicated by (15.49). Furthermore the averaged ratio of neutron width to spacing for the individual fine structure levels was not independent

[†] L. J. Cook, E. M. McMillan, J. M. Peterson and D. C. Sewell, *Phys. Rev.*, **75**, 7, 1949.
[‡] R. Serber, *Phys. Rev.*, **72**, 1008, 1947.
[§] S. Fernbach, R. Serber and T. B. Taylor, *Phys. Rev.*, **75**, 1352, 1949.
[¶] K. W. Ford and D. Bohm, *Phys. Rev.*, **79**, 745, 1950.
[‖] H. H. Barschall, *Phys. Rev.*, **86**, 431, 1952.

Nuclear reactions (detailed mechanisms)

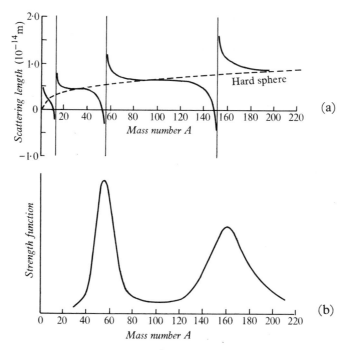

Fig. 15.12 (a) Slow neutron scattering length as a function of mass number. The dotted curve is the expectation for hard sphere scattering. The experimental points are not shown, but roughly follow the full curves, which are predictions for the scattering by a square well potential (Ford and Bohm). (b) Low energy neutron strength function, deduced from total cross-section measurements. The single peak shown for $A \approx 160$ in fact splits into two peaks because this is a region of permanent deformation and the potential is spheroidal.

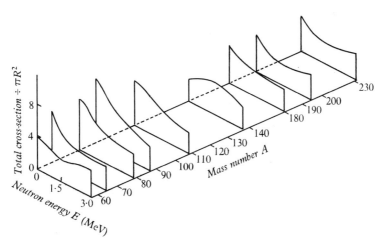

Fig. 15.13 Total cross-section of nuclei for neutrons as a function of energy E and mass number A (from Barschall, 1952).

405

of mass number as suggested by (15.40) but showed peaks which agreed in position with the anomalies in scattering length (Ref. 15.4 and Fig. 15.12(b)). The two curves in this figure show respectively the dispersive and absorptive effect of a well which is able to capture particles as an alternative to scattering them. The behaviour shown is quite unexpected if the incident neutron is strongly absorbed by the target nucleus, as assumed in the Bohr theory. The interpretation of the observations in terms of a long mean free path for the incident neutrons in the target was examined by Feshbach, Porter and Weisskopf† who were thus led to develop the low energy form of the optical model.

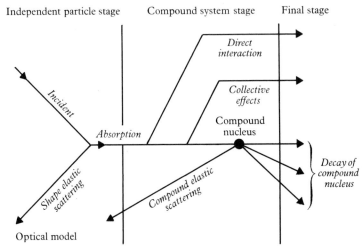

Fig. 15.14 Nuclear reaction scheme, according to Weisskopf (*Rev. mod. Phys.*, **29**, 174, 1957).

In uniting these two lines of evidence the optical model supposes that a nuclear reaction should be treated as a two-body problem, in which an incident particle sees the target nucleus as a limited region of complex potential

$$V(r) + iW(r) \quad (i = \sqrt{-1}). \tag{15.51}$$

The potentials $V(r)$ and $W(r)$ are energy dependent and at low (effectively negative) energies, V is the potential required by the nuclear shell model. Elastic scattering produced by the potential V is evidently of the single particle type and can account for the broad resonances seen in the results of Barschall and of Ford and Bohm. The imaginary part W of the potential attenuates the incident wave and is directly associated with the mean free path of the incident particle in the target nucleus; it is chosen to yield the mean free path required by the experiments at all energies. In the low energy region, W is often taken to describe compound nucleus formation but it may also account for direct interaction processes. The long mean free path

† H. Feshbach, C. E. Porter and V. F. Weisskopf, *Phys. Rev.*, **96**, 448, 1954.

associated with low values of W is at variance with the strong interaction theory of Bohr, but it may be understood because the Pauli principle will inhibit collisions which lead to occupied momentum states. In the optical model the single particle strength of shell model states spreads over a number of fine-structure compound states; the effect of this on experiments made with both high and low resolution has already been indicated (Fig. 9.7). The broad resonances, and their interpretation in terms of fine structure levels, are similar in nature to the giant dipole resonance of photodisintegration (Sect. 15.4.1).

The course of a nuclear reaction according to these ideas may be represented as in Fig. 15.14. There is a preliminary or *single particle* stage, in which the interaction of the incident wave with the potential V leads to *shape-elastic* or *potential scattering*. Part of the wave is also absorbed (potential W) to form a *compound system* towards which the first step may be the excitation of a nucleon from the target nucleus. This creates a two particle–one hole $(2p1h)$ state

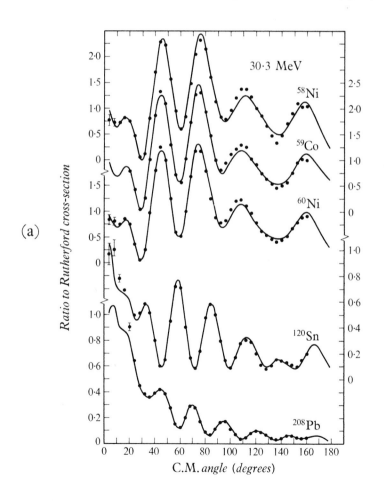

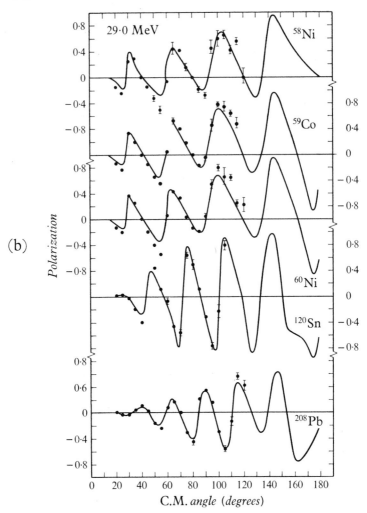

Fig. 15.15 Optical model analysis (Greenlees *et al.*, 1968). The figures relate to the elastic scattering of protons of energy ≈ 30 MeV by a range of nuclei. The points are experimental data and the curves are theoretical fits. (a) (p. 407) Differential cross-section. (b) Polarization.

which has sometimes been called a *doorway state*† since it leads both to direct interaction processes (Sect. 15.6) and to further particle-hole excitations (3p, 2h; 4p, 3h, etc.) which finally create the Bohr compound nucleus. Both the doorway state and the compound nucleus may decay back into the initial system (though on different time scales) and thereby contribute a component of *compound elastic scattering* which is not predictable in detail by the model itself.

Conventional optical model analysis, which has been successful for the

† Doorway states very probably enter into the formation of both giant dipole resonances and isobaric analogue states (Sect. **15.4**).

discussion of an extensive range of elastic scattering experiments, postulates that the potentials V and W have a different radial dependence, that they include a non-central component proportional to (**s. 1**) as in the shell model (Sect. 9.4) and that V is isobaric-spin dependent, i.e. different for neutron and proton scattering because of nuclear interactions as well as because of the Coulomb effect. From this potential a spin-dependent scattering amplitude may be obtained by numerical solution of the Schrödinger equation. From the scattering amplitude, differential cross-sections for the shape-elastic scattering and polarizations (Sect. 17.3.2) may be predicted; from the potential W, the total reaction cross-section may be deduced. Fits to cross-section and polarization data for 30 MeV proton scattering are shown in Fig. 15.15. The theoretical curves in this figure are actually derived from a *reformulated* version of the model due to Greenlees, Pyle and Tang.† This successfully reduces the number of arbitrary parameters in the conventional model by relating the potential to the fundamental nucleon–nucleon interaction.

15.6 Direct interaction processes

15.6.1 Insufficiency of the compound nucleus mechanism

The optical model does not concern itself with details of the processes which make up the inelastic part of the total cross-section, i.e. with nuclear reactions other than scattering. As shown in Fig. 15.14, some of these reactions proceed through the formation of a Bohr compound nucleus. Others, especially for incident particles of 10 MeV energy and above, exhibit features which are not easily described by the compound nucleus theory. The characteristics of these *direct interactions* are:

(i) Emission of excess particles of high energy in comparison with the number expected according to the statistical evaporation theory. These particles often show resolvable structure due to levels of the final nucleus (Fig. 15.9(a)). An associated effect is the observation of high cross-sections for (n, p) and (n, α) reactions in heavy target nuclei, from which the emission of charged particles should be severely hindered by the Coulomb barrier.

(ii) Forward peaking of the higher energy particles of the spectrum, in contrast with the symmetric angular distributions expected of evaporation particles (Fig. 15.9(b)).

(iii) Gradual and usually monotonic dependence of yield on bombarding energy. In the continuum region this is of course expected for averaged cross-sections (Fig. 15.8), but in the region of discrete levels a slowly varying 'background yield' may often be ascribed to a direct process.

These differences between direct and compound nuclear processes arise because the former are single-stage processes (eq. 15.5) in which the incident

† G. W. Greenlees, G. J. Pyle and Y. Tang, *Phys. Rev.*, **171**, 1115, 1968.

momentum of a tends to be transferred directly to b, while the latter are two-stage processes (eq. 15.4) in which the momentum is conveyed in the first instance to the compound nucleus C.

15.6.2 *Theory of direct interaction processes in general*

The Bohr theory of a compound nucleus reaction assumes that, in the resonance region of excitation at least, the mean free path of an incident nucleon in the compound nucleus is small. Although this may be true for energies near resonance, it is not necessarily true in the first interaction of the incident particle with the target nucleus, particularly for energies of the order of 10 MeV. The theories of direct interaction assume that in a scattering or reaction process the first event is a collision between the incident particle and a nucleon near the surface of the target nucleus. If a mean free path of the order of nuclear dimensions is assumed, the struck nucleon may emerge from the nucleus without the formation of a compound nucleus and the direct process, e.g. a (n, p) reaction, is complete.

In quantitative developments of this hypothesis, it is usually assumed that the direct interaction is confined to a surface layer of the target nucleus. Interactions within the nuclear core are not impossible, but there is a high probability of reflection of the resulting particles at the boundary and consequent enhancement of compound nucleus formation. For surface processes there is a fairly definite radius of interaction and, following Butler, Austern and Pearson[†] it is possible to give a simple semi-classical picture of the mechanism of these reactions. Let $\mathbf{k}_i$ and $\mathbf{k}_f$ (Fig. 15.16(a)) be the wave vectors for an incident and an emergent particle, both of which (for simplicity) traverse the surface of a nucleus of radius R without refraction. The direct interaction takes place at the point P, distant $\mathbf{r}$ from the centre of the nucleus. Then the change of linear momentum of the nucleus, i.e. the recoil momentum of the target is

$$(\mathbf{k}_i - \mathbf{k}_f)\hbar = \mathbf{q}\hbar \qquad (15.52)$$

where

$$\mathbf{q} = \mathbf{k}_i - \mathbf{k}_f \qquad (15.53)$$

(since $k_i = 1/\lambda_i = Mv_i/\hbar$). For the excitation of low-lying states of the final nucleus the emergent particle will carry a high momentum ($\mathbf{k}_i \approx \mathbf{k}_f$, $\mathbf{q} \approx 0$) and we should expect the angular distribution to be peaked in the forward direction (Fig. 15.16(b)). This expectation is modified because in the excitation of a discrete final state a certain definite angular momentum $l\hbar$ must be imparted to the nucleus. For the (n, p) reaction l is limited by the inequality[‡]

$$l_n + l_p \geqslant l \geqslant |l_n - l_p| \qquad (15.54)$$

[†] S. T. Butler, *Phys. Rev.*, **106**, 272, 1957; S. T. Butler, N. Austern and C. Pearson, *Phys. Rev.*, **112**, 1227, 1958.
[‡] Change of intrinsic spin direction is neglected.

where l_n and l_p are the orbital quantum numbers of the neutron in its final state after capture by the nucleus and of the proton in its initial state before interaction.

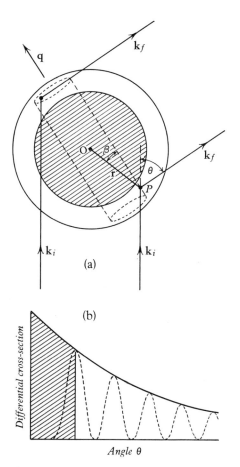

Fig. 15.16 Direct interaction (semi-classical picture). (a) Collision of an incident particle with a surface particle P. Conservation of angular momentum restricts the interaction to the surface of a certain cylinder; nuclear absorption confines it to large radii. (b) Angular distribution for a direct reaction. The full line is the classical expectation and the shaded area is excluded by the necessity for angular momentum conservation. The oscillations in the actual pattern arise from interference between the two rays shown in (a) (Butler, 1957).

Kinematically (Fig. 15.16(a)) the change of angular momentum is

$$\mathbf{l} = (\mathbf{k}_i - \mathbf{k}_f) \times \mathbf{r}\hbar \quad \text{or} \quad \mathbf{l} = (\mathbf{q} \times \mathbf{r})\hbar \tag{15.55}$$

so that

$$l = qr \sin \beta \tag{15.56}$$

where β is the angle between $\mathbf{r}$ and $\mathbf{q}$. For a given l-value, and a given angle

411

of scattering θ, which determines q through the equation

$$q^2 = k_i^2 + k_f^2 - 2k_i k_f \cos \theta = (k_i - k_f)^2 + 4k_i k_f \sin^2 \tfrac{1}{2}\theta \qquad (15.57)$$

the condition of angular momentum (15.56) restricts the possible points of direct interaction to the surface of a cylinder of radius l/q with axis in the direction $\mathbf{q}$. Scatterings from the nuclear interior are not likely to lead to direct interactions because of internal reflection, and the effective scattering elements are therefore the two spherical caps at the ends of the cylinder. Interference from the waves originating at these two areas leads to the maxima and minima in the angular distribution.

In the case that $l/q > R$ (15.56) shows that there will be no direct interaction. This may happen, for a given l, for small values of $\mathbf{q}$, i.e. for the forward direction. The larger the l-value concerned, the greater the angle at which the first scattering peak is observed in the angular distribution; for $l=0$ a forward peak may be seen. The angular distribution expected for a direct reaction, associated with the nuclear surface, is therefore as shown in Fig. 15.16(b). The forward peak associated with the high forward momentum of particles connected with low-lying residual states is forbidden (except for $l=0$) by the condition of angular momentum. The first peak in the angular distribution occurs at the angle for which

$$qR = l \qquad (15.58)$$

These conclusions are confirmed by detailed quantum mechanical calculations, which may be formulated at different levels of precision. One source of uncertainty in such calculations is the precise nature of the interaction experienced by the incoming particle, but for energies much greater than the binding energies of target nucleons it is reasonable to use the interaction with nucleons in a free state. This is known as the *impulse approximation*. The wave functions used to describe the ingoing and outgoing particles are *plane waves* in the simplest approach but *distorted waves* in a more realistic calculation. The distorting potentials may be obtained from optical model analysis (Sect. 15.5) of elastic scattering experiments.

The two general types of direct processes that have proved to be especially significant are (*a*) direct inelastic scattering reactions, and (*b*) transfer reactions.

15.6.3 *Direct inelastic scattering reactions*

These are processes of the type (e, e′) (p, p′), (d, d′) and (α, α′), and because of similarity of treatment the charge exchange reactions (n, p) and (p, n) may be included. Figure 15.17 shows the angular distribution of an inelastic scattering process; it may be used in the first instance to infer the angular momentum change in the reaction or a radius of interaction.

Direct inelastic scattering has been found to excite preferentially the collective states of nuclei, since these are structurally closely connected with the ground state and may have strong radiative transitions to it. The parameters of the collective model (e.g., the deformation β) may be obtained from the angular distribution. For strongly absorbed projectiles, such as α-particles, this may be regarded as a diffraction pattern for the deformed shape, just as the elastic scattering is analogous to the Fraunhofer pattern of a black disc (Sect. 14.2.2).

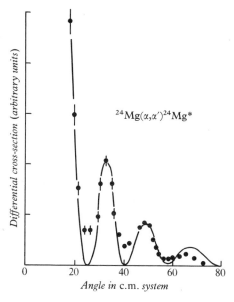

Fig. 15.17 Comparison of direct interaction theory with experiment for the inelastic scattering of 31 MeV α-particles by ^{24}Mg (1·37 MeV level) (Butler, 1957).

15.6.4 Transfer reactions

The theory of direct interactions outlined in Sect. 15.6.2 grew from the work of S. T. Butler and of A. B. Bhatia *et al.*† on the deuteron *stripping* reaction. In this reaction either a neutron (d, p reaction) or a proton (d, n reaction) is added to a target nucleus and evidently there will be some preference for the formation of single particle states. In the rather similar deuteron *pickup* reaction a neutron (d, t reaction) or proton (d, ^{3}He reaction) is removed from a target nucleus. In these cases there is some preference for exciting single hole states of the final nucleus. Many other *transfer* reactions are known e.g. (p, d), (p, ^{3}He), (^{3}He, p), (^{3}He, d) and the (p, 2p) reaction discussed in Sect. **9.3** and their basic features are similar; these will be illustrated for the (d, p) stripping reaction.

When accurate studies of the angular distributions of protons emitted in (d, p) reactions for deuterons of energy about 10 MeV became available in

† S. T. Butler, *Proc. roy. Soc.*, A, **208**, 559, 1951; A. B. Bhatia, K. Huang and H. C. Newns, *Phil. Mag.*, **43**, 485, 1952.

1949–50, the appearance of strong forward peaks was noted. Thus in the reaction

$$^{16}\text{O} + \text{d} \longrightarrow {}^{17}\text{O} + \text{p} + 1\cdot 92 \text{ MeV} \tag{15.59}$$

the differential cross-section (Fig. 15.18(a)) rises by a factor of 5 or more between 90° and the angle of the forward peak in the centre-of-mass system.

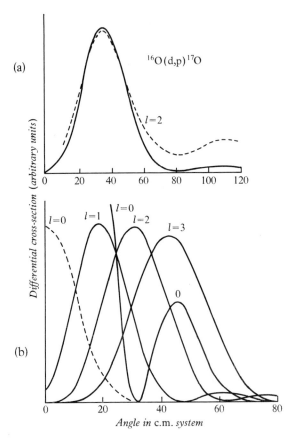

Fig. 15.18 Deuteron stripping reaction. (a) Comparison of experimental (dotted) and theoretical (full) proton angular distributions for ^{16}O bombarded by 8 MeV deuterons. (b) Theoretical angular distributions for different l-values in the stripping reaction. Two scales are used for the $l=0$ curve (Butler, 1951).

This behaviour cannot easily be understood on the basis of a compound nucleus mechanism, because very high order Legendre polynomials would be required to account for the peak, and these would not appear if the interaction took place within the nuclear radius at energies of the order of 10 MeV. If, however, the release of protons takes place well away from the centre of the nucleus these particles can more easily have a high orbital angular momentum and it is then possible to understand the forward peak. The fact that the deuteron is a loosely bound structure in which the two particles spend most of their time at a mutual distance greater than the range of the forces between them makes such a process possible.

In the approach of a deuteron to a nucleus one of the particles (in the case of reaction (15.59) the neutron) may collide with a nucleon in the nuclear surface while the other particle (the proton) is released at a distance from the nucleus of the order of the deuteron diameter. The proton continues with approximately its original forward momentum and with little or no inter-action with either the original or residual nucleus. The process has an obvious resemblance to the direct interaction mechanism illustrated in Fig. 15.16; thus referring to this figure, and considering a (d, p) reaction occurring by a direct process in the rim of the nucleus, the linear momentum carried into the nucleus by the absorbed neutron is $q\hbar$ where

$$q^2 = (k_d - k_p)^2 + 4k_p k_d \sin^2 \tfrac{1}{2}\theta \qquad (15.60)$$

and k_d, k_p are the wave numbers of the incident deuteron and emergent proton respectively. The angular momentum change is

$$\mathbf{1} = (\mathbf{q} \times \mathbf{r})\hbar \qquad (15.61)$$

where $|\mathbf{r}| \approx R$ is the radius of interaction. The neutrons will be absorbed at points over the nuclear surface between which interference is possible and maxima and minima in the proton angular distribution will arise as discussed in Section 15.6.2. From this pattern the angular momentum transferred to the target nucleus may be deduced and hence the *parity* change between the initial and final states. The l-value also sets limits on the spin I_f of the final state because of the selection rule

$$I_i + l + \tfrac{1}{2} \geqslant I_f \geqslant |I_i \pm l \pm \tfrac{1}{2}|_{\min} \qquad (15.62)$$

where I_i is the spin of the initial nucleus and $\tfrac{1}{2}$ is the spin number of the absorbed nucleon.

The theoretical angular distribution for a transfer reaction has been pre-sented in several equivalent forms, each of which takes account of the struc-ture of the incident deuteron and of the reduced width, which determines the relative ease of the transfer to a particular final state. In the plane wave approximation the angular dependence is specifically given by the squared spherical Bessel function $[j_i(qR)]^2$ in which it is assumed that the reaction is located at a radius R. In general, nuclear states are not of pure single-particle type and the observed cross-section may be expressed in the form

$$\left(\frac{d\sigma}{d\Omega}\right)_{obs} \propto S_{jl}\left(\frac{d\sigma}{d\Omega}\right)_{jl} \qquad (15.63)$$

where the suffixes jl refer to a single particle state, for which the transfer cross-section may be calculated using a distorted wave approximation to-gether with an appropriate final state wave function.† The quantity S_{jl} is the

† Use can be made of separation energies from (p, 2p) experiments and of elastic electron scattering data in setting up suitable potentials.

spectroscopic factor, which is essentially the reduced width (Sect. 15.2.5) for the reaction in units of the single-particle reduced width (Ref. 15.4).

The observed angular distributions thus yield the spectroscopic factor S_{jl} for several states of the final nucleus Y in the stripping reaction X(d, p)Y and these factors show the extent to which the states Y resemble the single particle configuration X + n. For nuclei with several particles (or holes) in a subshell the spectroscopic factors are directly related to the occupation numbers for the subshell. The spectroscopic factors may be calculated from any nuclear model which gives wavefunctions and provide a significant test of these models.

15.7 Nuclear reactions of heavy ions (Ref. 15.13)

The acceleration of ions such as ^{12}C, ^{16}O, ^{40}A, ^{79}Br in tandem electrostatic accelerators, linear accelerators and cyclotrons to energies ≈ 5–10 MeV per nucleon has opened a new field of nuclear studies. The special characteristics of heavy ion reactions are:

(i) Strong absorption in the nuclear surface so that there is good localization of a reaction.
(ii) High linear momentum for a given energy so that experiments requiring high recoil velocities, e.g. Doppler shift lifetime measurements, are helped.
(iii) High transfer of angular momentum to a compound nucleus.
(iv) Efficiency of Coulomb excitation processes owing to the high ionic charge.
(v) Efficiency of formation of highly excited compound nuclei.

In addition there has been extensive use of heavy ions in the synthesis of transuranic elements (Sect. 14.1.5) and they appear vital for the production of superheavy elements with $Z \approx 114$ (Ref. 14.8). Apart from such exotic uses heavy ions enter into the normal range of direct and compound nucleus reaction processes and have contributed much spectroscopic information particularly for neutron-deficient nuclei remote from the valley of stability which may be formed by the emission of several neutrons from a compound state formed by a heavy ion reaction. By proper choice of bombarding energy the evaporation of a particular number of neutrons can be favoured and the corresponding product nucleus emphasized. The spontaneously-fissioning isomers (App. 4) were discovered among the products of the bombardment of ^{238}U with ^{16}O and ^{22}Ne ions.

References

1. HUGHES, D. J. and SCHWARTZ, R. B. *Neutron Cross Sections*, Brookhaven National Laboratory Report BNL 325, 1958.
2. SCHIFF, L. I. *Quantum Mechanics* (3rd ed.), McGraw-Hill, 1968.
3. BLATT, J. B. and WEISSKOPF, V. F. *Theoretical Nuclear Physics*, Wiley, 1952.

4. JACKSON, D. F. *Nuclear Reactions*, Methuen, 1970.

5. KANE, J. V. 'Instrumentation for Nuclear Structure Analysis', *Physics Today*, **19**, July 1966.

6. RICHARDS, H. T. 'Charged Particle Reactions', in *Nuclear Spectroscopy*, Part A, ed. F. Ajzen-berg-Selove, Academic Press, 1960.

7. GOVE, H. E. 'Resonance Reactions, Experimental', in *Nuclear Reactions*, ed. Endt-Demeur, North Holland, 1959.

8. HUBY, R. 'Stripping Reactions', *Progr. nucl. Phys.*, **3**, 177, 1953.

9. LE COUTEUR, K. J. 'The Statistical Model', in *Nuclear Reactions*, ed. Endt-Demeur, North Holland, 1959.

10. HODGSON, P. E. *The Optical Model of Elastic Scattering*, Oxford University Press, 1963.

11. NEMIROVSKII, P. E. *Contemporary Models of the Atomic Nucleus*, Pergamon Press, 1963.

12. ROWE, D. J. *Nuclear Collective Motion*, Methuen, 1970.

13. NEWTON, J. O. 'Nuclear Spectroscopy with Heavy Ions', *Progr. nucl. Phys.*, **11**, 53, 1969.

16 Radioactive decay

The main phenomena of natural radioactive decay were established long ago and have been described in Chapter 2. Since 1935 many new α- and β-emitting bodies have been prepared, the former only among the heavy elements and the rare earths, but the latter throughout the periodic system. The factors determining the lifetime of these two types of unstable nuclei appear to be general and there is no important distinction between the naturally occurring and the artificially prepared species in this respect.

In Sect. **10.4** it was pointed out that all naturally occurring nuclei with $A > 150$ are unstable against α-emission, since the energy required to remove four nucleons from a nucleus can then be supplied by the combination of those four nucleons into a tightly bound α-particle. The fact that α-*emission* is a relatively rare phenomenon compared with α-*instability* is due to the exponential dependence of emission probability on decay energy (Sect. **16.2**). The α-emission observed in the rare earths is due to an increase of available energy resulting from closure of the $N = 82$ neutron shell. Coulomb effects also influence β-emission probabilities but there is no sharp dependence on transition energy and the main consequence is an inhibition of positron decay in heavy nuclei in favour of electron capture.

In Chapters 9 and 10 α- and β-decay were used as general evidence in favour of the single-particle shell model, and as examples of the application of the principles of nuclear stability. In this chapter we survey the data in rather more detail, with particular attention to the decay mechanism; α-decay will be seen to be essentially a nuclear reaction in reverse, but β-decay presents entirely new and (until recently) unexpected features.

16.1 Experimental information on α-decay

Observations of α-decay energies have been made since the discovery of radioactivity using methods of continually increasing precision, some of which have already been described in Chapter 7. The earliest measurements with ionization chambers and point counters established the lifetime for α-emission and the genetic relationships between the α-activity of given lifetime and that of its parent or daughter element (ch. 2). Range measurements with shallow ionization chambers (Bragg, 1906) demonstrated the approximate homogeneity in energy of the main α-activity from a given element. A correlation between the lifetime and energy of the α-particle emission was noticed by Geiger and Nuttall as early as 1911; more recent work has greatly enlarged the scope of this law. The resolution of α-particle energy measurements was much improved by the introduction of the magnetic spectrometer and many groups originally thought to be homogeneous have now been resolved into components.

The experimental data comprise (*a*) *energy–mass number* relationships, giving information on nuclear stability, (*b*) *energy–lifetime relationships*, giving information on decay mechanisms, and (*c*) *α-particle spectra*, giving information on nuclear states. Two types of spectra have been distinguished, namely those exhibiting *fine structure* and those containing *long-range α-particles*.

In discussing α-decay we must distinguish between the energy E_α of the emitted particle and the total decay energy. The latter is greater than E_α by the energy of nuclear recoil, and must be used when calculating mass changes.

16.1.1 *Dependence of α-decay energy on mass number*

The regularities of the ground-state α-decay energies in the heavy element region are illustrated in Fig. 16.1. This is based on data from artificially produced isotopes, including transuranics (Sect. 14.1.5) as well as from the naturally occurring series. A sequence of α-particle emissions (Fig. 2.9) finally

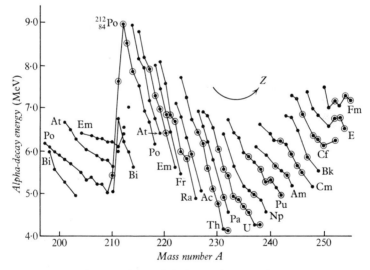

Fig. 16.1 Energy release in the α-decay of the heavy elements, showing effect of neutron shell closure at $N = 126$. The nuclides ringed are β-stable (adapted from Ref. 16.2, Hanna).

leads to a β-unstable nucleus by alteration of the ratio N/Z and one or two β-emissions then occur before α-transformations are resumed. The series thus contains a number of isotopes of certain elements. In Fig. 16.1 emphasis has been placed on the energy in relation to a given element, rather than to a particular series, by drawing appropriate lines; the variation of decay energy with A (or N) for a given Z is then apparent. For each mass number, the mass parabolas (Sect. **10.4**) show that one or more isotopes are β-stable and these isotopes are especially indicated in the figure. It is evident that the α-decay energy of the β-stable elements decreases as Z increases towards U ($Z=92$), in contrast with the prediction of the semi-empirical mass formula, but for $Z>92$ this trend is reversed. Other notable features in Fig. 16.1 are that

(i) E_α increases with Z for constant A,
(ii) E_α decreases as A increases for constant Z,
(iii)   If A is decreased until $N(=A-Z)$ falls below 128 the variation of E_α with A reverses abruptly over a limited region.
(iv)   A similar anomaly may be traced at $Z=84$ since the decay energies of bismuth isotopes ($Z=83$) are lower than expected and lead ($Z=82$) and thallium ($Z=81$) show no α-activity at all.

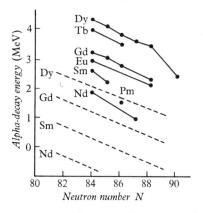

Fig. 16.2 Energy release in the α-decay of the rare earths showing effect of neutron shell closure at $N=82$. The dotted lines show the energies expected from the semi-empirical mass formula (Ref. 16.2, Hanna).

The variation of α-energy of β-stable elements with A, and effects (iii) and (iv) above are due to the closure of a neutron shell at $N=126$ and a proton shell at $Z=82$. A maximum in α-decay energy occurs when two loosely bound nucleons just above a closed shell are removed by the α-emission; thus $^{212}_{84}$Po (ThC′, $N=128$) and $^{213}_{85}$At ($N=128$) should have exceptionally high decay energies.† The effect of the 126-neutron shell for the polonium isotopes is also shown in Fig. 10.3.

The effect of the 82-neutron shell in the rare earth region is shown in Fig. 16.2. With the exception of ^{147}Sm the species shown are all artificially produced. The decay energies exceed those predicted by the semi-empirical mass formula for $N=84$ and above and no α-activity is observed for $N<84$.

† This is verified for ThC′, and seems likely for ^{213}At. The high decay energy means a short half-life.

The systematic behaviour of α-decay energies shown in Fig. 16.1, coupled with similar regularity in the energy dependence of half-lives (Sect. 16.1.2) has been of great use in the prediction of the properties, including the masses, of artificially produced transuranic elements.

16.1.2 Energy-lifetime relationships

The earliest energy-lifetime relationship was the celebrated law given by Geiger and Nuttall.† Using the apparatus shown in Fig. 16.3(a), these

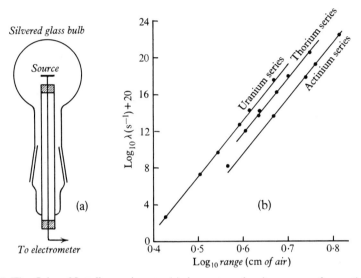

Fig. 16.3 The Geiger–Nuttall experiments. (a) Apparatus, showing source of α-particles in a vessel which can be evacuated. A potential of 700 V is applied to the inner silvered surface of the glass bulb. The ionization current falls off at the pressure at which the α-particle range becomes greater than the radius of the bulb. (b) Logarithmic plot of decay constant against range for α-emitting nuclei known in 1921 (Ref. 16.2, Hanna).

workers determined the mean range of the α-rays from a number of naturally occurring elements and plotted their results logarithmically against the measured decay constant. The results available in 1921 (Fig. 16.3(b)) fell on three straight lines of equal slope, one for each of the naturally occurring radioactive families, and demonstrated that for each of these families

$$\log \lambda = a + b \log R \qquad (16.1)$$

Since the range of an α-particle of a few MeV energy is connected with its velocity by the Geiger rule, eq. (5.60),

$$R \propto v^3 \qquad (16.2)$$

the Geiger–Nuttall rule may be written

$$\log \lambda = a + b' \log v \qquad (16.3)$$

† H. Geiger and J. M. Nuttall, *Phil. Mag.*, **22**, 613, 1911; H. Geiger, *Zeits. Physik*, **8**, 45, 1922.

It is now known that this rule is valid only for a rather limited number of even Z–even N nuclides.

Recent results are usually presented as diagrams such as Fig. 16.4, which relates the measured half-period and energy directly for nuclides of a given type.

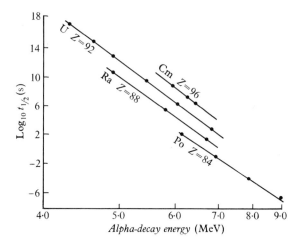

Fig. 16.4 Energy-lifetime relation for even–even α-emitting nuclei of indicated Z. The energy scale is linear in $E_\alpha^{-1/2}$ (Ref. 16.2, Hanna).

The lifetimes used are obtained from observed decay constants corrected for alternative decay modes such as β-emission or electron capture by determination of branching ratios. Regular behaviour in plots of this type is most marked in the decay of the ground states of even–even nuclei. If curves for odd Z elements are drawn between those for the even Z elements on this diagram, they too may be used as standards with which experimental results can be compared. In this way 'hindrance factors' for particle decays can be defined; it is for theories of nuclear structure to explain these factors. Thus in ^{235}U the partial decay constant for the ground-to-ground transition is about 1000 times smaller than that for an even Z–even N transition of the same energy.

16.1.3 *Fine structure of α-particle spectra*

Fine structure in α-ray spectra was demonstrated in the high resolution experiments of Rosenblum (1929) and of Rutherford and his students. Figure 16.5(a) shows a magnetic analysis of the α-particles emitted in the transition

$$\text{ThC} \longrightarrow \text{ThC}'' + \alpha + 6\cdot20 \text{ MeV} \tag{16.4}$$

and Table 16.1(a) gives the measured energies and intensities of the discrete groups. Low resolution experiments with simple ionization chambers would

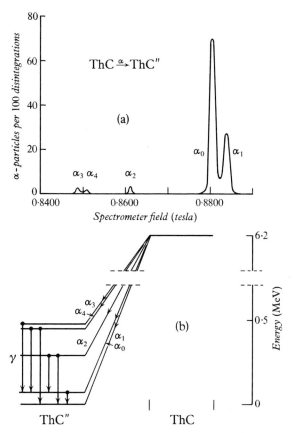

Fig. 16.5 Fine structure of α-particle spectra. (a) Magnetic analysis of groups from ThC ThC″ transition. (b) Level scheme, showing γ-ray transitions in final nucleus (Ref. 16.3).

not have separated groups α_0 and α_1, although the resolution required is now available with semiconductor counters (cf. Fig. 6.3(b)). These results are typical of many α-decays in which there are one or two closely spaced groups of comparable intensity and a few more widely spaced groups of lower energy and much lower intensity. This fine structure is due to the excitation of *levels of the residual nucleus* (ThC″ in the case of reaction (16.4)). The low intensity of groups α_2, α_3 and α_4 is due primarily to reduced penetrability of the potential barrier of ThC″ for the α-particles of lower energy, although nuclear structure effects are also important in affecting these intensities.

The excited residual nucleus ThC″ may lose its energy by γ-ray emission. The role of the γ-radiation as the electromagnetic spectrum of a nucleus has already been mentioned (Sect. 2.5.2 and Fig. 2.12). The γ-rays from many radioactive transitions were examined by Ellis (1922) using a magnetic spectrometer and relations of the form

$$h v_3 = h v_1 + h v_2 \qquad (16.5)$$

423

were found among the γ-ray transition energies deduced from conversion spectra, as would be expected from Fig. 16.5(b). A satisfactory correlation between the level schemes indicated by the γ-ray observations and by the later α-particle measurements has been established. The exact association between charged particle groups and the subsequent de-excitation spectra is now familiar throughout the whole range of nuclear spectroscopy and is extensively used in the investigation of nuclear reactions and level schemes.

TABLE 16.1(*a*) Fine structure of α-particle emission in the ThC→ ThC″ transition

Group	Relative abundance	α-energy MeV	Excitation energy in ThC″ MeV
α_1	27	6·082	0
α_0	70	6·043	0·040
α_2	1·8	5·761	0·328
α^4	0·1	5·619	0·472
α_3	1·0	5·600	0·492

TABLE 16.1(*b*) Long-range α-particle groups in the ThC′→ ThD transition

Group	Relative abundance	α-energy MeV	Excitation energy in ThC′ MeV
α_0	10^6	8·776	0
α_2	35	9·489	0·725
α_3	20	10·417	1·671
α_1	170	10·536	1·793

16.1.4 Long-range alpha particles

A second type of structure in α-particle spectra, of a different origin, was also observed and investigated by Rutherford and his students. This is the emission of a weak group or groups of α-particles of considerably longer range than the main group of particles in certain α-transformations. This phenomenon is to be clearly distinguished from the fine structure referred to in Sect. 16.1.3 because the energy differences are much larger and the intensities are extremely small, perhaps 1 in 10^5, in comparison with small energy differences and comparable intensities for the main fine-structure groups. The long-range α-particles are associated with disintegrations of an *excited state of the initial nucleus*, as illustrated in Fig. 16.6 for the ThC→ ThC′→ ThD transformation. A β-particle process can populate highly excited states of its product nucleus, because, in contrast with α-emission, the probability of β-decay is not an extremely sharp function of energy. The β-decay of ThC therefore leaves ThC′ not only in its ground state but also in three excited states at 0·725, 1·671 and 1·793 MeV (with partial β-spectra corresponding to each of these transitions). The ground state of ThC′ decays by α-emission to ThD emitting a single group of particles, and the excited states of ThC′ may radiate to the ground state in the normal way. The excited states are also unstable to α-emission and since the energy release is large the barrier penetration factor is so far reduced that α-decay can compete with radiative emission. In about two cases in 10^4 a long-range α-particle is observed instead of a γ-ray or internal conversion electron; the actual intensities for ThC→ ThC′→ ThD are shown in Table 16.1(*b*).

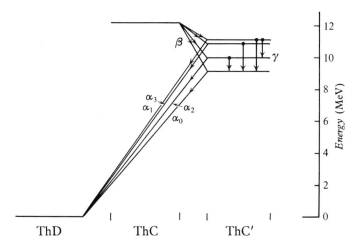

Fig. 16.6 Long-range α-particles in the ThC$\xrightarrow{\beta}$ ThC′$\xrightarrow{\alpha}$ ThD transition (Ref. 16.3).

The observation of long range α-particles has an important historical significance in connection with the estimation of radiative lives (Sect. 13.6.1).

The γ-ray spectra from the parent nucleus (ThC′ in the case discussed here) have also been shown to agree with the levels predicted from the energies of the long-range particles. For one group of long-range α-particles from RaC′ no associated γ-radiation could be found, although the corresponding conversion lines were already well known from the work of Ellis and the α-particle evidence suggested that the transition in RaC′ (of energy 1·414 MeV) should be strong. The phenomenon was interpreted by Fowler as a *total internal conversion* of the radiative transition, due to a spin change of 0→ 0 (Sect. **13.4**).

16.2 Theory of α-decay†

The partial width Γ_α for α-decay of an energetically unstable nucleus (or the decay constant $\lambda = \Gamma_\alpha/\hbar$) may be calculated using the method outlined in Sect. 15.2.5. We assume that an α-particle moves backwards and forwards in

† G. Gamow, *Zeits. Physik*, **51**, 204, 1928.

a simple potential well of radius R and depth U (Fig. 16.7) with uniform velocity v_α. The time τ_α between successive impacts on the barrier is

$$\tau_\alpha = \frac{2R}{v_\alpha} \tag{16.6}$$

and the probability of emission per unit time is therefore

$$\frac{1}{\tau_\alpha} T_0 = \frac{\Gamma_\alpha}{\hbar} \tag{16.7}$$

where T_0 is the barrier transmission coefficient. The internal velocity v_α is related to the velocity v with which the α-particle is observed when it has left the nucleus by the equation

$$\tfrac{1}{2}m_\alpha v_\alpha^2 = \tfrac{1}{2}m_\alpha v^2 + U = E_\alpha + U \tag{16.8}$$

where m_α is the mass of the α-particle and E_α its kinetic energy at a large distance from the nucleus. As discussed in Sect. **11.1** the α-particle leaves the barrier with zero kinetic energy and is accelerated by Coulomb repulsion to its final energy. Nuclear recoil is neglected in this calculation.

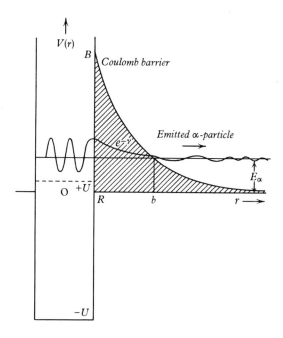

Fig. 16.7 Theory of α-decay. The wave representing the α-particle has a large amplitude within the nucleus and is attenuated exponentially in the region of negative kinetic energy ($R \leqslant r \leqslant b$). The emitted particle has total energy E_α, equal to the potential energy $zZe^2/4\pi\varepsilon_0 b$ at the point $r = b$. The internal nuclear potential is drawn at a level $-U$; for a heavy nucleus it is more realistic to take a positive value of internal potential, but the decay probability is not a sensitive function of this quantity.

The barrier transmission coefficient may be calculated as indicated in Sect. 14.2.3; the only difference in the present case is that the wave amplitude is large inside and small outside the nucleus, in contrast with the case of particle absorption. If we express T_0 as the product of a potential discontinuity factor at $r = R$ and a barrier penetration coefficient we have approximately

$$T_0 = \frac{4k'}{K} P_0 \qquad (16.9(a))$$

where k' is the (imaginary) wave number of the α-particle just outside the nuclear boundary at $r = R$ and K is the wave number within the potential well. If these wave numbers are expressed in terms of potentials we obtain

$$T_0 = 4 \sqrt{\frac{B - E_\alpha}{U + E_\alpha}} P_0 \qquad (16.9(b))$$

where $B = zZe^2/4\pi\varepsilon_0 R$ is the barrier height of the product nucleus.†

The barrier penetration coefficient P_0 represents the decay of *intensity* of the α-particle wave over the region of negative kinetic energies between $r = R$ and $r = b$, the point at which the α-particle leaves the Coulomb barrier. For s-wave emission the wave *amplitude* is a solution of the Schrödinger equation

$$\frac{d^2\psi}{dr^2} + \frac{2m_\alpha}{\hbar^2}\left(E_\alpha - \frac{zZe^2}{4\pi\varepsilon_0 r}\right)\psi = 0 \qquad (16.10(a))$$

and since $E_\alpha = \frac{1}{2}m_\alpha v^2 = zZe^2/4\pi\varepsilon_0 b$ this may also be written

$$\frac{d^2\psi}{dr^2} + \frac{m_\alpha zZe^2}{2\pi\varepsilon_0\hbar^2}\left(\frac{1}{b} - \frac{1}{r}\right)\psi = 0 \qquad (16.10(b))$$

For $R \leqslant r \leqslant b$ the coefficient of ψ is negative. We therefore assume a solution of the form

$$\psi \approx e^{-\gamma(r)}$$

where $\gamma(r)$ is a slowly varying function of r such that $d^2\gamma/dr^2$ may be neglected. From the physical nature of the problem we expect $d\gamma/dr$ to be positive.

Substituting in $(16.10(b))$ we obtain

$$\frac{d\gamma}{dr} = \frac{1}{\hbar}\sqrt{\frac{m_\alpha zZe^2}{2\pi\varepsilon_0}}\left(\frac{1}{r} - \frac{1}{b}\right)^{1/2}$$

and integrating this between the limits b and R (using a substitution $\cos^2 x = r/b$ and the limiting values $\gamma(b) = \gamma$, $\gamma(R) = 0$) we find

$$\gamma = \frac{zZe^2}{2\pi\varepsilon_0\hbar v}\left[\cos^{-1}\sqrt{\frac{R}{b}} - \sqrt{\frac{R}{b}\left(1 - \frac{R}{b}\right)}\right] \qquad (16.11(a))$$

---

† In Ref. 16.2 (Hanna) the internal potential U is taken positive, to represent the effect of the Coulomb force at small distances (cf. Fig. 11.1(d)). With this assumption the potential discontinuity factor becomes $4\sqrt{(E_\alpha - U)/(B - E_\alpha)}$ and the sign of U in (16.8) must be changed.

or, in terms of energies

$$\gamma = \frac{zZe^2}{2\pi\varepsilon_0 \hbar v}\left[\cos^{-1}\sqrt{\frac{E_\alpha}{B}} - \sqrt{\frac{E_\alpha}{B}\left(1 - \frac{E_\alpha}{B}\right)}\right] \qquad (16.11(b))$$

In the present approximation the barrier penetration factor for intensity is

$$P_0 = e^{-2\gamma}$$

and for the decay constant we obtain

$$\lambda = \frac{v_\alpha}{2R}T_0 = \frac{2v_\alpha}{R}\sqrt{\frac{B - E_\alpha}{U + E_\alpha}}\,e^{-2\gamma} \qquad (16.12)$$

The main dependence of the decay constant on α-particle energy arises from the exponential factor $e^{-2\gamma}$ in (16.12) and it is a good approximation to write

$$\log \lambda = \text{constant} - 2\gamma \qquad (16.13)$$

This is not the same as the empirical Geiger–Nuttall rule (16.3) but it exhibits the same sharp dependence of emission probability on energy. Observed decay constants are not very sensitive, in this model, to the internal potential U but depend mainly on the ratio E_α/B. For the heavy nuclei an increase of 1 MeV in E_α increases λ by a factor of about 10^5 while a 10 per cent increase in R (decrease in B) multiplies λ by 150.

In the limit of high barriers, $R \to 0$ and $B \to \infty$ effectively, and then

$$\gamma \longrightarrow \frac{zZe^2}{2\pi\varepsilon_0 \hbar v}\cdot\frac{\pi}{2} \qquad (16.14)$$

The factor determining α-emission is then

$$P_0 = e^{-2\gamma} \approx e^{-zZe^2/2\varepsilon_0 \hbar v}$$

the well-known *Gamow factor* already mentioned in Sect. 14.2.3.

The theory given so far is one-dimensional. If the α-particle carries angular momentum $l\hbar$, the centrifugal barrier appearing in (14.75) must be introduced into the penetrability calculation. The effect of this extra barrier on the emission probability is small in comparison with the effect of E_α or R; for a typical heavy nucleus, at the nuclear radius,

$$\frac{\text{Centrifugal barrier}}{\text{Coulomb barrier}} \approx 0{\cdot}002l(l+1)$$

in contrast with the situation for light nuclei, where centrifugal barriers are often larger than charge barriers. Formula (16.12) may be applied to estimate the 'decay constant without barrier' $v_\alpha/2R$ and the nuclear radius from the experimental data. If this is done values of the order of 10^{21} s^{-1} and $1{\cdot}5A^{1/3} \times 10^{-15}$ m are obtained (Sect. 11.3.2).

A more important use of the theoretical formula for α-decay is in the co-ordination of experimental data of the type displayed on energy-lifetime plots (Fig. 16.4), with the object of determining shell structure effects and hind-rance factors. The main points which emerge from this type of comparison are:

(i) *For even–even nuclei* ($I=0$ to $I=0$ with no parity change) transitions between ground states are well described by (16.12) and may be taken to represent normal α-decay. From (16.13) and (16.14) a relation of the form

$$\log t_{1/2} = a + \frac{b}{E_\alpha^{1/2}}$$

might be expected between half-life and decay energy and, as shown in Fig. 16.4, the isotopes of each element define a line on a plot of this type.

(ii) *For odd–odd nuclei* regularities in α-transition rates are less obvious than with even–even nuclei, but no transition is more probable than the 'normal' transition of even–even nuclei.

The notable success of the single-body theory of α-decay in accounting in a single formula for a range of lifetimes differing by a factor of 10^{25} should not divert attention from the fact that some of the assumptions of the theory are questionable. In particular the presence of preformed α-particles in the nucleus has been much discussed. Plausible evidence for the existence of α-clusters in the nuclear surface at least has been presented by Wilkinson.†

16.3 Experimental information on β-decay

Until 1957, when non-conservation of parity was discovered, study of the β-decay process was mainly a confirmation and interpretation of the charac-teristics of electron emission that had been established in the early days of radioactivity (ch. 2). We shall refer to the experiments of this period as the 'classical' experiments of β-decay although in fact there is little classical about the phenomenon. It has been responsible for the introduction of the radically new concept of the neutrino, and the objective demonstration of the existence of this particle may in a sense be said to complete the 'classical' period.

The emission of positive or negative electrons from nuclei, and the alter-native process of electron capture, determine the stability limits for nuclei throughout the periodic system as shown in Sect. **10.4**. The process changes the atomic number by ± 1, e.g. in the decay of radioactive sodium

$$^{22}_{11}\text{Na} \longrightarrow {}^{22}_{10}\text{Ne} + \beta^+ + \nu \tag{16.15}$$

stable neon is formed and the chemical identity both of the active nucleus and of its product have in some cases been verified. The electrons or positrons that arise in a typical β-decay such as (16.15), have a *continuous spectrum* of energies up to a definite limit determined by the mass change (Fig. 2.4) but

† *Proc. Rutherford Jubilee International Conference*, Manchester, 1961.

one β-particle only is emitted per nuclear disintegration. The homogeneous lines observed in many such spectra are due to internal conversion of γ-rays and are not directly connected with the β-decay process. When however, a β-transition leads to an excited state of its product nucleus, internal conversion lines may be used to identify the product, since they give accurate values of the X-ray energies of this body (Sect. **13.2**). The electrons emitted in β-decay have the same |e/m| as the ordinary atomic electrons and it has been established (by a search for X-rays) that the negative β-particles cannot form a new K-shell in a neutral atom. The β-particles are therefore identical with atomic electrons.

The techniques of β-spectroscopy have been outlined in Chapters 6 and 7; they provide information on nuclear energy changes of an accuracy comparable with or better than that obtained from heavy particle reactions. The data comprise:

(a) a survey of the *production and location of* β-*active species* in the N–Z diagram, establishing the limits of nuclear stability,
(b) *energy–lifetime relationships*, giving a general classification of decay data and offering information on nuclear structure, in particular on spin and parity changes,
(c) *detailed studies of spectrum shape*, and *correlations* between electrons and subsequent γ-radiation, giving information on nuclear decay schemes,
(d) 'modern' experiments (Sects. **16.7** and 21.3.3) including correlations between electrons and neutrinos, relating to the nature of the fundamental weak interaction.

16.3.1 Production of β-active nuclei

Many β-active bodies are found in the naturally occurring radioactive series. These are mainly negatron emitters because the nuclei concerned have an excess of neutrons. The lighter members of the thorium series provide well-known examples of naturally occurring β-emitters and of their relation to the associated α-decay (Fig. 2.9).

Both positron and negatron emitters are readily produced in nuclear reactions; (p, n) reactions such as

$$\ce{^{48}_{22}Ti} + p \longrightarrow \ce{^{48}_{23}V} + n + Q_1 \tag{16.16}$$

are usually endothermic (negative Q-value) and the product nucleus is generally positron or electron-capture active

$$\ce{^{48}V} \longrightarrow \ce{^{48}Ti} + \beta^+ + \nu + Q_2$$

or

$$\ce{^{48}V} + e_K^- \longrightarrow \ce{^{48}Ti} + \nu + Q_3$$

$$\left.\begin{array}{c}\\ \\ \end{array}\right\} \tag{16.17}$$

The radioactive decay reverses the neutron to proton change induced by the nuclear reaction and the observed positron energy may be calculated exactly from the energy threshold for the production of neutrons observed in re-action (16.16).

The (n, p) reaction produces negatron emitters, e.g.

$$^{32}_{16}S + n \longrightarrow {}^{32}_{15}P + p + Q_1 \tag{16.18}$$

which revert to the stable isobar with the emission of an electron

$$^{32}P \longrightarrow {}^{32}S + \beta^- + \bar{\nu} + Q_2 \tag{16.19}$$

and again the energy release in the forward and backward processes are found to agree.

Many other reactions produce unstable products, notably the (d, p), (d, n), (p, γ) and (n, γ) processes in which an extra nucleon is added. In practice the most prolific sources of negatron emitters for industrial and other purposes are made by neutron capture reactions in nuclear reactors, e.g. $^{59}Co(n, \gamma)^{60}Co$.

Addition products formed by heavy ion bombardments are, by contrast, neutron deficient and are either positron emitters or electron-capture bodies.

16.3.2 *Lifetimes of β-emitters*

The first attempt to classify β-emitting bodies was made by Sargent† who plotted the decay constant logarithmically against the maximum kinetic

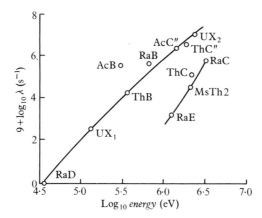

Fig. 16.8 Sargent diagram, showing relation between disintegration constant λ and β-decay (kinetic) energy for heavy radioactive nuclei.

energy in the β-spectrum for a number of naturally occurring nuclei. The Sargent diagram (Fig. 16.8) shows clearly a grouping of points about two lines which are now interpreted as 'allowed' and 'forbidden' lines, corresponding to different changes of nuclear spin as discussed in Sect. 16.5.2.

The discovery of artificial radioactivity has provided many hundreds of examples of β-active nuclei and the Sargent diagram may now be widely

† B. W. Sargent, *Proc. roy. Soc.*, A, **139**, 659, 1933.

extended. Since, however, according to (16.48) the decay constant may be written

$$\lambda \left(= \frac{0 \cdot 693}{t_{1/2}}\right) \propto f(Z, W_0) \qquad (16.20)$$

where Z is the atomic number of the product nucleus and W_0 is the maximum total energy of the β-spectrum in units of mc^2, a plot of log λ against log $(W_0 - 1)mc^2$ will obviously only give a straight line over a limited range of values of Z and W_0, for which $f(Z, W_0)$ may be considered nearly independent of Z and proportional to some power of $(W_0 - 1) \times mc^2$, the kinetic energy.

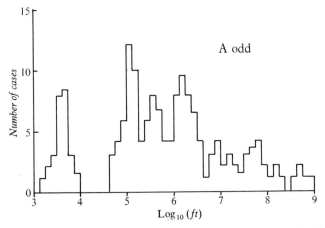

Fig. 16.9 Comparative half-life for β-transitions between ground states of odd-mass nuclei (Ref. 16.6).

Recent classifications are based on the *comparative half-life* $f(Z, W_0) \times t_{1/2}$ (or ft) for which use must be made of tabulated values of the function $f(Z, W_0)$ (Ref. 16.7); the half-life used must be obtained from the partial decay constant for the particular decay if there are competitive processes. The result of this classification for odd-mass nuclei is shown in Fig. 16.9.

   Well-defined groups in histograms such as this suggest that in such a group the nuclear factors in the decay probability are reasonably independent of Z and W_0. Such groups are seen

(i) *for log ft* $\approx$ 3–4. These are the especially probable transitions and are known as 'super-allowed' or favoured. Mirror nuclei and certain nuclei of mass $4n + 2$ fall into this class and it seems likely that the nuclear wave function remains practically unaltered as a result of the neutron $\leftrightarrow$ proton transition.

(ii) *for log ft* $\approx$ 4–5. These transitions are allowed in the sense of the selection rules (Sect. 16.5.2), but are unfavoured by comparison with the super-allowed group because of some change in nuclear wave function.

(iii) *for log ft* $\approx$ 6–10. This group includes the forbidden transitions (Sect. 16.5.2).

The connection between the *ft* groups and the selection rules for spin and parity changes can often be made with fair certainty from the predictions afforded by the single-particle shell model. Conversely β-decay data have supplied an important test of the predictions of the model for the spins of unstable nuclei (Ref. 16.8).

16.3.3 *The shape of β-spectra*

The theory of β-decay (Sect. **16.5**) predicts that the probability of emission of an electron with momentum between p and $p + \mathrm{d}p$ is

$$P(p)\,\mathrm{d}p \propto p^2 F(Z, p)(W_0 - W)^2\,\mathrm{d}p \tag{16.21}$$

where $F(Z, p)$ is a tabulated Coulomb correction factor known as the *Fermi function*. The energies W are measured in units of mc^2 and include the rest mass of the electron; p is measured in units of mc. A magnetic spectrometer set at momentum p will record an intensity proportional to $pP(p)$ since the slit width of the instrument is directly proportional to the momentum. If the momentum is varied over the spectrum then $P(p)$ can be obtained as function of p and from (16.21) a graph of the function

$$\sqrt{\frac{P(p)}{p^2 F(Z, p)}} \tag{16.22}$$

against total energy W should be a straight line, providing that proper correction has been made for electron screening of the nuclear charge. This *Fermi plot* (also known as a *Fermi–Kurie plot*, particularly when simplified forms of $F(Z, p)$ are used) extrapolates to the end point W_0 of the β-spectrum.

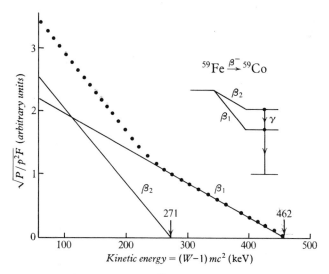

Fig. 16.10 Fermi plots of partial β-spectra of ^{59}Fe. There is also a third weak β-spectrum, not shown, leading to the ground state of ^{59}Co (Metzger, *Phys. Rev.*, **88**, 1360, 1952).

This is the standard analysis for determination of end points; values of $F(Z, p)$ are given in Ref. 16.7 and p is related to W by the relativistic equation

$$W^2 = 1 + p^2 \qquad (16.23)$$

Deviation of the Fermi plot from a straight line indicates either a forbidden transition, in which case the use of an appropriate correction factor produces a linear plot, or a complex spectrum involving transitions to more than one state of the residual nucleus, e.g.

$$
\begin{aligned}
{}^{59}\text{Fe} &\longrightarrow {}^{59}\text{Co} + \beta_1^- + \bar{\nu} + Q_1 \\
&\longrightarrow {}^{59}\text{Co*} + \beta_2^- + \bar{\nu} + Q_2
\end{aligned}
\qquad (16.24)
$$

for which the Fermi plot is shown in Fig. 16.10. In this case partial β-spectra with end points of 271 keV and 462 keV are found, and there is an associated γ-transition of energy $Q_1 - Q_2 = 191$ keV in the residual nucleus.

The measurement of the lifetime of the free neutron against β-decay and

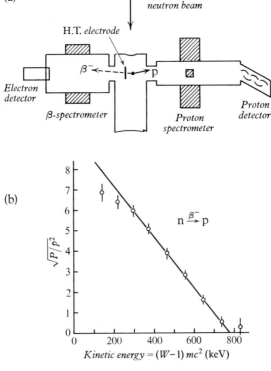

Fig. 16.11 Beta-decay of the neutron. (a) Apparatus for detection of recoil proton and decay electron from thermal neutrons decaying during the passage of a beam through a known volume. (b) Fermi–Kurie plot of the electron momentum spectrum, against kinetic energy (Robson).

the determination of the resulting β-spectrum (Robson†) are notable both for the technical elegance of the experiment and for the theoretical importance of the results. A slow neutron beam of 1.5×10^{10} neutrons s^{-1} from a reactor passed through a vacuum vessel (Fig. 16.11(a)) in which decay protons from the process

$$\text{n} \longrightarrow \text{p} + \beta^- + \bar{\nu} + 780 \text{ keV} \qquad (16.25)$$

were produced. The protons have initially a distribution of energies up to about 1 keV and they were accelerated by a voltage of 13 kV before detection by an electron multiplier. A lens-type magnetic spectrometer was used to focus protons of the expected energy from the decay region on to the first electrode of the multiplier. Electrons from the neutron decay were detected, also after magnetic analysis, by an anthracene scintillation crystal, and co-incidences between electron and proton signals were obtained by inserting a delay in the electron channel to allow for the time of flight of the slow protons. By careful determination of the volume from which protons were collected, the disintegration rate for a given neutron beam was found and the half-life of the neutron was calculated to be‡

$$t_{1/2} = 12.8 \pm 2.5 \text{ min.}$$

The momentum spectrum of the decay electrons was obtained in coincidence with the recoil protons and the Fermi plot (setting $F(1, p) = 1$) is shown in Fig. 16.11(b). The extrapolated kinetic energy endpoint is 782 ± 13 keV.

16.4 Neutrinos and antineutrinos

The continuous nature of the β-spectrum presented the first major problem for the theory of β-decay. The energy release in a nuclear transition is a definite quantity, given by the mass difference between the nuclei (A, Z) and $(A, Z+1)$ and this energy release accords well with the end point of the β-spectrum. This was first verified by considering the energies in the two branches of the ThC decay (Sect. 2.4.3); as shown below these agree if the upper limit of the β-spectrum is used.

$Q_\beta = 2.25$ ThC $Q_\alpha = 6.21$

β   α

ThC′   ThC″

$Q_\alpha = 8.95$ α $\beta^+ \gamma$ $Q_\beta = 2.37$

ThD $Q_\gamma = 2.62$

Total $\overline{11.20}$ MeV $\overline{11.20}$ MeV

Only one electron is emitted per β-disintegration, as may be seen by comparing successive α- and β-decays and the final nucleus cannot be left in a con-

† J. M. Robson, _Phys. Rev._, **83**, 349, 1951.
‡ The best recent value (Christenson _et al._, _Phys. Lett._, **26B**, 11, 1967) is $t_{1/2} = 10.80 \pm 0.16$ min.

tinuous series of residual states, since low-lying states are discrete and well-separated. The possibility that another absorbable type of radiation accompanied the decay electrons was eliminated in careful calorimetric experiments on RaE by Ellis and Wooster. They established that the energy of the absorbable radiation emitted agreed closely with the *mean* energy of the continuous spectrum rather than with the upper limit.† The unattractive possibility of the non-conservation of energy was seriously considered at this point.

A further difficulty arose when the neutron-proton model of nuclear structure became accepted. This predicts that the spins of odd-mass nuclei are always half-integral. The decay electron itself has half-integral spin, and β-decay changes only the charge and not the mass number of a nucleus. The conservation of angular momentum, as well as of energy, was therefore also questioned.

The way out of these difficulties was found by Pauli who suggested that a new light particle should be emitted in every β-transition. The simplest of all β-processes, the radioactive decay of the neutron, would then be represented by (16.25), and the converse process (which does not occur with free protons of course but is possible within a nucleus where energy is available) by:

$$\mathrm{p} \longrightarrow \mathrm{n} + \beta^+ + \nu - 1800 \text{ keV} \tag{16.26}$$

It is now customary following Fermi to describe the particle denoted by ν as a *neutrino* and the particle $\bar{\nu}$ as an *antineutrino* and to consider these as particle and antiparticle in the sense of the Dirac theory of the electron. We have indicated the emission of one or other of these particles in the equations for β-processes throughout this book.

The properties required of the neutrino are that it shall have:

(a) *zero charge*,
(b) *zero or nearly zero rest mass*, according to experiments on the shape of β-spectra near the upper limit (Sect. 16.5.2),
(c) *half-integral angular momentum* (component $\frac{1}{2}\hbar$ in order to conserve this quantity,
(d) *extremely small interaction with matter*, and hence essentially zero *magnetic moment* because of the failure of intensive experiments to show even the feeblest ionization caused by the passage of neutrinos through matter, and
(e) a definite *helicity*, in order to account for the 'modern' experiments on β-decay. Helicity is a two-valued quantity (± 1) which implies that the intrinsic spin momentum of a particle is either parallel or antiparallel to its direction of motion.

The neutrino differs from a light quantum in respects (c), (d) and (e), in the fact that it is not associated with an electromagnetic field, and because of the probable existence of a distinct antiparticle.

† Although the ThC branching is of historical interest, it may be noted that many cases of partial β-spectra and subsequent γ-radiation (e.g. process (16.24) now lead more convincingly to this conclusion.

Although the phenomenon of β-decay is itself the most compelling evidence for the existence of neutrinos, many attempts have been made to demonstrate the particle directly. The most convincing evidence for the fact that *linear momentum* may be carried away in β-decay, by a particle other than the electron and recoil nucleus, is to be found in beautiful photographs published by Szalay and Csikai. They introduced ^{6}He atoms into an expansion chamber and observed the process

$$^6_2\text{He} \longrightarrow \, ^6_3\text{Li} + \beta^- + \bar{\nu} \qquad (16.27)$$

Plate 15 shows that there is a large angle between the tracks of the electron and the recoil lithium nucleus as would be expected, owing to the finite neutrino momentum associated with all but the maximum energy electrons.

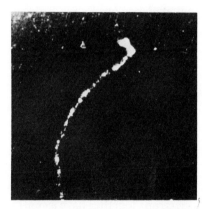

Plate 15 Expansion chamber photograph of the β-decay of ^{6}He, showing tracks of the decay electron and recoil nucleus (J. Csikai and A. Szalay, *Soviet Physics*, **8**, 749, 1959).

Another attractive possibility is to cause the neutrino to induce a β-process in the reverse direction. Nuclear reactors contain enormous quantities of β-decaying fission products and furnish an intense source of antineutrinos. In each β-decay, e.g. the decay of ^{131}I, a nuclear process of the type (16.19) occurs with the emission of the particle $\bar{\nu}$. If now these particles are used to bombard a suitable target, they might be expected to give rise to an observable yield of a radioactive product. The reaction tried† was

$$^{37}_{17}\text{Cl} + \bar{\nu} \longrightarrow \, ^{37}_{18}\text{A} + \beta^- \qquad (16.28)$$

and a search was made for 35 day ^{37}A decaying by the electron capture process

$$^{37}_{18}\text{A} + e^- \longrightarrow \, ^{37}_{17}\text{Cl} + \nu \qquad (16.29)$$

An extremely low, essentially zero, cross-section for the reaction was found.

† R. Davis, *Phys. Rev.*, **97**, 766, 1955.

This, however, does not necessarily mean that the neutrino does not exist, but only that the neutrino ν and the antineutrino $\bar{\nu}$ are different particles, like the positive and negative electron apart from charge. The true converse of reaction (16.29) is

$$^{37}\text{Cl} + \nu \longrightarrow {}^{37}\text{A} + \beta^{-} \qquad (16.30)$$

rather than the process shown in (16.28). A suitable source of neutrinos for initiating this reaction is the sun in which the nucleus ^{8}B is produced, among many others, by fusion reactions. Positron decay of ^{8}B yields a neutrino spectrum extending up to 14 MeV, and it is hoped that by working in a deep mine, cosmic ray background can be kept low enough to permit the unambiguous observation of neutrino-induced activity.

Evidence that the neutrino and antineutrino are distinct particles is also afforded by experimental evidence on double β-decay (Sect. **10.4**) between even-mass isobars. There are many examples of pairs of even–even nuclei which could decay in this way energetically, e.g.

$$^{124}\text{Sn} \longrightarrow {}^{124}\text{Te} + 2\beta^{-} + 2\bar{\nu} \qquad (16.31)$$

but the lifetime calculated for such a process is $\approx 10^{21}$ yr. If, however, the neutrino and antineutrino are the same particle, then the reaction can be written

$$^{124}\text{Sn} \longrightarrow {}^{124}\text{Sb} + \beta^{-} + \bar{\nu} \qquad (16.31(a))$$

$$\bar{\nu} + {}^{124}\text{Sb} \longrightarrow {}^{124}\text{Te} + \beta^{-} \qquad (16.31(b))$$

in which the neutrino emitted in the first stage is absorbed in the second stage. In this scheme the two electrons together have a constant energy and the lifetime is much shorter, $\approx 10^{17}$ years. Existing information, particularly for ^{130}Te (Sect. **10.4**), rules out the shorter lifetime. It therefore appears that (16.31(b)) is not a possible reaction and that it requires absorption of a ν particle rather than $\bar{\nu}$; there is then lepton conservation (Sect. 1.2.2).

A successful demonstration of an inverse β-process was made by Reines and Cowan† using a spectacular apparatus in which antineutrinos from a powerful reactor entered a liquid scintillator of volume $1\cdot4 \times 10^{3}$ litres. The reaction sought was the inverse of the neutron decay (16.25), i.e.

$$\bar{\nu} + \beta^{-} + \text{p} \longrightarrow \text{n} - 780 \text{ keV} \qquad (16.32)$$

using the protons of the liquid scintillator as target. Reaction (16.32) is equivalent, in terms of a Dirac theory for the electron and neutrino, to reaction (16.26) and also to the process

$$\bar{\nu} + \text{p} \longrightarrow \text{n} + \beta^{+} - 1800 \text{ keV} \qquad (16.33)$$

in which antineutrino absorption by a proton should release a neutron and a positron simultaneously if the antineutrino energy is above the threshold. In

† F. Reines and C. L. Cowan, *Phys. Rev.*, **113**, 273, 1959; F. A. Nezrick and F. Reines, *Phys. Rev.*, **142**, 852, 1966.

the experiment (Fig. 16.12) a 'prompt' signal was produced by the passage of the fast positron through the scintillator. The associated neutron (16.33) was slowed down by elastic nuclear collisions in the scintillator and was finally captured in a cadmium nucleus present in the scintillator in a suitable compound. The (n, γ) capture process produced a photon spectrum of total energy about 9 MeV per capture, which was detected by the liquid scintillator, with a delay time of up to 30 μs, corresponding to the slowing down process.

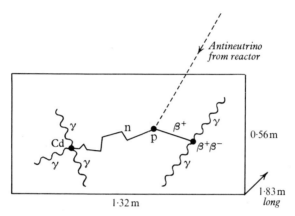

Fig. 16.12 Free antineutrino absorption cross-section. The diagram shows the principle of the liquid scintillator antineutrino detector in which a prompt pulse from the decay positron is followed by a delayed γ-pulse from capture of the associated neutron by a cadmium nucleus (Reines and Cowan).

The neutron and positron pulses were detected in delayed coincidence and the difference between the counting rate with the reactor on and off was taken to represent true antineutrino absorption events.

From the 'reactor associated' counting rate of 36 ± 4 events per hour the cross-section for the inverse β-decay was found to be

$$(11 \pm 2 \cdot 6) \times 10^{-48} \text{ m}^2 \tag{16.34}$$

The measurement of an effect with this extraordinarily small cross-section is a remarkable technical achievement. The result is consistent with theoretical expectation and the experimental verification of the Pauli neutrino hypothesis at last seems complete. In the same set of experiments the magnetic moment of the antineutrino was shown to be less than $10^{-9} \mu_B$.

16.5 The Fermi theory of β-decay

16.5.1 Formulation

Long before any direct attempt to verify the neutrino hypothesis had been made, Fermi† had embodied it in his theory of the β-process, which is the starting point for all theoretical studies of the subject and also the basis of the

† E. Fermi, *Zeits. Physik*, **88**, 161, 1934.

semi-empirical classification of experimental data (Sect. **16.3**). The Fermi theory does not explain the actual *rate* of β-decay, but it enables 'external' energy dependent factors to be removed from the experimental data.

The basic assumption of Fermi's treatment is that the light particles, the electron and neutrino, are created by the transformation of a neutron into a proton in a nucleus, or vice versa, in much the same way that a photon arises in an electromagnetic transition between energy levels of an atom or nucleus.

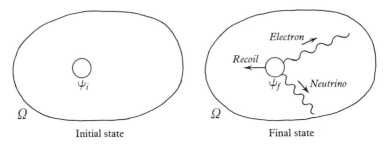

Initial state   Final state

Fig. 16.13 Beta-decay process. In the initial state a nucleon occupies a state with wave function ψ_i in a nucleus at $t=0$. In the final state a nucleon of opposite kind occupies a state with wave function ψ_f in the nucleus and electron and neutrino waves emerge. These waves are quantized within an arbitrary volume Ω.

The β-transition probability is obtained by applying the formulae of time-dependent perturbation theory (Ref. 3.1) to an initial system consisting of a single nucleon inside a volume Ω (Fig. 16.13). The final system is a different nucleon, an electron and a neutrino within the same volume; the nucleons in both initial and final systems may occupy states with wave functions ψ_i, ψ_f within a nucleus. The probability per unit time of the transition is

$$\frac{2\pi}{\hbar} |H_{if}|^2 \rho(E) \tag{16.35}$$

where H_{if} expresses the (unknown) potential or interaction relating the initial and final states, and 'causing' the transition, and $\rho(E)$ is the number of momentum states in the final system per unit energy range. In evaluating this it will be assumed that H_{if} is independent of the momenta of the electron and neutrino but that spin dependent factors are included in H_{if}. The electron and neutrino will in the present approximation be represented by plane waves. The mass of the recoil nucleus is large compared with the electron mass and it therefore takes up very little energy, but because of its momentum the neutrino momentum p_ν is not completely determined when the electron momentum p_e is given. If the total available energy (i.e. the upper limit of the β spectrum plus $m_e c^2$) is E_0 it is a good approximation to write

$$E_0 = E_e + E_\nu$$

where for the electron

$$E_e^2 = p_e^2 c^2 + m_e^2 c^4$$

and for the neutrino

$$E_\nu = cp_\nu$$

For a fixed electron energy E_e the total number of plane wave neutrino states in a volume Ω is

$$\Omega \int_0^{(E_0 - E_e)/c} \frac{4\pi p_\nu^2 \, dp_\nu}{(2\pi\hbar)^3} = \frac{\Omega 4\pi}{(2\pi\hbar)^3} \frac{(E_0 - E)^3}{3c^3} \qquad (16.36)$$

and the number of states per unit total energy interval is the derivative of this with respect to E_0. Allowing E_e to vary, it follows that the number of plane wave states available in the final system in the electron momentum range p_e to $p_e + dp_e$ is

$$\Omega \frac{4\pi p_e^2 \, dp_e}{(2\pi\hbar)^3} \times \Omega \frac{4\pi}{(2\pi\hbar)^3} \frac{(E_0 - E_e)^2}{c^3} \qquad (16.37)$$

This density of states will be used for $\rho(E_e)$ in (16.35) to predict the electron momentum spectrum, i.e. the probability per unit time $P(p_e) \, dp_e$ of emission of an electron in this momentum range. From (16.35) and (16.37) we thus obtain

$$P(p_e) \, dp_e = \frac{2\pi}{\hbar} |H_{if}|^2 \frac{4\pi p_e^2 \, dp_e}{(2\pi\hbar)^3} \Omega \frac{4\pi(E_0 - E_e)^2}{(2\pi\hbar)^3 c^3} \Omega$$

$$= \frac{|\Omega H_{if}|^2}{2\pi^3\hbar^7 c^3} p_e^2 (E_0 - E_e)^2 \, dp_e \qquad (16.38)$$

Apart from numerical factors, this formula contains:

(a) *the statistical factor* $p_e^2(E_0 - E_e)^2 \, dp_e$ which basically determines the shape of the β-spectrum. The distribution falls to zero for both high and low electron momenta; it would have been symmetrical if the electron and neutrino had been assumed of equal mass but for the known masses there are more electrons of low energy,

(b) *the factor* $|\Omega H_{if}|^2$ which relates the electron and neutrino emission to the associated nuclear transformation. The precise analytical form of the interaction cannot yet be predicted theoretically although its form is limited by relativistic laws. For the present we shall assume that this factor contains electron, neutrino and nuclear wave functions which are normalized in the volume Ω and that we may write

$$|\Omega H_{if}|^2 = g^2 |M|^2$$

where g is an arbitrary constant to be chosen to give the right strength for the β-decay and M contains details of the interaction so far unspecified. The general formula (16.38) then becomes

$$P(p_e) \, dp_e = \frac{g^2 |M|^2}{2\pi^3\hbar^7 c^3} p_e^2 (E_0 - E_e)^2 \, dp_e \qquad (16.39)$$

Frequently this is expressed differently by writing

$$p_e = p.m_ec \quad \text{and} \quad E_e = W.m_ec^2$$

so that

$$P(p)\,dp = \frac{m_e^5 g^2}{2\pi^3\hbar^7}\,|M|^2 c^4 p^2 (W_0 - W)^2\,dp$$

$$= \frac{G^2}{2\pi^3}\,|M|^2 p W (W_0 - W)^2\,dW \tag{16.40}$$

or, in terms of energies rather than momenta

$$P(W)\,dW = \frac{G^2}{2\pi^3}\,|M|^2 p W (W_0 - W)^2\,dW \tag{16.41}$$

where

$$G^2 = \frac{m_e^5 c^4}{\hbar^7}\,g^2$$

and W_0 is the total energy corresponding to the upper limit of the spectrum in units of m_ec^2.

16.5.2 Spectrum shape, decay rate and selection rules for β-decay

No account has been taken so far of the nuclear charge Ze of the product nucleus and when this is allowed for, the spectrum shape given by (16.40) or (16.41) differs according as positrons or electrons are emitted. Low-energy electrons are 'held back' and low-energy positrons are 'pushed forward' so that the momentum distributions are as illustrated in Fig. 16.14(a), and the energy distributions are as shown in Fig. 16.14(b). Analytically we write instead of (16.40):

$$P(p)\,dp = \frac{G^2}{2\pi^3}\,|M|^2 F(Z, p) p^2 (W_0 - W)^2\,dp \tag{16.42}$$

where $F(Z, p)$ is the Fermi function which corrects for finite nuclear charge. This is the formula on which the Fermi–Kurie plots of experimental data are based. The factor $F(Z, p)$ is essentially a barrier penetration factor and the non-relativistic approximation (14.76), with $\eta = Ze^2/4\pi\varepsilon_0\hbar v$ for positrons and $= -Ze^2/4\pi\varepsilon_0\hbar v$ for electrons may be used. In the case of a nucleus such as ^{64}Cu which undergoes dual decay the ratio of positrons to electrons of a given momentum is, using (16.42),

$$\frac{P_+(p)}{P_-(p)} = \frac{|M_+|^2}{|M_-|^2}\,\frac{F_+(Z, p)}{F_-(Z, p)}\,\frac{(W_0 - W)_+^2}{(W_0 - W)_-^2} \tag{16.43}$$

The variation of this ratio with p is thus mainly determined by the Coulomb factors, and this has been checked experimentally.

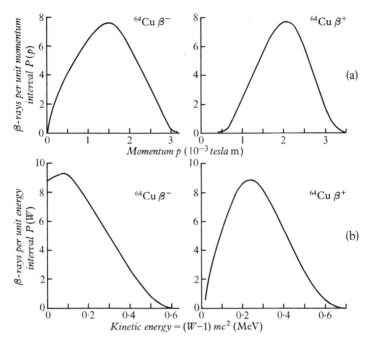

Fig. 16.14 Shape of β-spectra, illustrated by the case of ^{64}Cu, which shows dual decay. (a) Momentum spectrum for electrons and positrons. (b) Energy spectrum for electrons and positrons. The β$^-$ spectrum has a finite intensity at zero kinetic energy (Ref. 16.3).

The energy distribution of the electrons emitted in β-decay gives important evidence on the mass of the associated antineutrino. If the mass of this particle is zero the distribution (16.40) follows and this approaches the upper limit W_0 as $(W_0 - W)^2$, i.e. with a horizontal tangent (Fig. 16.15). On the other hand if the neutrino mass is finite, then near the upper limit the electron energy, electron momentum and neutrino energy ($\approx m_\nu c^2$) are all slowly varying and the neutrino may be treated non-relativistically, i.e.

$$p_\nu^2 = 2m_\nu E_\nu = 2m_\nu(E_0 - E_e) \qquad (16.44)$$

If this is used in (16.36), the density of states in this energy region becomes proportional to $(E_0 - E_e)^{1/2}$ or $(W_0 - W)^{1/2}$ and the electron momentum distribution $P(p)$ has a vertical tangent (Fig. 16.15). Accurate proportional counter and magnetic spectrometer studies of the tritium disintegration

$$^3\mathrm{H} \longrightarrow {}^3\mathrm{He} + \beta^- + \bar{\nu} \qquad (16.45)$$

(Curran, Angus and Cockroft, 1948; Langer and Moffatt, 1952) set an upper limit for the neutrino mass of

$$m_\nu < \frac{m_e}{2000} \qquad (16.46)$$

Similar limits have been obtained in proportional counter measurements with other low energy β-emitters. A direct determination of the energy balance in reaction (16.45)† is in agreement with these results.

If the mass of the neutrino is assumed to be zero, analysis of the shape of the spectrum near the end point can be used to check the suggestion of Weinberg‡ that in β-decay a neutrino may be absorbed from the Fermi level of a degenerate 'sea'. This would slightly extend the beta spectrum above its expected upper limit. No evidence for such an effect has been found.

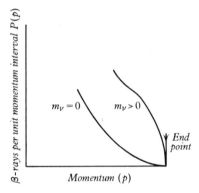

Fig. 16.15 Dependence of shape of β-spectrum near the upper limit on neutrino mass m_ν (not to scale).

Integration of formula (16.42) gives a value for the total probability per unit time of decay of the β-unstable nucleus. This is the decay constant and we write it, assuming the energy variation of $|M|^2$ to be small,

$$\lambda = \frac{\log_e 2}{t_{1/2}} = \frac{1}{\tau} = \int_0^{p_0} P(p)\,\mathrm{d}p$$

$$= \frac{G^2}{2\pi^3}\,|M|^2 f(Z, p_0) \tag{16.47}$$

or, if the Coulomb functions are given in terms of energy

$$\lambda = \frac{G^2}{2\pi^3}\,|M|^2 f(Z, W_0) \tag{16.48}$$

where $t_{1/2}$ is the half-life, τ the mean life for the decay and $f(Z, W_0)$ is an integral of the Fermi function over the spectrum and differs for positron emission, negatron emission and electron-capture. Abbreviating this integral to the symbol f we write

$$ft_{1/2} = \frac{2\pi^3}{G^2}\frac{\log_e 2}{|M|^2}\ \text{s} \tag{16.49}$$

where the quantity $ft_{1/2}$, usually written ft, is the comparative half-life used

† R. C. Salgo and H. H. Staub, *Nucl. Phys.*, **A138**, 417, 1969.
‡ *Phys. Rev.*, **128**, 1457, 1962.

in the systematization of experimental data (Sect. 16.3.2). The f-values used in this formula must take account of finite nuclear size and of the effect of the screening of the nuclear charge by atomic electrons. A small correction for radiative processes in the β-decay is also necessary, although this is not yet known precisely from theory (Sect. 21.3.4). For high energies, $E_0 \geqslant m_e c^2$, we have $p_e c \approx E_e$ and neglecting the Coulomb factors

$$\lambda = \int_0^{p_0} P(p) \, \mathrm{d}p \approx \int_0^{E_0} E^2 (E_0 - E)^2 \, \mathrm{d}E \propto E_0^5 \qquad (16.49(a))$$

The quantity $|M|^2$ depends theoretically on nuclear wave functions, taken together in a way which in turn depends on how the electron and neutrino are emitted. If these two particles emerge with their spins $\frac{1}{2}\hbar$ opposed we speak of *Fermi-type transitions*; if the spins of the light particles are parallel, giving a total spin angular momentum of $1\hbar$, we have *Gamow–Teller type transitions*.

Allowed transitions are those in which the light particles are emitted as an s-wave, and do not remove orbital momentum. The parity of the transforming nucleus consequently does not change. From the conservation of angular momentum we obtain the following *selection rules* for allowed β-transitions of a nucleus of spin I,

Fermi allowed: $\Delta I = 0$, no parity change,

Gamow–Teller allowed: $\Delta I = 0$, ± 1, no parity change, and in the latter case the transition $I = 0$ to $I = 0$ must be excluded since spin momentum must be carried away. Pure Fermi allowed transitions take place between such states, e.g.

$$^{14}\mathrm{O} \longrightarrow {}^{14}\mathrm{N}^* + \beta^+ + \nu \qquad (16.50)$$

while pure Gamow–Teller allowed decays are found when there is a spin change of 1, e.g.

$$^6\mathrm{He} \longrightarrow {}^6\mathrm{Li} + \beta^- + \bar{\nu} \qquad (16.51)$$

The examples (16.50) and (16.51) are in fact also *superallowed* according to the experimental ft values. This is because in these cases the initial and final nucleons occupy the same state of orbital motion and M has its maximum value. This is not in general so, as shown by the spread of ft values for the allowed group of transitions (Sect. 16.3.2).

For the allowed transitions in light nuclei at least we may write

$$|M|^2 = |C_F|^2 |M_F|^2 + |C_{GT}|^2 |M_{GT}|^2 \qquad (16.52)$$

where $|C_F|^2 + |C_{GT}|^2 = 1$ and M_F and M_{GT} are integrals over nuclear wave functions for the Fermi and Gamow–Teller interactions. Equation (16.49) may then be expressed in the form

$$ft = \frac{B}{(1-x)|M_F|^2 + x|M_{GT}|^2} \qquad (16.53)$$

where $x = |C_{GT}|^2$. The constants in this formula can be obtained from experimental results on nuclei such as ^{1}n, ^{3}H, ^{6}He and ^{14}O for which the quantities M_F and M_{GT} are known because of the simple nuclear structure. From these values the coupling constant g is found to be 1.410×10^{-62} J m^3 and the ratio $|C_{GT}/C_F|$ giving the relative strengths of the basic interactions governing Fermi and Gamow–Teller transitions, is 1.23.

If in a β-decay process the nuclear parity changes, as it would for a nucleon transition between an s-state and a p-state the light particles cannot be emitted as an s-wave. They originate effectively at a distance λ from the nuclear centre, where λ is the electron or neutrino wavelength, and the probability of emission is reduced by a factor of the order $(R/\lambda)^2$ in comparison with s-wave emission, where R is the nuclear radius. This is exactly analogous to the emission of high multipoles in radiation theory. The factor $(R/\lambda)^2$ is typically about $1/100$ for an average nucleus, and transitions of this type are known as *first forbidden*. When such weak processes are considered it is really unsatisfactory to use anything less than a fully relativistic treatment of the problem, and the selection rules become more complicated. For the non-relativistic case, however, the nuclear angular momentum and parity changes are summarized in the following table:

TABLE 16.2 Selection rules for β-decay (non-relativistic)

Types of transition	*Fermi rules*		*Gamow–Teller rules*	
	Spin ch.	*Parity ch.*	*Spin ch.*	*Parity ch.*
Allowed	0	No	$0, \pm 1$[a]	No
1st forbidden	$0, \pm 1$[a]	Yes	$0, \pm 1, \pm 2$[a]	Yes

[a] $0 \to 0$ transitions forbidden.

Forbidden transitions in general lead to β-spectra of 'non-allowed' shape and energy-dependent correction factors, predicted from a more detailed theory, must be introduced into (16.42).

16.6 Electron-capture decay

It has already been pointed out (Sect. **10.4**) that whenever positron decay can occur the capture of a K or L electron is an alternative process. This was first observed by Alvarez.† Electron capture is energetically favoured over positron emission by an amount $2mc^2$, because no positron rest mass has to be created, and the rest mass of the captured electron is added to the energy release. This energy is removed by a neutrino, e.g. the decay of ^{7}Be is written

$$^7_4\text{Be} + e^-_K \longrightarrow \, ^7_3\text{Li} + \nu + 0.86 \text{ MeV} \tag{16.54}$$

The neutrino energy is slightly less than that predicted by the mass change because the absorption of the K-electron of ^{7}Be by the nucleus leaves the

† L. W. Alvarez, *Phys. Rev.*, **52**, 134, 1937; **54**, 486, 1938.

product ^{7}Li atom neutral but with a K-shell vacancy. In the subsequent atomic rearrangement energy equivalent to the energy of the K-edge is radiated as X-rays or removed by Auger electron emission.

Equation (16.54) shows that electron capture is a two-body process, so that the neutrino would be expected to have a unique energy, in contrast with the distribution found in β-decay. This alters the formula corresponding to (16.48) for the decay constant since only the neutrino phase-space factor is required. The electron factor is replaced by a term representing the density of bound electrons at the nucleus and the decay constant for an allowed transition is expressed in the form

$$\lambda_K = \frac{G^2}{2\pi^3} |M|^2 f_K \qquad (16.55)$$

where (Ref. 16.3)

$$f_K \approx 2\pi(\alpha Z')^3 (W_0 + 1)^2 \qquad (16.56)$$

in which α is the fine structure constant ($=1/137$), W_0 is the *total* energy available for positron decay and Z' is the atomic number of the decaying nucleus. The energy of the shell from which the electron is taken is omitted in this approximate formula. In the decay of heavy neutron-deficient nuclei, electron capture is strongly preferred, not only because of the favourable energy release but also because a positron would have to penetrate a high potential barrier and the K-electron density in such nuclei is relatively large. In some cases when K-electron capture is energetically impossible, L-electron capture may still be possible because less energy is necessary to ionize the L shell than to ionize the K-shell of the product atom. The L-capture is always possible when K-capture takes place, and competition between the two has been observed (^{37}A). The ratio of decay constants for allowed electron-capture and positron emission is seen from (16.48) and (16.55) to be

$$\frac{\lambda_+}{\lambda_K} \approx \frac{f(Z, W_0)}{2\pi(\alpha Z')^3 (W_0 + 1)^2} \qquad (16.57)$$

which is independent of the nuclear factor $|M|^2$ so that the ratio is determined essentially by Coulomb effects. The observation of this ratio, which is known for a number of nuclei, is therefore a good test for β-decay theory.

Electron capture is usually observed by detection of the X-radiation from the product nucleus. For nuclei which can be studied in gaseous compounds, the proportional counter technique is very suitable and ratios λ_K/λ_+ of the order of 10^{-3} have been measured for light elements.† The special part played in this capture process by the electron density permits the decay constant to be influenced by chemical methods; it has for instance been found that for the process (16.54) the decay constants of ^{7}Be in the forms Be and

† J. L. Campbell, W. Leiper, K. W. D. Ledingham and R. W. P. Drever, *Nucl. Phys.*, **A96**, 279, 1967.

BeO differ by about one part in 10^4. A similar effect is found in low energy internal conversion (Sect. **13.2**).

The energy release in an electron capture such as (16.29) or (16.54) can be found by observing the energy of the recoil nucleus, which is typically a few eV, by means of a retarding potential. The energy can also be obtained by a time-of-flight method, using the K X-ray transition as a signal of zero time, and employing an electron multiplier as a detector of the slow recoil ion. It is also possible to obtain the disintegration energy by observing the energy distribution of a weak continuous radiation (inner bremsstrahlung) which always accompanies electron-capture. This is a radiative effect associated with the change of nuclear charge, and is involved in all β-decay processes.

16.7 Parity non-conservation and β-decay

Towards the end of 1956 an experiment was conducted† which had until then been thought unnecessary because there appeared to be no doubt about the result that would be obtained. Figure 16.16(a) shows the experimental arrangement; a source of ^{60}Co, a well-known β-emitter with the decay scheme shown in Fig. 16.16(b) was incorporated in the surface layers of a crystal of cerium magnesium nitrate, which possesses a strong internal magnetic field. The crystal was cooled by adiabatic demagnetization to 0·01 K and the ^{60}Co nuclei were then polarized (ch. 17) by applying a subsidiary magnetic field to the crystal. The polarization was checked and estimated by observing the anisotropy of gamma ray emission by means of counters A and B (Fig. 16.16(a), (c). The anthracene scintillation counter was then used to study the intensity of β-emission as a function of the direction of orientation (i.e. the direction of polarizing field), and the result shown in Fig. 16.16(d) was obtained. The asymmetry of β-emission and the anisotropy of γ-emission both disappeared as the crystal warmed up because of equalization of population of magnetic substates.

The demonstration of β-asymmetry in this experiment provided dramatic confirmation of the suggestion of Lee and Yang (Ref. 16.11) that parity might not be conserved in certain types of process, known as weak interactions, of which β-decay is one. A similar confirmation for another such process, the $\pi \to \mu \to$ e mesonic decay, was almost simultaneously forthcoming. To see why this experiment demonstrates parity non-conservation, we invoke the principle (ch. 17) that an asymmetric angular distribution always involves interference between amplitudes of opposite symmetry. The β-asymmetry therefore means that the transition between the ground state of ^{60}Co and the second excited state of ^{60}Ni (both of definite parity) can take place by emission of an electron–antineutrino pair in both odd (1^-) and even (1^+) parity states and that the two corresponding amplitudes interfere leading to an

† C. S. Wu, E. Ambler, R. W. Hayward, D. D. Hoppes and R. P. Hudson, *Phys. Rev.*, **105**, 1413, 1957.

angular distribution of electron emission of the form

$$a + b \cos \theta \tag{16.58}$$

with respect to the nuclear spin axis.

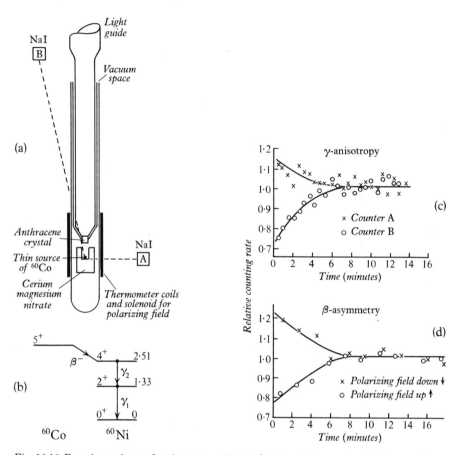

Fig. 16.16 Experimental test of parity conservation in β-decay. (a) Apparatus. (b) Decay scheme of ^{60}Co (Gamow–Teller decay). (c) Gamma-ray anisotropy obtained from counters A and B at different times as the crystal warms up. The difference between the curves measures the net polarization of the nuclei. (d) Beta-ray asymmetry shown by counting rate in the anthracene crystal for two directions of polarizing field (Wu *et al.*, 1957).

An alternative description of the β-asymmetry is that the measurement of the direction of emission of the β-particle in relation to the direction of the nuclear angular momentum vector defines the sense of a screw, and that one screw sense is preferred in β-decay to the other. In the particular case of ^{60}Co, and generally for emitters of negative electrons, the electrons prefer to come out in the direction opposite to the nuclear spin (Fig. 16.17(a)). That this circumstance violates parity conservation is easily understood in terms of the reflection principle, to which inversion plus rotation are equivalent. Nuclear spin is an axial vector but the linear momentum of an electron is a polar vector, and the reflection of such a system in a mirror leads to a different system (Fig. 16.17(a)).

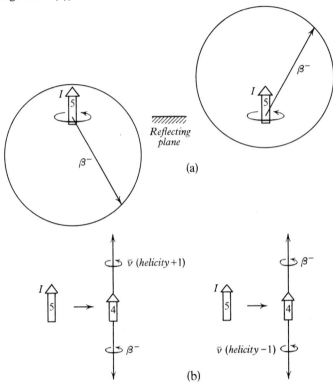

Fig. 16.17 Interpretation of ^{60}Co decay (allowed Gamow–Teller). (a) Reflection of the nuclear spin I (an axial vector) and the electron momentum (a polar vector) in a horizontal plane. The polar diagram is schematic only. (b) Explanation of the asymmetry in terms of a two-component neutrino theory. In order to change the nuclear spin by $\hbar$ the spins of the light particles must point along the nuclear axis. If the helicity of the antineutrino is prescribed only one of the two decay mechanisms shown is possible and the β-emission is asymmetric with respect to the axis of nuclear spin.

Another observable consequence of parity non-conservation in β-decay is that the electrons emitted from unaligned nuclei are longitudinally polarized. This may be seen simply in the case of an allowed Gamow–Teller transition (such as the ^{60}Co decay) when the electron–neutrino pair emerges along the axis of the nuclear angular momentum. Since this angular momentum has to change by one unit, the spins ($\frac{1}{2}\hbar$) of both electron and neutrino must also point along the nuclear axis. This has been proved in a number of elegant experiments in which the polarization is determined either by rotating the electron spin through 90° and conducting a scattering experiment designed to show left–right asymmetry, or by observing circular polarization of bremsstrahlung produced by the electrons as they slow down.

The intimate association of the electron and neutrino in β-decay makes it possible to account for the observed phenomena by attributing a parity non-conserving property to the neutrino and antineutrino. In the *two-component theory* of the neutrino it is specifically assumed that the neutrino has zero rest mass and that it is a left-handed particle, i.e. that its spin vector points in a direction opposite to its direction of motion. This is also expressed by saying that the neutrino has helicity -1: the antineutrino is a right-handed particle, with helicity $+1$. Experimental proof of this is discussed in Sect. 21.3.3.

The interpretation of the β-asymmetry experiment of Wu *et al.* in terms of the two-component theory of the neutrino is illustrated in Fig. 16.17(b). Many similar experiments have now been performed and the general effect of the discovery of parity non-conservation has been greatly to clarify our understanding of the β-decay process in particular and of weak interactions in general. Further discussion of the origin of the parity experiments and of their relevance to the weak interaction is given in Sect. 21.3.2.

References

1. SIEGBAHN, K. (ed.). *Alpha, Beta and Gamma Ray Spectroscopy*, North Holland, 1965.
2. RASMUSSEN, J. O. 'Alpha Decay' in Ref. 16.1, page 701; HANNA, G. C. 'Alpha Radioactivity', in *Experimental Nuclear Physics*, Vol. III, ed. E. Segrè, Wiley, 1959, p. 54.
3. EVANS, R. D. *The Atomic Nucleus*, McGraw-Hill, 1955.
4. DEUTSCH, M. and KOFOED-HANSEN, O. 'Beta-Rays', in *Experimental Nuclear Physics*, Vol. III, ed. E. Segrè, Wiley, 1959, p. 426.
5. KONOPINSKI, E. *Theory of Beta Radioactivity*, Oxford University Press, 1966.
6. FEENBERG, E. and TRIGG, G. 'Comparative Half-Lives', *Rev. mod. Phys.*, **22**, 399, 1950.
7. WAPSTRA, A. H. and NIJGH, G. J. *Nuclear Spectroscopy Tables*, North Holland Publ. Co., 1959; see also BHALLA, C. P. and ROSE, M. E. O.R.N.L. Report 2954 (1960).
8. MAYER, M. G., MOSZKOWSKI, S. A. and NORDHEIM, L. W. 'Nuclear Shell Structure and Beta-Decay', *Rev. mod. Phys.*, **23**, 315, 1951.
9. BYRNE, J. 'Radioactivity of Free Neutrons', *Contemp. Phys.*, **11**, 359, 1970.
10. RIDLEY, B. W. 'The Neutrino', *Prog. nucl. Phys.*, **5**, 188, 1956; LEWIS, G. M. *Neutrinos*, Wykeham Publications, London, 1970.
11. FRISCH, O. R. and SKYRME, T. H. R. 'Parity Non-Conservation in Weak Interactions', *Prog. nucl. Phys.*, **6**, 267, 1957.
12. BLACKETT, P. M. S. 'Non-conservation of Parity', *The American Scientist*, **47**, 509, 1959; see also The Rutherford Memorial Lecture, *Proc. roy. Soc.*, A, **251**, 293, 1959; P. Morrison, *Scientific American*, Vol. 196, April 1957.

17 Nuclear orientation

Nuclear spin affects the behaviour of nuclei in many phenomena and has already been introduced at several points in this book. In the present chapter we consider in rather more detail (*a*) the angular distribution of radiations arising from nuclei with predetermined spin directions, and (*b*) the production of systems of spin-oriented nuclei. The techniques described often permit the determination of the spin and magnetic moment of excited states, and thus supplement the methods described in Chapter 4 for determination of these properties for the ground state. It will be seen that many of the methods discussed in Chapter 4 will in fact lead to systems of oriented nuclei.

17.1 Definitions: polarization, orientation, alignment

The particles or quanta emitted from a radioactive source placed in a field free region are distributed isotropically in space. If the active nuclei have spin $\mathbf{I}$ (and magnetic moment μ_I), as defined in Sect. **4.1**, isotropic emission will result if the directions of the vectors $\mathbf{I}$, μ_I are randomly distributed, although the angular distribution of the radiation from a particular nucleus is not isotropic with respect to its spin axis (Sect. **3.5**). Quantum mechanically, isotropic radiation means that if a certain axis Oz is defined by a vanishingly small magnetic field (cf. Sect. **4.1**) then for the whole assembly of nuclei there are equal numbers with components of angular momentum $I(I-1)\ldots$ $-(I-1)$, $-I$ units of $\hbar$ along Oz. Thus for a single nucleus the probability $P(m)$ of the occurrence of any particular value $m\hbar$ of the resolved angular momentum† is independent of m and the system is unoriented (Fig. 17.1(a)). A given value of m defines one of the $(2I+1)$ magnetic substates, or independent spin states of the nucleus of spin I with respect to the axis Oz. If the magnetic field is zero, the *magnetic substates* all have the same energy and there is degeneracy with respect to m.

If $P(m)$ does depend on m then a system of nuclei is *oriented* (Fig. 17.1(b), (c)). If more spins point in one direction ($+Oz$) than in the opposite direction ($-Oz$), $P(+m)$ is not equal to $P(-m)$ and the system of nuclei is *polarized* (Fig. 17.1(b)); If $P(m)$ depends only on m^2 so that $P(m)=P(-m)$ the system is *aligned* (Fig. 17.1(c)); it has equal numbers of nuclei pointing in opposite directions and is therefore unpolarized but is nevertheless oriented with

---

† We shall omit the suffix I when it is not necessary to distinguish between nuclear and atomic or electronic spins.

respect to Oz since the populations of the spin states are not equal. The polarization P of a system may be defined as

$$P = \frac{1}{I} \Sigma m P(m) \tag{17.1}$$

and is then a number giving the excess of spins pointing in one direction. In the particularly simple case of a system of nuclei with spin $\frac{1}{2}$, of which N_1 are pointing 'up' ($m = +\frac{1}{2}$) and N_2 are pointing down ($m = -\frac{1}{2}$), we have

$$P(\tfrac{1}{2}) = \frac{N_1}{N_1 + N_2}, \qquad P(-\tfrac{1}{2}) = \frac{N_2}{N_1 + N_2}$$

and

$$P = \frac{2}{N_1 + N_2} (\tfrac{1}{2}N_1 - \tfrac{1}{2}N_2) = \frac{N_1 - N_2}{N_1 + N_2} \tag{17.2}$$

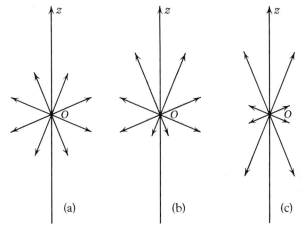

Fig. 17.1 Nuclear magnetic substates. Oz is an axis of quantization and the arrows represent allowed directions of orientation of a nuclear spin. The lengths of the arrows show the number of nuclei in an assembly with the particular orientation. (a) Unoriented assembly. (b) Oriented assembly (polarized). (c) Oriented assembly (aligned). The distributions are assumed to have axial symmetry about Oz.

In general, eq. (17.1) defines P to have a value between 1 and -1. Alignment as distinct from polarization is only meaningful for nuclei of spin greater than $\frac{1}{2}$; e.g. for spin 1, nuclei in the substate $m = 0$ form an aligned system. Orientation can arise in systems of nuclei of atoms bound in a solid lattice, in collision-free beams (as discussed in ch. 4), or in gases at low pressure. Although magnetic fields may be instrumental in the production of orientation, they are not essential to its existence. The detection of orientation depends on the nature of the system; oriented assemblies will usually show some anisotropy in subsequent emission of radiations but oriented beams are usually best studied in scattering or magnetic deflection experiments. In nearly all orientation phenomena the interaction between nuclear spin systems and their atomic environment is of great importance, both for the production and maintenance of their orientation.

17.2 Oriented nuclear systems

We consider first the most usual method of detecting nuclear orientation, and then the methods by which it has been produced.

17.2.1 Angular distribution of radiation from oriented nuclei

The semi-classical theory of radiation indicates that the radiation of multipolarity (L, M) emitted in a transition between nuclear states of spin $(I_i m_i)$

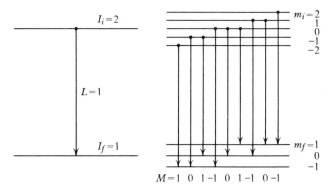

Fig. 17.2 Components of a radiative transition of multipolarity 1 between states of spin $I_i = 2$, $I_f = 1$.

and $(I_f m_f)$ has a definite angular distribution $F_L^M(\theta)$ with respect to the axis of quantization. The quantum numbers are related by the selection rules

$$I_i + I_f \geqslant L \geqslant |I_i - I_f|$$

$$m_i - m_f = M \qquad (17.3)$$

where $|M| \leqslant L$. The selection rule for M limits the number of components observable in the transition $I_i \longrightarrow I_f$ (Fig. 17.2). In optical transitions the spin states may be separated out clearly by application of a magnetic field, as in the Zeeman effect, and the different component transitions may be observed individually. In the nuclear case the separations obtainable are only of the order of $\mu_N B_i \approx 10^{-6}$ eV where B_i is the flux density of the internal field in the atom or lattice concerned. The transitions can therefore only be observed individually in special circumstances, such as those offered by the Mössbauer effect (Sect. 13.6.4). Usually all the line components are unresolvable experimentally, and the angular distribution observed must be obtained by adding the angular distributions of the components incoherently.

The probability of the vectors I_i, I_f combining to give a vector L, with particular magnetic quantum numbers, is given by a transformation amplitude or vector addition coefficient,† which may be written $\langle I_i I_f m_i m_f | LM \rangle$. The angular distribution arising from a given substate m_i of the initial level can then be written

$$W(\theta) = \sum_M \langle I_i I_f m_i m_f | LM \rangle^2 F_L^M(\theta) \qquad (17.4)$$

and if the initial substates have a weight $P(m_i)$, which depends on m_i, the total angular distribution is

$$W(\theta) = \sum_{M, m_i} P(m_i) \langle I_i I_f m_i m_f | LM \rangle^2 F_L^M(\theta) \qquad (17.5)$$

This must reduce to a constant if $P(m_i)$ is independent of m_i, i.e. if the initial levels are equally populated (random orientation). Thus, for example, if $I_i = 1$, $I_f = 0$, $L = 1$ (dipole radiation) we have, from the semi-classical theory (Sect. 3.5.2),

$$F_1^0 \propto \sin^2 \theta \qquad F_1^{\pm 1} \propto \frac{1 + \cos^2 \theta}{2}$$

and the vector addition coefficients are

$$\langle 010 \pm 1 | 1 \pm 1 \rangle = 1 \qquad \langle 0100 | 10 \rangle = 1$$

so that

$$W(\theta) = \sin^2 \theta + 2 \cdot \tfrac{1}{2}(1 + \cos^2 \theta)$$

$$= \text{constant, as expected.}$$

If the populations of the substates of the initial level are unequal, then even although the substates are not resolved, the resulting angular distribution is in general anisotropic. Such a distribution gives clear evidence of nuclear orientation and may provide information on I_i, I_f and L.

The functions $F_L^M(\theta)$ contain only Legendre polynomials of even order for pure multipole radiation so that we may also write

$$W(\theta) = a_0 + a_2 \cos^2 \theta + a_4 \cos^4 \theta + \cdots + a_{2L} \cos^{2L} \theta \qquad (17.6)$$

which again expresses the result already noted in Sect. **3.5** that no term in $\cos \theta$ of power higher than $2L$ appears. The distribution is symmetric with respect to the plane $z = 0$, a consequence of the conservation of parity, i.e. invariance under the inversion operation, for the electromagnetic interaction.

† Often known as a Wigner coefficient or Clebsch–Gordan coefficient. These coefficients are tabulated in Ref. 17.2.

17.2.2 *Orientation of nuclei in solids at low temperatures*

(*a*) *Principle.* The principle of all methods of obtaining nuclear orientation using low temperatures is to produce in a solid a sequence of energy levels characterized by different m_I values with respect to an axis of quantization. Unequal population of these levels by the nuclei of spin I arises if energy transfers between the spin system and the solid lattice establish thermal equilibrium, i.e. a Boltzmann distribution. The sequence of energy levels may be produced either by an externally applied magnetic field or by internal fields existing within the solid. The low temperatures required (≈ 0.01 K) are usually obtained by adiabatic demagnetization using a suitable crystal in

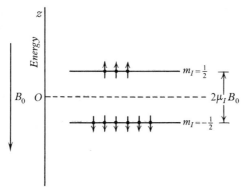

17.3 Separation of magnetic substates for a nucleus of spin $I=\frac{1}{2}$ and magnetic moment μ_I in a magnetic field B_0.

which the nuclei to be oriented are incorporated (Ref. 17.3). In such a system with a nucleus of spin $I=\frac{1}{2}$, a magnetic moment μ_I and a flux density B_0 (Fig. 17.3), the populations are

$$N_{1/2} \propto e^{-\mu_I B_0/kT} \qquad N_{-1/2} \propto e^{+\mu_I B_0/kT}$$

and from (17.2) the polarization is

$$P = \tanh\frac{\mu_I B_0}{kT} \tag{17.7}$$

This is appreciable only when $\mu_I B_0 \approx$ or $> kT$ and for μ_I equal to one nuclear magneton

$$B_0/T = 2.8 \times 10^3 \text{ tesla K}^{-1} \tag{17.8}$$

so that extremely low temperatures or high fields or both are necessary. This direct method of orientation has consequently been little used.

(*b*) *Orientation by hfs methods.* Gorter and Rose suggested independently in 1948 that the large internal magnetic fields ($B_i \approx 10$–100 T) existing in paramagnetic ions might be used in place of the large external field B_0 in (17.8) to produce nuclear orientation. If for instance the internal field is due to a single electron spin then at low temperatures this spin can be oriented by an external flux density of only 0·05 T because of the large electronic magnetic moment. The nuclear spin will follow the electron spin because of the hyperfine coupling and a nuclear polarization will exist in thermal equilibrium. This may be much larger at a given low temperature than is obtainable by the direct method because of the magnitude of B_i which increases the spacing between levels of different m_I. The internal fields concerned are those responsible for the hyperfine splittings in electron paramagnetic resonance (Sect. 4.3.2). The nuclei of diamagnetic elements may be oriented by hyperfine coupling in ferromagnetic metals.†

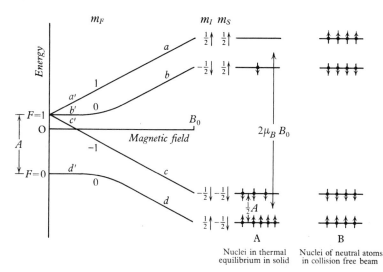

Fig. 17.4 Hyperfine structure of the ground state of an atom with $I=\frac{1}{2}$, $S=\frac{1}{2}$ in a magnetic field B_0. The figure is suitable for discussion of the hydrogen atom, or a paramagnetic ion with $S=\frac{1}{2}$. To the right of the diagram are shown the levels and nuclear spin directions at high fields ($>0·01$ T say). The spin symbols (A) represent the population of the levels in equilibrium at a temperature such that $kT \approx \frac{1}{2}A$. At B are shown the populations for a beam of neutral atoms.

For purposes of illustration we consider a simple system in which the internal field is due effectively to a single electron spin $S=\frac{1}{2}$ (cf. Sect. 4.3.2 and Fig. 4.3) and acts on a nucleus of spin $I=\frac{1}{2}$. The interaction energy will be written‡

$$Am_I m_S \tag{17.9}$$

where in this case $m_I = \pm\frac{1}{2}$, $m_s = \pm\frac{1}{2}$ and Oz is the axis of quantization defined by an external field. The energy levels of this system as a function of external field B_0 are shown in Fig. 17.4. At fields of about 0·05 T the vectors **I** and **S** are decoupled and the spin directions are shown in the figure. In states a, b,

† See for example I. A. Campbell *et al.*, *Proc. roy. Soc.*, A, **283**, 379, 1965.
‡ This is exactly as in eq. (4.11) except that it is customary to use A instead of a when the internal field may be due to more than one electron.

the electron spin is 'up', and in states c, d, it is 'down', and the energy separation between (ab) and (cd) is approximately $2\mu_B B_0$ where μ_B is the Bohr magneton. The energy separation between a and b, or between c and d, is $\frac{1}{2}A$ (putting $m_s = \frac{1}{2}$, $m_I = \pm\frac{1}{2}$ in (17.9)). At very small fields $F = I + S$ is a good quantum number and hyperfine pattern of states $a'b'c'd'$ of the type discussed in Chapter 4 arises.

In the Gorter–Rose method of orientation a sample is brought to a temperature T (≈ 0.01 K) such that $kT \approx \frac{1}{2}A$ in a residual field B_0 (Fig. 17.4). A nuclear *polarization* then exists because of preferential population of state d. In the *hfs* orientation methods proposed by Pound and by Bleaney single crystals in which anisotropic electric fields exist are used. The crystalline field provides an axis of quantization and causes splitting of the nuclear ground state so that no external field is required. Nuclear orientation results either from interaction between the electric field gradient and the nuclear quadrupole moment (Pound) or from an indirect interaction between the electronic and nuclear moments via the crystalline electric field (Bleaney). Since there is no external magnetic field, the energy in a particular state is unaltered by reversing both the nuclear and electronic spin and states with equal and opposite values of m_I have the same energy. These methods therefore only apply for $I > \frac{1}{2}$ and then give nuclear *alignment* rather than polarization in thermal equilibrium at temperatures of the order of 0.01 K. Since the angular distribution of γ-rays from a nuclear state involves only even powers of $\cos\theta$, aligned nuclei give the same anisotropic pattern as nuclei with the corresponding polarization, but asymmetries, such as those observed in the parity experiments (Sect. **16.7**), cannot be detected in this way. The angular distribution of γ-radiation with respect to the crystalline axis (or direction of applied field in the Rose–Gorter method) is usually characterized by an *anisotropy* ε where

$$\varepsilon = \frac{W(\frac{1}{2}\pi) - W(0)}{W(\frac{1}{2}\pi)} \tag{17.10}$$

and $W(\theta)$ is the intensity of radiation at an angle θ with the axis.

Figure 17.5 shows the cryostat used by Daniels *et al.*† in the first successful nuclear orientation experiment, and the results for ^{60}Co nuclei aligned in this way. The angular distribution of γ-radiation depends in a calculable way on the nature of the β-transition and on the spins of the states concerned in the γ-transitions. It also depends on the degree of alignment, i.e. the distribution of initial states with respect to m_I^2 and at a given temperature this is determined by the interaction constant A which involves the nuclear moment μ_I of the initial state and the internal fields. If the internal field can be calculated μ_I can be found; alternatively comparison can be made with results obtained with another isotope of known moment.

† J. M. Daniels, M. A. Grace and F. N. H. Robinson, *Nature*, **168**, 780, 1951; B. Bleaney, J. M. Daniels, M. A. Grace, H. H. Halban, N. Kurti, F. N. H. Robinson and F. E. Simon, *Proc. roy. Soc.*, A, **221**, 170, 1954.

Nuclear alignment experiments give information for single radiative transitions which is otherwise only obtainable for cascade transitions; the ratio of $E2$ to $M1$ components in a suitable transition may also be obtained. If the γ-radiation follows a β-transition the ratio of Fermi to Gamow–Teller type decays may be found.

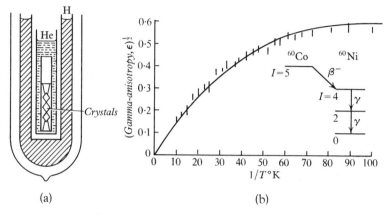

(a)   (b)

Fig. 17.5 Alignment of ^{60}Co nuclei. (a) Cryostat containing crystals of a Tutton salt (copper rubidium sulphate) surrounded by jackets of liquid helium and liquid hydrogen. The active cobalt nuclei are included in the crystal. (b) Gamma-ray anisotropy for the ^{60}Co decay plotted (as $\varepsilon^{1/2}$) against the inverse of the temperature. In this particular decay each of the two gamma rays is expected to have the same angular distribution. The curve corresponds to $I=5$, $\mu_I = 3\cdot5\pm0\cdot5$ nuclear magnetons for ^{60}Co (Bleaney *et al.*).

17.2.3 Orientation of nuclei by dynamical methods

These methods yield large polarizations in moderate magnetic fields (≈ 1 tesla) and temperatures (≈ 1 K); they have been especially useful in the production of polarized proton targets for accelerator experiments.

(*a*) *Microwave method* (Ref. 17.7). The energy separation between states (*ab*) and (*cd*) in Fig. 17.4 corresponds to a microwave quantum for magnetic fields of the order of 1 tesla. If sufficient power is supplied at the resonant frequency for one of the four possible transitions, the populations of the two states so connected can be equalized (saturation). This radically distorts the normal thermal distribution, and the combination of microwave power and thermal relaxation processes can lead to a much enhanced nuclear polarization. In practice the paramagnetic ion, with effective spin $S=\frac{1}{2}$, is generally neodymium and is contained within a lanthanum–magnesium nitrate crystal $La_2Mg_3(NO_3)_{12}24H_2O$ (LMN) whose water of crystallization supplies the protons to be polarized. The coupling between the Nd ion and the protons is weak and the states *a* and *b* in Fig. 17.4 are inverted, but the principle of saturation of a transition followed by relaxation still applies.

(*b*) *Optical pumping.* Kastler† has shown that free atoms, as distinct from

† A. Kastler, *Proc. phys. Soc. Lond.*, **67**, 853, 1954.

the atoms of condensed materials (which do not emit simple line spectra), can be oriented by absorption of resonance radiation, in the manner already discussed (Sect. 4.4.6) in connection with double resonance experiments. Since with hyperfine coupling the nuclear spin will follow the electronic spin, a nuclear orientation also exists, as in the Gorter–Rose method. In practice, this method requires the use of a vapour at a pressure of the order of 10^{-4} mmHg, or of a collision-free beam in order to minimize disorientation by collisions. The effect of small transverse magnetic fields may be serious, but can be minimized if a small axial field, in the direction of the incident light, is employed. If the pumping light is interrupted the oriented atoms will relax on the walls of a containing vessel. When the density of oriented atoms can be increased without destroying the orientation, as in the case of ^{3}He (Ref. 17.7, Baker *et al.*), this method is able to provide polarized targets.

17.3 Oriented particle beams (Ref. 17.6)

From the point of view of scattering and transmutation experiments, the production of oriented beams of bombarding particles is as interesting as alignment of target nuclei. We therefore consider briefly the means that have been used for providing sources of polarized nucleons, and the methods that are available for detecting polarization in particle beams. Since the nucleon spin is $\frac{1}{2}$, only polarization, as distinct from alignment, need be considered; for deuterons ($I = 1$) extra parameters are required.†

17.3.1 Magnetic separation of spin states

(*a*) *Atomic beam sources.* It was proposed by Clausnitzer, Fleischmann and Schopper‡ that a Stern–Gerlach type experiment with neutral hydrogen atoms could be used to provide a beam of atoms in which the nuclei were polarized. This can be understood from Fig. 17.4 which can be applied to the hydrogen atom ($I = \frac{1}{2}$, $J = S = \frac{1}{2}$). If a collision-free beam of neutral hydrogen atoms is passed through a uniform magnetic field B_0, the electron spins are aligned and the beam contains the four components *a, b, c, d*, in equal intensity, since thermal equilibrium is not established. The electronic magnetic moment is oppositely directed in states (*ab*) and (*cd*) and the beams (*ab*) can therefore be separated from (*cd*) by letting the magnetic field B_0 become non-uniform so that oppositely directed deflecting forces arise. The beam (*ab*) in the field B_0 has no net nuclear polarization but if this beam passes into a region of very low field, F becomes a good quantum number again because of the hyperfine coupling. Beam b' ($F = 1$, $m_F = 0$) becomes a mixture of the uncoupled states *b* and *d* and has no net polarization but beam a' is a pure state ($F = 1$, $m_F = 1$) and is fully polarized. The nuclear polarization of beams ($a'b'$) is therefore 50 per cent.

† S. E. Darden, 'Polarized spin one particles', *Amer. jnl. Phys.*, **35**, 727, 1967.
‡ G. Clausnitzer, R. Fleischmann and H. Schopper, *Z. Phys.*, **144**, 336, 1956.

A fully polarized beam can be obtained by use of the adiabatic passage method mentioned in Sect. 4.4.2. The incident beam, containing atoms in spin states *a* and *b* only, is exposed to the radiofrequency field as it passes through a region in which the transverse magnetic field varies smoothly from just below the nuclear resonance value to just above. This transfers atoms (by a 3-level quantum transition) from state *a* to state *c* and the resulting beam is 100 per cent polarized in nuclear spin. By use of a different frequency, atoms in state *b* can be transferred to *d* and this gives 100 per cent polarization of the opposite sign. Sources delivering up to 5 μA of polarized protons are available.

(*b*) *Lamb shift sources*. In these sources, which in fact also employ atomic beams, a state-selecting magnet is not necessary. The components (*c*, *d*) of the hyperfine pattern of hydrogen atoms excited to the metastable $2S_{1/2}$ state are in effect removed by a 'quenching' process. This takes place because at a certain magnetic field the $2S_{1/2}$ pattern of states crosses the similar hyperfine pattern of the $2P_{1/2}$ atomic state, which is separated from $2S_{1/2}$ at zero field by the Lamb shift. If an electric field is applied to the beam under these circumstances the $2S_{1/2}$ and $2P_{1/2}$ states become mixed and the rapid radiative decay of $2P_{1/2}$ transfers the atoms in the states (*c*, *d*) to the ground state. The beam then contains metastable atoms in states (*a*, *b*) and ground state atoms.

If now the beam is passed through argon gas a selective charge exchange process takes place in which the *metastable* neutral atoms preferentially become negative ions. This effectively rejects the ground state atoms and in weak magnetic field yields a beam with 50 per cent nuclear polarization, as in Sect. 17.3.1(*a*). Beams of 200 nA have been obtained, both for protons and deuterons and the negative ions are suitable for direct injection into tandem accelerators. The metastable neutral beam in Lamb shift sources is also produced by a highly efficient charge exchange process which takes place between H^+ ions of energy 500 eV and caesium atoms.

17.3.2 *Polarization by elastic scattering*

It is known from the success of the single-particle shell model (Sect. **9.4**) that the energy of a nucleon in a bound state depends on the relative orientation of its orbital and spin momenta. The same spin–orbit effect may be expected in unbound states, i.e. in the scattering of nucleons by complex nuclei. This introduces an asymmetry into the scattering of polarized nuclei, as may be seen from Fig. 17.6. Nuclei of given spin (**s**) orientation passing to right or to left of a nuclear centre *O* generate orbital vectors **l** of opposite sign. If the scattering depends on the sign of the scalar **s**.**l** the contributions to the scattering at a given angle from the two sides of the nucleus will be unequal. Interference between this one-sided scattering and any background scattering which may be present leads to an asymmetric angular distribution.†

---

† See the discussion of the high energy case given by Fermi, *Nuovo Cim. Suppl.*, Ser. 10, Vol. 2, 1955, p. 92.

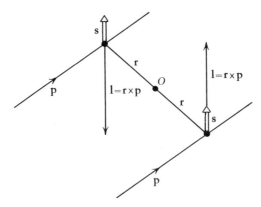

Fig. 17.6 Asymmetry of interaction of polarized nucleons with a nucleus, centre O.

Particles with spin 'up' are then preferentially scattered to the left or right and it may easily be seen (Fig. 17.7) that the elastic scattering of an unpolarized beam leads to scattered beams of equal intensity at a given angle of scattering but of opposite polarization. The existence of the polarization may be demonstrated by a second scattering, after which an asymmetry between the left and right scattered beams will be observed.

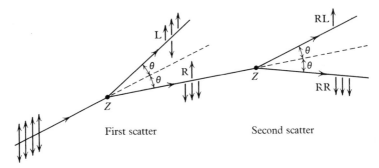

Fig. 17.7 Polarization in elastic scattering of unpolarized nucleons by a nucleus Z. After one elastic scattering the left (L) and the right (R) scattered beams have equal intensity but complementary polarization. If the right scattered beam is again scattered elastically the left (RL) and right (RR) beams have different intensity.

The first demonstration of the production of polarized beams was given by Heusinkveld and Freier† who studied the double scattering of 3·5 MeV protons by helium. The spin–orbit asymmetry arises in this case because the compound nucleus ^{5}Li has a ground state and first excited state which form a p-doublet with spins $\frac{3}{2}$ and $\frac{1}{2}$ respectively ($j = l \pm s$). If the energy of a proton incident upon a helium nucleus is such that one of these states is preferentially formed, interaction is stronger for one $\mathbf{s}.\mathbf{l}$ orientation than for the other, and a polarized beam results. The interference here is between the resonant $p_{3/2}$

† M. Heusinkveld and G. Freier, *Phys. Rev.*, **85**, 80, 1952.

or $p_{1/2}$ wave and the non-resonant s-wave background. The polarization was checked by a second scattering in helium, and was found to agree with that predicted from the angular distributions of elastic scattering, which can be analysed to give phase shifts for the s- and p-waves. The helium scattering can give proton polarizations of 100 per cent at certain angles and energies, and has been much used as a polarization analyser.

In the elastic scattering of protons of energy 10–20 MeV by heavier nuclei polarization arises because the potentials effective for left-sided and right-sided scattering are different owing to the different sign of the spin–orbit term $\mathbf{s}.\mathbf{l}$. Polarization of slow neutrons by passage through magnetized iron (as used in the determination of the magnetic moment of the neutron (Sect. 4.4.5)) also arises because of an interference effect, in this case between nuclear scattering and scattering from aligned atomic electrons.

17.4 Angular distribution and correlation experiments

17.4.1 Introduction

The observation of angular distributions and polarizations of nuclear radiations adds important information to that obtainable from total cross-sections or excitation functions. This information relates to

(*a*) the spin, parity and electromagnetic moments of nuclear levels, and
(*b*) the reaction mechanism.

There are two types of experiment from which this information is obtained; namely the *angular correlation experiment* (Sect. 17.4.2) in which two quanta appear in rapid succession in a nuclear decay scheme, and the *angular distribution, or differential cross-section experiment* in which a nuclear reaction takes place. This division of phenomena is not comprehensive; many processes can be described in either way, and many more complicated processes, such as triple correlations, have been studied. A common feature however is that either the emission of a quantum, or the incidence of a bombarding particle, defines a direction, or axis of quantization, and the distribution of intensity of emission of the second radiation round this axis can then be observed. If the second radiation is a particle the angular distribution is determined by the orbital angular momenta which it is permitted by selection rules to remove; if it is a photon, particular angular distribution patterns arise for each permitted multipole.

17.4.2 Angular correlation of successive radiations

Many excited nuclei decay by emitting a cascade of two or more gamma rays. This is especially noticeable when the state excited (I_2, Fig. 17.8(a)) has a spin which differs by several units of $\hbar$ from that of the ground state and levels of intermediate spin (I_1, Fig. 17.8(a)) lie between I_2 and the ground state I_0.

In such cases a succession of radiations of low multipolarity is more probable than a more energetic 'cross-over' transition of high multipolarity. In the particular case of a radiation γ_a followed by γ_b the observation of γ_a in a particular direction picks out a set of nuclei in states I_1 with an anisotropic distribution of spin directions. If this distribution of spins persists until γ_b is emitted and if γ_b is observed in coincidence with γ_a then an angular distribution of the form (17.6) is obtained.

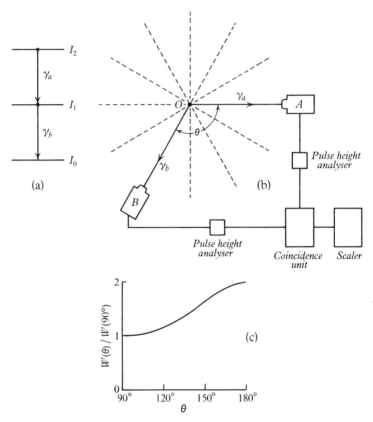

Fig. 17.8 Angular correlation of successive radiations from a radioactive nucleus at O. (a) Decay scheme. (b) Experimental arrangement of counters and coincidence unit. (c) Angular correlation $W(\theta) = 1 + \cos^2 \theta$.

That such a correlation exists may be seen in a simple case in a way indicated by Moon.† Let γ_a and γ_b both be electric dipole transitions and let $I_2 = I_0 = 0$, $I_1 = 1$ and let the intrinsic angular momentum of photons be disregarded. If OA (Fig. 17.8(b)) is the direction of γ_a the nuclear spins $I_1 = 1\hbar$ after this transition are aligned perpendicular to the direction OA. The photon γ_b removes this spin and is therefore emitted in a plane perpendicular

† P. B. Moon, *Artificial Radioactivity*, Cambridge University Press, 1949, p. 98.

464

to it with equal probability for all angles θ in this plane. A counter placed in this plane and operated in coincidence with counter A would show a coincidence rate independent of θ. Since however the direction of the intermediate angular momentum is actually random with respect to the plane OAB, an average for all such directions must be taken. If this is done by imagining the γ-ray distribution to rotate about the axis OA, the counters remaining fixed, it is clear that the coincidence rate will be a maximum for $\theta = 0$ and $\theta = \pi$.

In the general case we have for the relative population of the substates of the intermediate level I_1

$$P(m_1) = \sum_{m_2} \langle I_1 I_2 m_1 m_2 | L_a M_a \rangle^2 F_{L_a}^{M_a}(\theta) \qquad (17.11)$$

where

$$M_a = m_2 - m_1 \qquad (17.12)$$

A simplification results if we choose the axis of quantization to coincide with the direction of propagation of γ_a, since the orbital angular momentum of the photon has no component along this axis, and the intrinsic angular momentum only has components $\pm \hbar$. With this choice only the functions $F_{L_a}^1(\theta)$ and $F_{L_a}^{-1}(\theta)$ for $\theta = 0°$ appear in (17.11). The directional correlation between γ_a and γ_b is now easily seen to be (from (17.5) and (17.11)),

$$W(\theta) = \sum_{m_0 m_1 m_2} \langle I_1 I_2 m_1 m_2 | L_a \pm 1 \rangle^2 F_{L_a}^{\pm 1}(0)$$

$$\times \langle I_0 I_1 m_0 m_1 | L_b M_b \rangle^2 F_{L_b}^{M_b}(\theta) \qquad (17.13)$$

In the particular example quoted with $I_2 = I_0 = 0$, $I_1 = 1$ the transitions permitted when γ_a is observed at angle $\theta = 0°$ are $(I, m) = (0, 0) \rightleftarrows (1, \pm 1)$. The Clebsch–Gordan coefficients are unity (as always when one of the combining vectors is zero) and from (17.13) the angular correlation is given by

$$W(\theta) = F_1^{\pm 1}(0)[F_1^1(\theta) + F_1^{-1}(\theta)]$$

$$\propto 1 + \cos^2 \theta$$

in agreement with the type of variation predicted by the simple model.

Similar calculations may be made for more complicated cascades but the numerical evaluation of the correlation function becomes very tedious because of the large number of summations over magnetic quantum numbers which are necessary. Fortunately it is possible to avoid this through algebraic methods developed by Racah and it is now customary to write the general angular correlation function for transitions of pure multipolarity in the form

$$W(\theta) = \sum A_\nu P_\nu(\cos \theta) \qquad (17.14)$$

where

$$\nu = 0, 2, 4, \ldots$$

and

$$A_v = F_v(L_a I_2 I_1) F_v(L_b I_1 I_0) \qquad (17.15)$$

The functions F_v are tabulated for the simpler cases by Biedenharn and Rose† and depend on the multipolarity of the successive transitions and on the spins (but not parities) of the levels.

A typical experimental arrangement for the observation of angular correlations in cascade transitions is shown in Fig. 17.8(b). The detector pulses are combined in the coincidence unit and the number of coincidences $W(\theta)$ is determined as a function of the angle between the counter axes. The anisotropy

$$\varepsilon = \frac{W(\tfrac{1}{2}\pi) - W(0)}{W(\tfrac{1}{2}\pi)} = \frac{W(90°) - W(180°)}{(W(90°)}$$

can be compared with that predicted by (17.14), or a complete curve for $W(\theta)$ (Fig. 17.8(c)) may be obtained.

The angular correlation function involves only the spins of the nuclear levels concerned and not their parities. The parities of electric and magnetic radiation of the same multipolarity are however opposite, i.e. their electric vectors are in perpendicular planes with respect to the radiating moment and if counter B is replaced by a polarization-sensitive detector and the polarization at a given angle (other than 0° or 180°) determined, the parity change may be established.

It will be observed that the formulae developed in this section apply equally well if the first transition is an absorption rather than an emission process. The angular distribution of nuclear resonant scattering (Sect. 13.6.4) about the direction of the incident radiation is therefore given by (17.14) for the spin sequence $I_0 \longrightarrow I_1 \longrightarrow I_0$.

Directional correlation experiments of this type have contributed to the analysis of many decay schemes. The information obtained is only directly comparable with theory when the lifetime of the intermediate state is small compared with the times required for the nuclear moments (magnetic dipole or electric quadrupole) to alter their orientation in internal fields. The attenuation of angular correlation functions in cases $(\tau > 10^{-8}\text{ s})$ where this is not so may be used to study the electric fields existing in solids and liquids. If the internal fields are known the attenuation may be used to measure both magnetic moments (gyromagnetic ratio) and electric quadrupole moments of excited states. The gyromagnetic ratio may also be obtained by using external magnetic fields to cause appreciable spin precession within the lifetime of the state. It is even possible in certain special cases, to re-orient the excited nuclei by nuclear resonance methods (NMR). A review of some of these perturbed angular correlation (PAC) techniques is given in Ref. 17.8; this shows that gyromagnetic ratios have been measured for lifetimes down to 10^{-9} s in many cases.

† L. C. Biedenharn and M. E. Rose, *Rev. mod. Phys.*, **25**, 729, 1953, Table I. See also Ref. 17.10.

17.4.3 *Angular distribution in a two-stage reaction*

Extension of the methods of the preceding section to a two-stage nuclear reaction is straightforward and will not be discussed here. Tables of coefficients for the analysis of angular distributions are given in Ref. 17.10; they allow for coherence resulting from the partial wave decomposition of the incident plane wave.

A number of general results limit the complexity of patterns observed in angular distribution experiments. These are, for an unpolarized beam:

(*a*) Angular distributions are *isotropic* if $l = 0$ for the incident or emitted particle or if $\mathcal{J} = 0$ or $\frac{1}{2}$ for the compound state, since then the magnetic substates are equally populated.

(*b*) Anisotropic angular distributions are *symmetric* if only one compound state is involved or more states but of the same parity. They are *asymmetric* if states of opposite parity can be formed, at a given energy, for the same channel spin since there is then interference between waves with even and odd values of *l*.

(*c*) If only incoming waves of orbital angular momentum *l* contribute appreciably to a reaction, the angular distribution of the outgoing particles in the centre of mass system is an even polynomial in cos θ with an exponent not higher than $2l$. Multipole radiation of order L behaves in this respect as a particle of angular momentum L so that if any reaction dipole radiation appears the angular correlation cannot be more complicated than $a + b \cos^2 \theta$.

References

1. SIEGBAHN, K. (ed.), *Alpha, Beta and Gamma Ray Spectroscopy*, p. 997, North Holland, 1965.
2. WAPSTRA, A. H., NIJGH, G. J. and VAN LIESHOUT, R. *Nuclear Spectroscopy Tables*, North Holland, 1959.
3. KURTI, N. 'Spins and Cryogenics', *Contemp. Phys.*, **8**, 21, 1967.
4. BLIN-STOYLE, R. J., GRACE, M. A. and HALBAN, H. 'Oriented Nuclear Systems', *Progr. nucl. Phys.*, **3**, 63, 1953; BLIN-STOYLE, R. J. and GRACE, M. A. 'Oriented Nuclei', *Encyclopaedia of Physics*, vol. **42**, p. 555, Springer, 1957.
5. DANIELS, J. M. *Oriented nuclei; polarized targets and beams*, Academic Press, 1965.
6. DICKSON, J. M. 'Polarized ion sources and the acceleration of polarized beams', *Progr. nucl. tech. Instrum.*, **1**, 103, 1965; HAEBERLI, W. 'Sources of polarized ions', *Ann. rev. nucl. Sci.*, **17**, 373, 1967.
7. JEFFRIES, C. D. 'Dynamic Orientation of Nuclei', *Ann. rev. nucl. Sci.*, **14**, 101, 1964; 'Dynamic nuclear polarization in crystals', *Proc. roy. Soc.*, A, **283**, 471, 1965; S. D. BAKER, D. H. MCSHERRY and D. O. FINDLEY, *Phys. Rev.*, **178**, 1616, 1969.
8. ADLER, K. and STEFFEN, R. M. 'Electromagnetic Moments of Excited Nuclear States', *Ann. rev. nucl. Sci.*, **14**, 403, 1964.
9. FERGUSON, A. J. *Angular Correlation Methods in Gamma Ray Spectroscopy*, North Holland, 1965.
10. FERENTZ, M. and ROSENZWEIG, N., Ref. 17.1, page 1687.

E

The basic interactions
of nuclear physics

18 Nuclear forces

This book has so far been mainly concerned with the static properties of complex nuclei (Part C) and with the behaviour of such nuclei in nuclear reactions (Part D). Nuclear models are used in coordinating these phenomena since insufficient is known about nuclear forces to permit exact calculations for a many-body problem even if such calculations are technically possible.† It is of course necessary that nuclear models should be consistent with what is known about forces between nucleons, and it is in any case interesting to see how far one can proceed with detailed calculations for simple systems. In this chapter a brief account will be given of present-day knowledge of nuclear forces, and of their connection with nuclear properties. In surveying this field two approaches are useful:

(a) one may derive as much information as possible on internucleon forces by studying the properties, and particularly the regularity of properties, of complex nuclei, and

(b) one may examine the behaviour of two-nucleon systems, both in the bound state (the deuteron) and in scattering. This will involve experiments extending up to high energies at which meson production is important.

As a result of these investigations it is now possible to give an account of the main features of the force between nucleons. It would of course be intellectually more satisfactory to derive the law of nuclear force from a more fundamental theory based on the properties of mesons, in the way that the Coulomb law of force may be associated with a photon field. Some progress has been made in this direction.

The properties of the force between nucleons inferred from a study of complex nuclei are not necessarily those that would be expected for a pair of nucleons in free space. It remains to be proved that the internucleon force is unaffected by the presence of other nucleons near by, or that specifically nuclear many-body forces, depending on the presence of three or more nucleons, do not exist. This possible complication will be disregarded in this chapter.

It will be assumed that the fundamental constituents of a nucleus are neutrons and protons. The reasons why electrons cannot play a part in nuclear structure have already been discussed (ch. 9).

† An example of such calculations is to be seen in the work of Masterson and Lockett, *Phys. Rev.*, **129**, 776, 1963.

18.1 Information on nuclear forces from study of complex nuclei

18.1.1 Saturation property

Measurements of the size (ch. 11) and of the total binding energy (ch. 10) of nuclei show that to a fair approximation all nuclei with $A > 40$ have the same density and the same binding energy per nucleon; i.e. the nuclear radius may be written

$$R = r_0 A^{1/3} \tag{18.1}$$

with $r_0 = 1 \cdot 2$ fm, and the total binding energy is

$$B_A \propto A \tag{18.2}$$

for a nucleus of mass number A. These facts imply that the nuclear force *saturates*, i.e. that the number of nucleons with which a given nucleon interacts strongly is limited.

To see this we recall (Sect. **9.1**) that the kinetic energy of a Fermi gas of A nucleons is proportional to $A^{5/3}$ and the potential energy to $\frac{1}{2}A(A-1)$ if all pairs of nucleons interact attractively. The most stable state of such a nucleus will be one in which it has collapsed to a size of the order of the range of nuclear forces, and for a heavy nucleus the binding energy B_A will then follow the potential energy and increase as A^2. It is of course known from the scattering experiments and from calculations of the binding energy of simple systems (as first emphasized by Wigner) that the nuclear force is of short range (2×10^{-15} m, Fig. 18.1), and a collapsed nucleus will thus have too small a radius. It therefore becomes necessary to limit the number of attractive interactions within the nucleus.

One way of ensuring saturation is to assume that the potential energy of a pair of nucleons has the functional form known for the potential energy between the molecules of a liquid drop, a system which also shows saturation properties. This potential, sketched in Fig. 18.1, indicates an attractive force at extreme range, as is indicated by low energy scattering experiments, and a repulsive force at very short distances. This gives a minimum in the potential energy curve which may be associated with stable binding. The *repulsive core*, for which there is now good experimental evidence from scattering experiments at energies in the range 300–1000 MeV, prevents collapse and determines the nuclear volume.

In early discussions of the saturation property the idea of a repulsive core to the internucleon force did not appear reasonable. On the assumption of a Fermi gas of nucleons the highest occupied states in an average nucleus correspond to a kinetic energy of about 20 MeV. The observed binding energy of about 8 MeV then indicates that the potential well depth in Fig. 18.1 is of the order of 30 MeV and the height of the repulsive core, which has a much shorter range, must be much greater in order to be effective. A value of about 300 MeV is required, and this appeared incommensurate with what was known about nuclear energies. Moreover, a repulsive core could not be predicted from any type of meson field theory. Two alternatives considered were:

(a) the existence of *many-body forces*, and
(b) the possibility of *exchange forces*.

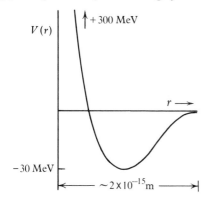

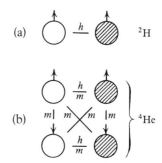

Fig. 18.1 The internucleon force (schematic). The curve shows the potential energy of two nucleons at a distance r between centres. Coulomb forces are neglected. Each nucleon has its own distribution of charge and magnetic moment.

Fig. 18.2 Nuclear bonds in the two-nucleon and four-nucleon system. The letters h, m refer to particular types of force, which are discussed in Sect. 18.1.3. The arrows represent spin directions.

Of these, Heisenberg chose to develop the latter hypothesis, since it could be based on theory, and since this type of force was already known in molecular problems. Although it now appears more reasonable to revert to the repulsive core explanation of saturation, there is also clear evidence from the angular distribution of high energy neutron–proton scattering (Sect. 18.2.7 and Fig. 18.10) that nuclear forces have important exchange characteristics. The way in which such forces might saturate will be discussed in Sect. 18.1.3. It may be noted that ordinary long-range forces, due for instance to electrostatic, magnetic or gravitational fields, are far too weak to account for nuclear binding energies within the observed nuclear size.

18.1.2 *Systematics of stable nuclei; charge independence*

If the distribution of stable nuclei among mass numbers (ch. 10) is examined it is seen that for light nuclei $N \approx Z \approx A/2$, which suggests that there is a tendency for neutrons and protons to pair off. Furthermore the large binding energy per nucleon for ^{4}He and the special stability of nuclei with $A = 4n$ indicate that the saturated nuclear system consists of two neutrons and two protons. This is in accordance with the Pauli principle according to which each proton or neutron can interact with one particle of the same type (but with opposite spin) and with two particles of the opposite type in the same state of orbital motion. It is to be expected that maximum binding energy for this system will arise, for attractive forces, when all the particles are in S-states of relative motion. If a fifth particle is added to this system it must enter a *p*-state and is less strongly bound, so that the average binding energy per nucleon for $A = 5$ is less than for $A = 4$. It will be seen from Fig. 18.2 that there are six 'bonds', four associated with (np) forces and one with (pp) and (nn) forces. In spectroscopic terminology the *S*-state interactions of two nucleons comprise

$$(a) \text{ the } {}^1S \text{ singlet configurations } {}^1(\text{np}), {}^1(\text{pp}), {}^1(\text{nn}) \qquad (18.3)$$

and

$$(b) \text{ the } {}^3S \text{ triplet configuration } {}^3(\text{np}) \qquad (18.4)$$

The evidence of the light nuclei is that the corresponding nuclear forces are attractive in each of these states. The neutron excess in heavy nuclei, which is required to balance the Coulomb repulsion of the protons, confirms that the ${}^1(\text{nn})$ force is attractive, but it is insufficiently attractive to lead to a stable di-neutron.

There is also good reason to suppose that for corresponding states of motion, e.g. in the singlet spin state, the (np), (pp) and (nn) forces are approximately equal. If this is accurately so, we characterize the forces as *charge independent*: if, however, only the equality of (nn) and (pp) forces can be established, the forces are *charge symmetric*. The similar number of nuclei with even Z, odd N and with odd Z, even N suggests charge symmetry (ch. 10), as does the approximate equality of neutron and proton separation energies. A closer test of these properties is, however, provided by the ground state energies and level systems of isobaric nuclei. Charge symmetry is well supported by these properties for odd mirror nuclei (Sect. 11.2.4) and charge independence, to a good approximation, by the corresponding properties for the even isobars (Sect. **12.1**). Under conditions of charge independence we describe the spin singlet and triplet configurations of the two-nucleon system by the isobaric spin (Sect. **12.1**) quantum numbers $T = 1$ and $T = 0$ respectively.

In a state of $T = 1$, but not of $T = 0$, a neutron or proton may be changed to the other particle without offending the exclusion principle. The concept of isobaric spin therefore permits a useful generalization of the Pauli principle in the case of charge independent forces. It is no longer necessary to specify nucleons of a particular kind in applying this principle; permitted states of two nucleons are those which are antisymmetrical for exchange of all coordinates, i.e. isobaric spin, ordinary spin and position. Obviously if the two nucleons are identical, the wave function of the pair is symmetrical in isobaric spin and the Pauli principle applies in its original form. Non-identical nucleons however are described by isobaric spin functions of both symmetries and there is no extra restriction on the permitted space-spin states.

18.1.3 Exchange forces

An exchange interaction can develop, according to quantum mechanics, when two interacting particles can exist in a state in which some common property can be shared. For the hydrogen molecular ion, for instance, the property is the orbital electron and the problem can be solved with high accuracy; for two nucleons Heisenberg assumed that the property shared would be the charge. The neutron–proton system is then regarded as a super-position of two states

$$\text{n}_1 + \text{n}_2 \qquad n'_1 + n'_2$$

which differ only because charge (e.g. owing to emission and absorption of a π-meson) has been exchanged. Because of the identity of these two systems in the sense that neither particle can be uniquely specified, the total energy is reduced below what it would be without exchange and binding results.

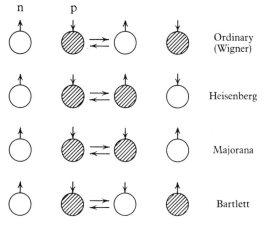

Fig. 18.3 Basic types of exchange interaction between nucleons. The properties which may be exchanged are charge (represented by shading) and spin (represented by arrows).

In order to see how an exchange force might produce saturation, consider a heavy nucleus in which only *ordinary* short range attractive forces operate. This nucleus collapses. If however only *exchange* forces operate, the inter-action between a pair of nucleons will be strong only if these nucleons occupy similar states of motion. Since the Pauli principle limits the number of particles in an assembly of fermions which may enter a given spatial state, the introduction of exchange forces immediately reduces the number of effective interactions. It can also be shown (Ref. 18.1) that some of the interactions that remain are repulsive. Together these effects create a saturated system.

Charge is not the only property that may be exchanged; spin is also a suitable property, as pointed out by Majorana. The possible types of ex-change for the $^1S_0$ neutron–proton system are shown in Fig. 18.3 together with the names by which the corresponding forces are known. It will be seen that the exchange of charge and spin (Majorana force) is exactly equivalent to an exchange of position and this force is sometimes known as a *space-*

exchange force. Similarly the Heisenberg charge exchange is equivalent to a joint change of spin and position.

The potential energy for a pair of nucleons is expressed by an integral in which occurs an operator P ($= \pm 1$) representing the exchange character of the force. The actual sign depends on the spin nature of the states involved. Taking $P = +1$ for an attractive force the exchange operators for the two-particle system are shown in Table 18.1.

TABLE 18.1 Exchange operators for the two-particle system

| | Even parity states | | Odd parity states | |
	triplet $S=1$	singlet $S=0$	triplet $S=1$	singlet $S=0$
Wigner	1	1	1	1
Heisenberg	1	-1	-1	1
Majorana	1	1	-1	-1
Bartlett	1	-1	1	-1

The Heisenberg and Bartlett forces are adequate to describe the deuteron (Fig. 18.2(a)), but cannot account for the increased binding energy in the α-particle in which all particles are in relative S-states. Majorana forces however are attractive in both singlet and triplet states and six attractive bonds arise (Fig. 18.2(b)).

With four varieties of nuclear force to employ it is possible to ensure saturation and to prevent collapse by choice of appropriate mixtures. The potential energy operator is written in the form

$$wP_W + hP_H + mP_M + bP_B$$

where w, h, m, b are distance-dependent potentials and P_W, P_H, P_M, P_B are the operators appearing in Table 18.1. The potentials may be chosen to fit the high energy scattering phase shifts, or may be derived from meson theory (Sect. **18.3**). Unfortunately potentials obtained in this way do not meet the saturation requirement, even when augmented by the spin–orbit terms which are also required (Sect. 18.1.4). The solution appears to lie in the addition of the repulsive core to the interaction.

18.1.4 The shell model; spin-orbit effect

The potentials of the single-particle shell model (ch. 9) are effective potentials for a complex system in interaction with a nucleon, but they are derivable in principle from the pure internucleon potentials. For the present purpose we recognize only that special features of the shell and optical model potentials probably have their origin in similar properties of the force between nucleons. Thus polarization phenomena (Sect. 17.3.2) are most easily interpreted in terms of the interaction between intrinsic spin and orbital motion which is assumed in the *j–j* coupling shell model. The internucleon force may be assumed to show the same property; polarization in high energy nucleon–nucleon collisions has in fact been observed.

18.2 Information on nuclear forces from the two-body problem

18.2.1 *The binding energy of the deuteron*

The deuteron has a binding energy of $2 \cdot 2245 \pm 0 \cdot 0002$ MeV and no stable excited states. It has an angular momentum of unity and both a magnetic dipole moment μ_d and an electric quadrupole moment. The magnetic moment is not equal to the algebraic sum of the magnetic moments of the proton μ_p and neutron μ_n, but it is sufficiently near to preclude the possibility of relative orbital motion of the two particles to a first approximation. The simplest structure to assume for the ground state of the deuteron is therefore a neutron and a proton in an *S*-state of orbital motion with parallel spins, i.e. a ${}^3S_1$ state. The ${}^1S_0$ state of the deuteron, in which the proton and neutron have antiparallel spins, is unbound.

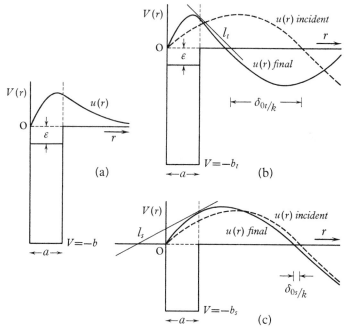

Fig. 18.4 The two-body system: (a) The deuteron, showing the *S*-state wave function $u = r\psi$ of relative motion of a neutron and proton interacting through a square well potential. For $r > a$ the kinetic energy of the particles is negative and the wave function decays exponentially to zero. (b) Neutron–proton scattering in the triplet (${}^3S$) state at low energy. The phase shift between the incident and final waves approaches π as the energy tends to zero ($k \longrightarrow 0$). (c) Neutron–proton scattering in the singlet (${}^1S$) state at low energy; the phase shift approaches zero as $k \longrightarrow 0$.

If an interaction potential for a neutron and a proton in this triplet state is prescribed, the binding energy of the deuteron may be calculated by solution of the Schrödinger equation, as in the corresponding problem of the hydrogen atom in atomic physics. It is simplest to assume a spherically symmetrical square-well potential function representing an attractive interaction whenever the neutron and proton approach within a distance a (Fig. 18.4(a)). The ground state of the deuteron is represented in Fig. 18.4 by a line drawn at energy $E = -\varepsilon$, the binding energy, and the repulsive core is neglected.

The solution of the wave equation is straightforward. If ψ is the wave function for the relative motion we have in general

$$\psi = R(r) Y_l^m(\theta, \phi) \tag{18.5}$$

but for an S-state there will be no angular dependence and

$$\psi = R(r) = \frac{u(r)}{r} \tag{18.6}$$

The wave function ψ (or u) determines the probability of finding a separation r between the neutron and proton in the deuteron. In terms of the function $u(r)$ the Schrödinger equation for a stationary state E for a nucleon of mass M moving in a potential field $V(r)$ in an S-state is

$$\frac{d^2 u}{dr^2} + \frac{2M}{\hbar^2} [E - V(r)] u = 0 \tag{18.7}$$

In the deuteron the masses of the two particles are nearly equal, each to the nucleon mass M, and they move about their centre of mass. Correction for this is obtained by using the reduced mass $\frac{1}{2}M$ in place of M in (18.7). The quantity $E - V(r)$ is the kinetic energy of the two particles, which has a non-classical, negative value for $r > a$. The solution of (18.7) is carried out in two stages:

I *Region $r > a$, $V(r) = 0$, $E = -\varepsilon$*
    Equation (18.7) becomes

$$\frac{d^2 u}{dr^2} - \frac{M}{\hbar^2} \varepsilon u = 0 \tag{18.8}$$

which has the solution

$$u = A e^{-\alpha r} \quad \text{with} \quad \alpha^2 = \frac{M\varepsilon}{\hbar^2} \tag{18.9}$$

assuming that $u(r)$ vanishes at $r = 0$ as it must if $|\psi|^2$, the particle density, is to remain finite.

II *Region $r < a$, $V(r) = -b$, $E = -\varepsilon$*
    Equation (18.7) becomes

$$\frac{d^2 u}{dr^2} + \frac{M}{\hbar^2} (b - \varepsilon) u = 0 \tag{18.10}$$

which has the solution

$$u = B \sin Kr \quad \text{with} \quad K^2 = \frac{M}{\hbar^2}(b - \varepsilon) \tag{18.11}$$

again assuming that $u(r)$ vanishes at the origin; K is the internal wave number. This solution is a particular case of the more general form (9.3). The two solutions (18.9) and (18.11) must be joined together at the potential boundary $r = a$. This may be done by requiring that $(1/u)(du/dr)$ (a measure of the particle density and current) shall be continuous at this point. This gives

$$\frac{1}{u}\left(\frac{du}{dr}\right)_{r=a} = K \cot Ka = -\alpha \tag{18.12}$$

or

$$\cot^2 Ka = \frac{\alpha^2}{K^2} = \frac{\varepsilon}{b - \varepsilon}$$

If now $b \gg \varepsilon$ it follows that

$$Ka \approx \frac{\pi}{2} \tag{18.13}$$

or, from (18.11),

$$ba^2 \approx \left(\frac{\pi}{2}\right)^2 \frac{\hbar^2}{M} \approx 1 \cdot 0 \text{ MeV} \times \text{barn} \tag{18.14}$$

The wave function $u(r)$ is shown in Fig. 18.4(a); the quantity $1/\alpha = 4 \cdot 3 \times 10^{-15}$ m is a size parameter for the deuteron. It is interesting to note that there is a high probability ($\approx 50\%$) of the particles in the deuteron being separated by more than the range of nuclear forces; the 'size' of the deuteron is determined by its binding energy, not by the force range.

Equation (18.13) and Fig. 18.4(a) show that in the deuteron the range of the forces is about equal to one-quarter of the de Broglie wavelength of relative motion. The wave function for $r < a$ then just turns over in a distance a to meet the rising wave function for $r > a$. The relation (18.14) between the range of the force and the depth of the potential well necessary to bind the deuteron is clearly independent of the precise form of the potential function so long as $b \gg \varepsilon$. The quantity ba^2 will remain finite, for a bound deuteron, in the zero range limit $a = 0$. For a well depth of ≈ 30 MeV the range of force required is about $1 \cdot 9 \times 10^{-15}$ m. The kinetic energy for this separation is therefore about 28 MeV; this is of course essentially the energy indicated by the uncertainty principle for a nucleon confined within the distance a.

No more information about the neutron–proton force than is contained in (18.14) can be obtained from the binding energy of the deuteron. Further knowledge of the force law must be obtained from a study of the unbound system, i.e. from neutron–proton scattering experiments. These establish that the neutron–proton force is *spin-dependent*, which may also be inferred from the absence of a bound state of the deuteron in which the nucleons have antiparallel spins (^1S).

18.2.2 *Low energy neutron–proton scattering*

If a slow neutron is captured by a proton, a γ-ray of energy 2·22 MeV is emitted and a deuteron is formed. The neutron is then forbidden classically to enter the region $r > a$ (Fig. 18.4(a)) in which its wave number is imaginary. If capture does not take place, and if there is no loss of energy in excitation processes, the neutron is elastically scattered by the proton and leaves the region $r < a$ with its original laboratory energy T_n and (small) channel wave number k (Sect. 14.2.1) given by

$$k^2 = \frac{M}{\hbar^2} \frac{1}{2} T_n \tag{18.15}$$

Within the region $r < a$ however, the wave number increases and is approximately equal to the internal value K calculated in Sect. 18.2.1.

The cross-section for elastic scattering may be calculated following the general method given in Sect. 14.2.2. For neutrons of energy 1 MeV, the reduced wavelength of relative motion is $\lambda = 1/k = 8\cdot8 \times 10^{-15}$ m which is much greater than a, the expected range of the neutron–proton force. A *zero range approximation* may therefore be used ($ka \ll 1$) and only s-wave scattering need be considered. When k is small we may use the concept of scattering length, as shown in Fig. 18.4(b) and (c). Outside the range of interaction the wave function is a sine wave of very long wavelength, but within the force range a it approximates to the internal wavefunction of the deuteron. The intercept of the tangent to the wavefunction at the point $r = a$ with the axis of r gives the scattering length l. From Fig. 18.4(b) this is positive for a bound deuteron, but if there is no bound state the internal wave function fails to turn over within the range of the force and the scattering length is negative (Fig. 18.4(c)). Specifically, from Fig. 18.4(b) and (c)

$$l = a - \frac{u(a)}{(du/dr)_{r=a}} \quad \text{for } k \longrightarrow 0 \tag{18.16}$$

The scattering length is determined by the phase shift δ_0 introduced into the asymptotic s-wave by the scattering potential. According to (14.40) and (14.47) this partial wave may be written

$$u(r) = r\psi = \frac{e^{i\delta_0}}{k} \sin(kr + \delta_0) \tag{18.17}$$

and δ_0 may be obtained by making the internal and external wave functions of the problem join smoothly at the potential boundary $r = a$ (Fig. 18.4(b) and (c)).

For the internal function we have from (18.11), for k small,

$$u(r) = B \sin Kr \qquad (18.18(a))$$

and for the external function from (18.17)

$$u(r) = C \sin (kr + \delta_0) \qquad (18.18(b))$$

Continuity of the quantity $(1/u)(du/dr)$ at $r = a$ gives

$$K \cot Ka = k \cot (ka + \delta_0)$$

$$= -\alpha \quad \text{from (18.12)}$$

so that

$$\tan (ka + \delta_0) = -k/\alpha \qquad (18.19)$$

for the s-wave interaction. The cross-section for scattering is then, from (14.63) and (14.66),

$$\sigma_{el}^0 = \frac{4\pi}{k^2} \sin^2 \delta_0$$

$$= \frac{4\pi}{k^2 + \alpha^2} (\cos ka + \frac{\alpha}{k} \sin ka)^2 \qquad (18.20(a))$$

by substituting from (18.19).

In the zero range approximation both αa and ka are small and $(18.20(a))$ becomes

$$\sigma_{el}^0 = \frac{4\pi}{k^2 + \alpha^2} \qquad (18.20(b))$$

A better estimate, to first order in αa, is

$$\sigma_{el}^0 = \frac{4\pi}{k^2 + \alpha^2} (1 + a\alpha) \quad \text{(Ref. 18.8)}$$

while from (18.16) and (18.12), if a may be neglected in comparison with the scattering length l,

$$\frac{1}{l} = -\frac{1}{u} \left(\frac{du}{dr}\right)_{r=a} = \alpha$$

so that

$$k \cot \delta_0 = -1/l$$

and

$$\sigma_{el}^0 \doteq \frac{4\pi}{k^2 + 1/l^2} \qquad (18.21)$$

In the *low energy limit*, $(k \longrightarrow 0)$

$$\delta_0 = -kl \quad \text{or} \quad (\pi - kl) \qquad (18.22)$$

and $\sigma_{\text{el}}^0 \longrightarrow 4\pi l^2$ which is the cross-section for scattering by a hard sphere of radius l.

The approximate expression $(18.20(b))$ may be expressed using (18.9) and (18.15) as

$$\sigma_{\text{el}}^0 = \frac{h^2}{\pi M} \frac{1}{\varepsilon + \frac{1}{2}T_n} = \frac{5\cdot 2}{\varepsilon + \frac{1}{2}T_n} \text{ barn} \qquad (18.23)$$

where ε and T_n are given in MeV. This formula indicates an energy variation of the neutron cross-section of the form shown in Fig. 18.5, curve (a), for energies up to 10 MeV. The experimental results obtained by the transmission method† are shown in curve (b) of this figure and are seen to be above the curve predicted by the simple theory‡ for energies for which T_n is not very large compared with ε. The difference is not removed by inserting the correction factors in $(18.20(a))$ for a reasonable value of a.

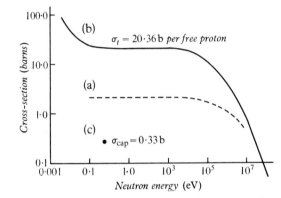

Fig. 18.5 Interaction of neutrons with protons. (a) Scattering cross-section according to theory based on triplet-state interaction only. (b) Experimental results for total cross-section. The cross-section rises at low energies because the two protons in the hydrogen molecule cannot then be treated as free (App. 5). (c) Capture cross-section of protons for thermal neutrons.

The reason for this discrepancy was first given by Wigner, who pointed out that although the $^1S_0$ state is unbound in the deuteron, its effect must be included in the scattering calculation. In a neutron–proton collision the probability of a triplet collision (parallel spins, $I = 1$) is three times that of a singlet collision (antiparallel spins, $I = 0$) because of the statistical factor $(2I + 1)$ and if the singlet state has a binding energy ε' the low-energy cross-section should be

$$\sigma_{\text{el}}^0 = \frac{4\pi}{k^2} \frac{1}{4} \{ 3 \sin^2 \delta_{0t} + \sin^2 \delta_{0s} \}$$

$$= \frac{h^2}{4\pi M} \left\{ \frac{3}{\varepsilon + \frac{1}{2}T_n} + \frac{1}{|\varepsilon'| + \frac{1}{2}T_n} \right\} \qquad (18.24)$$

$$= 4\pi \{ \frac{3}{4}l_t^2 + \frac{1}{4}l_s^2 \} \quad \text{in the low-energy limit.} \qquad (18.24(a))$$

† E. Melkonian, *Phys. Rev.*, **76**, 1744, 1949.
‡ The sharp increase at very low energies is due to a chemical binding effect (App. 5).

This gives a better account of the experimental results at low energies providing that ε' is small. The observed cross-section indicates that ε' is about 70 keV. This is clear evidence for a spin dependence of the neutron–proton force, leading to a different well-depth for the spin parallel and spin antiparallel interactions (Fig. 18.4). The neutron–proton scattering cross-section by itself does not prove that the ^1S state is *unbound* (l_s negative, Fig. 18.4(c)) but this may be inferred from the cross-section for the capture of slow neutrons by protons (Sect. 18.2.3) and from coherent scattering phenomena (App. 5). It must not be concluded that an observable singlet state exists in the continuum of positive energies of the (n, p) system; the singlet scattering length however is a necessary parameter of the neutron–proton interaction.

The zero range approximation conceals all information on the radial shape of the neutron–proton potential and of course on its range. It is therefore of interest to reformulate the *s*-wave phase shift in such a way that the extent of the approximations is more obvious. It is shown in Refs. 18.1, 18.2 and 18.3 that

$$k \cot \delta_0 = -\frac{1}{l} + \tfrac{1}{2}r_0 k^2 - P r_0^3 k^4 + \dots \tag{18.25}$$

where l is the scattering length already defined and r_0 is an *effective range* of the force, embodying the width and depth of the potential well. These two parameters, which differ for the singlet and triplet states, are all that can be obtained by analysis of the *s*-wave experiments. The detailed shape of the potential well appears in the coefficient P of the third term in (18.25) but this term is small at low energies. If it is neglected, we speak of the *shape-independent approximation*.

For the bound (triplet) state of the deuteron there is also an effective range formula:

$$\alpha = \frac{1}{l_t} + \tfrac{1}{2}\alpha^2 r_{0t} \tag{18.26}$$

where α gives the binding energy of the deuteron through eq. (18.9) and the suffix denotes the triplet interaction.

18.2.3 *Electromagnetic transitions in the neutron–proton system*

(*a*) *Capture of slow neutrons by protons.* It was found by Fermi in 1935 that slow neutrons live only for about 10^{-4} s in a large block of paraffin and that they disappear chiefly by the radiative capture process

$$^1\text{n} + {}^1\text{H} \longrightarrow {}^2\text{H} + \gamma + 2 \cdot 22 \text{ MeV} \tag{18.27}$$

The photons emitted in this capture have been detected and measurement of their energy has provided one of the most accurate determinations of the binding energy of the deuteron. Although capture *removes* thermal neutrons, the measured cross-section of 0·3b is small compared with the scattering cross-section.

It has already been noted that on account of the short range of nuclear forces, a neutron and proton with a relative energy of up to about 10 MeV interact mainly in the s-state of orbital motion. For a relative orbital angular momentum with $l \geqslant 1$ the separation of the two particles exceeds the range of the nuclear force. The probability of capture will also be proportional to $1/v$, i.e. to the time that the neutron spends within the range for forces and it follows that the capture cross-section, like that for other slow neutron reactions, will increase with decreasing velocity. The two particles may interact in either the $^1S_0$ or the $^3S_1$ state (since the P-state interaction is improbable at low energies) and the final state is the $^3S_1$ ground level of the deuteron. Of the two possible magnetic dipole transitions

$$^3S \longrightarrow \,^3S \quad \text{and} \quad ^1S \longrightarrow \,^3S \tag{18.28}$$

the former may be shown (Ref. 18.1, p. 604) to be forbidden and the capture process is therefore the latter, in which the neutron or proton undergoes a spin-flip and the $M1$ photon transports angular momentum $\hbar$. The angular distribution of the emitted photons will be isotropic because only S-states are involved.

The probability of the magnetic dipole process depends on the quantity $(\mu_p - \mu_n)$ and also on an integral over the wave functions of the singlet and triplet states. It follows from the observation of the photomagnetic capture that the magnetic moments of the neutron and proton are not equal and also that the singlet and triplet wave functions are determined by different potential wells, since otherwise the integral would vanish. This is further confirmation of the spin dependence of nuclear forces. The calculated cross-section for photomagnetic capture is sensitive to the sign of the binding energy of the singlet state. If comparison is made with the scattering cross-section we find for zero energy neutrons

$$\left. \begin{aligned} \frac{\sigma_{sc}}{\sigma_{cap}} &= 118 \quad (^1S_0 \text{ state bound}) \\ &= 71 \quad (^1S_0 \text{ state virtual}) \end{aligned} \right\} \tag{18.29}$$

The experimental value, Fig. 18.5, is $20·0/0·332 = 60$, confirming an unbound state.

Correction of the cross-section formula by the effective range theory leads to a value, unfortunately rather inaccurate, for the difference $r_{0t} - r_{0s}$, from which r_{0s} can be obtained using the more accurately known triplet range.

(b) *Photodisintegration of the deuteron.* The process converse to radiative capture is the disintegration of the deuteron by radiation, according to the scheme

$$^2H + \gamma \longrightarrow \,^1H + \,^1n - 2·22 \text{ MeV} \tag{18.30}$$

This reaction has been used in an accurate determination of the deuteron binding energy; the threshold energy for photoneutron or photoproton production is observed, using electron bremsstrahlung from an electrostatic generator of precisely defined energy.

The process (18.30) can be either *photomagnetic*

$$^3S \longrightarrow {}^1S \quad \text{(magnetic dipole)} \tag{18.31}$$

which is exactly inverse to the capture of slow neutrons by protons, or *photoelectric*

$$^3S \longrightarrow {}^3P \quad \text{(electric dipole)} \tag{18.32}$$

in which a transition to a *P*-state of the continuum of energies is induced by the electric vector of the electromagnetic wave, in analogy with the photoelectric effect in the atom. In the photomagnetic disintegration there is spin-flip, but no change of orbital motion; in the photoelectric disintegration the

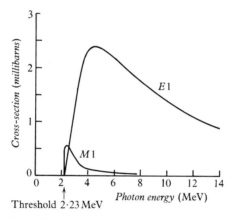

Fig. 18.6 Energy dependence of the photoelectric ($E1$) and photomagnetic ($M1$) cross-sections for disintegration of the deuteron. The threshold is at the deuteron binding energy (Ref. 18.3).

spin is unaltered but the neutron and proton are emitted with one unit of relative orbital momentum. The angular distribution of the magnetic dipole process is isotropic but that of the electric process has a normal dipole distribution with respect to the electric vector of the γ-radiation, i.e. $\sigma(\theta) \propto \sin^2 \theta$ where θ is the angle between the direction of the proton and the line of flight of the radiation.

The cross-sections for these processes as a function of energy can be calculated easily in the zero-range approximation and are shown in Fig. 18.6. The photomagnetic cross-section rises rapidly at the threshold to a maximum because it leads to the $^1S_0$ state whose energy is near zero for the n–p system. It is related simply, by arguments of detailed balancing, to the cross-section for capture of slow neutrons by protons, and increases linearly at first with the velocity of the disintegration particles. The photoelectric cross-section rises more slowly at first because a *P*-state is not favoured for slow particles; this is why the electric dipole absorption of slow neutrons is not observed. It reaches a maximum for $h\nu \approx 2\varepsilon$, where ε is the binding energy of the deuteron.

The deuteron photodisintegration has been studied experimentally, both for cross-section and angular distribution, over a wide range of energies. Measurements with radioactive sources have given information on the photo-magnetic transition near threshold, and similar work with γ-rays from nuclear reactions and with electron bremsstrahlung has extended up to 900 MeV. The photoelectric cross-section determines primarily the triplet effective range, r_{0t}, which enters as a correction to the zero range approximation, and the photomagnetic effect, as expected by analogy with its inverse process, yields the difference $r_{0t} - r_{0s}$.

The order of magnitude of the photoelectric cross-section is the area of the deuteron $\pi/4\alpha^2$ multiplied by the fine structure constant $\mu_0 e^2 c/2h = 1/137$, which represents the strength of the coupling between the electromagnetic field and matter.

18.2.4 Low-energy proton–proton scattering

Proton–proton scattering experiments can be carried out with greater accuracy than the comparable neutron–proton experiments at most energies. Analysis of the results is complicated by the Coulomb scattering and by the fact that allowance must be made for the identity of the two particles (cf. α-He scattering, Sect. 5.3.1). It is simplified in comparison with the neutron–proton case because the Pauli principle excludes certain states of motion for identical particles, so that either singlet or triplet states, but not both, occur for a given orbital state.

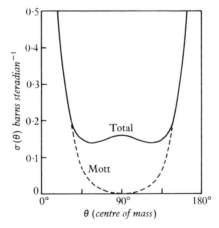

Fig. 18.7 Angular distribution for the scattering of 2·4 MeV protons by protons. Near $\theta = 90°$ (laboratory angle 45°) the scattering is mainly nuclear (Ref. 18.3).

The Coulomb scattering of identical particles was calculated wave-mechanically by Mott and is discussed in Ref. 18.3. Experimental results confirm this formula for energies of the order of 300 keV or less and thus verify that interference between waves corresponding to scattered and recoil particles occurs. At higher energies extra scattering is observed at large angles ($\approx 90°$ c.m. Fig. 18.7), and this must be attributed to the nuclear force. For energies up to about 10 MeV a phase shift analysis of angular distributions (Ref. 18.3) confirms that only s-wave interaction is important, as is expected because of the short range of the force ($ka \ll 1$). The Pauli principle then limits the interaction to the singlet state $^1S_0$ and there is only one phase shift δ_0 instead of the two required for the neutron–proton scattering (18.24).

The Mott scattering, which results from the long-range Coulomb force, involves both singlet and triplet states and all relative orbital momenta. There is consequently interference with the nuclear scattering and the final expression for the low-energy angular distribution for the centre-of-mass system may be written in the form

$$\sigma_{pp}(\theta) = \sigma_{Mott}(\theta) - A(\theta, \delta_0) \sin \delta_0 + \lambda^2 \sin^2 \delta_0 \qquad (18.33)$$

where λ is the reduced de Broglie wavelength for the relative motion ($= 2\hbar/m_p v$, where v is the incident velocity and m_p the proton mass) and $A(\theta, \delta_0)$ is a calculable coefficient. The last term is the nuclear s-wave scattering and the second term is due to Mott–nuclear interference. The presence of the interference term permits the sign as well as the magnitude of δ_0 to be obtained from the experimental results;[†] it is found to be positive (0·255 rad at 382·4 keV) corresponding to an attractive nuclear force. The interference is thus destructive for the s-wave and leads to a reduced cross-section, as already noted for the α-helium scattering (Sect. 5.3.1).

The effective range theory used to discuss the (np) interaction was developed by Jackson and Blatt to describe the variation of the $^1S_0$ phase shift for (pp) scattering with energy. The formula corresponding to (18.25) is

$$\frac{\pi \cot \delta_0}{\exp 2\pi\eta - 1} + h(\eta) = \frac{4\pi\varepsilon_0 \hbar^2}{m_p e^2} \left(-\frac{1}{l_s} + \tfrac{1}{2} r_{0s} k^2 - P r_{0s}^3 k^4 \ldots \right) \qquad (18.34)$$

in which $h(\eta)$ is a calculable function of the Coulomb parameter $\eta = e^2/4\pi\varepsilon_0 \hbar v$. The quantity dividing $\pi \cot \delta_0$ is part of the penetration factor for the potential barrier between the two protons, and appears when a fit is required between internal and external (Coulomb) wave functions at the nuclear boundary. In the case of neutron–proton scattering the s-wave phase shift is derived directly from the observed scattering cross-section, but owing to the strong Coulomb scattering, total cross-sections for the low energy (pp) interaction cannot easily be measured. The phase shift δ_0 is therefore obtained by analysis of the angular distribution (18.33) at a number of energies; (18.34) then yields values for the singlet parameters l_s and r_{0s}. The (pp)

† J. E. Brolley, J. D. Seagrave and J. G. Beery, *Phys. Rev.*, **135**, B1119, 1964.

scattering length l_s is found to be negative, corresponding to an unbound state for the two-proton system, in analogy with the singlet state of the (np) interaction (Fig. 18.4(c)).

18.2.5 Comparison of low-energy parameters (Table 18.2)

The parameters used in the effective range theory have the values in the following table according to the best experiments available in the energy range for which the *s*-wave interaction is predominant. The singlet scattering length for the (pp) interaction is obtained from the observed l_s ($-7·82$ fm) by correcting for the effect of Coulomb forces. Coherent scattering determinations of the (np) scattering lengths are described in Appendix 5. These parameters give a good account of most of the low-energy two-body phenomena, and contain all the information that can be extracted from such experiments. The existence of different singlet and triplet interactions, as has already been emphasized, is evidence for the spin dependence of nuclear forces. This is large enough, and in the right direction, to prevent stable binding of the singlet systems ¹(np), (the excited deuteron), ¹(pp) and ¹(nn). The di-neutron system has been investigated† by studying the energy distribution of the reaction products in such processes as ^{2}H(np)2n, ^{2}H(π, γ)2n and ^{3}H(^{3}H, α)2n.

The singlet scattering lengths for the (pp) and (nn) interactions support the hypothesis of charge symmetry of nuclear forces but the (np) singlet scattering length is different. Some difference would be expected according to the meson-exchange theory of nuclear forces (Sect. **18.3**) because charged mesons can contribute to (np) scattering but only neutral mesons to (pp) and (nn) and there is a mass difference (of an electromagnetic nature) between charged and neutral mesons. Because of the large correction necessary to the observed (pp) singlet scattering length to remove Coulomb effects, it is probably best to assess the validity of the full charge independence hypothesis by comparing effective ranges. Here the meson mass difference suggests a difference of 3 per cent between the (np) and (pp) singlet ranges, close to the observed value.

General evidence for charge symmetry or charge independence of the nucleon–nucleon interaction is discussed by Brink (Ref. 18.8, ch. 4). It seems likely that the neutron–proton interaction is slightly stronger than that between like nucleons.

18.2.6 Non-central forces

The central force assumption so far made in this chapter is unable to explain the nature of the electromagnetic moments of the deuteron. In a simple picture of a $^3S_1$ state of a neutron and a proton the magnetic moment should be the algebraic sum of the magnetic moments μ_p, μ_n of the two particles and

† E. E. Gross *et al.*, *Phys. Rev.*, **C1**, 1365, 1970.

TABLE 18.2 Parameters of the 2-Nucleon System

Parameter	Value (10^{-15} m) (np)	(pp)	(nn)	Method of determination	Reference
Triplet scattering length l_t	$5 \cdot 425 \pm 0 \cdot 004$	—	—	Low energy (np) scattering Coherent scattering	18.2.2 App. 5
Singlet scattering length l_s	$-23 \cdot 714 \pm 0 \cdot 013$	$-17 \cdot 0 \pm 0 \cdot 6$	-17 ± 1	Low energy (np) scattering Coherent scattering Low energy (pp) scattering Final state interactions	18.2.2 App. 5 18.2.4 18.2.5
Triplet effective range r_{0t}	$1.744 \pm 0 \cdot 003$	—	—	Binding energy of deuteron Photodisintegration	18.2.1 18.2.3
Singlet effective range r_{0s}	$2 \cdot 73 \pm 0 \cdot 03$	$2 \cdot 79 \pm 0 \cdot 02$	$2 \cdot 84 \pm 0 \cdot 03$	Low energy (np) scattering Low energy (pp) scattering Capture of slow neutrons by protons Final state interactions	18.2.2 18.2.4 18.2.3 18.2.5

General references: Ref. 18.4. R. Wilson, *Comments on Nuclear and Particle Physics*, **2**, 141, 1968. E. M. Henley, in *Isospin in Nuclear Physics*, ed. D. H. Wilkinson, North Holland Publ. Co., Amsterdam, 1969. H. P. Noyes and H. M. Lipinski, *Phys. Rev.*, **C4**, 995, 1971.

the electric quadrupole moment Q should vanish since there is only a symmetric charge distribution in an S-state. In fact the deuteron moment is smaller than that of the proton, so that the neutron moment must be negative but furthermore, in units of the nuclear magneton,

$$\mu_d = 0.857411 \pm 0.000019$$

whereas (18.35)

$$\mu_n + \mu_p = 0.87950 \pm 0.00010$$

and a quadrupole moment now known to be 2.82×10^{-31} m^2 was established in 1940 by the magnetic resonance method (ch. 4).

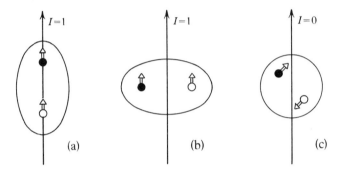

Fig. 18.8 Effect of tensor force on charge distribution in deuteron. (a) Ground state $(I=1)$ Q positive. (b) Ground state $(I=1)$ Q negative. (c) Virtual state $(I=0)$ $Q=0$.

These facts can be explained, without appreciable modification of the effective range theory, by the addition of an extra term to the central potential which creates a *tensor force*. Since the effects to be explained exist at low energies this force must be finite in the zero velocity limit. The tensor force resembles a magnetic interaction between two dipoles although its origin must be quite different; it results in a lower potential energy for the neutron and proton when the line joining them is parallel to the spin direction than when it is perpendicular (Fig. 18.8). The effect of such a non-central term, which can change markedly for different relative orientations of the spin axis and the interparticle axis, is to adjust the orbital motions to favour the low potential configuration (a), Fig. 18.8. The deuteron then becomes egg-shaped, and a positive electric quadrupole moment develops. At the same time the extra orbital motion contributes a magnetic moment which must be combined with the proton and neutron moments.

In quantum mechanical language, the tensor force mixes a fraction of ^3D-state wave function in with the predominantly S-state wave function of the deuteron ground state. Since total angular momentum and parity are good quantum numbers for a stable state the only possible admixture is $^3S_1 + {}^3D_1$. Detailed calculations show that if about seven per cent of the ground state wave function is effectively a D-state, then the electromagnetic moments of the deuteron are correctly predicted (Ref. 18.9).

Tensor forces are not effective in the singlet state (Fig. 18.8(c)) because there is no resultant spin and no preferred spin axis. It might therefore be possible to ascribe the whole of the spin dependence of nuclear forces to the existence of non-central effects in the triplet state. The tensor force is not the type of non-central force postulated in the single-particle shell model, which supposes a coupling between the spin and orbital motion of a single nucleon in a potential field. This *spin–orbit force* (which also does not exist in the singlet state of two nucleons) is velocity-dependent and vanishes in the low-energy limit. It is therefore not able to explain the electromagnetic moments of the deuteron, but it is necessary, in addition to the tensor force, to account for high-energy scattering phenomena in the two-body system. A force of this type, too, existing between nucleons and complex nuclei, is used to interpret the polarization observed in elastic scattering.

18.2.7 *High-energy experiments*

At energies above about 20 MeV partial waves with $l \geqslant 1$ become increasingly important in nucleon scattering, and effective range theory is no longer applicable. It might be expected that at these energies information about the

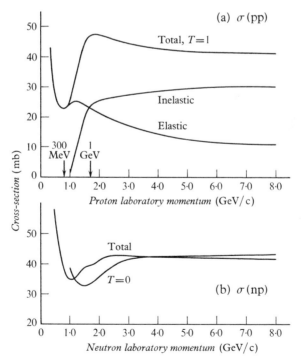

Fig. 18.9 Total cross-sections for the (pp) ($T=1$) and (np) interaction as a function of energy. The cross-section for the $T=0$ interaction is found by combining the (pp) and (np) data.

nuclear well shape might be obtained and that more precise tests of the hypothesis of charge independence might be forthcoming. Unfortunately the increased number of possible states of relative motion makes detailed analysis difficult, and the conclusions which have so far been drawn are of a rather general nature. Low energy experiments of high accuracy still provide better information on these points.

Nucleon–nucleon scattering has been studied at a large number of energies up to 70 GeV. Figure 18.9 shows some of the cross-sections for energies up to about 8 GeV; only monotonic variation has been seen beyond this. The increase in the (pp) total cross-section above 300 MeV is due to inelastic processes, namely pion production. The (pp) collision takes place wholly in the isobaric spin state $T=1$, but for (np) there is a mixture of $T=0$ and $T=1$ states and the cross-section for the pure $T=0$ state is obtained by combining the proton and neutron data (Example (19.14)). It is seen that meson production in the $T=0$ state is less marked and starts at a higher energy than for $T=1$; this will be discussed briefly in Sect. 19.2.5.

Differential cross-sections for the (np) and (pp) interactions have also been obtained over a wide range of energies. Polarization phenomena have been extensively explored, using at first double and triple scattering techniques, and later both polarized beams and polarized targets, as a result of which an important improvement in statistical quality has been obtained. A full discussion of this extensive work, and an enumeration of the experiments required to give complete information on the collision of two particles of spin $\frac{1}{2}$ is given by Lock and Measday (Ref. 18.7, Chapter 7) and by McGregor (Ref. 18.5). It essentially provides the data from which energy-dependent phase shift analyses may be constructed and these analyses, in extreme detail, now extend up to 750 MeV for both (pp) and (np) scattering. The most useful analyses so far however are those concerned with entirely elastic processes, i.e. up to about 400 MeV, since many extra parameters are needed to deal with inelastic channels.

The phase shifts of course join smoothly to the low energy values that have already been discussed. Thus, as in the case of low energy (np) scattering, there are both singlet and triplet spin states ($S=0$ and 1) for each partial wave. All of these are required to describe (np) scattering, but because of the Pauli principle the (pp) scattering, which is in a state of pure isobaric spin $T=1$, is confined to states of the type ^1S, ^3P, ^1D, ^3F, ...

Scattering in the isospin state $T=0$ is similarly confined to the states ^3S, ^1P, ^3D,... The phase shifts for (pp) scattering can be used successfully in describing the $T=1$ part of the (np) interaction, and this constitutes a high energy test of charge independence. A further important result of the (pp) analysis is that the $^1S_0$ phase shift, always positive at low energies corresponding to an attractive interaction, becomes negative at about 250 MeV. This can be taken to indicate a repulsive core in the nucleon–nucleon interaction of radius about 0·5 fm. The observation of polarization in (pp) scattering at all energies up to 10 GeV indicates the existence of a non-central force coupling spin to orbital motion in the triplet system (since polarization cannot arise in the zero spin singlet states).

The phase shifts for the two-nucleon system summarize a large body of experimental data. A successful analytical representation of the nucleon–nucleon interaction by means of a potential, with central and non-central terms, must give a good account of the best available phase shifts. If it does, then it may be used with some confidence in nuclear matter and nuclear reaction calculations. Several *phenomenological two-body potentials* containing empirically determined parameters now exist (e.g. Hamada–Johnston, Yale University, Tabakin). A more basic approach is to set up a specific model for

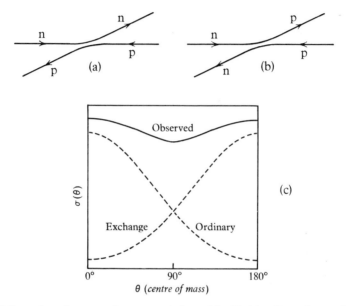

Fig. 18.10 Scattering of neutrons by protons (schematic) with (a) ordinary forces, (b) charge-exchange forces. The interpretation of an observed angular distribution as due to a mixture of ordinary and exchange interactions is shown in (c) (Ref. 18.3).

the nucleon–nucleon interaction from which a potential can be derived, which can be checked against the phase shifts. The *one-boson exchange potential* (Sect. **18.3**) is conceived in this way; a potential with an exchange character is indeed strongly suggested by the angular distribution observed in the scattering of high energy neutrons by protons (Fig. 18.10). In this scattering the backward peak in the c.m. system can most easily be understood if a charged pion is exchanged between the interacting nucleons. Such an exchange can give rise to a force capable of meeting saturation requirements (Sect. 18.1.3), as first suggested by Heisenberg. The existence of the repulsive core permits some latitude in the specification of the exchange force.

18.3 One-boson exchange potentials

18.3.1 Summary of knowledge of the nucleon–nucleon force

From the evidence presented in Sects. **18.1** and **18.2** it appears that the force between nucleons has the following properties:

(i) Charge symmetry, and most probably charge independence, to a good approximation.

(ii) Short range (≈ 2 fm).

(iii) Spin dependence.

(iv) An exchange component, although not so much as was required by the old saturation conditions.

(v) A repulsive core of radius ≈ 0.4 fm in the ^1S state and probably also in other states of relative motion.

(vi) A tensor component and probably a spin–orbit term in addition to the central forces.

The potential energy function for the central part of the force is typically as shown in Fig. 18.1.

The force necessary to account for the many relevant experimental phenomena is clearly a complicated one, as is obvious from the rather large number of parameters that appear in the phenomenological potentials (Ref. 18.8). In the boson exchange potentials, which stem from the fundamental work of Yukawa,† flexibility is afforded by the introduction of several types of exchange particle.

18.3.2 One-pion exchange

In electromagnetic theory the long-range Coulomb potential $V = e/4\pi\varepsilon_0 r$ is a solution of the equation

$$\nabla^2 V = 0 \tag{18.36}$$

which is itself the static limit of the wave equation

$$\nabla^2 V - \frac{1}{c^2}\frac{\partial^2 V}{\partial t^2} = 0 \tag{18.37}$$

The wave equation is derived from Maxwell's equations and describes the propagation of electromagnetic waves in free space. Any change in the potentials can only be conveyed to a distant point with a velocity c, and the relation between wavelength and frequency given by (18.37) is

$$c = \nu\lambda \tag{18.38}$$

In quantum theory it is assumed that although light is propagated as a wave, the energy is carried as discrete quanta or photons. Since these must travel

† H. Yukawa, *Proc. phys. math. Soc., Japan*, **17**, 48, 1935.

with velocity c, they must have no rest mass according to the theory of relativity and their energy and momentum are

$$E = h\nu$$

and

$$p = \frac{h}{\lambda} = \frac{h\nu}{c}$$

These relations have been verified in detail in studies of the Compton effect. In relativistic notation we relate the energy and momentum of a photon by the equation

$$E^2 - p^2 c^2 = 0$$

In order to obtain a force of shorter range than the Coulomb force between charges, we may, following Yukawa, modify (18.37) to the form

$$\nabla^2 U - \frac{1}{c^2} \frac{\partial^2 U}{\partial t^2} - K^2 U = 0 \tag{18.39}$$

where K is a constant and U is a potential function. Substituting a wave-like solution into this equation we obtain for the relation between wavelength and frequency

$$-\frac{1}{\lambda^2} + \frac{\nu^2}{c^2} - \frac{K^2}{4\pi^2} = 0 \tag{18.40}$$

and using the de Broglie expressions for the energy and momentum of a material particle (or photon) this gives

$$E^2 - p^2 c^2 - M^2 c^4 = 0 \tag{18.41}$$

where

$$M = \frac{\hbar K}{c} \tag{18.42}$$

This is the relativistic energy-momentum relation for a particle of rest mass M. Moreover, the static case of equation (18.39) is

$$\nabla^2 U - K^2 U = 0 \tag{18.43}$$

which gives for the corresponding potential

$$U = \frac{-f}{4\pi} \frac{e^{-Kr}}{r} \tag{18.44}$$

where f is a constant analogous to the charge e in electromagnetic theory. The potential energy of a nucleon in such a potential field will therefore be, by analogy with the electromagnetic case,

$$-\frac{f^2}{4\pi} \frac{e^{-Kr}}{r} \tag{18.45}$$

where f^2 is a *coupling constant*.[†] To obtain the range indicated by experiment, Yukawa proposed that K should be about 10^{15} m^{-1}, which implies the existence of a particle, according to (18.42), of mass equal to about 300 electron masses which should interact strongly with nuclei. The theoretical prediction of the existence of this particle and the subsequent discovery of the π-meson (pion) which has the predicted mass and nuclear interaction, together form one of the major joint achievements of theoretical and experimental physics of the twentieth century.

The exchange of a pion between a neutron and proton is analogous to the passage of an electron between a hydrogen atom and a proton in the binding of the hydrogen molecular ion. It is conveniently pictured in terms of Feynman diagrams (Sect. **21.1**), which show the equivalence of the two vertex reactions

$$\text{p} \rightleftarrows \text{n} + \pi^+$$

$$\text{n} \rightleftarrows \text{p} + \pi^-$$

for the exchange process. These processes do not conserve energy, but according to quantum mechanics, however, the energy $\Delta E = Mc^2$ necessary may be borrowed providing that it is returned within a time given by

$$\Delta E . \Delta t = \hbar$$

and the meson exchange can therefore be considered as a virtual process. In the time Δt the meson can travel a maximum distance $c \, \Delta t$ and if we identify this distance with the range of the nuclear force (Fig. 18.1) we obtain

$$a = c \, \Delta t = \frac{c\hbar}{\Delta E} = \frac{\hbar}{m_\pi c} = \frac{1}{K} \tag{18.46}$$

which is the value indicated by the potential (18.44). If we let $m_\pi \longrightarrow 0$ and $f^2 \longrightarrow \mu_0 c^2 e^2$ we obtain the Coulomb energy $(e^2/r)(\mu_0 c^2/4\pi)$ which can thus be regarded as arising from virtual photon exchange between charges. In this case the dimensionless coupling constant is the fine structure constant

$$\alpha = \frac{\mu_0 c^2}{4\pi} \frac{e^2}{\hbar c}$$

An important extension of Yukawa's theory was the prediction by Kemmer (1938) of the existence of a neutral pion (π^0). This offers an explanation of the force between like (as well as unlike) nucleons in terms of π^0 exchange

[†] The potential energy is sometimes written

$$-\left(\frac{f^2}{4\pi\hbar c}\right) \hbar c \frac{e^{-Kr}}{r}$$

The coupling constant $f^2/4\pi\hbar c$ is then dimensionless and is known as the pion–nucleon coupling constant. This usage is followed in Sect. 21.4.2.

resulting from the virtual processes

$$p \rightleftarrows p + \pi^0$$
$$n \rightleftarrows n + \pi^0$$

If all three pion exchanges are permitted it is straightforward to couple the corresponding fields to the nucleons in such a way that charge independence (modified because of the $\pi^+ - \pi^0$ mass difference) results.

18.3.3 Boson potentials

The pion–nucleon coupling constant may be deduced from pion scattering experiments, as will be described in Chapter 21. It can also be extracted from an analysis of nucleon–nucleon scattering data. The values agree, giving satisfactory confirmation to the whole idea of meson exchange. The coupling constant is large (Table 21.1) in comparison with the electromagnetic constant, but techniques have been developed to permit such a strong potential to be used in calculations. In developing a realistic potential it is of course necessary to check that the exchange particle has the properties required by the nucleon–nucleon force. This leads to the introduction of other particles each with its own coupling constant besides the pion to fulfil the following functions (Ref. 18.8):

(a) *π-exchange*. The pion has zero spin and odd parity (Sect. 19.2.2), and this means that the exchange particle must be in a *p*-state to conserve parity. The resulting force is attractive in both ^1S and ^3S states and is spin-dependent and non-central (tensor force). It constitutes the longer range (≈ 1.5–2 fm) part of the nucleon–nucleon interaction.

(b) *ω-, φ-, and ρ-exchange*. These are heavy mesons of spin 1 and odd parity (App. 8) whose role in nucleon structure has been probed by electron scattering experiments (Sect. 21.4.3). Their exchange between nucleons can produce a short-range repulsive core (≈ 0.15 fm) and also a spin–orbit force, both of which are required.

(c) *2π exchange*. The repulsive effect of the ω-meson is too large in the so-called 'middle range' of the interaction and exchange of either two pions, or of a new meson with similar properties, is required to allow for this.

There are at present several variants of the boson exchange potential, and some of them are in part phenomenological, i.e. a few empirically determined parameters are included. The development is sufficiently advanced for the potentials to be used in nuclear matter and nuclear structure calculations, as well as to predict the nucleon–nucleon phase shifts in some detail.

References

1. BLATT, J. M. and WEISSKOPF, V. F. *Theoretical Nuclear Physics*, Wiley, 1952.
2. SQUIRES, G. L. 'The Neutron–Proton Interaction', *Prog. nucl. Phys.*, **2**, 89, 1952.
3. EVANS, R. D. *The Atomic Nucleus*, McGraw-Hill, 1955.

4. HULTHEN, L. and SUGAWARA, M. 'The Two Nucleon Problem', *Encyclopedia of Physics*, **39**, 1, Springer, 1957.

5. MORAVCSIK, M. J. *The Two-Nucleon Interaction*, Oxford University Press, 1963; WILSON, R. *The Nucleon–Nucleon Interaction*, Interscience, 1963; MCGREGOR, M. H. 'Nucleon–Nucleon Scattering', *Physics Today*, **22**, No. 12, page 21, December 1969.

6. HUGHES, D. J. and SCHWARTZ, R. B. *Neutron Cross Sections*, Brookhaven National Laboratory Report BNL 325, 1958.

7. LOCK, W. O. and MEASDAY, D. F. *Intermediate Energy Nuclear Physics*, Methuen, 1970.

8. BRINK, D. M. *Nuclear Forces*, Pergamon Press, 1965; ELTON, L. R. B. *Introductory Nuclear Theory*, Pitman, 1965; HAFTEL, M. I., TABAKIN, F. and RICHARDS, K. C. *Nuclear Phys.*, **A154**, 1, 1970.

9. PRESTON, M. A. *Physics of the Nucleus*, Addison-Wesley, 1962.

19 Particle physics I (non-strange particles)

A satisfactory account of many of the phenomena of nuclear physics can be given in terms of neutrons and protons as fundamental constituents of nuclei. Relativistic phenomena, such as beta-decay, bring in the electron and the neutrino and, given the sources of current, electromagnetic theory is adequate to describe radiative processes. Our present knowledge that the proton, electron, neutrino and photon are merely the stable members of a large array of similar objects resulted from the discovery of the muon and pion and from the realization (ch. 18) that the short-range nuclear force and the electromagnetic properties of the nucleons cannot be described in terms of stable particles and classical fields. The interpretation of such properties is now a part, perhaps even a minor part, of the subject of *particle physics* which attempts to enumerate and to coordinate the so-called fundamental particles of nature.

The question whether any particle can in fact be regarded as fundamental or elementary is still open. Fortunately there is general agreement that the various types of particle, the leptons, mesons and baryons, are observable objects with measurable properties. It may be that some of these must be thought of as composed of even more 'fundamental' components or that some particles exist only in virtue of interactions between themselves. This is a concern at the moment of theoretical physics; in this book attention will be given only to the established behaviour of the particles, and its relation to nuclear physics.

Several important types of fundamental particle were established by visual techniques in cosmic ray observations. Properties of the particles have been explored and defined by accelerator experiments because of the enormous advantage in intensity up to the limit of machine energies. Beyond this limit interactions may once more have to be studied in cosmic ray laboratories.† Chapters 19 and 20 give a short account of the well-established properties of the particles and introduce the concept of strangeness which sharply distinguishes much of the present subject from conventional nuclear physics. The more important particles and their properties, including strangeness, are listed in Appendix 8; this list embraces the more familiar 'ordinary' particles shown in Table 1.2. Chapter 21 surveys some of the attempts that

† The contribution of elementary particle physics to our understanding of some of the complex phenomena of the cosmic radiation is discussed in Ref. 19.5.

have been made to present a coordinated picture of the particles and their interactions.

19.1 Cosmic radiation and mesons

19.1.1 Analysis of the cosmic radiation in the atmosphere

Cosmic radiation is the name given to the primary flux of high energy particles continuously incident on the earth from interstellar space, and also to the secondary radiations deriving from interactions of the primary particles in the atmosphere. The nature of the radiation was first investigated by determining the absorption in matter of the total radiation at sea level. A curve of the form shown in Fig. 19.1 is obtained for lead; the radiation may be divided

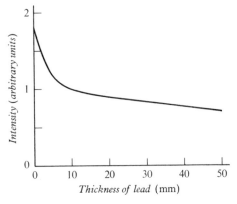

Fig. 19.1 Absorption in lead of cosmic radiation at sea level, showing 'soft' and 'hard' components (based on Auger *et al.*, *J. de Phys.*, **7**, 58, 1936).

into an easily absorbed (soft) component and a penetrating (hard) component with a mass attenuation coefficient much less than the least expected for gamma radiation (Fig. 5.17). Cloud chamber photographs showed that at sea level most of the ionizing particles carried a single electronic charge. The presence among these particles of some with positive charge and electronic mass, now known as positrons, was established by Anderson in 1932 (Plate 6); he observed the momentum of the positively charged particle before and after passage through a lead plate in which energy loss could be reliably estimated and sense of direction determined. It soon became clear that the soft component of the cosmic radiation contained positive and negative particles of electronic mass with energies ranging up to about 10^{11} eV and that these particles were produced in avalanches or showers which developed from the primary radiation in the atmosphere. The process of shower generation appeared to be proportional to Z^2 for the shower-producing medium and a typical shower produced in a lead plate in an expansion chamber is shown in Plate 16. The Z^2 dependence of shower production suggested the electromagnetic processes of bremsstrahlung and pair-production, and it is now known that the shower particles of the soft component are in the main the electronic descendants of high energy photons produced by π°-meson decay in the atmosphere. The primary particles are mainly protons.

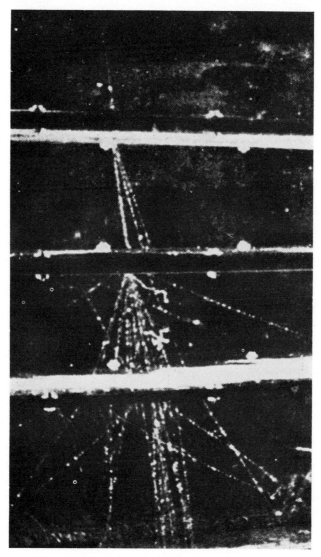

Plate 16 Electromagnetic shower produced by a cosmic ray particle entering a cloud chamber containing lead plates (D. K. Froman and J. C. Stearns, *Rev. mod. Phys.*, **10**, 133, 1938).

The penetrating component of the cosmic radiation was shown by expansion chamber photographs to consist of singly-charged particles which could pass through lead plates without producing an electromagnetic shower or a nuclear interaction. In 1936 Anderson and Neddermeyer presented conclusive evidence, based on range, momentum and ionization measurements, that some of these particles had electronic charge and a mass of about 200 times that of the electron. By 1939 many examples of such particles had accumulated and subsequent work has confirmed these observations; the penetrating component of the cosmic radiation at sea-level consists of the particles now known as muons, of both charges, which result from π or K-meson decay in the earth's atmosphere and which themselves decay into electrons and neutrinos (Sect. **19.3**). The hypothesis of a decaying particle in the cosmic radiation had already been invoked at an early stage in order to explain an apparent anomaly in the absorption coefficient for the penetrating radiation in matter, as determined by counter telescope experiments. This was found to depend not solely on the *mass* of material traversed, but also on the length of path, as would be expected for a particle with a lifetime of the order of a microsecond or so. The discovery of a meson in the cosmic radiation was hailed as a triumph for the Yukawa theory of nuclear forces (Sect. **18.3**) which had postulated the existence of a particle of comparable mass, but the difficulty remained that the cosmic ray mesons only rarely interacted with atomic nuclei according to the expansion chamber photographs. The Yukawa particle by its very nature must react strongly with nuclei and would not be expected to survive a passage through great thicknesses of the atmosphere or through many lead plates in a cloud chamber. Decay electrons from stopped mesons were however detected and the mean lifetime of the particle was determined by the delayed coincidence method to be $2 \cdot 16 \times 10^{-6}$ s. In 1946 a decisive experiment by Conversi, Pancini and Piccioni,[†] in which magnetic analysis was used, showed conclusively that when the mesons were brought to rest in iron, only the positive particles decayed radioactively, whereas when brought to rest in carbon, mesons of *both* signs decayed. Since the Coulomb field of a nucleus prevents slow positive mesons from approaching within the range of nuclear forces, the positive mesons should certainly decay; it is the negative mesons that are anomalous. These negative mesons must be presumed to fall into K-orbits (of radius about 1/200 that of the normal electronic orbit) in the carbon atom, and the experiment shows that they can remain in such orbits for a time of the order of the mean life of $2 \cdot 16 \times 10^{-6}$ s. Since the wave function of a particle in the K-shell actually overlaps the nucleus, this survival in carbon indicates that the negative mesons have only a small nuclear interaction. In heavier elements the orbits are smaller and nuclear absorption does indeed occur; the significant result however is that in light elements it does not predominate.

The conclusion of Conversi, Pancini and Piccioni seemed clearly at variance with the fact that mesons appeared to be produced by the cosmic radiation in the atmosphere with a cross section of the order of the geometrical area for nuclei of oxygen and nitrogen. This follows from the observed

† *Phys. Rev.*, **71**, 209, 1947.

attenuation coefficient. Suggestions were therefore made (Sakata and Inoue, 1946; Marshak and Bethe, 1947) that the mesons initially produced in the upper atmosphere were those connected with nuclear forces and that these mesons decayed radioactively into the mesons observed at sea-level. The strongly interacting mesons would be heavier than the sea-level mesons, and if produced from the primary cosmic radiation would have to be sought at high altitudes. The detailed constitution of the general cosmic radiation at such altitudes was not well known at the time of the two-meson hypothesis, but already in the late 1930s the exposure of nuclear emulsions on mountain tops had begun to reveal nuclear 'star' production (Plate 9). Such stars are now known to contain tracks of π-mesons in addition to the heavily ionizing tracks of protons and α-particles from nuclear fragmentation processes. The emulsions should therefore contain, in addition to tracks of primaries, tracks of mesons produced outside the sensitive volume.

19.1.2 Discovery of the π-meson and of π–μ–e decay (Ref. 19.1)

The nuclear emulsions first used for high altitude studies of cosmic radiation were unable to detect particles of minimum ionizing power. The sensitivity was improved by Messrs. Ilford in 1946–7 and in the latter year Perkins and also Occhialini and Powell showed that tracks of charged mesons (of mass estimated from grain counts and multiple Coulomb scattering observations) could be recorded. Some of these mesons produced nuclear disintegrations in elements of the emulsion at the end of their range, but about 10 per cent were found by Lattes, Muirhead, Occhialini and Powell[†] to emit a secondary particle. The subsequent introduction of emulsions sensitive to relativistic particles further revealed that the secondary particle disintegrated at the end of its range with the emission of an electron (Plate 17) in the way already known for the cosmic ray meson. The secondary particle appeared, from an examination of several events, to be emitted always with the same energy.

This remarkable and beautiful series of observations, made with equipment of extreme simplicity, immediately confirmed (and in point of fact preceded) the two-meson hypothesis and provided a dramatic clarification of many of the problems of cosmic ray physics. The parent particle is the π-meson (pion) which is produced copiously, in both π^+ and π^- forms,[‡] in the upper atmosphere as a result of nuclear interactions of the primary cosmic radiation. Both π^+ and π^--mesons enter emulsions exposed at high altitudes and the π^+-mesons usually slow down and decay, since they are kept away from nuclei, near the end of their range, by Coulomb repulsion. The π^--mesons interact strongly with emulsion nuclei producing stars and do not decay at rest; both types of π-meson however decay in flight and few reach sea level. The decay process yields a secondary particle (μ-meson or muon)

$$\pi^{\pm} \longrightarrow \mu^{\pm} + \nu \text{ (or } \bar{\nu}) \tag{19.1}$$

[†] *Nature*, **159**, 694, 1947.
[‡] The π^0 form (neutral meson) is also produced but this was not evident in the early experiments.

whose properties identify it with the particles of the penetrating component of the cosmic radiation, which thus arises from the decay of fast $\pi^{\pm}$-mesons.†

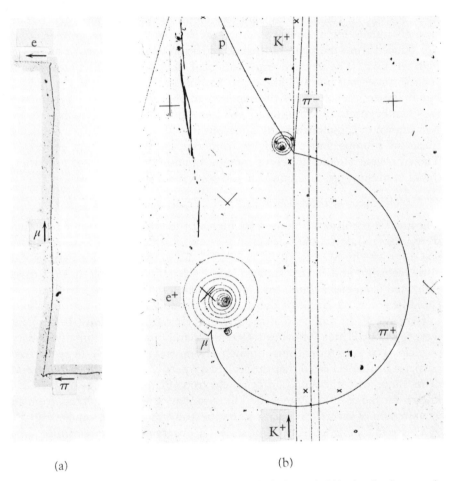

(a)   (b)

Plate 17 $\pi \to \mu \to$ e decay seen in (a) nuclear emulsion, (b) hydrogen bubble chamber in magnetic field. The emulsion tracks are from C. F. Powell, *Rep. progr. Phys.*, **13**, 350, 1950. The bubble chamber picture (D. C. Colley) shows the decay of a π^+ meson produced in the interaction $K^+ + p \to K^+ \pi^- \pi^+ p$.

In the decay of π^+-mesons at rest, the μ-meson is emitted with a unique energy of about 4·1 MeV (range 600 μm of emulsion) which is derived from the mass difference between the π and μ-particles. Since no identifiable particle or photon appears to be emitted in the π-decay it is assumed that a neutrino or antineutrino is created in the process (19.1) order to balance momentum. The muon itself decays with the emission of an electron, but the electron energy is not single-valued and the decay process is now assumed

† Muons also arise from the decay of K-mesons (Sect. **20.1**).

to be
$$\mu^{\pm} \longrightarrow e^{\pm} + \nu + \bar{\nu} \tag{19.2}$$

The majority of the positive muons disappear in this way; negative muons may in addition be captured by nuclei from the orbits of muonic atoms (Sect. 11.2.3).

The first observations of π–μ–e decay themselves yielded a large amount of information on the properties of the pion and muon. This has however been refined and extended by experiments with artificially produced particles as described in the following sections.

The pion exchange theory of nuclear forces implies that the forces between like nucleons are due to exchange of a neutral meson (Sect. 18.3.2). Possible decay products of such a particle, in particular high energy photons originating near nuclear disintegration stars, had been suspected for several years in cosmic ray studies. The neutral meson was however first identified conclusively as a result of accelerator experiments.

19.2 Pions

19.2.1 Artificial production of π-mesons from complex nuclei

Soon after the discovery of π-mesons in the cosmic radiation these particles were produced artificially by Gardner and Lattes† using a beam of 380 MeV

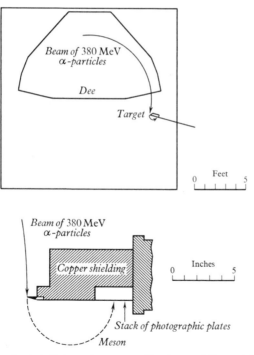

Fig. 19.2 Artificial production of charged π-mesons (Gardner and Lattes,† Berkeley 1948). The diagrams give a plan view of the cyclotron and details of the target assembly.

† Science **107**, 270, 1948.

α-particles from the 184″ synchrocyclotron at Berkeley. The experimental arrangement is shown in Fig. 19.2; the mesons emitted from a carbon target were detected in nuclear emulsions after deflection in the magnetic field of the cyclotron; this field was reversed so that mesons of both signs could be received. The π^--particles were identified by production of nuclear stars in the emulsion and the π^+-particles by the characteristic π–μ–e decay at the end of their range. Subsequent experiments at Berkeley showed that π-mesons could also be produced from nuclei by protons, neutrons and gamma radiation.

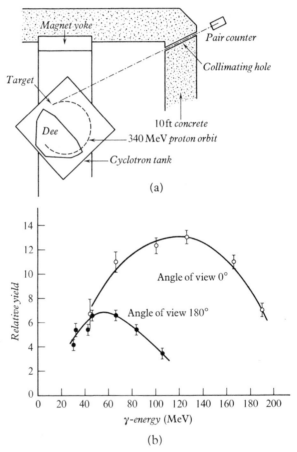

Fig. 19.3 Artificial production of neutral π-mesons (Bjorklund *et al.*, Berkeley 1950). (a) layout of experiment; (b) photon spectrum for two angles of emission with respect to the proton beam showing Doppler shift of energy.

In 1950 the neutral π-meson was produced artificially, also at Berkeley, by Bjorklund *et al.*† using the apparatus shown in Fig. 19.3(a). Protons from the synchrocyclotron struck a beryllium or carbon target, as in the charged pion

† R. Bjorklund, W. E. Crandall, B. J. Moyer and H. F. York, *Phys. Rev.*, **77**, 213, 1950.

experiments, and high energy gamma radiation from the target was analysed in a magnetic pair spectrometer. Conclusive evidence for the origin of the radiation in the decay of a neutral particle was obtained by observing the Doppler effect on the photon energy when the direction of motion of the mesons was altered by reversing the proton beam in the cyclotron. The change in the shape of the energy spectrum (Fig. 19.3(b)) is well predicted by the decay scheme assumed for the neutral pion (Sect. 19.2.2(*e*))

$$\pi^\circ \longrightarrow 2\gamma \tag{19.3}$$

19.2.2 Properties of π-mesons (π^+, π^-, π°)

The properties of the π-mesons, shown in Appendix 8, have been established as follows:

(*a*) *Mass and charge.* It may be shown that when the ratio of the momenta of a pion and a proton is equal to the ratio of their ranges in an absorber, then these particles have equal velocities and the common ratio is the mass ratio. In this way it is found (Ref. 19.2) that

$$m_{\pi^+} = 139 \cdot 6 \pm 0 \cdot 1 \text{ MeV}/c^2$$

A slightly more accurate value is obtained from a careful measurement of the energy release in the $\pi^+ \to \mu^+$ decay, using the muon mass (Sect. 19.3.1). The negative and neutral pion masses may be connected by observing the energy spectrum of photons from the reaction:

$$\begin{cases} \pi^- + p \longrightarrow \pi^\circ + n \\ \pi^\circ \longrightarrow 2\gamma \end{cases} \tag{19.4}$$

in a pair spectrometer. This leads to the result

$$m_{\pi^-} - m_{\pi^\circ} = 4 \cdot 606 \pm 0 \cdot 006 \text{ MeV } c^2$$

The π^--mass is obtained most accurately by bent-crystal measurement of X-rays from pionic atoms† (Sect. 19.2.4) and is given as

$$m_{\pi^-} = 139 \cdot 577 \pm 0 \cdot 013 \text{ MeV } c^2$$

This series of experiments establishes equality of the π^+ and π^- masses to a high degree of precision, as is required by general invariance principles (ch. 21) if these are particle and antiparticle. They also relate the pion and muon masses and help to indicate that the mass of the neutrino emitted in the π-μ decay is consistent with zero ($+0 \cdot 6$ MeV/c^2 at 90% confidence) as in the case of the neutrino emitted in beta decay (ch. 16).

All magnetic deflection and ionization measurements confirm that the pionic charge is equal to that of the electron.

(*b*) *Lifetime.* The first accurate measurement of the mean life of the charged pion was obtained by allowing artificially produced π^--mesons to

† R. E. Shafer, *Phys. Rev.*, **163**, 1451, 1967.

describe a known number of orbits in the magnetic field of the cyclotron in which the particles were produced. The circulating mesons could be intercepted by a detector after a suitable number of turns, and the mean life was deduced from the attenuation observed as a function of orbit number. More recently the attenuation due to decay in flight of a well-defined beam of pions has been studied; a typical apparatus is shown in Fig. 19.4. Cherenkov counters are used to eliminate unwanted particles which contaminate the

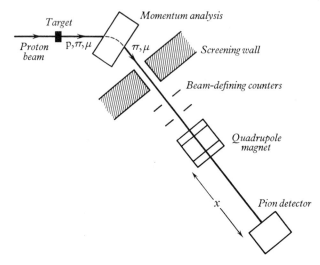

Fig. 19.4 Apparatus for determination of pion lifetime by observation of decay in flight. The counting rate in the detector (Cherenkov type, set to discriminate against muons) is found as a function of the distance x.

beam, and the experiment determines the fraction of the beam which survives along the flight path as a function of distance. This technique can be applied to both π^+- and π^--mesons. For π^+-mesons, which are not absorbed by nuclei when slowed down, the delayed coincidence method (Sect. 13.6.1) may be used; one signal indicates the arrival of a positive pion in a detector in which it stops and a delayed signal from the same or a similar detector indicates the appearance of the decay muon.

The results of these experiments show that the lifetimes of the π^+- and π^--mesons are equal within two standard deviations†; the accepted value is given in Appendix 8. The equality is expected, as in the case of the equality of masses, from invariance principles (ch. 21).

The neutral pion decays into two photons (para. (e), following) and early estimates of the lifetime were based on the alternative decay mode in which one photon and one electron–positron pair (Dalitz pair) are produced:

$$\pi^0 \longrightarrow \gamma + e^+ + e^- \tag{19.5}$$

† D. S. Ayres *et al.*, *Phys. Rev.*, **157**, 1288, 1967; *Phys. Rev. Lett.*, **21**, 261, 1968.

Paired tracks due to the charged particles have been seen to originate at a point slightly displaced from the centre of an event in a nuclear emulsion exposed to high energy particles and the gap has been interpreted as due to the flight of a π^0-meson before decay. Unfortunately the observed gap may be partly due to the close association of the e^+e^- pair at the beginning of its path which gives effectively a neutral system. Later work has exploited the relativistic dilation of the decay path of high energy π^0-mesons and has permitted the use of counter techniques. An alternative determination is possible using the decay of a K^+-meson at rest in nuclear emulsion. The decay scheme useful for this purpose is

$$K^+_{\pi 2} \longrightarrow \pi^+ + \pi^0 \qquad (19.6)$$

and the tracks of the K^+- and π^+-mesons locate the origin of the π^0-particle and also its line of flight which is opposite to that of the π^+ (Fig. 19.5). An

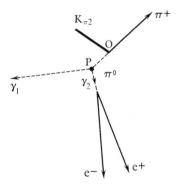

Fig. 19.5 Schematic diagram showing $K_{\pi 2}$ decay at point O, leading to π^0 decay at P. The lifetime of the π^0 determines the distance OP.

electron–positron pair produced by one of the π^0 decay photons can give an extrapolation back to the point of decay of the π^0 which is independent of the ionizing properties of the e^+e^- pair. The observed displacements are very small (0.05 μm) and the effect may be obscured by grain size, but a result of $(1.0 \pm 0.5) \times 10^{-16}$ s has been achieved from 232 events.[†] It is also possible, following a suggestion by Primakoff, to determine the π^0 lifetime from the cross-section for the photo-production of π^0-mesons in the Coulomb field of a heavy nucleus. The mean of all observations is $0.89 \pm 0.18 \times 10^{-16}$ s.

(c) *Spin, parity and statistics.* Strong evidence that the spin of the π^+-meson is 0 is obtained by comparing differential cross-sections for the forward and backward reactions:

$$p + p \rightleftarrows d + \pi^+ \qquad (19.7)$$

at corresponding energies and angles. It follows from the belief that nuclear

† P. Stamer *et al.*, *Phys. Rev.*, **151**, 1108, 1966.

processes are invariant under the operation of time reversal (App. 6) that the basic probabilities of the forward and backward reactions are equal provided that measurements are made without regard to polarization. Averaging over spin states must then be allowed for and this introduces the factors $(2s_p+1)$ $(2s_p+1)$ and $(2s_\pi+1)(2s_d+1)$ into the predicted cross-section ratio. The only unknown is s_π and the experimental data agree better with 0 than with 1; half-integral spin is excluded by the very occurrence of reaction (19.7). It is assumed that the π^- spin is the same as that of the π^+.

The spin of the neutral pion must be even if it decays into two photons, because of the transverse nature of the electromagnetic field. The two decay photons have z-components of angular momentum of $\pm\hbar$ with respect to their line of flight (Fig. 19.6 and Sect. 3.5.2), so that the total z-component is 0 or $2\hbar$. This is evidently consistent with a pion spin of zero although the value 2 is not excluded.

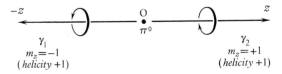

Fig. 19.6 Decay of π^0 (in rest system) into two collinear gamma rays γ_1, γ_2 of equal energy with opposite z-components of angular momentum. In the laboratory system the energies are Doppler shifted (see Fig. 19.3) (Bailey *et al.*, *Phys. Lett.*, **28B**, 287, 1968).

Mesonic parity is deduced from the phenomena observed when π^--mesons are slowed down and captured in deuterium. The processes

$$\begin{cases} \pi^- + d \longrightarrow n+n \\ \pi^- + d \longrightarrow n+n+\gamma \end{cases} \tag{19.8}$$

are highly favoured above the process

$$\pi^- + d \longrightarrow n+n+\pi^0. \tag{19.9}$$

The initial state for these reactions is a π^--meson in an 'atomic' s-orbit round the deuteron and this system has the parity of the meson. In the first of the reactions (19.8) the two neutrons must occupy a ^1S, ^3P or ^1D state (Sect. **12.1**) at the energy available in the reaction. Of these states only the $^3P_{(1)}$ state can satisfy angular momentum conservation with spin zero for the pion and spin one for the deuteron. The P-state has odd parity and this is therefore the conclusion for the π^--meson. If the π^0-meson also has odd parity the process (19.9) will be retarded because it will be necessary for the two neutrons to occupy a P-state and for the π^0-meson to be emitted as a p-wave ($l=1$) in order to conserve angular momentum and parity. The emission of p-waves is not favoured at the energy available because of the low mass and long wavelength of the pion.

The even spin of the pions indicates that they are Bose particles, and, as with photons, their number is not conserved in interactions.

(*d*) *Isobaric spin (isospin)*.† The evidence so far presented suggests that pions occur in three charge states with identical spin and parity and nearly the same mass. This may be expressed concisely by ascribing isobaric spin $T = 1$ (Sect. **12.1**) to the pion so that the three pions $\pi^+ \; \pi^- \; \pi^0$ are distinguished by the third component of isospin $T_z = +1, -1$ and 0. It is possible to consider the π^+- and π^--mesons as antiparticles and the π^0-meson as its own antiparticle. This is quite different from the case of the Σ-hyperon (App. 8) which also exists in three charge states and has $T = 1$. These particles however are fermions and *each* has its own antiparticle. An *s*-wave fermion particle–antiparticle pair has odd parity; the *s*-wave $\pi^+\pi^-$ system is of even parity.

(*e*) *Decay schemes*. The main decay modes of the pions are intimately involved in the experiments by which the existence of these particles was established and most of them have already been discussed. The π^+-meson at rest decays with a mean life of about $2 \cdot 5 \times 10^{-8}$ s by the process

$$\pi^+ \longrightarrow \mu^+ + \nu_\mu \tag{19.10}$$

(Sect. 19.1.2 and Plate 17) in which the muon is emitted with a unique energy of 4.17 MeV. It is longitudinally polarized because of parity non-conservation in the decay (Sects. **16.7** and 21.3.2). The energy release is consistent with zero mass for the muon-associated neutrino ν_μ. Negative pion decay is not observed at rest because of the Coulomb attraction of this particle to nuclei and strong nuclear absorption but the decay

$$\pi^- \longrightarrow \mu^- + \bar{\nu}_\mu \tag{19.11}$$

takes place in flight. An antineutrino is indicated in this reaction in accordance with the principle of lepton conservation (Sect. 1.2.2) and the assignment of particle rather than antiparticle status to the μ^-.

In 1958 it was discovered‡ that electron decay of the pion, by the process:

$$\pi^\pm \longrightarrow e^\pm + \nu_e \text{ (or } \bar{\nu}_e) \tag{19.12}$$

takes place at a rate about 10^4 times less than the rate of the main $\pi \rightarrow \mu$ decay. Since the energy release in reaction (19.12) is much greater than that in (19.10) or (19.11) owing to the mass of the muon, more states of motion are available to the final particles in the electron process (cf. ch. 16) and it might be expected that this process would be favoured. This however ignores the constraint placed on the decay by the helicity of the neutrino. Figure 19.7 shows that in the $\pi^+ \rightarrow e^+$ decay the zero spin of the pion requires that the helicity of the positron shall be the same as that of the neutrino, i.e. -1 (left handed). The evidence from beta decay (Sect. **16.7**) however is that positrons are

---

† *Nomenclature.* In high energy physics the isobaric spin quantum number is usually written *I*; in this book the low energy usage *T*, with component T_z will be continued. Nuclear angular momentum quantum numbers are usually written $\mathscr{J}$ (rather than *I*) and this usage will be adopted in Chapters 19, 20, 21. The angular momentum component will be written $m_z\hbar$.

‡ T. Fazzini *et al.*, *Phys. Rev. Lett.*, **1**, 247, 1958.

emitted with a longitudinal polarization corresponding to $+v/c$ and in the relativistic limit these particles would have helicity $+1$. The $\pi^+ \to e^+$ decay would then be forbidden if it were governed by the same type of interaction as nuclear beta decay; in fact the finite rest mass of the electron prevents the relativistic limit from being reached and the reaction is permitted but with a severe retardation. The $\pi \to \mu$ decay is retarded for the same reason but by a much smaller factor because of the lower value of v/c. The ratio of $\pi \to e$ to $\pi \to \mu$ decay calculated with the use of these factors agrees with observation.

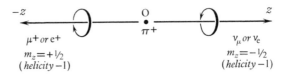

Fig. 19.7 Decay of π^+-meson at rest into collinear muon or positron and neutrino, showing helicities required by conservation of angular momentum. The μ^+, e^+ helicity is opposite to that favoured.

In 1962 the weak decay mode

$$\pi^+ \longrightarrow \pi^0 + e^+ + \nu_e \qquad (19.13)$$

was observed.† This process is analogous to a $0^+ \to 0^+$ Fermi transition in nuclear beta decay (ch. 16) and its rate can be predicted from the known beta decay coupling constant. The result is $1 \cdot 07 \times 10^{-8}$ of the $\pi \to \mu$ decay, and this agrees with observation. Both the existence and the rate of this decay mode are clear predictions of the *conserved vector current* hypothesis in the theory of the weak interaction (Sect. 21.3.4).

The decay of the neutral pion into two photons has already been discussed (para. (*b*) above) in connection with the determination of the lifetime of this particle. The alternative decay mode in which a positron–electron pair is emitted instead of one photon was suggested by Dalitz, and has been observed with a branching ratio of 1 in 80; more rarely still, two electron pairs are observed. These characteristic decays often identify π^0 production in high energy interactions. Neutral pions, with a lifetime of 10^{-16} s in their rest system, always decay in flight and the photons emitted may be drastically Doppler-shifted in energy. For a pion of velocity v the energy E_L of one of the photons observed in the laboratory system is obtained by a Lorentz transformation from the rest system of the pion, in which each photon has an energy $\frac{1}{2}m_\pi c^2$. This gives

$$E_L = \gamma[E + vp] \qquad (19.14)$$

where E and p are the pion rest frame energy and momentum. Substituting

† R. Bacastow *et al.*, *Phys. Rev. Lett.*, **9**, 400, 1962.

for these quantities for a rest frame angle θ we obtain:

$$E_L = \gamma[\tfrac{1}{2}m_\pi c^2 \pm \tfrac{1}{2}m_\pi cv \cos\theta]$$

$$= \tfrac{1}{2}m_\pi c^2 \gamma \left[1 \pm \frac{v}{c}\cos\theta\right] \tag{19.15}$$

Since $\gamma = (1-v^2/c^2)^{-1/2}$ this means that as θ varies the laboratory energies vary between

$$\frac{1}{2}m_\pi c^2 \sqrt{\frac{1+v/c}{1-v/c}} \quad \text{and} \quad \frac{1}{2}m_\pi c^2 \sqrt{\frac{1-v/c}{1+v/c}}$$

The decay in the pion rest frame is isotropic and this leads to a uniform distribution of energies between the two limits in the laboratory system. Observed distributions exhibiting these features are shown in Figs. 19.3(b) and 19.8.

19.2.3 Interaction of π-mesons with nucleons

Although fast π-mesons can be used for nuclear reaction studies (Ref. 19.8, ch. 11), there has until now been an overwhelming concentration of effort on the pion–nucleon interaction. This is in any case basic to an understanding of pion-induced effects in complex nuclei. We therefore consider in this section only the pion–proton and pion–deuteron systems.

(a) *Absorption of stopped π⁻-mesons.* It has been calculated that a fast charged pion slows down from an energy of about 100 MeV to about 1 keV in liquid hydrogen in 10^{-9} s by the ordinary processes of energy loss. From the energy of 1 keV, a π⁻-meson will lose further energy and will finally become bound (with negative energy) in a 1s orbit to a proton in a further time of about 10^{-12} s. From the 1s orbit the pion may either decay or, more probably, react with the proton as follows:

$$\begin{cases} \pi^- + p \longrightarrow \pi^0 + n \\ \pi^0 \longrightarrow 2\gamma \end{cases} \tag{19.16}$$

or

$$\pi^- + p \longrightarrow n + \gamma \tag{19.17}$$

These reactions were studied at Berkeley by Panofsky, Aamodt and Hadley†
using a high-pressure hydrogen gas target, and a pair spectrometer to detect the gamma radiation; the results of this work and of a later experiment are shown in Fig. 19.8. The photon spectrum shows an apparently homogeneous peak of energy about 130 MeV which is ascribed to reaction (19.17) and is consistent with the energy expected from the known masses. It also confirms that the π⁻-meson is a boson. The lower peak of mean energy 70 MeV is broader and represents the energies of photons from the π⁰-meson decaying

† *Phys. Rev.*, **81**, 565, 1951.

in flight as discussed in Sect. 19.2.2(*e*). It may be noted that this particular photon spectrum confirms the existence of the π^0 as a parent radiating particle; the alternative reaction not involving a π^0, i.e.

$$\pi^- + p \longrightarrow n + 2\gamma \qquad (19.18)$$

would yield a continuous photon spectrum. The neutron energy in reactions (19.16) and (19.17) is about 1 MeV and 10 MeV respectively, and emission in *s*-states is therefore favoured. The π^0-meson has a wavelength of 5×10^{-15} m and *s*-state emission for this particle is also indicated. The initial state has

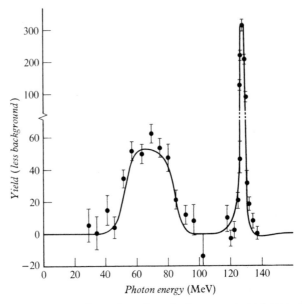

Fig. 19.8 Gamma ray spectrum from absorption of π^--mesons in hydrogen, showing homogeneous peak from radiative capture and Doppler broadened peak due to decay of π^0-mesons in flight. Note change of scale. (K. M. Crowe and R. H. Phillips, *Phys. Rev.*, **96**, 470, 1954.)

spin and parity $\mathcal{J} = \frac{1}{2}^-$ since the π^- is captured from the *K*-shell and the pion has odd parity. In reaction (19.17) the final neutron state has $\mathcal{J} = \frac{1}{2}^+$; the photon must therefore carry odd parity and unit angular momentum.

The absorption of a π^--meson from a bound state of negative energy by a proton is formally almost identical with *charge exchange scattering*, in which a pion of positive energy interacts with a proton according to the equation:

$$\pi^- + p \longrightarrow \pi^0 + n \qquad (19.19)$$

It was pointed out by Anderson and Fermi that observations of the so-called 'Panofsky ratio' for negative pion capture in hydrogen by reactions (19.16) and (19.17),

$$P = \frac{\sigma(\pi^- p n \pi^0)}{\sigma(\pi^- p n \gamma)} = 1 \cdot 56 \pm 0 \cdot 07$$

would permit the charge exchange scattering of negative pions to be related to meson photoproduction. This is because the radiative capture cross-section $\sigma(\pi^- p n \gamma)$ can be transformed into the photodisintegration cross-section $\sigma(\gamma n p \pi^-)$ by the principle of detailed balancing (App. 6) and this cross-section is essentially given by that for the photoproduction of π^--mesons from deuterium. There is now satisfactory agreement between the observed charge exchange cross-section extrapolated to zero energy and that deduced from photoproduction data.

Panofsky, Aamodt and Hadley also first studied the reactions:

$$\pi^- + d \longrightarrow 2n + \gamma \qquad (19.20)$$

$$\pi^- + d \longrightarrow 2n \qquad (19.21)$$

$$\pi^- + d \longrightarrow 2n + \pi^0 \qquad (19.22)$$

which are possible when negative pions are absorbed in deuterium. This experiment, which proves that the pion has odd parity, has already been discussed. The reaction (19.22) is not observed; a typical photon spectrum for

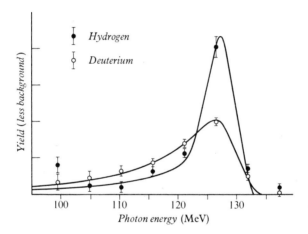

Fig. 19.9 Gamma ray spectrum from absorption of π^--mesons in deuterium (ϕ), with spectrum of radiative capture in hydrogen (ϕ) for comparison (J. A. Kuehner, A. W. Merrison and S. Tornabene, *Proc. Phys. Soc. (Lond.)*, **73**, 551, 1959).

reaction (19.20) with a spectrum for $(\pi^- p n \gamma)$ for comparison, is shown in Fig. 19.9. The occurrence of the reaction (19.21) was inferred by comparing the rate of photon production from mesonic atoms formed in hydrogen and deuterium; the rate of non-radiative capture, relative to that of radiative capture in deuterium is $2 \cdot 35 \pm 0 \cdot 35$. From this ratio, assuming the cross-section for photoproduction of π^+-mesons from protons (at threshold) it is possible to deduce the cross-section for the non-radiative capture and hence by the principle of detailed balancing the cross-section for the process

$$n + n \longrightarrow \pi^- + d. \qquad (19.23)$$

This is found to agree well with the measured cross-section for the charge symmetric reaction

$$p + p \longrightarrow \pi^+ + d \tag{19.24}$$

(*b*) *π-meson–nucleon scattering (total cross-section)*. The total and differential cross-sections for the *scattering of pions by protons* have been measured accurately over a wide range of energy using momentum-analysed beams from both internal and external targets of accelerators. The detecting systems are generally based on scintillation counter telescopes, and Cherenkov counters are also included to exclude unwanted particles. Lead radiators are placed in front of scintillators for conversion of decay photons to electron pairs when

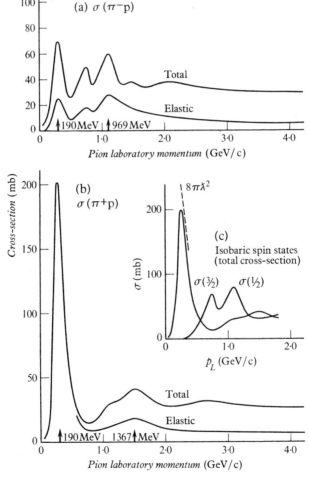

Fig. 19.10 Total and elastic cross-sections for scattering of pions by protons (a) π^+, (b) π^-, (c) in states of pure isospin. The maximum cross-section for a $\mathcal{J} = \frac{3}{2}$ resonance $(8\pi\lambda^2)$ is shown (from data in Refs. 19.3 and 19.7, see also Ref. 19.8, Ch. 5).

the production of π^0-mesons is studied. Total cross-sections are measured by the attenuation technique using a liquid hydrogen absorption cell.†

The total and elastic cross-sections for $\pi^{\pm}$-meson scattering by protons are shown in Fig. 19.10. At energies below about 300 MeV the scattering is due to the elastic processes:

$$\pi^+ + p \longrightarrow \pi^+ + p \tag{19.25}$$

$$\pi^- + p \longrightarrow \pi^- + p \tag{19.26}$$

and the charge exchange scattering:

$$\pi^- + p \longrightarrow \pi^0 + n \tag{19.27}$$

which increases the total π^--cross-section above the value for elastic scattering only. Each of these reactions shows a major resonance for a pion energy of about 200 MeV, and other resonances at higher energies. From general collision theory (eq. (15.36)) the maximum cross-section for elastic scattering when this is the only process possible (as for the π^+p system at 200 MeV) is

$$\sigma_t = 4\pi\lambda^2 \frac{2\mathcal{J}+1}{(2s_1+1)(2s_2+1)} \tag{19.28}$$

where $\mathcal{J}$ is the spin of the resonant state and s_1, s_2 are the spins of the interacting particles. Inserting the values for the π^+p case we find

$$\sigma_t^+ \leqslant 2\pi\lambda^2(2\mathcal{J}+1) \tag{19.29}$$

and the magnitude of the resonant cross-section requires $\mathcal{J}=\frac{3}{2}$ (Fig. 19.10) which indicates formation by p-wave pions. This state, to which we shall see that the isobaric spin of $T=\frac{3}{2}$ is assigned, may be regarded as an excited state of the nucleon; it can also be described as a *baryon resonance* and has received the special name $\Delta(1236)$ where the figure in brackets represents the mass of the resonance in MeV/c^2. The formation process predicts even parity.

(c) *Differential cross-section.* At about 120 MeV only s and p waves are effective in pion–proton scattering and in a phase shift analysis (ch. 14) only the phases corresponding to $s_{1/2}$, $p_{1/2}$, and $p_{3/2}$ initial states are required. The differential cross-section may be expressed as

$$\frac{d\sigma}{d\Omega} = A + B\cos\theta + C\cos^2\theta \tag{19.30}$$

in which the highest power of $\cos\theta$ is consistent with a p-wave reaction and the odd power of $\cos\theta$ arises from s–p interference. The coefficients A, B and C, or the three equivalent phase shifts for a given reaction may be determined by cross-section measurements at three angles at a given energy.

† An experiment of this type is described in Sect. 20.3.2.

Since there are three types of scattering, represented by eqs. (19.25), (19.26) and (19.27), analysis of the observations† starts with nine apparently independent pieces of information. If it is assumed that the π-nucleon interaction is charge independent then these nine pieces of data reduce, as will be shown in the following section, to six real phase shifts, three associated with each of the isobaric spins $T=\frac{1}{2}$ and $T=\frac{3}{2}$ possible for the pion–nucleon system. These phase shifts are denoted by the symbol δ with suffixes as follows:

$$T = \tfrac{1}{2} \quad s\text{-wave } \delta_1, \, p\text{-wave } \delta_{11}, \, \delta_{13}$$

$$T = \tfrac{3}{2} \quad s\text{-wave } \delta_3, \, p\text{-wave } \delta_{31}, \, \delta_{33}$$

The phase shift δ_{33} corresponding to the resonant state $T=\frac{3}{2}$, $\mathcal{J}=\frac{3}{2}$ passes through 90° at the resonant energy of 195 MeV. By observing the interference between Coulomb and nuclear scattering at small angles, it has been found that δ_{33} is positive, corresponding to an attractive force whereas in the s-wave, δ_3 is negative and the force is repulsive. The s-wave phase shifts are important since they determine the cross-section for the absorption of a bound negative pion from a mesonic atom formed with a proton. Direct absorption of a meson of energy ≈ 100 MeV by a proton has a small cross-section, so that the approximation of real rather than complex phase shifts is a good one.

(d) *Isobaric spin analysis.* It has been seen (Sects. **12.1** and 18.1.2) that there is good evidence for the approximate charge independence of nuclear forces. If these forces arise from a meson-exchange process (Sect. 18.3.2) then the basic pion–nucleon interaction may also be charge independent. Isobaric spin should then be a good quantum number, i.e. isobaric spin should be conserved in pion–nucleon reactions just as angular momentum is conserved. For the nucleon ($T=\frac{1}{2}$) and pion ($T=1$), isospin states with $T=\frac{1}{2}$ and $T=\frac{3}{2}$ can be formed as in the case of ordinary spin states, and with appropriate third components. As far as phase shift analysis (ch. 14) is concerned these states should be independent with scattering amplitudes $f_{1/2}(\theta)$ and $f_{3/2}(\theta)$ which depend only on total isospin and not on its third component. The isobaric spin states‡ $|T, T_z\rangle$ can be constructed from pion–nucleon states by the use of Clebsch–Gordan vector coupling coefficients:

$$|T, T_z\rangle = \sum (\text{C.G. coefficient}) \, |\text{nucleon, pion}\rangle \qquad (19.31)$$

Specifically, substituting for the coefficients, we arrive at the following table of normalized orthogonal states, in which $|p\pi^+\rangle$ stands for a state with the quantum numbers of the proton and π^+-meson.

† H. L. Anderson, E. Fermi, R. Martin and D. E. Nagle, *Phys. Rev.*, **91**, 155, 1953.

‡ The Dirac symbol $|\ \rangle$ represents a *state vector* in quantum mechanics, e.g. an ordinary wave function ψ or a column vector $\binom{a}{b}$. For the nucleon, $|p\rangle$ and $|n >$ represent eigenstates of the operators $\hat{t}^2$ and $\hat{t}_z$ (Sect. 12. 1) and may alternatively be written $|\frac{1}{2}, \frac{1}{2}\rangle$ and $|\frac{1}{2}, -\frac{1}{2}\rangle$. For more complex systems, the form $|T, T_z\rangle$ represents a simultaneous eigenstate of the operators T^2 and T_z. The state written $|\text{nucleon pion}\rangle$ however is not necessarily the eigenstate of any operator.

TABLE 19.1 *Isospin states for the pion–nucleon system*

$$|\tfrac{3}{2},\tfrac{3}{2}\rangle = |p\pi^+\rangle$$

$$|\tfrac{3}{2},\tfrac{1}{2}\rangle = \sqrt{\tfrac{1}{3}}\,|n\pi^+\rangle + \sqrt{\tfrac{2}{3}}\,|p\pi^0\rangle$$

$$|\tfrac{3}{2},-\tfrac{1}{2}\rangle = \sqrt{\tfrac{1}{3}}\,|p\pi^-\rangle + \sqrt{\tfrac{2}{3}}\,|n\pi^0\rangle$$

$$|\tfrac{3}{2},-\tfrac{3}{2}\rangle = |n\pi^-\rangle$$

$$|\tfrac{1}{2},\tfrac{1}{2}\rangle = \sqrt{\tfrac{2}{3}}\,|n\pi^+\rangle - \sqrt{\tfrac{1}{3}}\,|p\pi^0\rangle$$

$$|\tfrac{1}{2},-\tfrac{1}{2}\rangle = \sqrt{\tfrac{2}{3}}\,|n\pi^0\rangle - \sqrt{\tfrac{1}{3}}\,|p\pi^-\rangle$$

This shows how the isospin substates can be constructed from mixtures of the nucleon–pion states. Conversely we can express the nucleon–pion states as mixtures of isospin states, e.g.

$$|p\pi^+\rangle = |\tfrac{3}{2},\tfrac{3}{2}\rangle$$

$$|p\pi^0\rangle = \sqrt{\tfrac{2}{3}}\,|\tfrac{3}{2},\tfrac{1}{2}\rangle - \sqrt{\tfrac{1}{3}}\,|\tfrac{1}{2},\tfrac{1}{2}\rangle$$

$$|p\pi^-\rangle = \sqrt{\tfrac{1}{3}}\,|\tfrac{3}{2},-\tfrac{1}{2}\rangle - \sqrt{\tfrac{2}{3}}\,|\tfrac{1}{2},-\tfrac{1}{2}\rangle$$

$$|n\pi^+\rangle = \sqrt{\tfrac{1}{3}}\,|\tfrac{3}{2},\tfrac{1}{2}\rangle + \sqrt{\tfrac{2}{3}}\,|\tfrac{1}{2},\tfrac{1}{2}\rangle$$

$$|n\pi^0\rangle = \sqrt{\tfrac{2}{3}}\,|\tfrac{3}{2},-\tfrac{1}{2}\rangle + \sqrt{\tfrac{1}{3}}\,|\tfrac{1}{2},-\tfrac{1}{2}\rangle$$

$$|n\pi^-\rangle = |\tfrac{3}{2},-\tfrac{3}{2}\rangle$$

In the scattering of positive pions by protons the incident wave contains a factor $|\tfrac{3}{2},\tfrac{3}{2}\rangle$ and the scattered wave the factor $f_{3/2}(\theta)|\tfrac{3}{2},\tfrac{3}{2}\rangle$. In the π^--proton scattering however the incident system is a mixture of isospin states and the scattered wave contains the factor

$$\sqrt{\tfrac{1}{3}}f_{3/2}(\theta)\,|\tfrac{3}{2},-\tfrac{1}{2}\rangle - \sqrt{\tfrac{2}{3}}f_{1/2}(\theta)\,|\tfrac{1}{2},-\tfrac{1}{2}\rangle$$

If we now return to pion–nucleon states this becomes:

$$\sqrt{\tfrac{1}{3}}f_{3/2}(\theta)\{\sqrt{\tfrac{1}{3}}\,|p\pi^-\rangle + \sqrt{\tfrac{2}{3}}\,|n\pi^0\rangle\} - \sqrt{\tfrac{2}{3}}f_{1/2}(\theta)\{\sqrt{\tfrac{1}{3}}\,|n\pi^0\rangle - \sqrt{\tfrac{2}{3}}\,|p\pi^-\rangle\}$$

$$= |p\pi^-\rangle\left\{\tfrac{1}{3}f_{3/2}(\theta) + \tfrac{2}{3}f_{1/2}(\theta)\right\} + |n\pi^0\rangle\left\{\tfrac{\sqrt{2}}{3}f_{3/2}(\theta) - \tfrac{\sqrt{2}}{3}f_{1/2}(\theta)\right\}$$

This gives the immediate physical interpretation that for π^-p the elastic scattering amplitude is $\tfrac{1}{3}(f_{3/2} + 2f_{1/2})$ and the charge-exchange amplitude is $\sqrt{2}/3(f_{3/2} - f_{1/2})$. For the π^+p system there is only elastic scattering with amplitude $f_{3/2}$. The amplitudes are functions of energy as well as of angle.

If the isobaric spin state $T = \tfrac{3}{2}$ is dominant, $f_{1/2}$ may be neglected and the differential cross-sections for the three reactions at any angle should stand in the ratio $(1/3)^2 : (\sqrt{2}/3)^2 : 1$, i.e. as $1:2:9$. The cross-sections integrated over angle should exhibit the same ratio and the total (π^+p) to (π^-p) cross-section should be $9:(2+1)$, i.e. $3:1$. This is consistent with the cross-section values near 200 MeV shown in Fig. 19.10. If the total cross-sections for the two isospin states are extracted from the data the variation with energy is as shown in Fig. 19.10(c).

The angular distributions for the three reactions being considered are not in fact identical and a closer description requires s and p amplitudes (or phase shifts) in both the $T=\frac{3}{2}$ and $T=\frac{1}{2}$ isobaric spin states, i.e. $f_{1/2}$ cannot be entirely disregarded. As long as inelastic reactions are neglected however only six real phase shifts are required. That these phases give an adequate account of nine pieces of data in the experiment of Anderson *et al.* is good evidence for the hypothesis of charge independence† in the pion–nucleon system. The extraction of the phase shifts of course does not shed light on the mechanism of the pion–nucleon scattering, for which some kind of dynamical model must be invoked.

19.2.4 Bound pions (pionic atoms)

As in the case of muonic atoms (Sect. 11.2.3) atoms may be formed in which a π^--meson replaces an electron. The pion occupies a set of levels whose energies are approximately m_π/m_e times those of an electron, and whose wave functions extend over a region scaled down in size by the same factor. Such atoms were observed at Rochester in 1952 and recently accurate measurements of the pionic X-ray energies have been made using silicon and germanium detectors.‡ The energies of pionic levels can be calculated from the appropriate wave equation with a correction for the fact that the nucleus is not a point charge. For the higher pionic states in light nuclei this is a small correction and accurate determination of the transition energies (e.g. $4f \rightarrow 3d$ in titanium, $E = 87.622 \pm 0.001$ keV) give the pion mass (Sect. 19.2.2).

As the pion cascades through the sequence of atomic levels, the probability of nuclear absorption, which is always much greater than in the similar muonic atom, increases and the K X-rays ($2p \rightarrow 1s$) have only been observed with precision for elements with $Z \leq 12$. The nuclear capture from the $1s$ state is a rapid process and the line width is correspondingly large (≈ 5 keV for $Z = 12$). Since the strong interaction produces energy shifts, the pionic X-rays are less useful for nuclear size determination than the muonic radiations. The sign of the energy shift is consistent with a repulsive potential in the s-wave pion–nucleon interaction.

19.2.5 π-meson production

(a) *Photoproduction.* The bremsstrahlung from electron synchrotrons has been used to create charged and neutral mesons by the processes

$$\gamma + p \longrightarrow \begin{cases} \pi^+ + n \\ \pi^0 + p \end{cases} \qquad (19.32)$$

---

† Other evidence includes the existence of charge multiplets in nuclear spectra (Sect. **12.1**), and in elementary particle physics (App. 8), the properties of stable nuclei (Sect. **18.1**), the behaviour of the two-body system (Sect. **18.2**), and the existence of selection rules for nuclear reactions and gamma ray emission, which are based on isobaric spin conservation.

‡ D. A. Jenkins *et al.*, *Phys. Rev. Lett.*, **17**, 1, 1966.

in hydrogen, and

$$\gamma + n \longrightarrow \begin{cases} \pi^0 + n \\ \pi^- + p \end{cases} \tag{19.33}$$

in deuterium. The mesons may be detected in counter telescopes or spark chambers, using lead sheets to convert photons from π^0-decay into electron pairs.

The total cross-section for π^0-meson production as a function of energy (Fig. 19.11) shows a peak at a photon energy of about 300 MeV, which yields a pion–nucleon energy corresponding to the $\mathcal{J} = \frac{3}{2}^+$, $T = \frac{3}{2}$ resonance seen in scattering. Other peaks corresponding to higher pion–nucleon resonances are also seen and from the angular distributions of pion production it is possible to infer spin-values for the state concerned, which act as short-lived intermediate systems in reactions (19.32) and (19.33).

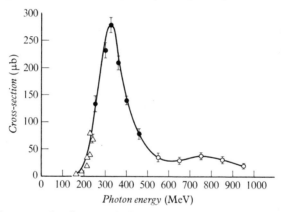

Fig. 19.11 Total cross-section for neutral pion photo-production from protons (reprinted by permission from *Nuclei and Particles* by E. Segrè, © 1964, W. A. Benjamin, Inc., Menlo Park, California.

(b) *Nucleon–nucleon collisions.* The Yukawa one-pion exchange theory of nuclear forces (Sect. 18.3.2) envisages the sharing of a virtual pion between two nucleons, e.g.

$$p + p \rightleftarrows p + p + \pi^0$$

The pion cannot be observed directly for nucleons at rest, but if the energy of the system is sufficiently increased, production of real pions takes place. The combined inelastic cross-section for the three processes

$$\begin{align} p + p &\longrightarrow p + p + \pi^0 \\ p + p &\longrightarrow p + n + \pi^+ \tag{19.34} \\ p + p &\longrightarrow d + \pi^+ \end{align}$$

(together with the contribution from multiple pion processes at high energies) is shown in Fig. 18.9(a); the data are obtained from work with synchrotrons and synchrocyclotrons using counter detection as in the case of photoproduction. These inelastic reactions cause an increase in the total (elastic + inelastic) cross-section for the (pp) interaction as shown in Fig. 18.9(a) above the pion production threshold at about 300 MeV. Below this energy the interaction is wholly elastic. For incident energies near 600 MeV (lab.) one of the pions in reactions (19.34) and a nucleon can form the $\mathcal{J} = \frac{3}{2}^+$ $T = \frac{3}{2}$ excited state $\Delta(1236)$. The reactions may then be supposed to proceed as follows:

$$
\begin{matrix}
(a) & p+p \longrightarrow n+\Delta^{++} \\
\begin{pmatrix} T & \frac{1}{2} & \frac{1}{2} & & \frac{1}{2} & \frac{3}{2} \\ \mathcal{J} & & {}^1D_2 & & \frac{1}{2} & \frac{3}{2} \end{pmatrix} \\
(b) & \Delta^{++} \longrightarrow p+\pi^+ \\
(T & \frac{3}{2} & & & \frac{1}{2} & 1)
\end{matrix}
\qquad (19.35)
$$

in which the values of isobaric spin and angular momentum are indicated. The two protons together have $T_z = 1$ so the initial state is of isospin 1. At bombarding energies up to about 1000 MeV only s-wave emission of the nucleon and Δ will be important and the corresponding system (nΔ) will have spin-parity 1^+ or 2^+. Because of the Pauli exclusion principle only the state 2^+ (spectroscopically ${}^1D_2$) is then possible for the two protons in the incident system and from the general theorems on reactions (ch. 15) the maximum cross-section for such a process is

$$
\sigma_{\text{inel}}^{\text{max}} = 2 \frac{2l+1}{(2s_1+1)(2s_2+1)} \pi \lambdabar^2
\qquad (19.36)
$$

where the 2 arises from identity of the particles, $l = 2({}^1D_2)$, $s_1 = s_2 = \frac{1}{2}$ and $\lambdabar$ is the reduced de Broglie wavelength of the incident proton in the c.m. system. This gives a value of 7 mb compared with the observed 25 mb for all reactions (19.34). Processes other than Δ production with an accompanying s-wave nucleon must therefore contribute. A full discussion of the development of this *isobar model* for pion production in nucleon–nucleon collisions is given in Ref. 19.8, Chapter 8.

In the collision of neutrons with protons the single pion production processes are

$$
\begin{aligned}
n+p &\longrightarrow n+p+\pi^0 \\
n+p &\longrightarrow d+\pi^0 \\
n+p &\longrightarrow n+n+\pi^+ \\
n+p &\longrightarrow p+p+\pi^-
\end{aligned}
\qquad (19.37)
$$

and existing results on the total cross-section are shown in Fig. 18.9(b). The initial system has $T_z = 0$ and this can arise equally from isospin states with $T = 0$ and $T = 1$. On the assumption of charge independence the $T = 1$ contribution is given by Fig. 18.9(a) and the cross-section in the isospin state $T = 0$ can be deduced. This is given in Fig. 18.9(b) and it is seen that there is no strong Δ production in the $T=0$ state, in agreement with the known isobaric spin ($\frac{3}{2}$) of the Δ.

(c) *Meson–nucleon collisions; the ρ-meson.* The processes

$$\pi^{\pm}+p \longrightarrow p+\pi^{\pm}+\pi^{0}$$
$$\pi^{\pm}+p \longrightarrow n+\pi^{+/0}+\pi^{+/0} \qquad (19.38)$$
$$\pi^{-}+p \longrightarrow n+\pi^{+}+\pi^{-}$$

have the especial interest that two mesons appear in the final state, and information on the meson–meson interaction may be obtained. Such reactions have been extensively studied by track chamber methods in which the mass of the meson pair is calculated from the observed momenta. It is found that $\pi^{+}\pi^{-}$ states show a grouping at mass 765 MeV whereas the $\pi^{+}\pi^{+}$ states ($T=2$) do not. The width of this state is ≈ 100 MeV so that its life is $\approx 10^{-23}$ s, but despite this it has been possible to assign quantum numbers $\mathcal{J}=1$, $T=1$ (odd parity), by analysis of production cross-sections. The state is obviously basic to the meson–meson interaction and it is customary to describe it as the ρ-meson; it occurs in three charge states ρ^{+}, ρ^{-} and ρ^{0}.

(d) *Antiproton–proton collisions: the ω-meson.* The antiproton $\bar{p}$ was discovered at Berkeley† in a counter study of the reaction

$$p+p \longrightarrow p+p+p+\bar{p} \qquad (19.39)$$

for which there is a threshold of 5·6 GeV. The antineutron $\bar{n}$ was identified later in a similar experiment. In terms of the Yukawa theory the pion is coupled strongly to nucleons and antinucleons and in particular a pion may be regarded as a virtual nucleon–antinucleon pair, e.g.

$$\pi^{+} \rightleftarrows p+\bar{n}$$

It follows that the most likely mode of antinucleon annihilation is by pion production and reactions of the type:

$$p+\bar{p} \longrightarrow n\pi^{+}+n\pi^{-}+\pi^{0} \qquad (19.40)$$

are extremely convenient for analysis of pion–pion interactions. Figure 19.12 shows the results of a celebrated bubble chamber experiment with antiprotons‡ in which the distribution of effective mass§ of various combinations of pions was explored. In the case of 3 pions, the state with charge zero shows a marked peak at a mass of 790 MeV but no such peak appears in the states with charge 1 or 2. The favoured state therefore has $T=0$. The width of the

† O. Chamberlain *et al.*, *Phys. Rev.*, **100**, 947, 1955; B. Cork *et al.*, *Phys. Rev.*, **104**, 1193, 1965.
‡ B. C. Maglic, L. Alvarez, A. H. Rosenfeld and M. L. Stevenson, *Phys. Rev. Lett.*, **7**, 178, 1961.
§ The effective mass of a group of particles is the total energy in their c.m. system. If E_i, $\mathbf{p}_i$ are the total energy and momentum of the individual particles then the effective mass is defined by

$$M^{2}c^{4} = (\textstyle\sum E_{i})^{2} - c^{2}(\textstyle\sum \mathbf{p}_{i})^{2}$$

where the momenta are added vectorially. In high energy physics it is customary to simplify equations such as this by writing $c=\hbar=1$.

peak is much smaller than in the case of the ρ-meson and is determined by the experimental resolution. Analysis of the decay data (App. 7) indicates that the state has odd parity and spin 1; it is known as the ω-meson.

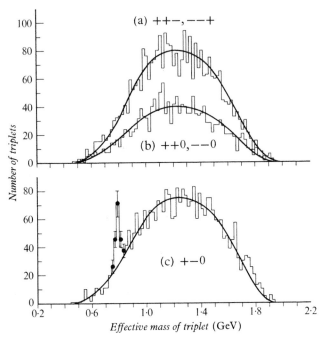

Fig. 19.12 Discovery of the ω-meson, in a bubble chamber study of the reaction $p + \bar{p} \longrightarrow 2\pi^+ + 2\pi^- + \pi^0$. The distributions show the effective mass of all combinations of 3 pions, i.e. the tendency for any grouping of specific energy to form. Distributions (a) and (b) relate to total charge 1 and 2 and are essentially 'phase space' or statistical plots. Distribution (c), for charge zero, shows the ω peak at 787 MeV (B. C. Maglic *et al.*).

(*e*) *The η-meson.* The η-meson is also seen by analysis of pion triplets, but in the reaction

$$\pi^+ + d \longrightarrow p + p + \pi^+ + \pi^- + \pi^0 \qquad (19.41)$$

Figure 19.13 shows the mass distribution obtained in such an experiment, and a small peak at ≈ 550 MeV can be seen. From detailed analysis of this and other production reactions this peak (the η-meson) has been shown to correspond to $\mathcal{J} = 0$, $T = 0$ (parity odd). This particle decays electromagnetically by the processes

$$\begin{aligned} \eta &\longrightarrow 2\gamma \\ \eta &\longrightarrow \pi^0 + \pi^0 + \pi^0 \qquad (19.42) \\ \eta &\longrightarrow \pi^+ + \pi^- + \pi^0 \end{aligned}$$

The $3\pi^0$-decay implies that isospin is not conserved in the η-decay, because three π^0-mesons cannot form a $T = 0$ state (Ref. 19.3). Since the decay is too fast to be a weak process, this confirms that the electromagnetic interaction is involved (ch. 21). The ω (and ρ)-mesons, which decay by the strong interaction, conserve isospin and $3\pi^0$ decay of the ω is not observed.

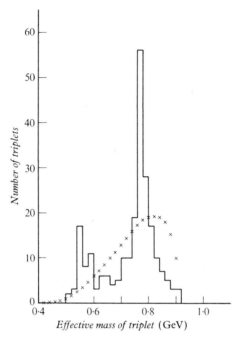

Fig. 19.13 Discovery of the η-meson, in a deuterium-filled bubble chamber study of the reaction $\pi^+ + d \longrightarrow p + p + \pi^+ + \pi^- + \pi^0$. The peak near 800 MeV in the effective mass distribution is due to the ω-meson. The crosses show the expected statistical or phase space distribution. The small peak near 550 MeV is due to the η (A. Pevsner *et al.*, *Phys. Rev. Lett.*, **7**, 421, 1961).

19.3 Muons

The discovery of the muon in the cosmic radiation and the subsequent observation of the $\pi \rightarrow \mu$ decay as a source of cosmic ray muons at sea level have already been discussed (Sect. **19**.1). As for the pion, precise knowledge of muon properties comes from accelerator experiments, particularly those done with synchrocyclotrons. These reveal the fact that the muon, which occurs in the two charge states μ^+ and μ^- only, is in all known respects so far essentially a heavy electron.

19.3.1 Properties of the muon (App. 8)

(a) *Mass and charge.* A direct determination of the mass of the μ^+-meson was made by comparing it with the π^+ mass in a magnetic deflection experiment using the field of the Berkeley synchrocyclotron. Pions originating in an internal target and muons from the decay of pions stopped in the same target were received after magnetic deflection in a nuclear emulsion detector in which their ranges were measured. In the same experiment the π^+ mass was compared with that of the proton. As in this case, when the momentum ratio is equal to the ratio of ranges, the two particles have the same velocity and the momentum ratio then gives the ratio of rest masses.

The most accurate method is based on the radiations from muonic atoms (Sect. 11.2.3). The $3d \to 2p$ mesonic transition in phosphorus has an energy which is very close to the K-absorption edge of lead (Fig. 5.17). Near the edge the attenuation coefficient varies very rapidly with energy, but it is known with high accuracy from X-ray measurements with a crystal spectrometer. A precise measurement of the attenuation then determines the μK transition energy to about 1 part in 10^4 and from this the μ^- mass to similar accuracy.

Although no experiment specifically shows that the muon carries the electronic charge, this conclusion is highly probable because of the agreement of mass values obtained by different methods.

(b) *Lifetime.* The original cosmic ray measurements have been greatly refined; the most recent work essentially uses the signal from a muon production process $\pi^+ \to \mu^+$ in a scintillator to start a timing circuit and the subsequent decay electron pulse $\mu^+ \to e^+$ to stop it. A precision of 0·1 per cent has been achieved. The μ^+ and μ^- lifetimes have been compared and are equal, also to 0·1 per cent, as predicted theoretically by invariance properties.

(c) *Decay modes; parity non-conservation.* Plate 17 shows the $\pi \to \mu \to e$ decay as registered in a nuclear emulsion and in a bubble chamber. The decay electrons of positive muons from accelerator produced pions have been studied using bubble chambers and magnetic spectrometers and a spectrum is shown in Fig. 19.14(a). Since this is continuous as in the case of β-decay (ch. 16) the decay process must be at least of a three-body nature, and since no electromagnetic radiation other than bremsstrahlung has been detected, the decay plausibly involves neutrinos, e.g.

$$\mu^+ \longrightarrow e^+ + \nu_e + \bar{\nu}_\mu \tag{19.43}$$

and

$$\mu^+ \not\longrightarrow e^+ + \gamma \tag{19.44}$$

From the lifetime and the energy release in the decay of unpolarized muons the basic probability of the reaction, or coupling constant, can be calculated as in the case of β-decay. The value found is $g^\mu = 1·435 \times 10^{-62}$ J m^3 which is extremely close to the β-decay value $g^\beta = 1·410 \times 10^{-62}$ J m^3. The significance of this result, which emphasises the fact that both β and muon decay are examples of the basic weak interaction, will be discussed in Chapter 21.

The upper limit of the electron spectrum (≈ 54 MeV) corresponds with prediction from the masses of the muon and electron on the assumption that at the upper limit the two neutrinos move in the opposite direction to the electron. If the neutrinos have a definite helicity as suggested by the discovery of parity non-conservation, then with the scheme shown in Fig. 19.14(b), the neutrinos cannot be identical, since this collinear decay would be forbidden by the Pauli principle. Originally it was proposed that a particle–anti-particle pair should be emitted, but now it is known that there is the further difference, shown by suffixes in eq. (19.43), that one neutrino is electron-associated and the other is muon-associated.

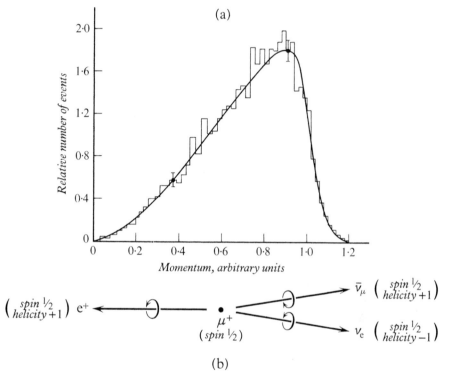

Fig. 19.14 (a) Momentum spectrum of positrons from μ^+ decay observed in a liquid hydrogen bubble chamber in a magnetic field (R. J. Plano, *Phys. Rev.*, **119**, 1400, 1960). (b) Helicities in muon decay. Note that in contrast with the pion decay (Fig. 19.7), the e^+ particle may now carry positive helicity. The collinear decay takes place with high probability.

Parity non-conservation (PNC) in $\pi \to \mu \to e$ decay was established almost simultaneously with non-conservation in β-decay. If there is PNC in the pion decay then the muon will be longitudinally polarized (cf. Sect. **16.7** and Fig. 19.7). If also there is PNC in the muon decay there will be an asymmetry in the direction of electron emission from the muon with respect to its polarization, just as in electron emission from polarized nuclei. The detection of this asymmetry by Garwin, Lederman and Weinrich† not only established the

† *Phys. Rev.*, **105**, 1415, 1957; see also *Phys. Rev.*, **118**, 271, 1960.

presence of PNC in the ($\pi\mu e$) process (para. (*e*) below) but also provided a powerful and elegant method of investigating the properties of the muon. The helicity of the electrons from decaying muons has been determined by observation of the circular polarization of bremsstrahlung; it agrees with the assumption shown in Fig. 19.14(b).

(*d*) *Spin of the muon; isobaric spin; parity.* All direct evidence, such as decay schemes and hyperfine effects in muonic atoms, support the assignment of a spin quantum number $\frac{1}{2}$ to the muon. Direct evidence is obtained from the properties of muonium, an association of a positive muon with a negative electron which is formed when muons slow down in matter (Ref. 19.4). The muonic atom $\mu^+ e^-$ has a resultant magnetic moment which may be calculated in a way analogous to the calculation of the Landé factor for an ordinary atom. The moment may be observed by a precision experiment using the muon decay asymmetry to indicate the spin direction. The result is only consistent with $s_\mu = \frac{1}{2}$.

Isobaric spin is a quantum number associated with strongly interacting particles and is not defined for muons, electrons or neutrinos. The spin of the muon and its close analogy to the electron indicate that it is a fermion. This means that, like the electron and unlike the pion, it is subject to conservation laws (ch. 21). Production of muons by an electromagnetic interaction for example yields a muon *pair* and an intrinsic parity (odd) is only meaningful for this particle–antiparticle system.

(*e*) *Magnetic moment of the muon; the (g-2) experiment.* If the muon can be described by the Dirac theory of the electron, it should have a magnetic moment of one muon magneton, namely $e\hbar/2m_\mu$ corresponding to a g-factor of 2 for spin $\frac{1}{2}$. The asymmetry of the muon decay permits extremely accurate measurements of the moment and particularly of any anomalous part of it, which may be specified by the quantity $(g-2)/2$. The muon moment, in units of the muon magneton, is $\frac{1}{2}g$.

In the experiments of Garwin *et al.* (Fig. 19.15(a)) muons from π^+-mesons decaying in flight near a synchrocyclotron target were brought to rest in a carbon target. If the muons maintain the longitudinal polarization with which they are endowed by the π^+ decay their spins are aligned antiparallel to the direction of incidence (Fig. 19.7). A magnetic field B perpendicular to the plane of incidence then causes a precession of the muon moment with the Larmor frequency $g_\mu \mu_\mu B/h$ in the plane of incidence and an alteration in the positron counting rate in the detectors, because of the non-isotropic distribution of the decay. The detectors are set to respond at a certain fixed time after the arrival of the muon in the carbon target and the response obtained as a function of magnetic field is shown in Fig. 19.15(b). The precessional frequency observed in the first experiments already proved that the muon g-factor was extremely close to 2 as in the case of the electron.

The rapid and extreme refinement of this result has been one of the outstanding achievements of physics in the last 10 years. From the many elegant

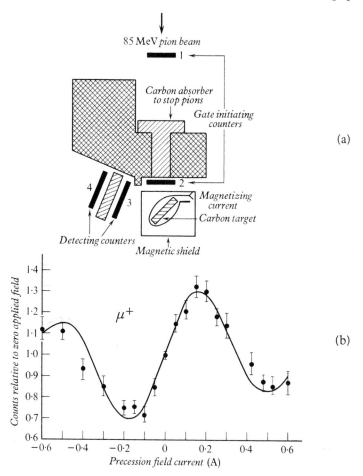

Fig. 19.15 Parity non-conservation in muon decay. (a) Apparatus of Garwin *et al.* to detect precession of muon magnetic moment in a magnetic field using asymmetric decay as indicator of spin direction. (b) Variation of electron counting rate in fixed detector as function of magnetic field.

experiments we select here only the latest and most precise, due to Bailey, Farley *et al.*† at CERN and specifically directed at the anomalous part of the moment. This essentially compares the Larmor frequency of the muon in a given magnetic field with the orbital frequency of the particle in the same field. These are equal if $g_\mu = 2$, but if $(g\text{-}2)$ has a small but finite value then a beam of longitudinally polarized muons in a circular orbit will gradually build up misalignment between the spin direction and the direction of motion. This situation has been realized by the use of a storage ring (Fig. 19.16) which is filled with fast muons produced by pion decay in the same orbit. When these muons decay in flight, the relativistic transformation from the rest system ensures that the forward-going electrons have the highest energy, and

† J. Bailey *et al.*, *Phys. Lett.*, **28B**, 287, 1968; *Progr. nucl. Phys.*, **12**, 43, 1970.

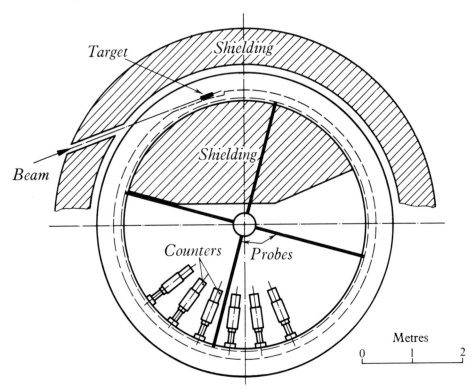

Fig. 19.16 Muon storage ring. Protons strike a target within the magnetic field and muons created by $\pi \longrightarrow \mu$ decay are stored. The probes are for measuring the magnetic field (Bailey *et al.*).

the detector circuits are set to respond only to these. Since the angular distribution of emission from muon decay is not isotropic, the observed intensity of electron emission is modulated by the muon precession due to the anomalous moment. Figure 19.17 shows the spectacular result obtained for the counting rate as a function of time after the injection of pions into the ring. From this experiment, which has probably not yet reached the limit of accuracy obtainable

$$a_\mu = \tfrac{1}{2}(g_\mu - 2) = 116616 \pm 31 \times 10^{-8}$$

with no difference within errors between results for μ^+ and μ^-. This is to be compared with calculation (Sect. 21.2.2)

$$a_\mu = 116564 \times 10^{-8}$$

and with a similar result for the electron (Sect. 21.2.2). The agreement between these figures confirms other evidence that the muon behaves essentially as a heavy electron but above all verifies the theory of electrodynamics.

The same experiment confirms the relativistic time dilation for muons in a circular orbit and gives an extremely accurate estimate of the muon lifetime.

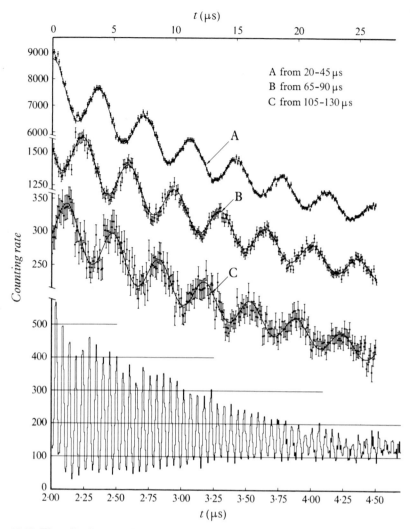

Fig. 19.17 Time distribution of decay electron events in muon storage ring. Lower curve gives the rotation frequency of muon (period 50 ns) (Bailey *et al.*).

19.3.2 *Interactions of muons with nuclei*

(a) *Electromagnetic interaction.* The passage of muons through matter is analogous to that of electrons except that radiative processes are much diminished because of the greater mass. There is cosmic ray evidence that fast muons, observed at a depth of many metres underground, can produce nuclear disintegration stars and this has been interpreted as an electromagnetic process in which the field of the moving muon may be regarded as equivalent to a continuous distribution of photons (cf Coulomb excitation, Sect. 13.6.2). These virtual quanta may be absorbed by nuclei with resulting

531

disintegration. Muon beams of energy up to several hundred MeV have been used for nuclear scattering experiments and the results agree with those obtained by electron scattering.

Slow negative muons form muonic atoms and slow positive muons may form muonium (Ref. 19.4) which is analogous to positronium. Normal decay takes place from both these systems, and (weak) nuclear interaction is also possible in the case of the mesonic atoms. The negative muon may also bind two atoms to form a muonic molecule.

Pair production of muons ($\mu^+\mu^-$) by high energy bremsstrahlung has been observed and is in agreement with theory.

(b) *Weak interaction; the muon neutrino.* The chief weak process associated with the muon is its own decay (Sect. 19.3.1(c)) but the capture of negative muons by nuclei from the lower orbits of mesonic atoms is also possible. The basic process within the nucleus is

$$\mu^- + p \longrightarrow n + \nu_\mu \tag{19.45}$$

and this has in fact been observed for hydrogen although comparison with theory is complicated by atomic processes. In heavier nuclei e.g.:

$$\begin{aligned} \mu^- + {}^3\mathrm{He} &\longrightarrow {}^3\mathrm{H} + \nu_\mu \\ \mu^- + {}^{12}\mathrm{C} &\longrightarrow {}^{12}\mathrm{B} + \nu_\mu \\ \mu^- + {}^{16}\mathrm{O} &\longrightarrow {}^{16}\mathrm{N} + \nu_\mu \end{aligned} \tag{19.46}$$

the processes are analogous to β-decay between the nuclei, but because of the larger energy release, many excited states of the residual nuclei may be produced. The basic probability for the capture reaction may be estimated from the yield of product nuclei and is similar to that already encountered in β-decay and muon decay. Such nuclear capture reduces the observed mean life of muons stopping in heavy material.

It will be noted that so far reactions involving muons have usually been written to indicate the appearance of a muon-associated neutrino ν_μ. The particle was conclusively distinguished from the electron-associated neutrino ν_e of β-decay in a fundamental experiment[†] at the Brookhaven AG-Synchrotron. The apparatus, shown in Fig. 19.18, was arranged to allow neutrinos from the decay of pions in flight to pass through extensive shielding to an interaction area. The production processes

$$\pi^\pm \longrightarrow \mu^\pm + \nu_\mu \text{ (or } \bar{\nu}_\mu) \tag{19.47}$$

provided the presumed muon-associated neutrinos and the interactions sought in the material of spark chambers (weighing 10 tons), were

$$\begin{aligned} \nu_\mu + n &\longrightarrow p + \mu^- \quad \text{or} \quad p + e^- \\ \bar{\nu}_\mu + p &\longrightarrow n + \mu^+ \quad \text{or} \quad n + e^+ \end{aligned} \tag{19.48}$$

† G. Danby *et al.*, *Phys. Rev. Lett.*, **9**, 36, 1962.

Clear evidence, later confirmed with better statistics at CERN, was obtained to show that the muon-associated neutrinos do not produce electron reactions. This established the existence of two types of neutrino, each with its own antiparticle. All experimental evidence so far confirms the identity of these two types of neutrino with the single exception of the property of muon or electron association.

The existence of two types of neutrino immediately explains the non-occurrence of the decay mode:

$$\mu^+ \longrightarrow e^+ + \gamma \qquad (19.49)$$

This does indeed occur with neutrino emission in accordance with the scheme

$$\mu^+ \longrightarrow e^+ + \nu_e + \bar{\nu}_\mu + \gamma \qquad (19.50)$$

and is known as radiative decay, but cannot occur without neutrinos unless ν_e and $\bar{\nu}_\mu$ are particle and antiparticle; and this is ruled out by the experiment of Danby *et al.*

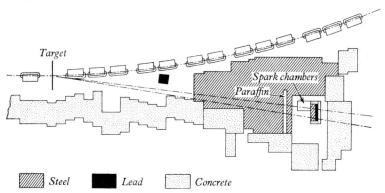

Fig. 19.18 The muon–neutrino experiment. Pions produced by 15 GeV protons at the target in the accelerator decay in flight. Neutrinos from the decay reach the 10-ton aluminium spark chambers through a 13·5 m steel wall (G. Danby *et al.*).

(c) *Muonic atoms.* These have been discussed (Sect. 11.2.3) in connection with the determination of the charge radii of nuclei. Such systems have much intrinsic interest and offer possibilities of the determination of nuclear moments by hyperfine interaction effects (Ref. 19.4).

References

1. POWELL, C. F. 'Mesons', *Rep. progr. Phys.*, **13**, 350, 1950.
2. BARKAS, W. H. 'Masses of the Metastable Particles', *Ann. rev. nucl. Sci.*, **15**, 67, 1965.
3. KÄLLEN, G. *Elementary Particle Physics*, Addison-Wesley, Reading, Mass., 1964.
4. BURHOP, E. H. S. 'Exotic Atoms', *Contemp. Phys.*, **11**, 335, 1970.
5. WOLFENDALE, A. W. *Cosmic Rays*, Newnes, London, 1963; ROSSI, B. *Cosmic Rays*, McGraw-Hill, New York, 1964.
6. SEGRÈ, E. *Nuclei and Particles*, Benjamin, New York, 1964.
7. GALBRAITH, W. 'Hadron-nucleon cross sections at high energies', *Rep. prog. Phys.*, **32**, 547, 1969.
8. LOCK, W. O. and MEASDAY, D. F. *Intermediate Energy Nuclear Physics*, Methuen, London, 1970.

20 Particle physics II (strange particles)

The discovery of the muon and pion and the understanding of their properties seemed for a time to offer the possibility of explaining the main phenomena of low energy nuclear physics and of the cosmic radiation. However even before the elucidation of the $\pi \to \mu \to e$ decay had been completed, isolated tracks deriving from new types of particle had been seen in cloud chambers exposed to the cosmic rays. These were in fact due to what are now called K-mesons and hyperons, and their discovery opened an exciting field of still unknown extent in which new laws have been found to operate. The study of these *strange particles* now centres experimentally on large accelerators where the properties of the particles themselves and the laws that govern their interactions are still being established; the present chapter and that following survey the status of this rapidly developing subject.

20.1 The discovery of K-mesons and hyperons

As part of the search for the Yukawa particle, cosmic ray physicists from 1938 onwards devoted increasing attention to the showers of penetrating particles occasionally observed at sea level. These had to be distinguished from the larger background of electron–positron cascade showers and by 1946 the most powerful technique was that of Rochester and Butler at Manchester in which a counter-controlled cloud chamber containing a lead plate was operated in a strong magnetic field. Plate 18 shows one of the pictures obtained in 1946 and 1947[†] which contained tracks which were ascribed to new types of particle. Because of the characteristic appearance of some of these tracks the particles concerned were named V-particles.[‡]

Although the first results of Powell's work on the $\pi \to \mu \to e$ sequence were only about to appear, Rochester and Butler established with certainty from the kinematics of two events that the unusual tracks must originate from a particle considerably heavier than the π-meson. The first photograph (Plate 18) showed the decay of an uncharged particle (now known to be $K^0 \to \pi^+ + \pi^-$) and the second that of a charged particle ($K_\mu^+ \to \mu^+ + \nu$). To confirm and extend the original observations the cloud chamber and magnet were taken to the 2687 m high Pic du Midi in the French Pyrenees and by 1950 more

[†] G. D. Rochester and C. C. Butler, *Nature*, **160**, 855, 1947.
[‡] This terminology survived until about 1953 when the present nomenclature came into general use.

534

V-particles were being recorded, not only by the Manchester group but also by other laboratories. It soon became clear that there existed two uncharged V-particles, the V_1^0 which could decay into a proton and pion

$$V_1^0 \longrightarrow p + \pi^- \quad \text{(now known as } \Lambda^0\text{)} \tag{20.1}$$

as well as the original particle with the decay

$$V_2^0 \longrightarrow \pi^+ + \pi^- \quad \text{(now known as } K^0\text{)} \tag{20.2}$$

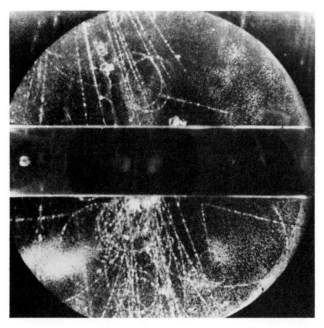

Plate 18 Discovery of strange particles. A K^0-meson produced by a cosmic ray interaction in a lead plate in an expansion chamber decays (bottom right of picture) into $\pi^+ + \pi^-$ (G. D. Rochester and C. C. Butler, *Nature*, **160**, 855, 1947).

The masses were found to be greater than the sum of the product masses by the Q-value (≈ 40 MeV for V_1^0, ≈ 200 MeV for V_2^0) and the lifetimes deduced from the distances travelled from the presumed area of origin in the lead plate, seemed to be about 10^{-10} s.

In the early 1950s many events of these types and a variety of new ones were observed in cloud chambers and nuclear emulsions exposed to the cosmic rays and finally in a diffusion cloud chamber traversed by 1·4 GeV negative pions from the Brookhaven Cosmotron (proton synchrotron). This last work indicated from the decays observed that one production reaction was†

$$\pi^- + p \longrightarrow \Lambda^0 + K^0 \tag{20.3}$$

† See Plate 20.

The Cosmotron work also revealed that the K^0-particles included a variety with a lifetime of the order of 10^{-8} s in accordance with a theoretical prediction by Gell-Mann and Pais that a beam of this form of meson might consist of a mixture of short-lived and long-lived types (see Sect. 20.4.4). Long-lived charged mesons decaying into two or three charged particles were discovered in nuclear emulsions by a Bristol group. Emulsion work (Danysz and Pniewski) also showed that the Λ^0 particle could bind itself to a nucleus, replacing a nucleon and forming a short-lived hyperfragment (Sect. 20.5.5).

The accepted nomenclature for the elementary particles, reached in 1953 after examination of the accumulated cosmic ray data, is as given in Sect. 1.2.1, namely there are recognized *leptons, mesons* and *baryons*. The heavier baryons (*hyperons*) which decay ultimately to nucleons are labelled Y and the heavier mesons which decay to pions and muons are labelled K. There may be individual names within these classes, e.g. the hyperons include Λ^0 (V_1^0), Σ, Ξ and Ω particles and the K-mesons include θ and τ particles (see Sect. 20.6.1). A general survey of particle classification will be given in Chapter 21.

20.2 Strangeness and isobaric spin; hypercharge

The key to the classification of the K-mesons and hyperons is the property described by the somewhat ill-chosen name *strangeness*. This arose from the early observations that the cross-section for the production of the strange particles by pions or protons from accelerators was a few per cent of the geometrical area of the target nucleus (usually a proton). On the other hand the decay times of the particles were $10^{-8} - 10^{-10}$ s which is very long on a nuclear time scale. These two facts are inconsistent in the same way that the long life of a negative muon near a nucleus (Sect. 19.1.1) is inconsistent with the properties necessary for the Yukawa particle. In both cases we expect the production and decay process to be connected by the law of detailed balancing (App. 6) so that particles produced with high probability should decay rapidly.

The way out of this difficulty was suggested by Pais, who proposed that the new particles when produced by the *strong* interaction process should always arise as seen in the Cosmotron experiments, e.g. as in (20.3) or as

$$\pi^- + p \longrightarrow \Sigma^- + K^+ \tag{20.4}$$

and that they should decay via the *weak* interaction, which has a much smaller basic probability. This *associated production hypothesis* was made specific by Gell-Mann and Nishijima who introduced a *strangeness* quantum number $S = 0, \pm 1, \ldots$ for each of the new particles and asserted that associated production (e.g. (20.3), (20.4)) can proceed rapidly because the nucleon and pion have strangeness zero and the Λ^0 (Σ^-) and K^0 (K^+) have strangeness -1 and $+1$ respectively so that this new quantity (for which the quantum number is additive) is conserved in the process.

In the decay processes

$$\Lambda^0 \longrightarrow p + \pi^-$$
$$K^0 \longrightarrow \pi^+ + \pi^- \tag{20.5}$$

the final products are of zero strangeness and this property is not conserved. Violation of strangeness conservation in particle physics indicates a weak process and a relatively long decay time.

Appendix 8 gives the strangeness numbers S for the new particles as deduced from their production mechanisms; all particles shown in Table 1.2, among which appear the nucleons and pions, are non-strange, with $S = 0$.

The Appendix table is built up from the definition that the Λ^0-particle has strangeness -1. Since the reaction (20.3) proceeds with high probability we infer from the hypothesis $\Delta S = 0$ that the K^0-meson has $S = +1$. Similarly, since it is found that the process

$$\pi^+ + n \longrightarrow \Lambda^0 + K^+ \tag{20.6}$$

occurs frequently we deduce $S = +1$ for the positive kaon. The reaction $\pi^- p \rightarrow \Sigma^- K^+$ (20.4) occurs and so does $\pi^+ p \rightarrow \Sigma^+ K^+$ but $\pi^- p \rightarrow \Sigma^+ K^-$ has never been observed so $S = -1$ for both Σ^- and Σ^+. Particles and their antiparticles, as shown in Appendix 8, have opposite strangeness.

Strangeness changing reactions, with $\Delta S = \pm 1$, are not absolutely forbidden, but are of low probability, occurring only in the weak interactions. There is a similar situation with isobaric spin, for which too the conservation law is not absolute and is violated in both weak and electromagnetic processes. The isobaric spins of strange particles (App. 8) can be inferred by arguments analogous to those used in assigning strangeness, remembering that $2T + 1$ gives the number of charge states of a particle with isobaric spin T, and that the actual charge of a non-strange particle is

$$Q = T_z + \frac{B}{2} \tag{20.7}$$

where B is the baryon number, i.e. the number of nucleons into which the particle ultimately resolves, with antinucleons counted negatively. For the strange particles we may use the reaction (20.3) and the fact that only one Λ-particle is known ($T = 0$) to deduce that the isobaric spin of the K^0 is $\frac{1}{2}$ or $\frac{3}{2}$; since four different charge states of the K-particle are not found we assume it to have $T = \frac{1}{2}$. Similar considerations enable the table to be completed; antiparticles have T_z opposite to that for particles.

It can now be seen that the strangeness number and the third component of the isobaric spin vector are related for the strange mesons ($B = 0$) by the relation

$$Q = T_z + \frac{S}{2} \tag{20.8}$$

and if we combine (20.7) and (20.8) into one formula, then the expression

$$Q = T_z + \frac{S}{2} + \frac{B}{2}$$
(20.9)

covers all particles, strange or non-strange. The quantity $S+B$ is called the *hypercharge*. This formulation shows that strangeness is not an independent, new quantity; it is related to a combination of charge, third component of isobaric spin and baryon number, each of which is regulated by conservation laws. The conservation of total isobaric spin is not so far implied but analysis of the yields of the πp reaction to different strange-particle charge states indicates that total i-spin is in fact a good quantum number in this case, although the evidence is less compelling than in the case of pion–nucleon scattering. For the discussion of strange particle properties we will assume that both strangeness and i-spin are rigorously conserved in strong inter-action processes.

20.3 Experimental studies in particle physics

Most of the recent data on particle properties have been obtained from bubble chamber or counter and spark chamber experiments in large synchrotron laboratories. All types of interaction can be studied in a bubble chamber and there is little control over the relative prominence of different types of event because the chamber cannot be triggered. In contrast, counter and spark chamber experiments are deliberately designed to study selected types of interaction and these have usually been simple processes such as total cross-sections, elastic scattering angular distributions and specific decay modes. These experiments produce results with higher statistical accuracy than those made with the bubble chamber, but the latter has so far proved best for the study of complicated events involving several particles in the final state.

The general principles of the bubble chamber and spark chamber were outlined in Chapter 6. Two typical investigations will now be described to illustrate the techniques involved and the type of information obtained. Although reactions involving strange particles have been chosen for these illustrations the techniques have also been used for studying many of the processes described in Chapter 19.

20.3.1 *Bubble chamber experiments* (Ref. 20.3)

Plate 19 shows tracks in a bubble chamber of particles originating from the process

$$K^- + p \longrightarrow K^0 + \Xi^- + \pi^+$$
(20.10)

The two strange particles themselves decay

$$K^0 \longrightarrow \pi^+ + \pi^-$$
$$\Xi^- \longrightarrow \pi^- + \Lambda^0 \qquad\qquad (20.11)$$
$$\Lambda^0 \longrightarrow \pi^- + p$$

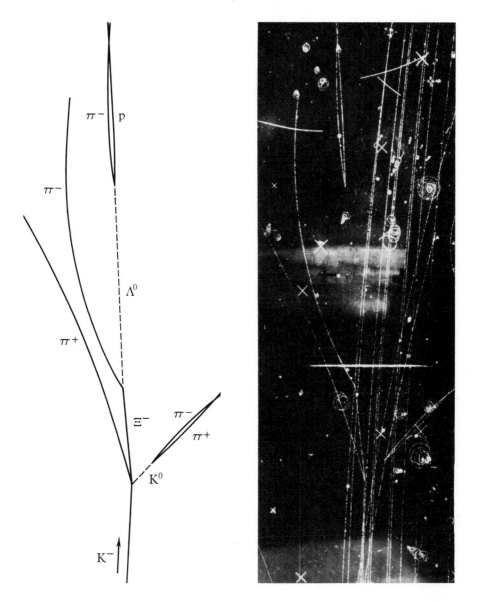

Plate 19 Hydrogen bubble chamber picture of a K^-p event (Birmingham collaboration experiment).

This spectacular event is in one of the rare inelastic channels of the K^-p system. Of more frequent occurrence are the simpler groups containing only one strange particle

$$
\begin{aligned}
K^- + p &\longrightarrow \overline{K}^0 p \pi^- \\
K^- + p &\longrightarrow \overline{K}^0 p \pi^- \pi^0 \\
K^- + p &\longrightarrow \overline{K}^0 n \pi^- \pi^+ \\
K^- + p &\longrightarrow \Lambda^0 \pi^+ \pi^-
\end{aligned}
\tag{20.12}
$$

To study these reactions a beam of K^--mesons is produced by allowing the external beam of a proton synchrotron to strike a copper target of length about 0·1 m. At a typical accelerator 10^{11} protons (per pulse, at a repetition rate of 0·3–0·5 Hz) can produce at an experimental site some distance from the copper target about 1000 K^--mesons, 10^5 π^--mesons and 10 antiprotons in a solid angle of 10^{-3} sr with a momentum p of 1 GeV/c and a momentum spread $\Delta p/p$ of 1 per cent. The production reactions yield a wide range of momenta and the required value is selected using a bending magnet. A specified type of particle is selected by an electrostatic separator in which a field gradient of 5 MV m^{-1} is maintained for the necessary distance along the particle trajectory. An essential feature of the beam transport system is magnetic quadrupole focusing, by which the beam is held together over large distances. With K-mesons, the distances should not be too large because of the short lifetime of the particle, although the distance for a given loss by decay increases linearly with the particle momentum.

The K-mesons produce tracks and events in a liquid hydrogen bubble chamber in a magnetic field. These are photographed by three cameras with known optical properties; a film demagnification of a factor of 10 is typical. The picture taking rate is high since the bubble chamber is capable of rapid recycling. In order to match the rate of data production from large accelerators, computer-controlled measuring apparatus† has been developed which accurately and rapidly presents track coordinates in digital form. A geometry programme computes momenta from curvatures and reconstructs the events in space and a kinematics programme tests energy and momentum balance in accordance with a series of hypotheses about the event. At this point difficulty may arise if neutral particles without visible decay products are emitted. In the K^-p interaction however the channel $\overline{K}^0 p \pi^-$ containing only one neutral particle is straightforward for study and a typical photograph may be as shown in Fig. 20.1.

When sufficient measured events have been accumulated questions relating to the physical nature of the process may be posed and the records of the events held on magnetic tape may be displayed by computer to answer these questions. In the case of a three-body final state, the *Dalitz plot*‡ is a con-

---

† Two well-known types are the F.S.D. (Flying Spot Digitizer) and P.E.P.R. (Precision Encoding and Pattern Recognition).

‡ A discussion of the technique of Dalitz plots is given in Appendix 7.

venient and powerful method of representing the data, since it immediately shows whether the reaction is dominated only by statistical factors or whether there is some preferred association of particles at a particular energy. In one form of this plot the effective mass (squared) of one pair of particles is plotted

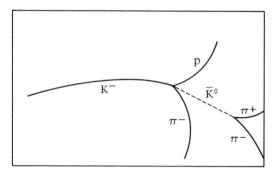

Fig. 20.1 Sketch of bubble chamber photograph of tracks corresponding to the interaction $K^-p \longrightarrow \overline{K}^0 p\pi^-$. The $\overline{K}^0$ particle decays inside the chamber. This type of event is classified as 201 where the figures give the number of tracks originating at the production vertex (2), the number of these tracks with a kink (0) and the number of associated Vee particles (1).

for each event against the effective mass (squared) of another pair. The effective mass of particle 1 and 2 is defined by the equation (cf Sect. 19.2.5(*d*))

$$M_{12}^2 c^4 = (E_1 + E_2)^2 - (\mathbf{p}_1 + \mathbf{p}_2)^2 c^2 \tag{20.13}$$

where E_i, $\mathbf{p}_i$ are the total energy and momentum of particle i. In the particular example the clustering of points shows a preference for a mass $M = 892 \cdot 4$ MeV in the $\overline{K}^0 \pi^-$ pair, which is evidence for an excited state K^* of the $K\pi$ system at this energy. The spin of the excited state may then be investigated by calling for an angular distribution of the decay products of the K^* in their centre of mass system. If this is anisotropic, spin zero can be excluded. A large number of excited or *resonant* states of mesons and baryons have been found in this way. Since the information required is normally obtained from observations of particles in the final state of the reaction, such experiments are known as *production processes*.

Plate 20 is a further example of the high information content of bubble chamber pictures; it shows the associated production of Λ^0 and K^0 particles in the π^-p interaction (20.3).

20.3.2 Counter/spark chamber experiments

An alternative method of studying resonant states is to employ a *formation process*. If a particle or state can decay in a given mode, then it can be formed by reversing this mode providing that the correct energy is available in the centre of mass system. If the c.m. energy can be varied, then the state may be seen as a resonance in the total or partial cross-section for the reaction. As in

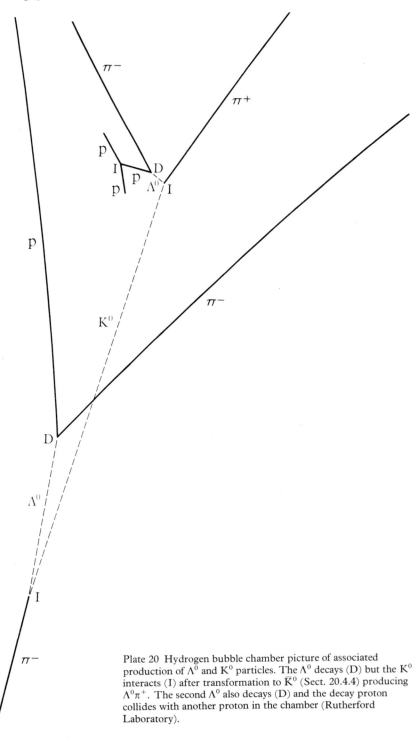

Plate 20 Hydrogen bubble chamber picture of associated
production of Λ^0 and K^0 particles. The Λ^0 decays (D) but the K^0
interacts (I) after transformation to $\bar{K}^0$ (Sect. 20.4.4) producing
$\Lambda^0\pi^+$. The second Λ^0 also decays (D) and the decay proton
collides with another proton in the chamber (Rutherford
Laboratory).

low energy nuclear physics, if the state has a spin $\mathscr{J}$ the elastic scattering cross-section at a Breit–Wigner resonance may be written for incident wavelength λ

$$\sigma_R = 4\pi\lambda^2 \frac{2\mathscr{J}+1}{(2s_1+1)(2s_2+1)} x \qquad (20.14)$$

where s_1 and s_2 are the spins of the bombarding particle and target nucleon and $x = \Gamma_{\text{el}}/\Gamma$ is the 'elasticity' of the reaction, i.e. the fraction of all processes at a given energy which take place in the elastic channel. These resonances may be conveniently sought as peaks in excitation functions, using counter telescope detection methods.

The discovery of the $\Delta(1236)$ baryon resonance, already described in Sect. 19.2.3, is an example of the use of a formation process. As a further example, involving strange particles, we consider again the K^-p interaction.

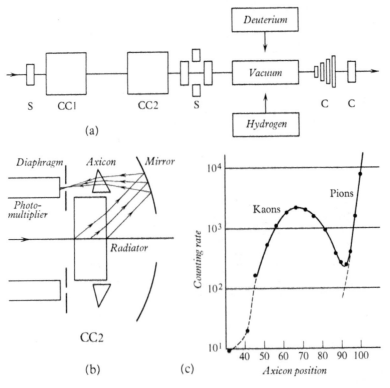

Fig. 20.2 Transmission method of measuring total cross-sections for π-nucleon and K-nucleon interactions. (a) Experimental layout, showing beam defining scintillators S, Cherenkov detectors CC1 and CC2 and transmission counters C determining solid angle. Liquid hydrogen targets are compared with vacuum to determine the transmission. (b) Detail of differential Cherenkov counter CC2 which, with appropriate setting of the 'axicon' lens, gives a signal for kaons but not for pions of the same momentum. (c) Velocity scan obtained by moving axicon. (From Carter *et al.*, *Phys. Rev.*, **168**, 1457, 1968, and Bugg *et al.*, *Phys. Rev.*, **168**, 1466, 1968).

This can lead to intermediate states of the same strangeness which correspond to an excited lambda (Λ^*) or sigma (Σ^*) particle. These may also be described as baryon resonances Y_0^*, Y_1^* where the suffix denotes the isobaric spin, and each may decay back into the K^-p system. Since $s_1 = 0$, $s_2 = \frac{1}{2}$ for this system the maximum cross-section at resonance might be expected from (20.14) to be

$$\sigma_R = 4\pi\lambda^2(\mathcal{J}+\tfrac{1}{2})x \tag{20.15}$$

as in the case of π^+ p elastic scattering (eq. (19.29) where $x=1$). In the K^- p system however there are two possible isobaric spin states, with $T=0$ and $T=1$ and if the resonance is in only one of these states the value σ_R given in (20.15) must be halved; it may still however affect the total cross-section. A suitable experiment is illustrated in Fig. 20.2 in which a beam of K^--mesons is attenuated in a liquid hydrogen absorption cell of length l: the total cross-section is determined, after corrections, by the formula (which may be deduced from eq. (5.6))

$$\sigma = \frac{1}{Nl} \log \frac{n_0}{n_l} \tag{20.16}$$

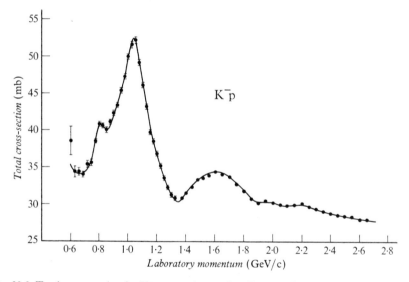

Fig. 20.3 Total cross-section for K^--proton interaction (Bugg *et al.*, 1968). By combining these data with similar observations for K^--neutron, obtained from deuterium experiments, it is possible to extract cross-sections for the K-nucleon interaction in the states of pure isospin $T=0$ and $T=1$.

The beam-preparation and transport for such an experiment requires equipment of the type discussed in Sect. 20.3.1. In the case of a counter experiment however it is not necessary always to exclude particles of correct momentum but unwanted mass from the counter system because the desired particles may be selected by a Cherenkov counter, preferably of the ring-focus, differential type (Fig. 20.2) in which a given velocity range may be chosen by adjusting the angle of the light collected by setting a so-called 'axicon' lens. It is possible with such a counter to obtain a 0·1 per cent resolution in velocity and to maintain high efficiency of counting in this range, which may also be adjusted by altering the refractive index of the transparent medium (gas or liquid) used in the counter. The selected particles are defined by a set of plastic scintillator telescopes (Fig. 20.2) and an array of these determines the solid angle within which the beam transmitted through the absorption cell is detected; the total cross-section is obtained by an extrapolation to zero subtended solid angle. Figure 20.3 shows results obtained over a range of incident momenta in this way for the K^-p system in which several resonances

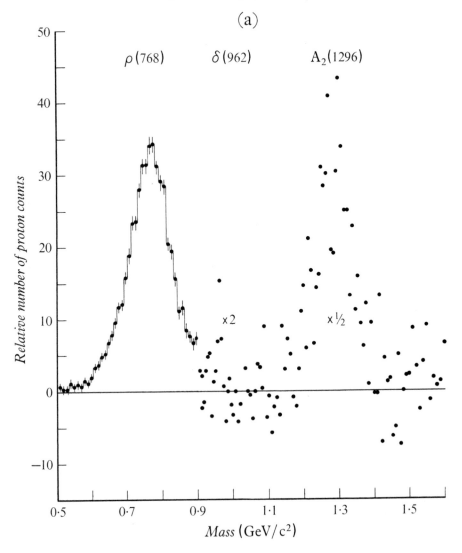

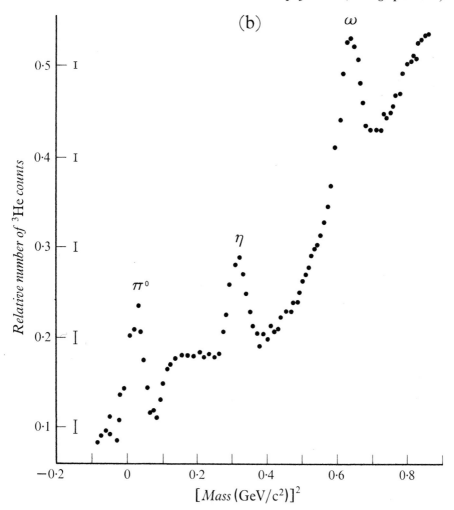

Fig. 20.4 Missing mass spectrometer. (a) Spectrum of mesons X^- produced in the reaction $\pi^- + p \longrightarrow p + X^-$. Statistical errors not shown above 0.9 GeV/c^2 (Focacci *et al.*). (b) Spectrum of neutral mesons X^0 produced in the reaction $p + d \longrightarrow {}^3\text{He} + X^0$ (Brody *et al.*).

appear; similar data for the pion–nucleon interaction are shown in Fig. 19.10. In cases where the resonance is wholly in the elastic scattering, its spin may be determined from the height of the resonance peak ((20.15) with $x = 1$). If the elasticity x is not known, as is usually the case, it is necessary to measure the angular distribution of elastic or inelastic processes at the resonances to find the spin by phase shift analysis (ch. 14). Sonic or wire spark chambers (Sect. **6.3**) triggered by scintillation and Cherenkov counters, and acting as position defining counters, provide a convenient means of investigating these distributions. Spark coordinates are obtained in digital form and are processed by an on-line computer.

Counter/spark chamber systems can also be used in production type experiments, and this is important for those baryon resonances, e.g. Y_1^* (1385) or Y_0^* (1405) which do not decay into the πp or Kp systems and therefore cannot be studied in formation processes. In such cases the *missing mass spectrometer* as used by Focacci *et al.*† is a powerful technique; it essentially studies production, e.g. in the reaction

$$\pi^- + p \longrightarrow p + X^- \tag{20.17}$$

and measures the momentum and angle of the product proton. On the assumption of the reaction type the mass of X^- may be obtained. The trajectory of the proton is defined by a spark chamber system triggered by scintillation counters. Figure 20.4 shows results obtained for the reaction (20.17) and for neutral mesons observed in the reaction‡

$$p + d \longrightarrow {}^3\text{He} + X^0 \tag{20.18}$$

Again this type of investigation is familiar in low energy nuclear physics, and also in atomic physics (as the Franck–Hertz experiment).

20.4 Properties of strange particles (Ref. 20.1)

The strange particles have the directly measurable properties of charge, mass, and decay lifetime and the inferred properties of intrinsic spin, strangeness, isobaric spin and relative parity. The intrinsic parity of weakly-decaying strange particles cannot be deduced from their decay products because of the non-conservation of parity in the decay process. The determination of strange particle properties cannot properly be separated from their production and decay. Appendix 8 lists the quantum numbers and measurable properties of these particles.

20.4.1 Mass (Ref. 20.2)

The mass of a strange particle is determined from the kinematics of a production or decay process.

The K^+-meson has a partial decay mode to three charged pions (τ-decay) and observations of the ranges of pion triplets originating from a K^+ stopped in nuclear emulsion gives an accurate mass value. The neutral kaon decays into two charged pions (θ-decay) and the energies of these particles from K^0-mesons decaying in flight have been measured with a spark chamber plus magnetic spectrometer system. The K^0 mass follows from the equations for energy and momentum balance. The masses of the charged and neutral kaons may be connected by observation of the reaction

$$K^- + p \longrightarrow \overline{K}^0 + n \tag{20.19}$$

† M. N. Focacci *et al.*, *Phys. Rev. Lett.*, **17**, 890, 1966.
‡ H. Brody *et al.*, *Phys. Rev. Lett.*, **24**, 948, 1970.

in a hydrogen bubble chamber, in which the neutron energy is obtained from an observable n–p collision.

The Λ^0-hyperon mass is obtained by observing the momenta of the decay products ($\pi^- + p$) and the angle between their tracks, when Λ^0-particles are produced and decay in flight in a nuclear emulsion stack. The Q-value is 37·6 MeV. The Σ-mass may be obtained in a similar way, and also from production reactions, while the masses of charged and neutral sigma-particles are connected by the process

$$\Sigma^- + p \longrightarrow \Sigma^0 + n \tag{20.20}$$

The Ξ mass and Ω^- mass are found from the energies of decay products.

20.4.2 Spin and magnetic moment (Ref. 20.3)

The spin of a particle may sometimes be inferred by simple arguments based on its decay properties. Thus in the production process

$$\pi^- + p \longrightarrow \Lambda^0 + K^0 \tag{20.21}$$

one may, following Adair,[†] consider Λ^0-particles emitted at $0°$ or $180°$ to the incident pion beam. This condition ensures that there is no component of angular momentum in the beam direction so that the magnetic quantum number along this direction is $\pm\frac{1}{2}$ only. The angular distribution of Λ-decay products (e.g. $\pi^- p$) in the Λ rest frame in relation to the beam direction then has a simple form determined by the Λ spin. The observed isotropic distribution is consistent with $s_\Lambda = \frac{1}{2}$.

The Λ, Σ and Ξ particles all appear to have spin $\frac{1}{2}$ and are therefore fermions; they possess a magnetic moment. Since they decay by a weak process the decay products of *polarized* hyperons are asymmetrically distributed, as in the case of nuclear beta-decay. This can be used to indicate the spin precession of an initially polarized hyperon beam (from a suitable production reaction) in a strong magnetic field, whence a value for the magnetic moment of the hyperon may be obtained.

The K^0-meson is known to exhibit a $2\pi^0$ decay mode and since the pions are bosons this system must be space symmetric, i.e. the spin is even. Arguments based on the decay angular distribution can be given to indicate spin zero for the K^0. An Adair type analysis shows this to be the case.

20.4.3 Parity (Ref. 20.3)

Because of parity non-conservation in weak decay processes it is impossible to relate the parity of the Λ-particle to that of the πp system. If the orbital angular momenta involved in reaction (20.21) are known it is possible to infer the relative parity (pKΛ). This is however more clearly indicated (as odd) by the fact of formation of the hypernucleus $^4_\Lambda$He (Sect. 20.5.5). By

† R. K. Adair, *Phys. Rev.*, **100**, 1540, 1955.

convention the Λ-particle, like the proton, is taken to have even parity and the parity of the kaon is then odd. The Σ parity can be related to that of the Λ by analysis of the electromagnetic decay process $\Sigma \rightarrow \Lambda e^+ e^-$.

20.4.4 K^0-mesons (Ref. 20.4)

In Sect. **20.2** it was pointed out that because of the requirements of strangeness and isobaric spin conservation in the production process $\pi p \rightarrow \Lambda K$ the K-meson should be ascribed isobaric spin $\frac{1}{2}$. Since K^+, K^- and K^0 mesons are known it might be expected that the isobaric spin would be 1 as in the case of the pions. The K^+ and K^- mesons have the same mass, lifetime and decay schemes and thus behave as particle and antiparticle. To accommodate them within an i-spin $\frac{1}{2}$ classification it is necessary to postulate two $T = \frac{1}{2}$ doublets one containing K^+ and K^0 and the other K^- and $\bar{K}^0$ where $\bar{K}^0$ is the antiparticle of K^0; these two doublets would have strangeness $+1$ and -1 respectively.

The suggestion of the $\bar{K}^0$ was made by Gell-Mann and Pais[†] who pointed out that this particle must be distinct from the K^0, contrary to the case of the π^0 meson which is its own antiparticle. The K^0 and $\bar{K}^0$ are coupled by decay to pions:

$$K^0 \longrightarrow \pi^+ + \pi^- \longleftarrow \bar{K}^0 \qquad (20.22)$$

but only weakly. As far as strong processes are concerned the transition $K^0 \rightleftarrows \bar{K}^0$ is evidently forbidden by strangeness conservation and moreover identity of these two particles would lead to double Λ^0 production from nucleons by the process:

$$\begin{aligned} \mathrm{N} &\longrightarrow \Lambda^0 + K^0 \\ \bar{K}^0 + \mathrm{N} &\longrightarrow \Lambda^0 \\ \hline \mathrm{N} + \mathrm{N} &\longrightarrow 2\Lambda^0 \quad \text{if } K^0 = \bar{K}^0 \end{aligned} \qquad (20.23)$$

and this is not observed.

The existence of two neutral kaons leads to striking predictions which have been experimentally verified. In order to understand them we must use not only the parity operation P (Sect. **3.1**) but also a *charge conjugation* operation C by which particles are changed into their antiparticles. Since we are taking the parity of K^0 mesons to be odd the effect of the two operations on the neutral kaon states $|K^0\rangle$, $|\bar{K}^0\rangle$ is as follows

$$\begin{aligned} P|K^0\rangle &= -|K^0\rangle & P|\bar{K}^0\rangle &= -|\bar{K}^0\rangle \\ C|K^0\rangle &= |\bar{K}^0\rangle & C|\bar{K}^0\rangle &= |K^0\rangle \\ CP|K^0\rangle &= -|\bar{K}^0\rangle & CP|\bar{K}^0\rangle &= -|K^0\rangle \end{aligned} \qquad (20.24)$$

The last line shows that the K^0 and $\bar{K}^0$ states are not eigenstates of the joint operator CP, because the wave function is changed by this operation

† *Phys. Rev.*, **97**, 1387, 1955.

from one to the other. Such eigenstates can, however, be constructed by setting up

$$|K_1^0\rangle = \frac{1}{\sqrt{2}} (|K^0\rangle - |\bar{K}^0\rangle)$$

$$|K_2^0\rangle = \frac{1}{\sqrt{2}} (|K^0\rangle + |\bar{K}^0\rangle)$$

(20.25)

for which it is easy to see that

$$CP|K_1^0\rangle = +|K_1^0\rangle \quad \text{(eigenvalue } +1, \text{ symmetric)}$$
$$CP|K_2^0\rangle = -|K_2^0\rangle \quad \text{(eigenvalue } -1, \text{ antisymmetric)}$$

(20.26)

Since the state $\pi^+\pi^-$ transforms as

$$CP|\pi^+\pi^-\rangle = +|\pi^+\pi^-\rangle$$

(20.27)

only the K_1^0 state should be able to decay into *two pions*.† By similar arguments, which will not be given here (see Ref. 20.5, p. 437) it is found that *three pion* decay is essentially confined to the K_2^0 state.

At this point it is useful to refer to the experimental situation, which is that when K^0 mesons are produced (e.g. by $(p\bar{p})$ annihilation) in a hydrogen bubble chamber of say 30″ diameter, two pion decays ('neutral V's') are seen to originate within the chamber at various distances from the annihilation vertex.‡ The momentum of the kaons is known from the energy release in the annihilation and the mean lifetime of the decaying neutral particles can then be found. Figure 20.5 shows the results obtained, which give $\tau = 0.843 \pm 0.013 \times 10^{-10}$ s for the lifetime of what we must, in terms of the previous discussion, describe as the K_1^0 (or sometimes K_S^0, S for short) meson. Neutral particles with a lifetime of 10^{-9} s or more would have a good chance of escaping from the chamber. If however an experiment is arranged to examine neutral decays at large distances from the point of production, e.g. an accelerator target, then a fall-off with distance is observed§ corresponding to a mean lifetime of 5×10^{-8} s and the decay modes do not include the two pion process. This lifetime can therefore be ascribed to the K_2^0 (or K_L^0, L for long) meson, and it differs from the K_1^0 lifetime because the requirement for invariance under the CP operation excludes $\pi^+\pi^-$ or $2\pi^0$ decay in this case.

We therefore arrive at the picture that in a production process such as $\pi^-p \rightarrow \Lambda^0 K^0$ the K^0-meson is produced strongly and strangeness is a good quantum number. It cannot however decay strongly but rather through much weaker couplings so that it is reasonable to represent it as an appropriate superposition of wave functions corresponding to the systems K_1^0 and K_2^0.

† Recent experiments (see Sect. 21.3.2) modify this conclusion.
‡ L. Kirsch and P. Schmidt, *Phys. Rev.*, **147**, 939, 1966 is a recent example.
§ Landé *et al.*, *Phys. Rev.*, **103**, 1901, 1956.

From eq. (20.25) such a wave function is

$$|K^0\rangle = \frac{1}{\sqrt{2}}(|K_1^0\rangle + |K_2^0\rangle) \qquad (20.28)$$

and similarly for $\bar{K}^0$. The K^0 beam from the accelerator target is therefore initially half of K_1^0 type and half of K_2^0 type. The short-lived K_1^0 particles decay within 0·1 m or so and the beam at larger distances consists of K_2^0.

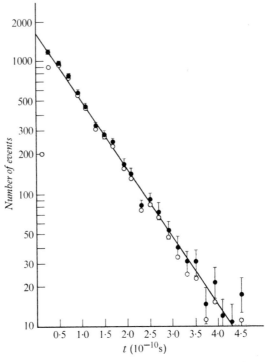

Fig. 20.5 Determination of lifetime of K_1^0-meson from distances of 2π-decays from origin of K^0 (Kirsch and Schmidt). The line indicates $\tau = 0.843 \times 10^{-10}$ s.

Noting that

$$|K_2^0\rangle = \frac{1}{\sqrt{2}}(|K^0\rangle + |\bar{K}^0\rangle) \qquad (20.29)$$

we see that half this beam is of strangeness $+1$ and half of strangeness -1. At times before this final state is reached it may be shown (Ref. 20.4) that the strangeness (or $K^0/\bar{K}^0$ ratio) oscillates with a frequency determined by the $(K_1^0 - K_2^0)$ mass difference.

Suppose now that the K_2^0 beam interacts with matter. The result of this will be different for the K^0 and $\bar{K}^0$ components of the beam and the phase relation between them will be altered, so that we must replace eq. (20.29) by

$$\frac{1}{\sqrt{2}} \{ a|K^0\rangle + b|\bar{K}^0\rangle \} \tag{20.30}$$

where a and b depend on the precise interactions. But this is no longer a pure K_2^0 beam; in fact by returning to the $K_1^0 K_2^0$ representation we see that it is

$$\tfrac{1}{2} \{ (a-b)|K_1^0\rangle + (a+b)|K_2^0\rangle \} \tag{20.31}$$

which means that K_1^0 mesons have been regenerated and the fast two-pion decay of this particle should again be observable (Fig. 20.6) after the absorber

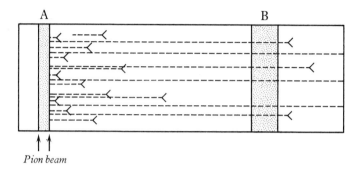

Fig. 20.6 K_1^0 regeneration following passage of K_2^0 beam through matter, due to preferential absorption of K^0. The symbols represent 2π decays of the K_1^0 produced at A and regenerated at B (Ref. 20.4, Fig. 1.1).

(or regenerator). This has been confirmed by experiment, and a mass difference $m_{K_2} - m_{K_1} = 3 \cdot 7 \times 10^{-6}$ eV between the K_1^0 and K_2^0 particles has been deduced from a detailed analysis.

20.5 Strong interactions of strange particles

20.5.1 Production

No satisfactory dynamical theory of strong interactions has yet been given, but useful relations between production amplitudes, e.g. for the processes $\pi^+ p \to \Sigma^+ K^+$, $\pi^- p \to \Sigma^- K^+$ and $\pi^- p \to \Sigma^0 K^0$ can be given (Ref. 20.1, p. 50) on the assumption of isobaric spin conservation.

The production of strange particles in nucleon–nucleon collisions is closely related to their production in the pion–nucleon interaction. Figure 20.7 shows a possible process in which a proton collides with a virtual pion emitted by a target proton; such processes, taking place at distances between the nucleons of the order of the pion Compton wavelength are known as *peripheral* and can be treated theoretically with some plausibility in terms of the pion–nucleon *coupling constant* (Sect. 21.4.2) and the cross-section for the $\pi p \to YK$ reaction. Figure 20.7 is an example of a *Feynman diagram* and there are precise rules for obtaining cross-sections from such useful pictorial representations. A similar discussion may be given of photoproduction of strange particles.

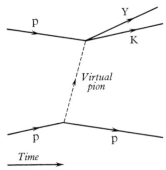

Fig. 20.7 Space-time (Feynman) diagram showing collision of an incident proton with a virtual pion emitted at lower vertex by a target proton. At the upper vertex a hyperon and kaon are produced by πp interaction.

One of the outstanding events in recent years has been the observation† of the particle Ω^-. This was predicted to exist by Gell-Mann and Ne'eman from their classification of elementary particles (ch. 21) and was expected to be of strangeness -3. It could therefore only be sought in reactions of the type

$$K^- + p \longrightarrow \Omega^- + K^+ + K^0 \tag{20.32}$$

and it was identified by a spectacular sequence of decays involving both Ξ and Λ^0 hyperons.

20.5.2 *Scattering of kaons by nucleons* (Ref. 20.6)

As in the case of pion–nucleon scattering we may consider total, elastic and charge exchange scattering of K^- and K^+ particles by nucleons. Experimental data are less precise than for the πp interaction but it is clear that the $K^+ p$ total cross-section is relatively featureless as a function of energy compared with that for $K^- p$. This is because of the different strangeness which requires that the only two-body inelastic reaction of the K^+ is

$$K^+ + n \longrightarrow K^0 + p \tag{20.33}$$

while the K^- offers many more possibilities

$$
\begin{aligned}
K^- + p &\longrightarrow \bar{K}^0 + n \\
K^- + p &\longrightarrow \Lambda^0 + \pi^0 \\
K^- + p &\longrightarrow \Sigma^\pm + \pi^\mp \\
K^- + p &\longrightarrow \Sigma^0 + \pi^0
\end{aligned}
\tag{20.34}
$$

† V. E. Barnes *et al.*, *Phys. Rev. Lett.*, **12**, 204, 1964.

For low energies, at which inelastic cross-sections are small, the K-nucleon cross-sections can be fitted by the assumption of pure s-wave scattering, as in the case of n–p scattering (Sect. 18.2.2) and scattering lengths and effective ranges can be obtained. The K^+p system is a pure state of $T = 1$, but K^-p can exist in states of $T = 0$ and 1. Of reactions (20.34) that leading to $\Lambda^0\pi^0$ involves only the $T = 1$ state; the others may proceed from both types of initial state.

20.5.3 Bound kaons (kaonic atoms)†

When a K^- meson stops in matter it may enter an atomic orbit as in the case of π and μ mesons. In all mesonic atoms there is a tendency for the initial capture state to lead to a 'circular orbit' state $(l = n-1)$. The important difference in the case of K^- mesons is the strength of the interaction as a result of which an absorption by a nuclear proton producing one of the reactions (20.34) may take place from circular orbits with very high quantum numbers, i.e. a long way from the nuclear centre. The cut-off of kaonic X-rays of a particular atomic sequence (e.g. $5g \rightarrow 4f$) at a certain Z-value of the atomic target shows the onset of nuclear absorption from the particular orbit (e.g. $5g$) and provides a sensitive probe of the nuclear density distribution at large radii. Reactions of the kaon with nuclear neutrons can be identified by the appearance of the processes

$$K^- + n \longrightarrow \Lambda^0 + \pi^-$$
$$K^- + n \longrightarrow \Sigma^0 + \pi^- \tag{20.35}$$

Present results suggest that the neutron distribution in nuclei may extend beyond the proton distribution in, for instance, a nucleus such as Gd.

20.5.4 Resonance production (Ref. 20.7)

The first hyperon resonance Y* was observed in 1960 by Alston *et al.*‡ using the bubble chamber technique described in Sect. 20.3.1. The process was:

$$K^- + p \longrightarrow \Lambda^0 + \pi^+ + \pi^- \tag{20.36}$$

and a Dalitz scatter plot (Fig. 20.8) showed concentrations of events with a pion energy of ≈ 290 MeV. These correspond with the decay of an excited lambda hyperon Y_1^* built from either $\Lambda^0\pi^+$ or $\Lambda^0\pi^-$ and therefore of $S = -1$ and $T = 1$; the third charge state $\Lambda^0\pi^0$ should also exist. The mass is 1385 MeV and the width is about 40 MeV which means that the Y_1^* decays by the strong interaction.

The Y_1^* (1385) resonance has a mass less than that of the K^-p system (1432 MeV) and must therefore be examined in a production process. Many additional hyperon resonances are known however with higher mass which

† C. E. Wiegand and D. A. Mack, *Phys. Rev. Lett.*, **18**, 685, 1967; C. E. Wiegand, *Phys. Rev. Lett.*, **22**, 1235, 1969; see also Ref. 19.4.
‡ *Phys. Rev. Lett.*, **5**, 520, 1960.

may be seen in formation experiments, i.e. as peaks in total cross-section as a function of energy (cf. Fig. 20.3). They can occur in both $T=0$ and $T=1$ states, which are obtained from the observed K^-p and K^-n cross-sections by suitable combination. The spin and parity of resonant states may be obtained from the angular distributions of the decay products by application of the resonance analysis techniques of low energy nuclear physics. The method suggested by Adair (Sect. 20.4.2) has been successful in some cases.

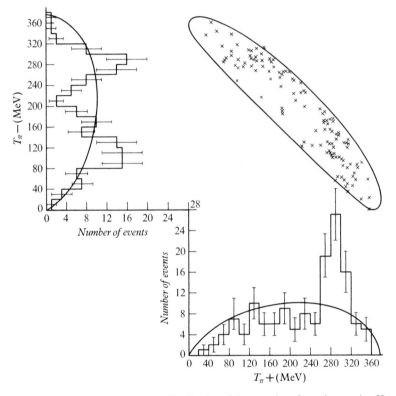

Fig. 20.8 Centre of mass kinetic-energy distribution of the two pions from the reaction $K^-p \longrightarrow \Lambda^0\pi^+\pi^-$ observed in a hydrogen bubble chamber. Concentrations along the lines $T_\pi \approx 290$ MeV correspond to the production of the Y_1^* (1385) resonance. This is an example of a Dalitz plot; (App. 7); if phase space dominated the reaction the density of points in the populated area would be uniform. The solid lines drawn on the histograms are phase space curves (Alston *et al.*).

For parity determination use is often made of the interference between resonant and background amplitudes. Alternatively the polarization of an outgoing spin-$\frac{1}{2}$ particle may be observed. A particularly important resonance is the K^* (890) seen in reactions

$$K^+ +p \longrightarrow K^{*+} +p; \qquad K^* \longrightarrow K+\pi$$
$$K^- +p \longrightarrow K^{*-} +p \qquad\qquad\qquad (20.37)$$
$$K^- +p \longrightarrow \bar{K}^{*0} +n$$

These reactions have been analysed using the Dalitz plot (App. 7) (and the Adair method) and a spin 1 is suggested with $T = \frac{1}{2}$ and two values of strangeness $S = \pm 1$ in the two charge states as for the K-meson. The K*, like the non-strange ρ- and ω-mesons, is a particle which may have relevance in the understanding of nucleon structure and the strong interactions. The ρ- and ω-particles are themselves examples of meson resonances and the Δ (1236) is a non-strange baryon resonance.

20.5.5 Hypernuclei

Low energy Λ^0 hyperons cannot react inelastically with nucleons, either because of lack of energy or because of conservation of strangeness. Elastic scattering is possible, and if the Λ^0-nucleon force is attractive there is also the possibility of the formation of a bound state which may survive for a time comparable with the free Λ^0 lifetime of $\approx 10^{-10}$ s. This is much longer than the lifetime of a typical nuclear electromagnetic transition and a system consisting of a complex nucleus with a Λ^0-particle attached, in its lowest bound state, may therefore be quasi-stable. Evidence for the existence of such *hypernuclei* was first given by Danysz and Pniewski (1953) who found a fragment originating in a cosmic ray star in emulsion which showed a striking form of decay (Plate 21).

The motion of a Λ^0-particle in a complex nucleus is not restricted by the Pauli exclusion principle, and the particle may even be bound more strongly than a nucleon to the rest of the nucleus. Many light hypernuclei† have now been studied, mostly in emulsion experiments with K^- beams from accelerators, and from the decay energies observed it is possible to infer the binding energy of the Λ-particle. Figure 20.9 shows the results obtained, from which it can be seen that for light nuclei the binding energy is proportional roughly to A, and does not fluctuate rapidly as in the case of nucleons (Fig. 10.1). This is a result of the switching-off of the Pauli principle, which also leads to the 'stability' of some nuclei such as $^8_\Lambda$Be which are heavy-particle unstable when the Λ^0 is replaced by a nucleon. There is evidence that the Λ^0-nucleon force is spin-dependent.

Hypernuclei decay mesonically in accordance with the $\Lambda \rightarrow p\pi^-$ or $n\pi^0$ processes, or non-mesonically, in which case a virtual pion causes a nucleon transition, e.g.

$$^8_\Lambda \text{Be} \longrightarrow {}^3\text{He} + {}^4\text{He} + \text{n} \tag{20.38}$$

This latter process is sometimes described as analogous to internal conversion of electromagnetic transitions, and analysis of the ratio of mesonic to non-mesonic decay is dependent on the spin of the Λ^0-particle; a value of $\frac{1}{2}$ is in agreement with observation.

---

† A hypernucleus is denoted by the symbols $^A_\Lambda$X where X is the chemical symbol and A is the total number of baryons.

A particularly interesting case of hypernuclear decay occurs with $^4_\Lambda$He formed by stopping K^--mesons in a helium bubble chamber

$$K^- + {}^4He \longrightarrow {}^4_\Lambda He + \pi^- \tag{20.39}$$

If the spin of $^4_\Lambda$He is zero, the orbital momenta of the initial and final systems in this process are the same and consequently the initial and final parities are

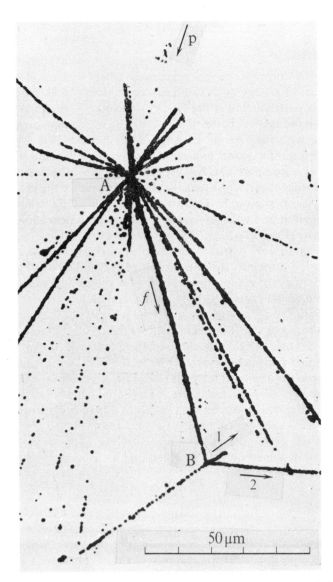

Plate 21. Discovery of hypernuclei. Among the products of a cosmic ray star recorded in nuclear emulsion at a height of 28 300 m is a fragment stable for a time greater than 3×10^{-12} s, which finally disintegrates at B (M. Dansyz and J. Pniewski, *Phil. Mag.*, **44**, 348, 1953).

the same. Taking the lambda parity to be even, the parity of the K^- is the same as that of the pion, namely odd. The assumption that the spin of $^4_\Lambda$He is zero is plausible because of the preponderance of the decay mode

$$^4_\Lambda\text{He} \longrightarrow {}^4\text{He} + \pi^0 \qquad (20.40)$$

in which S-states only are involved for spin zero.

The formation of Σ-hypernuclei is in principle possible but has not been observed, probably because the Σ-nucleon reactions are fast. The observation of X-rays from a presumed Σ^- hyperonic atom with a $_{19}$K nucleus has been reported.†

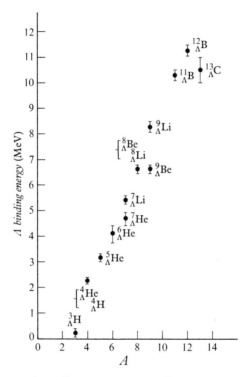

Fig. 20.9 Binding energy of the Λ^0 in light hypernuclei. A is the total number of baryons in the hypernucleus $^A_\Lambda$X formed from the parent nucleus $^{A-1}$X (from Gajewski *et al.*, *Nucl. Phys.*, **B1**, 105, 1967).

20.6 Weak interactions of strange particles

Strange particles are produced through the strong interaction but decay via the weak interaction with or without change of strangeness, or electromagnetically. All decay modes of the charged K-mesons have the same lifetime, and so do the alternative modes of the various hyperons.

† C. E. Wiegand, *Phys. Rev. Lett.*, **22**, 1235, 1969. For later work see Backenstoss *et al.*, *Phys. Lett.*, **33B**, 230, 1970.

20.6.1 Decay of charged kaons

The main modes of decay of both K^+ and K^- with lifetime $1 \cdot 23 \times 10^{-8}$ s are

Process		Fraction of decays
$K \longrightarrow \mu + \nu_\mu$	$(K_{\mu 2})$	$0 \cdot 63$
$K \longrightarrow \pi + \pi$	$(K_{\pi 2}, \theta)$	$0 \cdot 21$
$K \longrightarrow \pi + \pi + \pi$	$(K_{\pi 3}, \tau)$	$0 \cdot 08$
$K \longrightarrow \mu + \nu_\mu + \pi$	$(K_{\mu 3})$	$0 \cdot 03$
$K \longrightarrow e + \nu_e + \pi$	(K_{e3})	$0 \cdot 05$

The neutrinos are particle or antiparticle according to the status of the associated lepton. The τ-mode has two forms, e.g. $K^+ \to \pi^+ \pi^+ \pi^-$ or $\pi^+ \pi^0 \pi^0$.

The tau and theta modes of the K^+ non-leptonic decay originally presented a difficulty of interpretation which led to the discovery of the non-conservation of parity in weak processes† (tau-theta problem). The mass and lifetime of the particles decaying in these two ways appeared to be identical, but the parities of the final system were different. In the theta decay, because of the zero spin of the pions, the spin-parity of the parent meson must be determined by the relative orbital momentum of the decay particles (which together have even parity), i.e. it must be 0^+, 1^-, 2^+. In the tau decay we consider the final system to be built from a similar pair of identical pions, which as bosons must form a $0^+, 2^+, 4^+, \dots$ state, together with a third pion in a relative S-state. This adds no angular momentum, but changes the parity so that the τ-decaying parent would be of character $0^-, 2^-, \dots$ which is inconsistent with the result from the θ mode. Similar arguments apply if one builds the τ-decay from a $\pi^+ \pi^-$ pair and a third pion. The assumption of a relative S-state is verified by observation of the energy distribution of the decay pions. It then follows that either the tau and theta parents are different particles, or that *parity is not conserved* in the decay. We now know that the latter is correct, that the K-meson has $J^P = 0^-$ (Sect. 20.4.3) (and the pionic states opposite parities) but that the weak interaction does not conserve this quantity.

Detailed analysis of the tau decay by Dalitz, in terms of the triangular diagrams which have proved so useful in discussing 3-particle decays, established the K-spin to be zero. Figure 20.10 is a Dalitz plot for the K^+ tau decay, based on emulsion measurements of ranges. The theory of this representation is outlined in Appendix 7; the uniformity of the distribution of the points is consistent with spin zero for the kaon.

The difference between the partial lifetime for two pion decay of the charged kaon ($5 \cdot 9 \times 10^{-8}$ s) and for the similar decay of the K_1^0 (10^{-10} s) suggests that the process

$$K^+ \longrightarrow \pi^+ + \pi^0 \tag{20.41}$$

is inhibited. Since the K^+ has spin zero, the final state $(\pi^+ \pi^0)$ is spatially

† T. D. Lee and C. N. Yang, *Phys. Rev.*, **104**, 254, 1956.

symmetric. It is also symmetric in isospin† and must therefore have $T=2$ since $T_z=1$. If therefore there is a *selection rule* $|\Delta T|=\frac{1}{2}$ the retardation of the K^+ decay can be understood. This rule appears to be approximately true in other processes, e.g. hyperon decays.

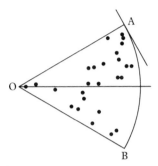

Fig. 20.10 Decay of the K^+ meson in the $\tau(3\pi)$ mode, as originally analysed by Dalitz (App. 7). The area OAB shown must contain points representing the energies of all observed τ-decays and uniformity of the distribution is consistent with a K-meson spin of zero. The data are those available in 1954 (R. H. Dalitz, *Phys. Rev.*, **94**, 1046, 1954).

Leptonic decays of kaons, such as

$$K^+ \longrightarrow \mu^+ + \nu_\mu \tag{20.42}$$

exhibit the same features of helicity due to parity non-conservation that have already been discussed for pion decay.

20.6.2 Decay of hyperons

The main decay schemes are

$$
\begin{array}{lll}
\Lambda^0 \longrightarrow p+\pi^- & \Sigma^+ \longrightarrow p+\pi^0 & \Xi^0 \longrightarrow \Lambda^0+\pi^0 \\
\Lambda^0 \longrightarrow n+\pi^0 & \Sigma^+ \longrightarrow n+\pi^+ & \Xi^- \longrightarrow \Lambda^0+\pi^- \\
& \Sigma^- \longrightarrow n+\pi^- & \Omega^- \longrightarrow \Xi+\pi \\
& \Sigma^0 \longrightarrow \Lambda^0+\gamma &
\end{array}
$$

Most of these decays except the electromagnetic process $\Sigma^0 \to \Lambda^0+\gamma$ seem to have lifetimes of the order of 10^{-10} s which enables the decays to be studied in detail in bubble chambers. It has already been noted that the Λ- and Σ-particles are polarized by their production processes and also show the asymmetric decay pattern characteristic of parity non-conservation. Most of the decays obey the selection rule $\Delta S=1$ and direct decay of the cascade particle Ξ to $n\pi$ is apparently excluded because of $\Delta S=2$. Fast decay through the strong interaction, with $\Delta S=0$, is impossible for unexcited baryons, in contrast with resonant states, because of their mass values.

† We recall that for fermions such as neutrons and protons the total wave function is antisymmetric in space, spin and isospin. For bosons, the wave function is symmetric. The state $T=1$ has opposite symmetry in the two cases. See Ref. 20.5, p. 428.

The ratio of the two modes of decay of the Λ-particle may be calculated to be 2/1 in favour of the $p\pi^-$ products if the $\Delta T = \frac{1}{2}$ rule is assumed; this eliminates $T = \frac{3}{2}$ components of the final state since $T_\Lambda = 0$.

Leptonic decay modes such as

$$\Sigma^+ \longrightarrow \Lambda^0 + e^+ + \nu_e$$
$$\Sigma^- \longrightarrow n + e^- + \bar{\nu}_e \qquad (20.43)$$
$$\Lambda^0 \longrightarrow p + e^- + \bar{\nu}_e$$

are permitted by the rule $\Delta S = 0, 1$; the strangeness-changing decays are at least one order of magnitude less probable than nuclear beta decays.

References

1. ADAIR, R. K. and FOWLER, E. C. *Strange Particles*, John Wiley, New York, 1963.
2. BARKAS, W. H. 'Masses of the Metastable Particles', *Ann. rev. nucl. Sci.*, **15**, 67, 1965.
3. TRIPP, R. D. 'Spin and Parity Determination of Elementary Particles', *Ann. rev. nucl. Sci.*, **15**, 325, 1965.
4. KABIR, P. *The CP Puzzle; Strange decays of the neutral kaon*, Academic Press, New York, 1968.
5. KÄLLEN, G. *Elementary Particle Physics*, Addison-Wesley, Reading, Mass., 1964.
6. GALBRAITH, W. 'Hadron–nucleon total cross-sections at high energies', *Rep. progr. Phys.* **32**, 547, 1969.
7. HILL, R. D. 'Resonance particles', *Scientific American*, January 1963; ADAIR, R. K. 'Problems in the Analysis of Final-state Resonances', *Rev. mod. Phys.*, **37**, 473, 1965.
8. KENYON, I. R., 'Bubble chamber film analysis', *Contemp. Phys.*, **13**, 75, 1972.

21 Theories

In the preceding two chapters the properties of the particles have been surveyed. It remains to comment briefly on attempts which are being made to reduce this diverse family to some sort of order, and to set it within a framework of general physical theory. The difficulty in such a procedure is that consideration of elementary particles brings us to a situation where basic laws are unknown. As in nuclear structure physics, but for somewhat different reasons, it is necessary to proceed in terms of plausible models, which may be tested by comparison with experiment. Even this is a difficult task, because in processes involving the strong interaction (Sect. **21.4**) only approximate calculations can be made. By contrast the electromagnetic and weak interactions (Sects. **21.2, 21.3**) are fairly well understood, except for the question of CP conservation (Sect. 21.3.2).

The dominant features of the strong interaction are the existence of resonant states and the occurrence of forward (and backward) peaks in angular distributions of reaction processes at high energies. Although these two properties are certainly connected there has been, at least until recently, less success in understanding the dynamics of scattering than in setting up schemes of classification. In the classification of strongly interacting particles it will be found that the interactions are such that the particles actually lose their identity, so that the term 'elementary' becomes inappropriate and it is generally better to speak of *sub-nuclear* bodies.

In this chapter we consider mainly the systematics of sub-nuclear particles and the constraints placed on interactions between them by conservation laws. Classification is a necessary preliminary to understanding and it usually has also a merit and a convenience of its own.

21.1 Classification

21.1.1 The sub-nuclear particles and their interactions

The general nomenclature for the particles discussed in Chapters 19 and 20 was given in Sect. 1.2.1; it was based essentially on rest mass. The interactions between these particles were introduced in Sect. 1.2.2. It is now appropriate to put the kaons and pions together with the baryons into a group of *strongly interacting* particles called *hadrons*, and of course to recognize that the muons belong to the *weakly interacting* group of *leptons*. The *electromagnetic interaction* is represented by the photon, which is the exchange particle of the

Maxwell field, in the sense that the electromagnetic force between two charged particles may be thought to arise as a result of transfer of virtual photons between the two. Similarly the strong interaction may be represented by the exchange of virtual hadrons (Sect. **18.3**); for the leptons the corresponding quantum remains to be discovered. The exchange of a field quantum between two interacting particles may conveniently be represented by a space-time graph or Feynman diagram (Fig. 21.1).

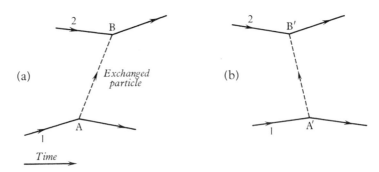

Fig. 21.1 Feynman diagrams representing an interaction between two particles 1, 2 (e.g. nucleons) resulting from particle exchange (e.g. a pion). In (b) the exchange particle is moving backwards in time so that it is absorbed before it is emitted; it is interpreted as an *antiparticle* moving forwards. At A, B, A′ and B′ momentum is conserved but not energy and the time of propagation of the exchanged particle and the energy imbalance are related by $\varDelta t\, \varDelta E \approx \hbar$. The range of the force between particles 1 and 2 is inversely proportional to the mass of the exchanged particle (Sect. 18.3.2). Conservation of both energy and momentum at the vertices may be assumed if the exchanged particle is taken to have a mass different from its physical value; it is then said to be '*off the mass shell*'.

Two diagrams are necessary for Lorentz invariance (Ref. 21.4, Burkhardt p. 105) but in later sections only one will be shown for a given process.

The nomenclature can thus be summarized as follows:†

    (*a*) *the photon*
    (*b*) *the leptons:*   neutrinos
                      electrons
                      muons
    (*c*) *the hadrons:*   pions   } mesons
                      kaons
                      nucleons } baryons
                      hyperons

The hadrons enter into a great variety of *resonant* states which must be treated as particles but decay through the strong interaction (App. 8) and have lifetimes so small that the decay probability is most conveniently specified and measured as a width Γ. All particles have antiparticles, which are identical with the particle in the case of the photon, π^0-meson and η-meson. The

† Although mass increases downwards in this list it should be noted that some of the meson resonances now known have greater mass than the nucleon.

kaons and hyperons have non-zero strangeness. All hadrons are members of charge multiplets in which the number of states or individual particles is given by $2T+1$ where T is the isobaric spin.

If a detailed ordering of the hadrons in terms of rest mass is attempted, excluding resonances, one arrives at a diagram of the type shown in Fig. 21.2 in which groupings of charge, hypercharge (Sect. **20.2**) and isobaric spin are given. These quantities are linked by the relation already given in Sect. **20.2**.

$$Q = T_z + \frac{Y}{2} \quad \text{with} \quad Y = B + S \quad\quad (21.1)$$

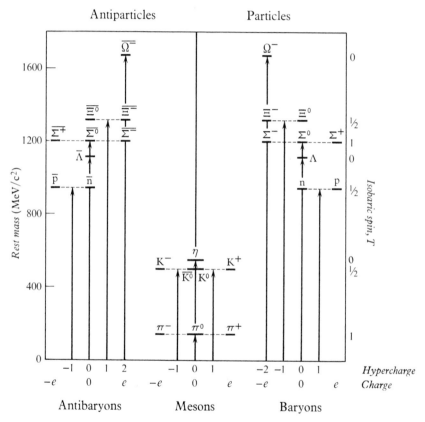

Fig. 21.2 Classification of mesons and baryons in terms of rest mass (vertically) and charge and hypercharge (horizontally) (Diagram due to Dr. L. Riddiford.)

where Q is the charge number, B the baryon number, S the strangeness and Y the hypercharge. All these numbers (unless zero) are opposite for particle and antiparticle. The particles shown (App. 8) are either stable or decay with measurable lifetimes through the weak interaction except for those such as π^0, Σ^0 and η to which fast electromagnetic decay is open.

21.1.2 Invariance and conservation laws (Ref. 21.1)

(*a*) *Symmetry*. The importance of invariance properties in physics may be approached through the readily appreciated property of symmetry. In all phenomena in which order can be discerned, e.g. in crystal structures, there will be certain points, lines or planes in space with respect to which symmetrical patterns may be recognized. Figure 21.3 is a reproduction of some wallpaper designs which exhibit these features. Reflection symmetry (sometimes not quite exact) is seen in innumerable works of art throughout many ages; it is clearly an important element in aesthetic appreciation.

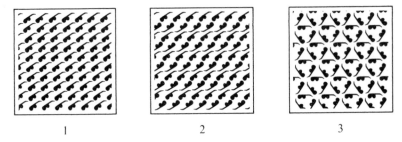

1   2   3

Fig. 21.3 Wallpaper designs deriving from a simple basic unit (1), each showing a particular type of symmetry. Case (3) contains a threefold axis of symmetry. (From a larger figure given by A. V. Shubnikov and quoted by A. E. Kitaigorodskiy, *Order and Disorder in the World of Atoms*, Longman, New York, 1967.)

The geometrical symmetries of ordinary life relate generally to very complex objects, and the physicist is concerned to seek for the origin of these symmetries in much simpler systems, ultimately of molecular, atomic, nuclear or sub-nuclear nature. Figure 21.4 for instance exhibits such a simpler symmetry; the plane figures there drawn are the same for rotations of 60° and 120° respectively about an axis through the midpoint of the figure perpendicular to the plane of the paper. There is now great significance in these

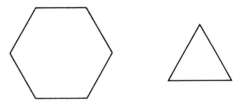

Fig. 21.4 The hexagon and triangle have respectively a sixfold and threefold axis of symmetry.

angles (except for these particular plane figures) and one may usefully enquire what happens for different rotations, including in particular infinitesimal ones. If a physical system is essentially unchanged by such a process it is said to be *invariant* under a *transformation*; in the case of Fig. 21.4 there is invariance for 60° and 120° rotations as a result of the symmetry of the figures.

The terms invariance and symmetry are thus closely related, and are used frequently to mean the same thing.

The concept of infinitesimal transformations is important in that it permits one to make a transition from the complicated objects of classical physics to the elementary systems which demand a quantum mechanical description. Moreover it is possible to discuss in the same way other types of transformation, e.g. displacement in space, or development in time as well as rotation. In quantum mechanics such transformations are effected by the application of a suitable operator $U(\rho)$ to a wave function ψ describing the state (for instance) of a particle:

$$U\psi_\alpha(x) = \psi_\alpha(x-\rho) \tag{21.2}$$

where α labels the state concerned and ρ is the displacement (or transformation) generated by the operator U. By expanding ψ_α in a Taylor series we obtain, if ρ is small,

$$\psi_\alpha(x-\rho) = \psi_\alpha(x) - \rho\,\frac{\partial}{\partial x}\,\psi_\alpha(x) + \frac{\rho^2}{2}\,\frac{\partial^2}{\partial x^2}\,\psi_\alpha(x) - \cdots$$
$$= e^{-\rho(\partial/\partial x)}\psi_\alpha(x)$$
$$= e^{-i\rho p/\hbar}\psi_\alpha(x)$$

where the substitution $p = -i\hbar\partial/\partial x$ relating momentum p to its quantum operator has been made. It follows that

$$U(\rho) = e^{-i\rho p/\hbar} \tag{21.3}$$

or, for small displacements,

$$U(\rho) = 1 - i\rho p/\hbar \tag{21.4}$$

It will be noted that U is *unitary*, i.e. $|U|^2 = 1$ as is indeed necessary if the transformation is only to displace and not otherwise to modify the system. In eq. (21.3) the dynamical variable p is known as the *generator* of the transformation specified by ρ. The parameter ρ is here a scalar, but for more complicated transformations it may be a vector in n-dimensional space. If the displaced system, resulting from the operation U, is to be the same as the initial system it may be shown (Ref. 21.1, Schiff) that the momentum p is a *constant of the motion*. In other words if the system is invariant with respect to displacement in space, linear momentum is conserved. Other conservation laws, of an *additive* nature, and valid in classical as well as quantum physics, may be established; thus we have

Invariance with respect to	leading to	*Conserved quantity*
displacement (x)		linear momentum (p_x)
angular displacement (θ)		angular momentum
time (t)		energy (E)

These symmetries are in fact part of the more general symmetry described by the proper† Lorentz transformation of special relativity. From this transformation one defines the invariant mass m of a particle or system, given by

$$m^2c^4 = E^2 - \mathbf{p}^2 c^2 \qquad (21.5)$$

as a constant of the motion with total energy E and momentum $\mathbf{p}$.

In this discussion we have considered unitary transformations of quantum states. Although the wave functions concerned play a vital role they are not observable quantities. Physical measurements determine properties such as static and transition moments, lifetimes and production cross-sections in which two states are connected by an operator whose *matrix element* gives the observed quantity. It is the squares of the matrix elements corresponding to physical operators which are essentially invariant under the appropriate unitary transformations. The symmetry properties of the operator however depend on the physical interaction which it represents, and (except in the case of energy, momentum and angular momentum) may differ between electromagnetic, weak and strong processes. Such processes will thus take place only between states which jointly match the required interaction symmetries or invariance properties; restrictions imposed in this way are *selection rules*. In Sects. **2, 3, 4** of this chapter we shall be mainly concerned with the invariance properties of the interactions themselves.

(b) *Group concepts and conservation laws.* A powerful treatment of invariance properties may be based on the mathematical theory of groups. A group G is defined by a set of *elements g* and a *law of combination* which obey the following conditions:

(i) if g and g' are elements of G then gg' is also an element.

(ii) the set contains an *identity* element g_I such that

$$gg_I = g_I g = g$$

(iii) for every element g in G there is an *inverse* element g^{-1} such that

$$gg^{-1} = g_I$$

(iv) the law of combination is associative, i.e.

$$(gg')g'' = g(g'g'')$$

This is a purely abstract concept, but if quantities of a special kind such as numbers or matrices can be found which behave in this particular way, then they are said to form a *representation* of the group. Evidently the positive and negative integers . . . $-2, -1, 0, 1, 2, . . .$ form a group representation (with 0 as the identity element) if the law of combination is ordinary addition. In a matrix representation the combination is (matrix) multiplication. Groups are especially appropriate to describe transformations since two transformations such as displacement or rotation may be combined to give a third. The unitary

† 'Proper' has the meaning that space reflection and time reversal (see (c) below) are excluded.

operators U effecting transformations contain the *generators* of the corresponding group and these are multiplied as n-dimensional vectors with displacement parameters.† For each group there is a representation, known as *regular*, in which the number of elements is equal to the number of independent generators. There is also a *singlet* representation for which an element transforms into itself. In high energy physics a close correspondence is established between group representations and particle multiplets.

The simplest unitary group, $U(1)$, is that which contains transformations (sometimes called *gauge transformations*) which essentially add only a phase factor to particle wave functions. Invariance under such transformations leads to the following additive conservation laws:

$$
\begin{array}{lll}
\text{Charge number} & \Delta(Q-\overline{Q}) = 0 & \\
\text{Baryon number} & \Delta(B-\overline{B}) = 0 & (21.6) \\
\text{Lepton number} & \Delta(L-\overline{L}) = 0 & \\
\text{Hypercharge number} & \Delta(Y-\overline{Y}) = 0 & (21.6(a))
\end{array}
$$

where the symbols $\overline{Q}$, $\overline{B}$, $\overline{L}$ and $\overline{Y}$ indicate that in applying these laws antiparticles are counted negatively. The rules (21.6) appear to be absolute but (21.6(a)) is not valid for the weak interactions. These rules are inferred from observations on many reaction processes of which the following sequence is typical

$$
\begin{array}{llll}
\pi^- + p \longrightarrow & \Sigma^- + K^+ & \text{Yes} & \\
\longrightarrow & n + \pi^+ & \text{No} & (\Delta Q \neq 0) \\
\longrightarrow & \pi^- + K^+ & \text{No} & (\Delta B \neq 0) \quad (21.7) \\
\longrightarrow & e^- + p & \text{No} & (\Delta L \neq 0) \\
\longrightarrow & \Sigma^+ + K^- & \text{No} & (\Delta Y \neq 0)
\end{array}
$$

There is also an additive conservation law for total isobaric spin **T**. The invariance here is not of a geometrical nature except by analogy. It corresponds with the constancy of length of the vector **T** under rotations which change the component T_z in a fictitious charge space. In physical terms the symmetry derives from the charge independence of nuclear forces; it is broken by the electromagnetic interaction. The simplest isobaric spin transformations change a neutron into a proton, or vice versa, and these may be expressed by the operation of a 2×2 matrix on a two component wave function. The set of unitary matrices which effect these transformations is known as the group $U(2)$.

In order to describe an invariance more comprehensive than that covered by $U(2)$ we may consider the group $U(3)$, which can be represented by a set of 3×3 matrices. The three component wave functions on which these matrices

---

† If, for instance, the transformations U cause a rotation of coordinate axes then the displacements are angular and the generators are angular momentum operators, which give the observable angular momentum components of the system.

operate can be constructed to represent not only the neutron and proton but also a particle with hypercharge such as the Λ.

The groups $U(2)$ and $U(3)$ may be reduced to the *special unitary groups* $SU(2)$ and $SU(3)$ by imposing the condition of unit determinant on the U matrices. In applications to high energy physics, this removes from the group the transformations corresponding to baryon conservation (Ref. 21.8(*b*)). The groups $SU(2)$ and $SU(3)$ then have respectively three and eight generators of infinitesimal transformations (generally $n^2 - 1$ for $SU(n)$) and these generators have to be *traceless* matrices. Classifications arise when one determines the particle states of given baryon number, spin and parity which transform among themselves under these matrix operations.

The group $SU(3)$ has had outstanding success in classifying the states of the oscillator potential (especially in nuclear physics, in which the basic quantities are oscillator quanta) and in the classification of the hadrons, to which we return in Sect. **21.4**. It will be seen that whereas in $SU(2)$ the generators are components of a vector, in $SU(3)$ they are related similarly to an eight-component family, or octet.

(*c*) *Discrete transformations.* The parity operation is not a continuous transformation, such as those which lead to additive conservation laws; the eigenvalues of the parity operator P are ± 1. Similar discrete transformations may be discussed which transform particles into antiparticles and vice versa (charge conjugation, C), and which reverse the direction of time (time reversal, T). Invariance with respect to these operations may be visualized as invariance or symmetry under a type of mirror reflection (Ref. 21.2). These three operations are not independent since it may be shown that Lorentz invariance requires invariance under the joint operation $C \times P \times T$. Physical systems however need not be invariant under the operations separately, as has been particularly demonstrated for the weak interaction (Sect. **21.3**). When there is invariance, a useful quantum number may be defined for parity and charge conjugation (but not for time reversal, which turns initial into final states and therefore has no eigenstates).

(*d*) *Symmetry breaking.* It is now useful to return to Fig. 21.2, showing the ground-state hadrons and to observe the significance of isobaric spin T and hypercharge in labelling families. The mass differences between the $2T+1$ members of a family of given T are much less than those arising from changes of hypercharge. The former splittings $\Delta M/M$ are of the order of the fine structure constant and can reasonably be ascribed to the Coulomb interaction which breaks isobaric spin symmetry ($SU(2)$); the latter must be due to the breaking of a higher symmetry ($SU(3)$) by the strong interaction itself. The hadronic states at present observed, including resonances, are all in the spectroscopic sense likely to be fine structures of a basic ground state, determined by some unknown but hyperstrong interaction.

The validity of the conservation laws (or symmetry operations) with respect to the three interactions is shown in Table 21.1 which also includes

orders of magnitude of the lifetimes of particles decaying through these interactions and of the cross-sections for their production. Dimensionless coupling constants derived from these lifetimes or cross-sections are also included.

TABLE 21.1 Conservation laws

| | | Interaction | |
Conserved quantity or symmetry operator	Electromagnetic	Weak	Strong
Baryon number, B	Yes	Yes	Yes
Electric charge, Q	Yes	Yes	Yes
Isobaric spin, T	No	No	Yes
Hypercharge, Y	Yes	No	Yes
Parity, P	Yes	No	Yes
Charge conjugation, C	Yes	No	Yes
Time reversal, T	Yes	Yes?	Yes
Characteristic lifetime (s)	10^{-15}	10^{-8}	10^{-24}
Characteristic cross-section (mb)	10^{-4}	10^{-12}	10
Dimensionless coupling constant	$1/137$[a]	$1 \cdot 2 \times 10^{-5}$[b]	15[c]

[a] See Sect. 21.2.1.　　[b] See Sect. 21.3.1.　　[c] See Sect. 21.4.2.

21.1.3 Scattering amplitude and dispersion relations

Although it is likely that interactions between particles are intimately bound
up with their nature, it is nevertheless useful to discuss collisions in the way
introduced in Chapters 5 and 14 but with a relativistic generalization of both

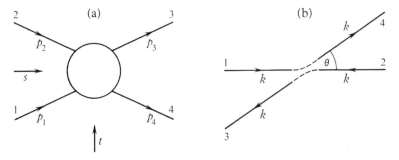

Fig. 21.5 (a) Notation for four-vectors p_i in collision between particles 1 and 2 producing particles 3 and 4. The variable s is the square of the centre of mass energy and $-t$ is the square of the
momentum transfer. (b) Elastic collision in centre of mass system, momentum k.

the kinematics and the scattering amplitude $f(\theta)$. For the kinematics of the
two particle reaction

$$1 + 2 \longrightarrow 3 + 4 \tag{21.8}$$

shown in Fig. 21.5(a) we use the four vectors p defined by

$$p^2 = p_0^2 - \mathbf{p}^2 \tag{21.9}$$

where p_0 is the total energy (E) of the particle, of mass m, and $\mathbf{p}$ its ordinary momentum.† In the initial and final states, when the particles are free, we have the 'on the mass shell' condition

$$p^2 = m^2 \qquad (21.10)$$

For the interaction shown, conservation of energy and momentum give

$$p_1 + p_2 = p_3 + p_4 \qquad (21.11)$$

Following Mandelstam‡, we may combine these four-vectors to form the invariant quantities

$$
\begin{aligned}
s &= (p_1 + p_2)^2 = (p_{10} + p_{20})^2 - (\mathbf{p}_1 + \mathbf{p}_2)^2 \\
t &= (p_1 - p_4)^2 = (p_{10} - p_{40})^2 - (\mathbf{p}_1 - \mathbf{p}_4)^2 \\
u &= (p_1 - p_3)^2 = (p_{10} - p_{30})^2 - (\mathbf{p}_1 - \mathbf{p}_3)^2
\end{aligned}
\qquad (21.12)
$$

These quantities are not independent because using (21.10) and (21.11) we have

$$s + t + u = \sum m^2 \qquad (21.13)$$

Their physical significance is best seen re-drawing Fig. 21.5(a) for the centre of mass system and considering (Fig. 21.5(b)) the elastic scattering of two identical particles of mass m in this system in which $\mathbf{p}_1 = -\mathbf{p}_2$ $(=k$ say$)$ and each energy $p_{10} = \sqrt{m^2 + k^2}$. It then follows that

$$
\begin{aligned}
s &= (2\sqrt{m^2 + k^2})^2 = 4(m^2 + k^2) \\
t &= -\{(k - k \cos \theta)^2 + k^2 \sin^2 \theta\} = -2k^2(1 - \cos \theta) \\
u &= -\{(k + k \cos \theta)^2 + k^2 \sin^2 \theta\} = -2k^2(1 + \cos \theta)
\end{aligned}
\qquad (21.14)
$$

from which we see that s is the square of the *total energy* of the interacting particles in the c.m. system and $-t$ is the *momentum transfer* (squared) between particles 1 and 4. Similarly, $-u$ may be regarded as an *exchange momentum transfer* (squared) between the incident particle 1 and the emergent particle 3. It will be noted that $s^{1/2}$ gives the effective mass (Sects. 19.2.5 and 20.3.1) of the particles $1 + 2$ or $3 + 4$.

The differential cross-section for the process (21.8) with spinless particles may be written, in terms of s and t, as

$$d\sigma = |F(s, t)|^2 \, d\Omega \qquad (21.15)$$

where s determines the available energy and t the angular coordinate. The *scattering amplitude* $F(s, t)$ reduces in the non-relativistic limit to the quantity $f(E, \theta)$ used in Chapter 14.

† In high energy physics it is customary to use a system of units in which $\hbar = c = 1$, so that $E^2 - \mathbf{p}^2 = m^2$. This system will be used in the remainder of this chapter when the physical meaning is clear. It should be noted that some authors use the convention $p^2 = \mathbf{p}^2 - p_0^2$ so that $p^2 = -m^2$ for a single particle.
‡ *Phys. Rev.*, **115**, 1741, 1959.

The scattering amplitude F is normally considered to be an *analytic* function of s and t, i.e. to be uniquely defined and differentiable for the whole range of these variables, including the non-physical values which occur below thresholds, e.g. for $s < 4m^2$ for the elastic scattering discussed above. Physical properties of the scattering amplitude, based on experience, are

(a) *unitarity*, the fact that the sum of the probabilities of producing all possible final states from any initial state must equal unity. This leads (Ref. 21.3) to the optical theorem for forward scattering ($t = 0$) which was introduced in Sect. 14.2.2.

(b) *crossing*, which relates the amplitude for a phenomenon in which a particle is absorbed to that for the reaction in which the corresponding antiparticle is created. Thus the processes

$$1 + 2 \rightarrow 3 + 4 \quad (s\text{-channel}) \tag{21.16}$$

$$1 + \bar{4} \rightarrow \bar{2} + 3 \quad (t\text{-channel}) \tag{21.17}$$

$$1 + \bar{3} \rightarrow \bar{2} + 4 \quad (u\text{-channel}) \tag{21.18}$$

where $\bar{2}, \bar{3}, \bar{4}$ represent the antiparticles of 2, 3, 4 are each physically possible, although with different ranges of the variables (s, t), which do not overlap.

The analytic property of the scattering amplitude permits one to formulate the consequences of the *principle of causality* which states that a scattered wave cannot precede the incident wave in time. If this is so, the amplitudes must obey a *dispersion relation*, which connects the real part of a complex amplitude with its imaginary part. Dispersion relations (first derived by Kramers and Kronig for light) do not depend on specific interaction processes and are applicable even when the detailed dynamical situation is not known, e.g. in a discussion of strong interactions. A dispersion relation (Ref. 21.4) has the typical form

$$\text{Real part } F(\omega) = \frac{1}{\pi} \text{ Principal part of } \int_{-\infty}^{\infty} \frac{\text{Im } F(\omega') \, d\omega'}{\omega' - \omega} \tag{21.19}$$

where ω is the total (laboratory) energy of an incident particle (1). In rendering this suitable for comparison with experiment we first eliminate the negative energies by use of crossing relations which bring in the observed imaginary part for antiparticle scattering. Then concentrating on forward scattering, we use the optical theorem to replace imaginary parts by total cross-sections. Finally we must deal separately with the region of integration $m > \omega > -m$, where m is the mass of the scattered particle, because this is an unphysical region of total energy. In this region however there may be total mass values ($s^{1/2}$) which correspond to bound states of the system $1 + 2$ and such *poles* can dominate the scattering amplitude in the physical region. The contribution from a pole, if calculated in lowest order by perturbation theory, is called a Born term. If the s channel (or direct) process is represented by a Feynman diagram (Fig. 21.6(a)) and if the mass of the intermediate particle in this process is M, then the direct Born term has the form

$$(\text{const.}) \frac{f^2}{s - M^2} \qquad (21.20)$$

where f^2 is a coupling constant (Sect. 21.4.2) which measures the strength of the interaction. This constant is split between the two vertices as shown in Fig. 21.6(a).

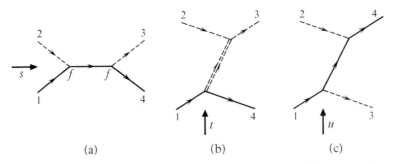

(a)   (b)   (c)

Fig. 21.6 Virtual particles in scattering process $1+2 \longrightarrow 3+4$: (a) Direct channel particle; (b) Crossed channel particle (see also Fig. 21.1); (c) Exchange channel particle.

For the particular case of pion–proton scattering the initial masses are m_π and m_p and there is a nucleon pole when $s = M^2 = m_\mathrm{p}^2$. Using (21.12) this gives

$$(\omega_\pi + m_\mathrm{p})^2 - \mathbf{p}_\pi^2 = m_\mathrm{p}^2 \quad \text{or} \quad \omega_\pi = -\frac{m_\pi^2}{2m_\mathrm{p}} = -10 \text{ MeV} \qquad (21.21)$$

which corresponds to a *negative* kinetic energy. Extrapolation of experimental data to the pole leads to a determination of the coupling constant f^2. The intermediate particle is in general virtual and does not have its physical mass. At the pole the particle becomes real, but at an unphysical value of the total energy. Pion-nucleon dispersion relations, including pole terms, are given in detail in Ref. 21.4.†

In addition to the direct s-channel process we can also consider the crossed channels, Figs. 21.6(b) and (c). In these channels also the intermediate particle must have quantum numbers to match the two vertices and there is a pole in t or u at the mass of the exchanged particle, which is at an unphysical value of the momentum transfer (or scattering angle). If the interaction at the vertex in diagrams such as Fig. 21.6 cannot be represented entirely by a coupling constant a momentum-dependent form factor G is introduced. The Fourier transform of G is related to spatial distributions of charge and magnetic moment in the case of electron–proton scattering (Sect. 21.4.3).

† They are also discussed by T. E. Ericson and M. P. Locher, *Nuclear Physics*, **A148**, 1, 1970 in connection with the application of dispersion relations to nuclear reactions.

21.2 The electromagnetic interaction

21.2.1 General features

The electromagnetic interaction, specified by laws associated with the names of Coulomb, Ampère and Faraday, is well known in classical physics. It is most powerfully expressed in Maxwell's equations for field amplitudes. The forces of the electromagnetic field ($\propto e^2/r^2$) are of *long range*, in contrast with the short range forces which give rise to nuclear binding, and are therefore observable as a property of matter in bulk. An electromagnetic process is the principal mode of de-excitation of the excited bound states of free atoms and nuclei.

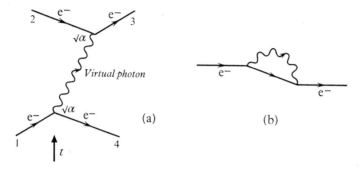

Fig. 21.7 The electromagnetic interaction: (a) Electron–electron scattering, described by one-photon exchange. The real mass of the photon is zero and there is a pole at $t = 0$. (b) Emission and re-absorption of a virtual photon by an electron giving rise to self-energy.

The electromagnetic field is quantized in terms of the photon, and the force between two charges e may be thought of as arising because of the mechanical effect of the emission and absorption of virtual photons† (Fig. 21.7(a)). The long range character of the force is ensured if the rest mass of the photon is zero (Sect. 18.3.2); the strength of the interaction is measured by the dimensionless fine structure constant

$$\alpha = \frac{\mu_0 c^2}{4\pi} \frac{e^2}{\hbar c}$$

also known as the electromagnetic coupling constant. In terms of the Feynman diagram (Fig. 21.7(a)) we place a factor $\sqrt{\alpha}$ or e at each electron–photon–electron vertex. The strength of the electromagnetic interaction is

† It must be noted that a charged particle may also absorb the virtual photon which it emits (Fig. 21.7(b)) and this is essentially an interaction of the particle with itself, whose effect, leading to an increase in effective mass, must be included in all electromagnetic calculations. The virtual photon may also be regarded as a virtual electron–positron pair and this leads to the appearance of additional charges in the field. These effects result in infinite integrals and progress has only been made by the introduction of *re-normalization* techniques to justify the use of the normally observed mass and charge. The emission of virtual photons is the source of the Lamb shift of atomic energy levels (Sect. 21.2.2).

compared with that of other interactions in Table 21.1 in terms of typical lifetimes and interaction cross-sections. These are based essentially on the coupling constants, which are also shown, in dimensionless form, in the table.

On the assumption of the validity of re-normalized mass and charge it is now possible to give a satisfactory account of all basically electromagnetic phenomena to a level of precision at which corrections due to other interactions may be felt. In particular, using Dirac theory for the particles (spin $\frac{1}{2}$) we may describe (a) the motion of electrons and positrons in prescribed fields, and (b) the interaction between particles, i.e. *quantum electrodynamics*, of which the electron–electron scattering illustrated in Fig. 21.7(a) is a simple example. The differential cross-section $d\sigma$ for a particular type of interaction between a point particle and a field, or between particles, is obtained according to the well-tried rules of quantum mechanics and since the coupling constant α is small compared with unity, the perturbation method may be used with confidence. In this way we arrive at an expansion of the total interaction in powers of α (or e^2) which enables us to set up the following classification;†

(a) *First order processes* ($d\sigma \propto e^2$). These include emission and absorption of radiation by nuclear and atomic states, and the photoelectric effect, and have been discussed in Chapters 5 and 13. For a free particle the emission or absorption of *real* photons is forbidden by the conservation of energy and momentum.

(b) *Second order processes* ($d\sigma \propto e^4$). These include electron–electron scattering, two-quantum annihilation of positrons and the Compton effect. They are described by diagrams of the type shown in Fig. 21.7(a), in which there are two interactions between electrons and real or virtual photons.

(c) *Third order processes* ($d\sigma \propto e^6$). These include pair creation and bremsstrahlung which are processes that do not take place in free space because energy and momentum cannot then be conserved. Interaction with the Coulomb field of a charged particle, e.g. through a virtual photon, is required and this counts as one of the three electromagnetic interactions involved (Fig. 21.8). There is a virtual electron in the intermediate state.

If electromagnetic processes are studied at high values of the momentum transfer variable $-t$ ($=q^2$) it may be possible to detect a breakdown of the theory. Such experiments probe any possible structure of the interacting particles over a distance of the order of $\hbar/q$; they may be conducted either directly, using high energy electrons or muons, or indirectly by examining properties such as the magnetic moment of the electron to a high degree of precision. Present efforts in this very basic field are outlined in Sects. 21.2.2 and 21.2.3.

The electromagnetic interaction is obviously charge dependent. As a result, energy differences arise between the members of charge multiplets whether they be nuclear states (Sect. **12.1**) or elementary particles (e.g. $\Sigma^+ - \Sigma^\circ - \Sigma^-$). In terms of isobaric spin, the interaction depends on T_z and,

† It should be noted that Feynman diagrams lead to matrix elements or probability amplitudes. Cross-sections or radiative widths are proportional to the square of these quantities.

in analogy with angular momentum selection rules we expect electromagnetic processes to be governed by the isospin rule $\Delta T = 0, \pm 1$. All other quantities are conserved (Table 21.1) including parity (as is known from the validity of selection rules in atomic physics, Ref. 21.10). Neutral particles such as π^0 ($\to 2\gamma$) and Σ^0 ($\to \Lambda^0 + \gamma$) decay electromagnetically since these processes involve no change of strangeness. Weak decay processes such as $\mu \to e + \nu + \bar{\nu}$ may also be accompanied by radiation (radiative decays) with a probability $\approx 1/137$ times the non-radiative decay. Decays such as $\Sigma^+ \to p + \gamma$ are forbidden because the change $\Delta S = 1$ is required. Since the electromagnetic interaction need not conserve isospin, decay of the η-meson to $3\pi^0$-mesons is permitted through this interaction although it is forbidden to the ω-meson which decays strongly. Two photon decay of the η is also permitted.† The η decays have been used to examine the possibility of C-invariance violation in electromagnetic processes (Sect. 21.3.2). Violation of T-invariance (and parity) is excluded by the absence of an observable electric dipole moment of the electron, the muon and the proton.

The photon quantum numbers for spin and parity are 1^- and particles with the same character, e.g. the ρ-meson, may be involved as virtual states in photon-induced reactions.

21.2.2 The electromagnetic interaction at low energy

The Dirac theory of the electron and positron does not take account of the emission of virtual photons or virtual electron–positron pairs. Corrections to Dirac predictions of static electromagnetic quantities arise as a power series in α/π and test the presence of virtual photons of high momentum.

(a) Spectroscopic quantities. The basic *fine structure constant* α may be obtained from the displacement of 10969 MHz between the $2^2P_{3/2}$ and $2^2P_{1/2}$ terms of the hydrogen atom due to spin-orbit interaction. This interval does not involve s-states and is relatively little affected by virtual photon emission. An alternative method of high accuracy is based on the Josephson effect, i.e. the tunnelling of particle-hole pairs between two superconductors, with the appearance of an alternating current of frequency $2eV/h$ where V is the applied potential difference. By combining the value of e/h thus obtained with other accurately known quantities, α may be deduced. From all methods of high precision the best value of 1969 is

$$\alpha^{-1} = 137 \cdot 03608 \pm 1 \cdot 9 \text{ parts per million} \qquad (21.22)$$

The *Lamb shift* is a displacement of about 1058 MHz between the $2^2S_{1/2}$ and $2^2P_{1/2}$ terms of the hydrogen atom, which were predicted by Dirac theory to be of equal energy. It arises because the emission of virtual photons disturbs the electron motion with greater effect in s-states than in p-states.

† The decay of an odd parity system to two photons (of intrinsic odd parity) is discussed by Yang, *Phys. Rev.*, **77**, 242, 1950. For photon parity see Ref. 21.1: Sakurai (p. 19), Marshak and Sudarshan (p. 79).

The observed value of $1057{\cdot}90 \pm 0{\cdot}010$ MHz agrees exactly with theory and may be taken to verify the validity of electromagnetic theory down to a distance of about 10^{-15} m.

(*b*) *Anomalous parts of magnetic moments.* The Lamb shift arises quantitatively in large part from an alteration of the electron *g*-factor to a value†

$$g_e = 2 \left(1 + \frac{1}{2}\frac{\alpha}{\pi} - 0{\cdot}3285\,\frac{\alpha^2}{\pi^2} + 0{\cdot}55\,\frac{\alpha^3}{\pi^3} \cdots \right) \qquad (21.23)$$

or

$$\tfrac{1}{2}(g_e - 2) = 1\,159\,644 \times 10^{-9} \qquad (21.24)$$

The correction $a_e = \tfrac{1}{2}(g_e - 2)$ has been measured accurately by Wilkinson and Crane and by Wesley and Rich‡ by comparing the orbital and spin precession frequencies of polarized electrons in a magnetic field and there is now essentially no difference between theory and experiment.

The muon $(g_\mu - 2)$ experiment has already been described (Sect. 19.3.1). The predicted *g*-factor is

$$g_\mu = 2 \left(1 + \frac{1}{2}\frac{\alpha}{\pi} + 0{\cdot}7658\,\frac{\alpha^2}{\pi^2} + 2{\cdot}55\,\frac{\alpha^3}{\pi^3} \cdots \right) \qquad (21.25)$$

and the result of the CERN storage ring experiment at a level of accuracy of 270 parts per million, agrees with this within one standard deviation.§ Because of the higher mass of the muon, this result provides a more severe test of the electromagnetic interaction. The accuracy approaches the level at which effects of virtual pion pairs, i.e. of the strong interaction, may be seen.

21.2.3 *The electromagnetic interaction at high energy*

In high energy experiments, phenomena associated with high momentum transfers in the electromagnetic interaction may be studied directly, by suitable choice of bombarding energy and angle of observation.

(*a*) *Electron–nucleon scattering.* The scattering of electrons of energy 200 MeV by a nuclear charge distribution was discussed in Sect. 11.2.1. Such experiments, which have been extended to higher energy, reveal the charge and magnetic moment distribution of the neutron and proton arising from virtual pion emission. These distributions have a spatial extent of the order of the Compton wavelength of the pion, but they arise because of the strong interaction and do not provide a clear test of quantum electrodynamics at small distances. The electromagnetic structure of the nucleon is discussed in Sect. 21.4.3 and a diagram for the electron scattering process is given in Fig. 21.16(a).

---

† S. J. Brodsky and S. D. Drell, *Ann. rev. nucl. Sci.*, **20**, 147, 1970.
‡ *Phys. Rev.*, **130**, 852, 1963; *Phys. Rev. Lett.*, **24**, 1320, 1970.
§ Recent theoretical work revises the coefficient of the α^3 term in eq. (21.25) and gives improved agreement with experiment.

(b) *Wide-angle pair production by photons and bremsstrahlung by electrons.*
These two processes are similar in origin as may be seen from Fig. 21.8 which
shows typical diagrams, of the so-called Bethe–Heitler type, which may be
connected by the principle of crossing symmetry. Other diagrams exist
representing processes which involve the photon in a strong interaction and
do not test quantum electrodynamics as clearly as is desired, but such processes can be eliminated experimentally by a suitable choice of angle of
observation, which also determines the momentum transfer in the process.

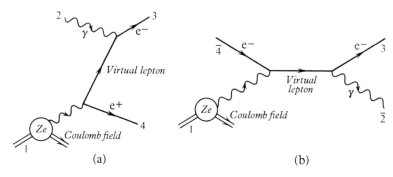

Fig. 21.8 The electromagnetic interaction: (a) Pair production by photon in field of nucleus Ze.
(b) Bremsstrahlung by electron in field of nucleus Ze.

In each diagram (Fig. 21.8(a), (b)) there is a virtual lepton whose energy
and momentum are off the mass shell. In fact the momentum transfer variable
t has opposite sign in the two cases shown.† These two experiments‡ thus
permit a wide range of the momentum transfer variable to be used; so far in
experiments corresponding to a lepton mass of the order of 1 GeV/c^2 no
deviation of cross-sections from the predictions of quantum electrodynamics
has been established. The corresponding length is $\approx 10^{-16}$ m.

(c) *Colliding beam experiments.* It is not possible conveniently to reach a
high centre of mass energy in a collision between an incident electron and a
target electron at rest. The c.m. energy increases only as the square root of
the incident energy. If however beams of electrons of equal velocity collide,
in a storage ring, this limitation is removed. Such experiments have now
been made at Stanford for the (e⁻e⁻) system with beams of energy 556 MeV
and provide the best test so far of the point like nature of the electron ($r \leqslant$
4×10^{-17} m). A diagram for this process is given in Fig. 21.7(a), p. 575.

The (e⁻e⁺) collision has been studied in storage rings at Orsay and
Novosibirsk. The diagrams possible resemble Figures 21.8(a), (b) with
photon lines changed to leptons and the virtual lepton to a virtual photon.

† This may be seen by considering the special case in which the photons are at right angles to the
electrons and each particle and photon has the same energy E. The virtual electron then has
energy 0 or $2E$ and momentum $\sqrt{2}E$ in the two cases and $t=(0-2E^2)$ or $(4E^2-2E^2)$ follows.
‡ The Compton scattering of photons in principle gives similar information to the bremsstrahlung
experiment, but the momentum transfer is smaller because the target electron is at rest.

Again the squared momentum transfer t has opposite sign in the two cases, which differ physically in that one scattering is represented as an annihilation followed by pair creation. A large range of t is accessible to experiment and evidence is found for a resonance at an energy in the intermediate state corresponding to the mass of the ρ-meson. This is basically a manifestation of the strong interaction and yields the form factor of the pion.†

21.3 The weak interaction

21.3.1 Summary

The weak interaction is responsible for the leptonic decay of strange and non-strange particles and for the non-leptonic decays of strange particles. Both leptonic and non-leptonic modes can be regarded as four-fermion processes (Fig. 21.9(a), (b)) because a pion is equivalent to a virtual nucleon–antinucleon pair.

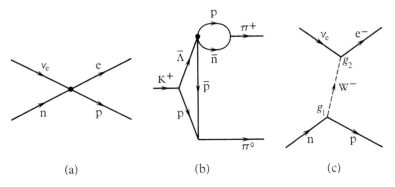

(a)   (b)   (c)

Fig. 21.9 Weak interactions: (a) Four-fermion point interaction (representing beta decay of the neutron by absorption of a neutrino). (b) Non-leptonic decay of the K^+-meson showing a four-fermion vertex at which a weak coupling constant applies. (c) Beta decay with an interaction of finite range, which might be associated with a boson W^-.

The strength of the weak interaction is measured by a coupling constant g which in practice is obtainable from the observed lifetime of some suitable particle such as the muon which decays by the weak process only. Its value is $g = g^\mu = 1.4354 \pm 0.0003 \times 10^{-62}$ Jm³. A dimensionless form for comparison with other interactions (Table 21.1) may be obtained by dividing this by the rest energy of the proton, $m_p c^2$ and by the cube of the proton Compton wavelength $(\hbar/m_p c)$; the number so obtained is 1.2×10^{-5}. Following Elton (Ref. 1.9, p. 265) we may compare this with the strong interaction by calculating the quantity ba^3 (Sect. 18.2.1) in the same units; the result is 27.6.

Figure 21.9(a) implies a point (contact) interaction but there may be an exchange particle W for the interaction in which case the diagram of Fig. 21.9(c) is appropriate and is obviously analogous to the similar diagram for

† R. Wilson, *Physics Today*, **22**, No. 1, 47, 1969.

the strong and the electromagnetic interaction. In this diagram the 'weak charge' to be associated with each vertex is not simply $\sqrt{g}$ because of the difference between the coupling of the W-particle to hadrons and to leptons. Furthermore, perturbation theory shows that the amplitude contains the mass m_W of the exchanged particle, which is unknown. If the true vertex factors are g_1, g_2 (as shown in Fig. 21.9) then the experimental coupling constant is defined by

$$\frac{g}{\sqrt{2}} = \frac{g_1 g_2}{m_W^2} \qquad (21.26)$$

The weak interaction conserves energy, momentum and angular momentum; also charge, baryon number and lepton number. It is invariant as far as is known, under the CPT transformation. It is *not* invariant to the parity and charge conjugation operations although until recently it was thought to be so to the joint process CP. This is no longer true for K_2^0 decays. Neither isobaric spin nor strangeness is meaningful for leptons, but when hadrons are involved the weak interaction connects only states between which the changes of isobaric spin and strangeness are limited. In particular we find examples of

$$\Delta S = 0, \pm 1; \quad \Delta T = 0, \pm\tfrac{1}{2}, \pm 1 \quad \text{and} \quad \Delta T_z = \pm\tfrac{1}{2}, \pm 1$$

i.e. for the weak interaction ΔS, ΔT, ΔT_z may be non-zero.

The basic properties of the weak interaction are well displayed by nuclear beta decay which is discussed in Chapter 16. In the following sections a brief account is given of the main steps towards an understanding of the interaction together with a survey of some experiments designed specifically for this purpose.

21.3.2 Parity and charge conjugation

The apparent inconsistency of the $K_{\pi 2}$ (θ) and $K_{\pi 3}$ (τ) modes of decay of the K^+-meson, indicating opposite parities for identical kaons led Lee and Yang to the conclusion that parity might not be conserved in the weak interaction. They further observed that none of the 'classical' experiments of β-decay would have revealed such an effect, and they suggested new experiments by which such a violation of reflection symmetry could be examined. The two crucial experiments by which this hypothesis was first tested have already been described. They are the observation of

(*a*) an asymmetry in the β-decay of polarized nuclei (Wu *et al.*, Sect. **16.7**)
(*b*) an asymmetry in the β-decay of the muon in μ-decay of the pion (Garwin *et al.*, Sect. 19.3.1)

Other observations which also verify parity non-conservation (PNC) in β-decay are (Ref. 21.5)

(c) the longitudinal polarization (helicity) of electrons and neutrinos emitted in β-decay

(d) the circular polarization of gamma radiation following β-decays to excited states,

while the asymmetry of the leptonic decay of hyperons (Sect. 20.6.2) verifies that PNC is characteristic of the interaction rather than of a certain group of particles.

The asymmetries observed in PNC experiments arise because of interference between parity conserving and parity non-conserving transitions

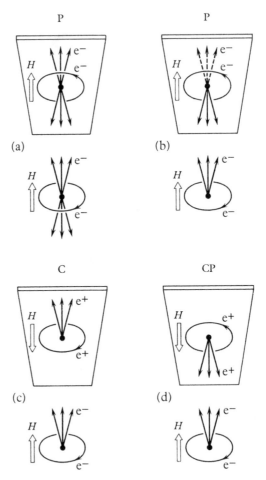

Fig. 21.10 Beta decay of polarized nuclei (see text). H shows the direction of a magnetic field due to a current loop, used for polarizing the nuclei. The upper diagrams represent the result of reflecting the process shown in the lower diagrams in various types of 'mirror' as discussed by Wigner, *Scientific American*, Vol. 213, p. 28, 1965. (a) Parity mirror (P), parity-conserving decay; (b) Parity mirror (P), parity-non-conserving decay; (c) Charge conjugation mirror (C), parity-non-conserving decay; (d) CP mirror, parity-non-conserving decay.

between two states (e.g. nuclear energy states) of well-defined parity. The interpretation of experiment (*a*) above may be followed from Fig. 21.10 which shows the angular distribution of electrons in relation to the direction of an external magnetic field which polarizes the magnetic moment of the emitting nuclei, i.e. sets up a preferred spin direction. This field is produced by a current in a coil which is represented by an electron moving in the direction shown. In Fig. 21.10(a) (lower) a symmetric distribution is shown and this would be the case if parity were conserved; by space reflection (*P*) in a mirror, as in the upper part of the diagram, the situation is unchanged. If, however, the emission of electrons is asymmetric, as shown in Fig. 21.10(b), then *P* reflection creates a different system because the reflection changes the direction of the decay arrows but not the direction of the magnetic field or nuclear spin. The experiment of Wu *et al.* proved that the electron distribution was in fact asymmetric and parity non-conservation follows. Stated in a different way, beta decay does not reflect in the *P*-mirror; the actual decay distribution remains as shown by the dotted arrows in Fig. 21.10(b). It will be noted that a reflection is equivalent to the parity operation defined in Sect. **3.1** followed by a rotation, and since rotational invariance is not in question (because of conservation of angular momentum) reflection experiments test parity conservation.

If instead of space reflection (*P*) the charge conjugation reflection (*C*) is imposed, a decay distribution will behave as shown in Fig. 21.10(c). Electrons are replaced by positrons and both the magnetic field and nuclear moment change sign, so that the nuclear angular momentum stays in the same direction. The antinucleus however is now emitting positrons in the same direction as the original nucleus emitted electrons with respect to the spin direction and this result does not exhibit 'reflection' type of symmetry.

Symmetry may be restored if both *P* and *C* operations are applied to the decay. This is shown in Fig. 21.10(d) in which it is seen that under *CP* reflection a nucleus emitting electrons say antiparallel to the nuclear spin turns into an antinucleus emitting positrons parallel to the spin. This is consistent with experiment,† and it may be concluded that the β-decay weak interaction is invariant under *CP*. This conclusion might be thought to extend to the non-leptonic weak decays but in 1964 evidence to the contrary was presented.‡ In a spark chamber experiment with beams of K^0-mesons at the Brookhaven laboratory evidence for a 2π decay mode of the K_2^0-meson

$$K_2^0 \longrightarrow \pi^+ + \pi^- \quad \text{(1 in 500 of all charged decays)} \quad (21.27)$$

was obtained. This violates *CP* conservation because the K_2^0 particle (Sect.

---

† If the spin change in the ^{60}Co decay is examined it may be inferred that the emitted electrons are left handed. Experiments with antimatter cannot yet be performed, but it is possible to compare the electrons and positrons from the decay of unpolarized negative and positive muons (Sect. 19.3.1(*c*)). The experiments of Culligan *et al.* (*Nature*, **180**, 751, 1957) and of Macq *et al.* (*Phys. Rev.*, **112**, 2061, 1958) confirm that the electrons are left-handed and the positrons right-handed. This proves violation of *C*-symmetry, but is consistent with *CP* invariance.

‡ J. H. Christenson, J. W. Cronin, V. L. Fitch and R. Turlay, *Phys. Rev. Lett.*, **13**, 138, 1964.

20.4.4) has CP odd and the $\pi^+\pi^-$ system in an S-state is obviously even with respect to C, P and therefore CP. Only the K_1^0 meson, which has CP even, should show 2π decay. So far however such effects seem confined to kaons and the notion of restoration of symmetry to the polarized ^{60}Co experiment by the assertion of a mirror image anticobalt decay may still be possible. The origin of CP violation is not yet understood and is being actively investigated by detailed study of the branching ratios to charged and neutral pions in the K_2^0 decay. New types of interaction have been postulated but another possibility, since the effect is weak, is that there is a breakdown of C-invariance in the electromagnetic interaction which could be conveyed to weak processes by coupling to virtual photons. This would show up typically as an asymmetry between π^+ and π^- emission in the electromagnetic decays.

$$\eta^0 \longrightarrow \pi^+ + \pi^- + \gamma$$
$$\eta^0 \longrightarrow \pi^+ + \pi^- + \pi^0$$

(21.28)

Experiments, in several laboratories, are still inconclusive on this question. The violation of CP in the weak interaction means also a violation of time reversal invariance. This can in fact be tested by observing the correlation between neutrino (or recoil) direction and electron direction in the decay of a polarized nucleus. Results for ^{19}Ne, obtained by Calaprice et al.[†] show no evidence for T non-conservation within the limits of error. This remains one of the topics of current interest and importance.

21.3.3 *Nature of the weak interaction in beta decay*

So far we have mainly been concerned with the phenomenology of weak decays, and have not embodied the results into any expression suitable for handling by quantum mechanics; in other words we have no way of writing down the matrix element H_{if} of the weak interaction between initial and final states (cf. eq. 16.35). One way to start is to enquire what matrix elements are allowable which contain the wave functions of the four fermions concerned in a weak interaction and which are either scalars or pseudoscalars[‡] with respect to the Lorentz transformation. The squares of these matrix elements would be Lorentz invariants.

It turns out that there are five ways of constructing overall scalars and pseudoscalars from the nucleon and lepton wavefunctions. Within these matrix elements the nucleon and lepton wavefunctions themselves behave as scalar (S), vector (V), axial vector (A), tensor T and pseudoscalar (P) quantities under the Lorentz transformation. It is conventional to describe the overall interaction matrix elements as of S, V, T, A or P character according to the behaviour of the nucleon wavefunctions.[§] There are thus 10 independent

[†] *Phys. Rev.*, **184**, 1117, 1969.
[‡] A pseudoscalar is a quantity which is odd under the parity operation but is otherwise of a scalar nature.
[§] The original Fermi formulation of the theory of β-decay was essentially a vector interaction (V).

matrix elements, each with its own coupling constant C_S, C_V, C_T, C_A, C_P (scalar, parity conserving) and C'_S, C'_V, C'_T, C'_A, C'_P (pseudoscalar, parity non-conserving) and it is the task of experiment to determine these. They are normalized so that the total interaction is represented in strength by the coupling constant g (Section 21.3.1). Each interaction implies certain selection rules for the initial and final nuclear states, and by proper choice of these states, particular interactions can be isolated. Thus, Fermi transitions (Sect. 16.5.2) might be governed by the S and V interactions and Gamow–Teller transitions by A and T: the P interaction turns out to be not important for ordinary nuclear β-decay.

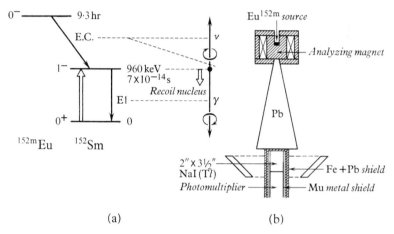

Fig. 21.11 Neutrino helicity experiment (see text): (a) Decay scheme, and angular momenta of opposed neutrino and photon, whose frequency is Doppler shifted by the velocity of the recoil nucleus. (b) Resonant scattering arrangement with solid iron magnet for determining sign of circular polarization of forward moving quanta (Goldhaber, Grodzins and Sunyar).

Detailed theory shows that a clear decision between interaction types can be obtained if the angular correlation between the electron and neutrino directions in an allowed β-decay is observed. These experiments have been done and now exclude the S and T types of interaction. Before this situation was reached however, and because of the difficulty of such experiments, an alternative method of considerable elegance was devised by Goldhaber, Grodzins and Sunyar.† This is essentially a determination of the *helicity of the neutrino* emitted in the electron-capture process for 9·3 hour europium-152

$$^{152}\mathrm{Eu} \xrightarrow{\ ^c\bar{\kappa}\ } {}^{152}\mathrm{Sm}^* + \nu \tag{21.29}$$

followed by photon emission from the excited product nucleus, which has a lifetime of 7×10^{-14} s:

$$^{152}\mathrm{Sm}^* \longrightarrow {}^{152}\mathrm{Sm} + \gamma \ (960 \ \mathrm{keV}) \tag{21.30}$$

† *Phys. Rev.*, **109**, 1015, 1958. See also L. Grodzins, *Progr. nucl. Phys.*, **7**, 163, 1959.

The spins involved in these transitions are shown in Fig. 21.11(a) and for the electron-capture process we can evidently write, assuming capture from the K-shell,

$$\text{Spins } \mathcal{J}: \quad \tfrac{1}{2}(e_K) + 0(\text{Eu}) \longrightarrow 1(\text{Sm*}) + \tfrac{1}{2}(\nu)$$
$$\mathcal{J}_z: \quad \pm\tfrac{1}{2} \quad 0 \qquad \pm 1, 0 \quad \pm\tfrac{1}{2}$$

(21.31)

If however the neutrino is characterized by a definite helicity $\mathcal{H}$ and if we let this particle define the z-axis then $\mathcal{H} = +1$ requires $\mathcal{J}_z(\nu)$ to be $+\tfrac{1}{2}$ and $\mathcal{J}_z(\text{Sm*})$ will be limited to 0 or -1, while for $\mathcal{H} = -1$ we have $\mathcal{J}_z(\text{Sm*}) = +1$ or 0. The excited nucleus is thus spin polarized.

If the photon emitted in process (21.30) is observed along the z-axis, in a direction opposite to the neutrino, conservation of angular momentum requires it to take away spin 1 with a component ± 1, and it is circularly polarized. For $\mathcal{J}_z(\text{Sm*}) = +1 \rightarrow \mathcal{J}_z(\text{Sm}) = 0$ we have $\Delta\mathcal{J}_z = -1$ and the photon is left-circular ($\mathcal{H} = -1$) and for $\mathcal{J}_z(\text{Sm*}) = -1$ the polarization is right-handed ($\mathcal{H} = +1$). The sign of circular polarization is therefore the same as the neutrino helicity.

The sign of circular polarization was determined by the inefficient but conventional method of transmission through magnetized iron (Fig. 21.11(b)). The selection of quanta emitted in a direction opposite to that of the neutrino was accomplished by imposing a resonant scattering condition (Sect. 13.6.4). In order that the radiation shall be scattered in this way by the samarium oxide annulus (Fig. 21.11(b)) it is necessary that recoil losses in emission and absorption shall be fully compensated by the Doppler shift due to the recoil of the ^{152}Sm* nucleus against the neutrino emission. This in turn requires forward emission from the moving nucleus, before the nucleus slows down. In the experiment all these conditions were satisfied because of the fortunate characteristics of the decay scheme, and the unambiguous result was obtained that the neutrino has *negative helicity*.

If now one returns to theory, this result, coupled with knowledge of the electron helicity in β-decays,† implies that the Gamow–Teller interaction is of *axial vector* (A) character. This is confirmed by the latest (eν) correlation experiments, which also indicate that in Fermi transitions the interaction is *vector* (V).

The magnitude of the longitudinal polarization effects suggests that one is dealing with lepton states which try to be as wholly left-handed or as wholly right-handed as possible. It is then possible to combine the parity conserving and parity non-conserving coupling constants into just one coupling constant (for each interaction) which implies such states and to speak of the beta interaction as of type V and A, with constants g_V and g_A.

Comparison between lifetimes for Fermi and Gamow–Teller transitions shows that g_V and g_A are not equal, and from observations on the decay of polarized neutrons, the amplitudes of V and A must be taken in opposite

† Obtainable from observation of circular polarization of bremsstrahlung.

phase. We therefore now state the general beta interaction to be of type

$$(V - 1 \cdot 2A) \tag{21.32}$$

Both the vector and the axial vector interactions give maximum non-conservation of parity, i.e. they carry the sense of a left-handed or right-handed screw.

21.3.4 The universal Fermi interaction (Ref. 21.5)

(*a*) *Universality.* If a coupling constant is extracted from the observed rate of weak interactions it is found that very similar results are obtained for the processes

(*a*) nuclear β-decay   (Sect. 16.5.2)
(*b*) muon decay   (Sect. 19.3.1)
(*c*) muon capture   (Sect. 19.3.2)
(*d*) $\Sigma \rightarrow \Lambda$ decay   (Sect. 20.6.2, eq. 20.43)

Process (*b*) is wholly leptonic, processes (*a*), (*c*), (*d*) are semi-leptonic since hadrons are involved, and all these processes are non strangeness-changing ($\Delta S = 0$). There are also strangeness-changing semi-leptonic and non-leptonic decays of K-mesons and hyperons. These processes with $\Delta S = 1$ have a rate of a few percent only of the $\Delta S = 0$ transitions.

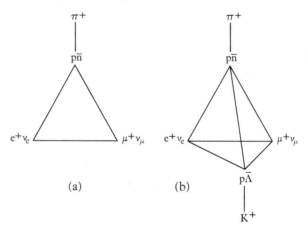

Fig. 21.12 Weak interactions: (a) Non-strange particles (Puppi triangle). Each pair shown may be connected with the pairs at the other vertices by a weak interaction. The nucleon vertex shows how the pion may enter into the weak interaction via a nucleon–antinucleon pair. (b) Addition of an extra vertex of positive strangeness showing a connection of the kaon with the weak interaction.

The similarity of the strengths of the $\Delta S = 0$ processes led to the suggestion that they might all be manifestations of a general basic property of particles, known as the *universal Fermi interaction* which involves in each case four fermions. A schematic representation of this suggestion is shown in Fig. 21.12(a) which places the appropriate pairs of weakly interacting fermions at the corners of a triangle (generally known as the *Puppi triangle*). This figure may be extended to strangeness-changing decays by converting it to a tetrahedron with particles equivalent to the K^+ $(=p\overline{\Lambda})$ at the fourth vertex (Fig. 21.12(b)). We also note that the pion is coupled (strongly) to the nucleon vertex because of the equivalence $\pi^+ = p + \overline{n}$. This fact immediately introduces a problem because the leptonic vertices $(e^+ v_e)$ and $(\mu^+ v_\mu)$ have no such coupling so it might be thought that the strength of weak decays involving nucleons would be strongly affected in comparison with μ-decay by the possibility of the emission of virtual pions. There is in fact a plausible hypothesis which suggests why this is not so.

(b) *The conserved vector current hypothesis (CVC).* This hypothesis is based on an analogy between the vector part of the weak interaction and electromagnetism. It is a well-known fact that the charge of the proton is equal to the charge of the positron. The positron has no couplings to the strong interaction, but the proton is emitting and re-absorbing virtual pions (Fig. 21.13). It then follows that the pion must carry exactly the proton

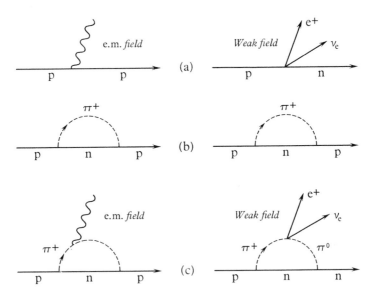

Fig. 21.13 Coupling of a proton to the electromagnetic and weak interaction fields; (a) proton without meson cloud, coupled to fields, (b) proton having emitted virtual meson could be uncoupled, (c) restoration of coupling by ascribing the properties of charge and weak decay to the pion.

charge in order that no observable change in this quantity shall result from the strong interaction. We then have *conservation of charge.* Similarly (Fig. 21.13), if the pion carries with it the power of β-decay to the right extent, the nucleon beta-decay will not be affected by the strong couplings which we know to exist. We then have *conservation of weak charge.* When these principles are expressed precisely† it is found that it is only the vector (V) part of the weak interaction which exhibits this exact analogy with electromagnetic currents. There is in fact reason to believe that more than analogy is involved here and that the currents representing nuclear beta decay, and the electromagnetic current, form part of an $SU(3)$ representation (Sect. 21.1.2).

The *CVC* hypothesis may be tested by observing the beta decay of the pion. The process $\pi^+ \to \pi^0 + e^+ + \nu_e$ in fact takes place to the extent (1 in 10^8 of normal $\pi \to \mu$ decay) required by the hypothesis. More precisely, the coupling constant g^μ obtained from muon decay, i.e.

$$g^\mu = 1\cdot4354 \pm 0\cdot0003 \times 10^{-62} \text{ J m}^3 \qquad (21.33)$$

should be equal to the coupling for the vector part of nuclear beta decay. The best value‡ of this quantity is at the moment

$$g_v^\beta = 1\cdot4102 \pm 0\cdot0012 \times 10^{-62} \text{ J m}^3 \qquad (21.34)$$

These two values are close, but their difference (1·8 per cent) is well outside the combined experimental error. It now seems likely that this difference is real and that it is a result of a connection between normal beta decay and the strangeness-changing processes of the weak interaction.

(*c*) *The Cabibbo hypothesis.* The formal treatment of the weak interaction, proceeding again in analogy with electromagnetism, sets up an energy of interaction in the form of the coupling constant times a product of two currents, or more precisely the product of a current $\mathscr{J}$ with itself. This current has the symbolic form

$$\mathscr{J} = j_e + j_\mu + \mathscr{J} + S \qquad (21.35)$$

where, $j_e, j_\mu, \mathscr{J}$ and S refer respectively to electrons, muons, hadrons ($\Delta S = 0$) and hadrons ($\Delta S = 1$). The product of $\mathscr{J}$ with itself may be thought of as representing the scattering of one particle by another, with the exchange of some property, and by taking appropriate pairs of currents we get terms corresponding to the known physical processes, e.g.

$$\begin{array}{ll} j_e j_\mu & \text{—muon decay} \\ j_e \mathscr{J} & \text{—nuclear beta decay} \\ j_\mu \mathscr{J} & \text{—muon capture} \\ j_e S, j_\mu S & \text{—}\Delta S = 1 \text{ leptonic decays of strange particles} \\ \mathscr{J} S & \text{—non-leptonic decays of strange particles} \end{array}$$

So far this adds nothing new. Cabibbo§ however proposed that the hadronic currents $\mathscr{J}$, S should not be coupled in equally. Specifically he writes

$$\mathscr{J} = j_e + j_\mu + \mathscr{J} \cos \theta + S \sin \theta \qquad (21.36)$$

---

† R. P. Feynman and M. Gell-Mann, *Phys. Rev.*, **109**, 193, 1958; Ref. 21.5 (C. S. Wu).
‡ To isolate the vector part of the weak interaction, decays between analogous nuclear states with $\mathscr{J} = 0^+$ are studied. These occur, for instance in the nuclei ^{14}O, ^{14}N; ^{26}Al, ^{26}Mg; and several other pairs linked by positron emission.
§ N. Cabibbo, *Phys. Rev. Lett.*, **10**, 531, 1963.

and the coupling constant for nuclear beta decay then becomes $g \cos \theta$ where $g = g^{\mu}$. The $\Delta S = 1$ leptonic decays of the baryons (and of the mesons) have coupling constants proportional to $\sin \theta$. The connection of the $\Delta S = 0$ and $\Delta S = 1$ currents through θ is made plausible by the existence of unitary symmetry multiplets (Sect. 21.4.4) among the hadrons. The close association between the particles of a multiplet implies that the form of the weak decay interaction is a property of the whole family of particles and not exclusively of any one individual member. In other words, if nuclear beta decay appears to have a smaller coupling constant than it should according to the universal Fermi interaction, that is because some of the weak charge is associated with particles Λ, Σ, Ξ which together with the nucleons form a multiplet.

The leptonic decays of the baryons are less accurately measured than those of the mesons and it is the latter that are normally used for determinations of the Cabibbo angle θ from a $|\Delta S| = 1$ transition. Thus from the probability of the decay

$$ K^+ \longrightarrow \pi^0 + e^+ + \nu \tag{21.37} $$

dependent on $g_V^K = g^{\mu} \sin \theta$ one finds $\sin \theta = 0\cdot221 \pm 0\cdot004$. From $g_v^{\beta} = g^{\mu} \cos \theta$ the corresponding value is $0\cdot186 \pm 0\cdot004$. The difference between these two values is probably to be ascribed to theoretical corrections which are not yet fully understood but which must certainly enter into the decay rates because of coupling between the charges and electromagnetic field. The closeness of the two estimates of θ however generally supports the Cabibbo hypothesis, which unifies a considerable area of weak interaction physics.

21.4 The strong interaction

21.4.1 *General features*

The strong interaction governs the mutual influence of the hadrons (pions, kaons, nucleons and hyperons). It was first interpreted by Yukawa, who described the short-range internucleon force in terms of one-meson exchange (Sect. 18.3.2 and Fig. 21.1). Following Yukawa, we now seek information on the strong interaction in nucleon–nucleon scattering over a wide range of energies, in the form factors of nucleons and pions and in photoproduction experiments. Hypernuclei and baryon and meson resonances are also manifestations of this same interaction, and it is in the realm of strongly interacting particles that the great increase in the number of observable states has taken place. We focus attention, in the spirit of Sect. **21.1**, on (a) the classification of states, with particular reference to quantum numbers arising from invariance under transformations, and on (b) the dynamics of hadron processes, leading to the determination of coupling constants from experimental cross-sections. These two aspects are not independent; any interaction which couples two well-defined states must conform to the necessary selection rules and to the underlying invariance properties.

Strong interactions between elementary particles are responsible for the total cross-sections as a function of energy shown in Fig. 21.14, from which one can see a tendency to asymptotic limits. The elastic scattering contribution to these total cross-sections has a forward peaked angular distribution characteristic of diffraction scattering. Inelastic collisions have been studied mainly with the object of identifying resonant states and determining the channels through which they decay.

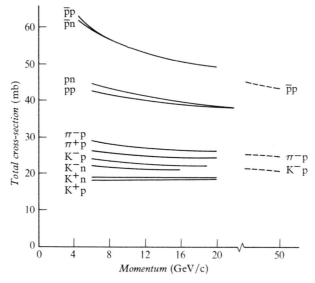

Fig. 21.14 Total cross-sections for strong interactions at high energy (S. J. Lindenbaum, quoted by R. J. Eden, Ref. 21.4; the figures for 50 GeV/c are given in a preliminary communication from the Serpukhov laboratory, see Denisov *et al.*, *Phys. Lett.*, **36B**, 415, 1971).

The strong interaction like all other known processes conserves baryon number B. There is also good evidence that it conserves parity, i.e. is invariant to the operation P (except for minor effects due to the weak interaction, Sect. 21.4.6). The equivalence of cross-sections for forward and backward nuclear reactions (e.g. $^{25}Mg + p \rightleftarrows ^{24}Mg + d$) is evidence for invariance under time reversal T, and from the CPT theorem we infer invariance under charge conjugation C.

Evidence for isobaric spin conservation is clearly seen in pion and kaon scattering by nucleons and in the existence of isospin multiplets and isospin selection rules in both nuclear and particle physics. Ordinary charge conservation is a fact of experience and the phenomenon of associated production shows that the strong interaction also conserves hypercharge. We shall see in Sect. 21.4.4 that isospin and hypercharge may be brought together into a deeper symmetry scheme. The symmetry is broken by Coulomb forces in isobaric spin multiplets and by the strong interaction itself in the more general symmetry so that mass differences arise for example between the neutron and the proton and between the nucleons and strange particles as shown in Fig. 21.2.

Summarizing, the strong interaction is a short-range force between hadrons (range $\approx 10^{-15}$ m corresponding to one-pion exchange, but with other exchanges possible), subject to invariances which lead to the selection rules:

$$\Delta B = \Delta Q = \Delta Y = 0$$

$$\Delta \pi \text{ (parity)} = 0$$

$$\Delta T = \Delta T_z = 0 \qquad (21.38)$$

21.4.2 Coupling constant

The basic pion–nucleon interaction is represented at the πNN vertices of the Feynman diagrams for nucleon–nucleon scattering (Fig. 21.1) and for pion–nucleon scattering (Fig. 21.6(a)). The coupling constant $f^2/4\pi\hbar c$ is analogous to the electromagnetic coupling constant $\mu_0 c^2 e^2/4\pi\hbar c$. It may be determined, as indicated in Sect. 21.1.3, by analysis of pion–nucleon scattering experiments. These comprise the variation of total cross-section for $\pi^+ p$ and $\pi^- p$ with energy, and angular distributions, also as a function of energy. A recent analysis[†] gives $f^2/4\pi\hbar c = 0.081 \pm 0.002$. This constant is also often quoted as $g^2/4\pi\hbar c \approx 15$ where

$$f^2 = g^2 \left(\frac{m_\pi}{2m_p} \right)^2 \qquad (21.39)$$

The pion–nucleon coupling constant determines the strength of the nucleon–nucleon force according to the pion exchange theory (Sect. **18.3**) and the observed potentials are consistent with the value quoted above. The constant f^2 may also be obtained directly by analysing neutron–proton scattering experiments, which are free from Coulomb effects. It was pointed out by Chew[‡] that in this case the best technique of extrapolation would be from a study of the variation of differential cross-section with angle (i.e. with momentum transfer) at a fixed centre of mass energy. If a dispersion relation is written in terms of the momentum transfer variable a pole occurs at the point $t = m_\pi^2$ corresponding to single pion exchange (Fig. 21.1). From (21.14) the c.m. angle for momentum k is given by

$$\cos \theta = 1 + \frac{m_\pi^2}{2k^2} \qquad (21.40)$$

The differential cross-section for n–p scattering at fixed energy, extrapolated suitably to this non-physical angle, yields the coupling constant. In an experiment with neutrons of 350 MeV energy, Ashmore *et al.*[§] obtained $g^2/4\pi\hbar c = 14.3 \pm 1.0$ which is consistent with the values obtained from pion scattering.

[†] J. Hamilton and W. S. Woolcock, *Rev. mod. Phys.*, **35**, 737, 1963.
[‡] *Phys. Rev.*, **112**, 1380, 1958.
[§] *Nuclear Physics*, **36**, 258, 1962.

The energy in such experiments should be high so that the pole approaches the physical region, but not so high that multiple pion exchange becomes important.

In strong interaction experiments in which the poles lie near to the physical region, the effect of the pole can be seen in sharp forward (or backward) peaks in the angular distribution. It is sometimes useful to think of these in terms of the peripheral model which postulates the collision of an incident particle with a virtual pion emitted by the target particle. Feynman diagrams provide a natural representation for such a process, which emphasizes the long range part of the interparticle force. An example of the peripheral mechanism for the reaction $p + p \to p + Y + K$ is given in Fig. 20.7; the cross-section in this case depends on the pion–nucleon coupling constant as well as on the cross-section for the reaction at the upper vertex.

The forward† peak seen in elastic scattering processes is at least partly analogous to a classical diffraction phenomenon (Ref. 21.4, Burkhardt, p. 64). The peripheral model suggests in addition that forward *and* backward peaks will be seen in all processes, elastic and inelastic, if a known particle can be exchanged with low momentum transfer. Thus the backward peak observed in K^+p elastic scattering is not seen in K^-p because of the absence of a suitably doubly charged baryon (Fig. 21.15). Neutron–proton scattering, in which a pion may be exchanged, shows both forward and backward peaks (Sect. 18.2.7).

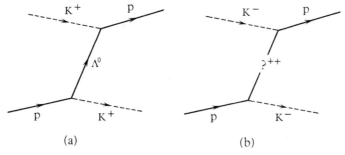

(a)　　　　　　　　　　　(b)

Fig. 21.15 Backward peaks in elastic scattering: (a) K^+p in which baryon (Λ^0) exchange is possible and a backward peak is found. See also Fig. 21.6(c). (b) K^-p in which baryon exchange is impossible because it requires a doubly charged strange particle and a backward peak is not seen.

21.4.3 *Electromagnetic structure of the nucleon*‡

The emission and reabsorption of virtual pions by a nucleon at rest (cf. Fig. 21.7(b) for the analogous electromagnetic case) creates the anomalous part of the magnetic moment. The field of virtual pions extends out to a distance

---

† Forward is here taken to mean the direction of the incident particle in the laboratory or c.m. system. Backward is then the direction of the target particle in the c.m. system.
‡ See R. Wilson, *Physics Today*, **22**, No. 1, 47, 1969; R. R. Wilson and J. S. Levinger, *Ann. rev. nucl. sci.*, **14**, 135, 1964; Ref. 21.6, vol. 1.

of the order of the pion Compton wavelength ($\approx 10^{-15}$ m) and gives the nucleon a finite charge structure. This may be explored by electron scattering experiments (Sect. 11.2.1) which probe the structure down to a distance $\hbar/q$ for momentum transfer q. Such experiments in principle reveal any substructure of the nucleon.

The scattering process may be represented in a diagram as shown in Fig. 21.16(a) and if the scattering potential is spherically symmetrical the differential cross-section for the process can be expressed as in eq. (11.6) with $Z = 1$. The *form factor* written in eq. (11.6) as F must however be more closely examined in the case of the proton and at the electron energies required to explore the nucleon charge distribution, the recoil of the heavy particle may not be neglected. Taking these points into account we find instead of (11.6) a formula due to Rosenbluth in which there are now two form factors, G_E and G_M, relating to the charge and magnetic moment distributions respectively. The form factors are functions of the momentum transfer variable t (often written $-q^2$) and essentially describe the interaction of the virtual photon with the nucleon at the upper vertex in Fig. 21.16(a). In the limit of zero momentum transfer ($t = 0$) we have for the proton $G_E = 1$, $G_M = 2\cdot79$ and for the neutron $G_E = 0$, $G_M = -1\cdot91$, values familiar as the charge and magnetic moment in units of e and μ_N. In the case of electron scattering by

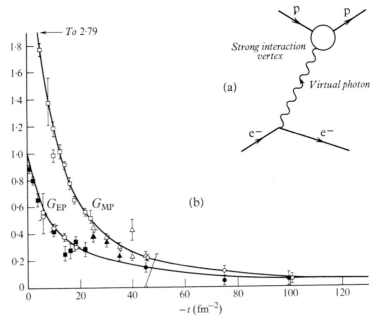

Fig. 21.16 Electron–proton scattering: (a) Space time diagram, showing exchange of virtual photon between electron and nucleon vertex. At this vertex the virtual pion cloud illustrated in Fig. 21.13(c) exists. (b) Experimental results for proton form factors. The units of ($-t$) are such that $\hbar = 1$, i.e. fm^{-2} (from Ref. 21.6, Vol. 1, p. 354).

heavy nuclei, for which recoil may be neglected, the nuclear charge distribution is given by a Fourier transform of the form factor. For the nucleon, because of recoil effects, and because of the emergence of both G_E and G_M it is more convenient to present experimental results in terms of invariant form factors than as charge or magnetic moment distributions. The experimental results for the proton are shown in Fig. 21.16(b) as a function of four-momentum transfer $(-t > 0)$.

An understanding of the nucleon form factors cannot reasonably be obtained by perturbation theory because of the large value of the strong interaction coupling constant (g^2) but dispersion theory calculations which relate the form-factors for low momentum transfer to pion–pion and pion–nucleon scattering amplitudes have been successful. From such calculations it was clearly established that the nucleon structure could not be wholly described in terms of a particle as light as a pion. The prediction was therefore made that a heavier meson must exist.† Moreover two such mesons would be necessary so that their effects could be added to describe the proton and subtracted in the case of the neutron.‡ Since the problem is to describe an electromagnetic structure (albeit that it is produced by the strong interaction) the mesons ought to have the same quantum numbers as the photon (1^-).

The discovery of the ρ- and ω-mesons (Sect. 19.2.5(c), (d)) provided evidence for the particles theoretically required. The ρ, with $T = 1$, is an isobaric vector particle existing in three states and capable of exerting a different effect in the proton and neutron. The ω, with $T = 0$, is an isobaric scalar and its effect is the same in the proton and neutron. Together these particles, with perhaps some contribution from still heavier meson resonances, are adequate to account for our present knowledge of the electromagnetic structure of the nucleons. The η-meson has spin 0 and is not concerned in this structure.

Analyticity implies that the form factors exist for positive values of $t(-t < 0)$. This is a region which has been explored by the study of the proton–antiproton annihilation process yielding electrons, which is described by applying the crossing principle to Fig. 21.16(a) but data of adequate accuracy are not yet available. The role of the ρ-meson is however seen both in photoproduction experiments and in electron–positron colliding beam studies (Sect. 21.2.3), from which a clear resonance at the ρ-meson mass has been established.

21.4.4 Unitary symmetry [$SU(3)$] and quarks

The outstanding feature of the spectroscopy of the hadrons is the existence of narrow resonant states up to at least 3 GeV, as shown for instance by the missing mass spectrometer (Sect. 20.3.2, Fig. 20.4), and by the existence of isospin multiplets of particles of given strangeness. The narrow states suggest

---

† W. R. Frazer and J. R. Fulco, *Phys. Rev. Lett.*, **2**, 365, 1959; Y. Nambu, *Phys. Rev.*, **106**, 1366, 1957.
‡ Neutron form factors are obtained from electron scattering in deuterium.

that we are concerned with a 'fine structure' pattern. The usefulness of the isospin quantum number results from invariance under transformations of a type described by the special unitary group $SU(2)$ which was introduced briefly in Sect. 21.1.2(b). It was there mentioned that the reduction of the group $U(2)$ to $SU(2)$ by imposing the requirement of unit determinant removes the transformations corresponding to baryon conservation and leaves three infinitesimal generators which are by definition traceless 2×2 hermitian matrices. Under the operations of this group representation the nucleon states $|p\rangle$, $|n\rangle$ (or antinucleon states $|\bar{p}\rangle$, $|\bar{n}\rangle$) transform between each other. Of particular interest are the states of two such particles which similarly transform among themselves. These may be obtained (Ref. 21.8(b)) as products of the basic two-component state vectors,† giving rise to a multiplet of states each with the same baryon number, spin and parity. Thus in the discussion of isobaric spin (Sect. **12.1**) we already noted for the case $B=2$ the formation of a charge triplet and a charge singlet, the latter to be identified with the stable deuteron. For the case $B=0$ a nucleon must be combined with an antinucleon and this process may be represented symbolically in the form (omitting $|\ \rangle$ brackets for clarity, and separating the trace from the traceless part)

$$\begin{pmatrix} p \\ n \end{pmatrix} \times (\bar{p}\bar{n}) \longrightarrow \frac{p\bar{p}+n\bar{n}}{\sqrt{2}} \begin{pmatrix} 1 & 0 \\ 0 & 1 \end{pmatrix} \qquad \text{(singlet)} \qquad (21.41)$$

$$+ \begin{pmatrix} \dfrac{p\bar{p}-n\bar{n}}{2} & p\bar{n} \\ n\bar{p} & \dfrac{p\bar{p}-n\bar{n}}{2} \end{pmatrix} \qquad \text{(triplet array)} \quad (21.42)$$

The singlet corresponds to the η-meson ($T=0$, $\mathcal{J}=0^-$) and the triplet array to the pions ($T=1$, $\mathcal{J}=0^-$), which can similarly be written in array form, with normalizing factors:

$$\begin{pmatrix} \dfrac{\pi^0}{\sqrt{2}} & \pi^+ \\ \pi^- & -\dfrac{\pi^0}{\sqrt{2}} \end{pmatrix} \qquad (21.43)$$

These multiplets, within which particle states transform into each other under the $SU(2)$ group operations, correspond with the singlet and triplet (regular) group representations.

The $SU(2)$ group cannot accommodate the hypercharge quantum number and to bring it in we move to $SU(3)$. This group includes $SU(2)$, i.e. isobaric spin invariance. The eight infinitesimal generators (3×3 matrices) operate

† The proton–neutron isospin doublet is conveniently represented by the two component vector $\binom{p}{n}$ where the eigenvectors are $|p\rangle = \binom{1}{0}$, $|n\rangle = \binom{0}{1}$.

on three-component wavefunctions which describe three basic quantities known as *quarks*. In a first approach to this problem Sakata (1957) suggested that these should be the observable proton, neutron and Λ-particle (or their antiparticles) and that multiplets containing other observed particles should be constructed as linear combinations of these. Particle multiplets corresponding to this representation would be produced by combining suitable numbers of particles and antiparticles according to the group rules (Ref. 21.8(*b*)). Thus for states with $B = 0$ we may form particle–antiparticle states to fill a 3×3 (traceless) array:

$$\begin{pmatrix} p \\ n \\ \Lambda \end{pmatrix} \times (\bar{p}\bar{n}\bar{\Lambda}) \longrightarrow$$

$$\left\{ \begin{matrix} \frac{1}{3}(2p\bar{p} - n\bar{n} - \Lambda\bar{\Lambda}) & p\bar{n} & p\bar{\Lambda} \\ n\bar{p} & \frac{1}{3}(-p\bar{p} + 2n\bar{n} - \Lambda\bar{\Lambda}) & n\bar{\Lambda} \\ \Lambda\bar{p} & \Lambda\bar{n} & \frac{1}{3}(-p\bar{p} - n\bar{n} + 2\Lambda\bar{\Lambda}) \end{matrix} \right\} \quad (21.44)$$

which could be identified with known spin-zero mesons

$$\left\{ \begin{matrix} \dfrac{\pi^0}{\sqrt{2}} + \dfrac{\eta}{\sqrt{6}} & \pi^+ & K^+ \\ \\ \pi^- & -\dfrac{\pi^0}{\sqrt{2}} + \dfrac{\eta}{\sqrt{6}} & K^0 \\ \\ K^- & \bar{K}^0 & -\dfrac{2\eta}{\sqrt{6}} \end{matrix} \right\} \quad (21.45)$$

The neutral mesons are now written as

$$\pi^0 = \frac{p\bar{p} - n\bar{n}}{\sqrt{2}}, \qquad \eta = \frac{p\bar{p} + n\bar{n} - 2\Lambda\bar{\Lambda}}{\sqrt{6}} \quad (21.46)$$

and there is in addition the symmetrical neutral combination or singlet.

$$\eta' = \frac{p\bar{p} + n\bar{n} + \Lambda\bar{\Lambda}}{\sqrt{3}} \quad (21.47)$$

The matrix array contains a total of eight states, and is known as an octet; it corresponds with the regular representation of $SU(3)$. Since the mesons are formed from fermion particle–antiparticle pairs they will have odd parity, as observed. A similar spin 1 meson octet is predicted and has been found. Unfortunately the Sakata model, which constructs hyperons from the system $\Lambda N \bar{N}$, also predicts entities such as $\bar{\Lambda} pp$, a strange baryon of charge 2, and these have never been seen.

In 1961–2 Gell-Mann and Ne'eman† pointed out that the baryons also formed an octet array

$$
\left\{
\begin{matrix}
\dfrac{\Sigma^0}{\sqrt{2}}+\dfrac{\Lambda}{\sqrt{6}} & \Sigma^+ & p \\[2ex]
\Sigma^- & -\dfrac{\Sigma^0}{\sqrt{2}}+\dfrac{\Lambda}{\sqrt{6}} & n \\[2ex]
\Xi^- & \Xi^0 & -\dfrac{2\Lambda}{\sqrt{6}}
\end{matrix}
\right\}
\qquad (21.48)
$$

and suggested that this also corresponded with an octet representation of $SU(3)$. Here however we have $B=1$ and to create both the meson and baryon octets from quarks and antiquarks in the simplest way it is necessary to assume that a baryon is an even-parity state of three quarks. It is then possible, using the formula

$$
Q = T_z + \frac{Y}{2} \qquad (21.49)
$$

to deduce the properties of the basic quarks q_1, q_2, q_3. Thus letting this formula be centred ($Y=0$, $T_z=0$) on the central member of the array, the quarks would have the following quantum numbers (and antiquarks the opposite in each case)

	Q	B	Y	T_z	S
q_1	$\frac{2}{3}$	$\frac{1}{3}$	$\frac{1}{3}$	$\frac{1}{2}$	0
q_2	$-\frac{1}{3}$	$\frac{1}{3}$	$\frac{1}{3}$	$-\frac{1}{2}$	0
q_3	$-\frac{1}{3}$	$\frac{1}{3}$	$-\frac{2}{3}$	0	-1

We then see the composition of the proton as $q_1 q_1 q_2$, the π^+ as $q_1 \bar{q}_2$ and the K^+ as $q_1 \bar{q}_3$.

The transformations of the $SU(3)$ matrix representation mix the particles of a given multiplet, because of the underlying quark structure, so that each can be regarded as to some extent a superposition of all others. Within an $SU(3)$ multiplet all particles should have the same mass, spin and parity, although the quite large actual mass differences between the submultiplets with different Y show that the symmetry is broken by the strong interactions. If however the basic quark mass is very large then the symmetry breaking is a relatively small correction to a much larger binding energy generated perhaps by a superstrong interaction, and the ground states of all particles effectively belong to the lowest energy level of the appropriate quark system. This picture has obvious resemblances to the nuclear shell model, and higher hadronic resonances can be associated with quark orbital momenta.

† M. Gell-Mann, *Phys. Rev.*, **125**, 1067, 1962. Y. Ne'eman, *Nuclear Physics*, **26**, 222, 1961.

Once the role of $SU(3)$ in hadron classification is allowed, the number of states expected in baryon and meson multiplets may be predicted. For baryons for instance $q_1 \times q_2 \times q_3$ indicates 27 objects which split into a singlet, two octets and a decuplet according to the group properties. For mesons $q_1 \times \bar{q}_2$ gives 9 objects splitting into a singlet and octet. The pictorial representations of some of these multiplets, which are logical extensions of the singlet/triplet structure of $SU(2)$, are shown in Fig. 21.17. In this figure hypercharge is plotted against T_z the third component of isospin as abscissa. Other orientations of these patterns however are also meaningful, e.g. a rotation of the octet through $60°$ still leaves an octet (cf. Sect. 21.1.2) but shows groupings of equal charge as a function of another combination of particle parameters so far not used. Experiment verifies the spin-parity assignments shown in Fig. 21.17 and determines the mass differences, from which follow the decay routes, all of which involve the weak interaction except the π^0, η and Σ^0 decays.

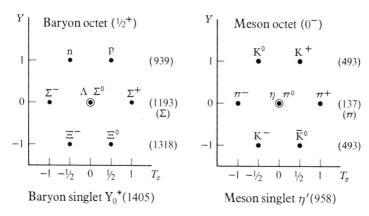

Fig. 21.17 The $Y–T_z$ diagram of the baryon octet $\frac{1}{2}^+$ and the meson octet 0^-. With each octet is also associated a singlet which arises from the combination of 3 basic fields $[3 \times 3 = 8 + 1]$. For the baryon octet $\frac{1}{2}^+$ the singlet is the Y_0^* (1405); for the 0^- mesons it is the η' (958). The figures in brackets give masses in MeV/c².

The baryon decuplet or family of 10 and the 1^- meson octet in the unitary symmetry classification can be populated by known resonant states. Figure 21.18(a) shows the baryon decuplet with spin parity $\frac{3}{2}^+$ and indicates the masses involved. The lightest grouping is the $SU(2)$ quartet of states corresponding to the $\mathcal{J} = \frac{3}{2}$, $T = \frac{3}{2}$ pion–nucleon resonance, denoted by Δ; this exists in four charge states. The triplet is the hyperon resonance Y_1^* which may also be thought of as containing excited Σ particles. The doublet is a further hyperon resonance equivalent to excited cascade particles, and finally and most spectacular of all is the celebrated Ω^-. This isobaric singlet, whose charge follows unambiguously from (21.49) once its place in the decuplet is admitted, was predicted in 1962 well before its discovery in 1964 (Sect. 20.5.1). There are now a convincing number of Ω^- events in bubble chamber photographs and its appearance was both a triumph and a stimulus for the theory. The mass splittings in the baryon decuplet are approximately equal; this is predicted by detailed theory which was thus able to specify the Ω^- mass.

The quarks themselves form a triplet pattern as shown in Fig. 21.18(*b*). The octet and decuplet patterns can be created by geometrical superposition of these triangular diagrams.

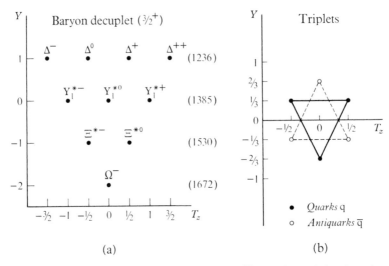

(a)   (b)

Fig. 21.18 (a) The Y–T_z diagram of the baryon decuplet $\frac{3}{2}^+$. The figures in brackets give masses in MeV/c². (b) The Y–T_z diagrams for the 3 quarks and 3 antiquarks.

These ideas can be carried further by assuming that each of the three quarks has spin $\frac{1}{2}$ (and therefore two spin states) and that the strong interactions are approximately invariant with respect to unitary transformations on these six states. This generalization, which is known as $SU(6)$ symmetry, or the quark model, leads to predictions of surprising accuracy both for spin and parity of the multiplets and for dynamic processes. The ratio of the proton to neutron magnetic moment for instance is given as $-\frac{3}{2}$ compared with the experimental value of $-1\cdot47$. The asymptotic limit of total cross-sections follows by assuming that quarks interact in pairs as 'black' objects. A nucleon–nucleon collision then involves 3×3 such interactions but a meson–nucleon collision only 3×2, so that the ratio of cross-sections might be $\frac{3}{2}$. This is close to the value found experimentally (Fig. 21.14).

The objective existence of quarks has not been demonstrated conclusively despite intensive search based on the properties of fractional charge,† on a presumed large mass, or on the relative immunity of the quark to interactions except with other quarks. The mass of these particles seems to be $\geqslant 5 \text{ GeV}/c^2$

† Fractional charge is ascribed by Cairns, McCusker *et al.* (*Phys. Rev.*, **186**, 1394, 1969) to particles producing certain tracks of low ionization density in cosmic ray showers.

from work carried out so far and the binding of three such particles to produce the mass of the nucleon (≈ 1 GeV/c^2) must involve forces of a quite exceptional strength. The search for quarks may however be in principle fruitless; if the ultimate structure is that already revealed by the recognizable elementary particles, quarks become no more than an alternative method of description, of obvious theoretical, but no practical significance.

21.4.5 Regge trajectories

Some progress in the classification of hadrons has also been made in terms of a theory of scattering developed by the Italian physicist T. Regge. The general results of this theory are highly relevant to the description of interactions but this will not be considered here (cf. Ref. 21.9); for the purposes of classification we note only that it is proposed that there should be an analytic

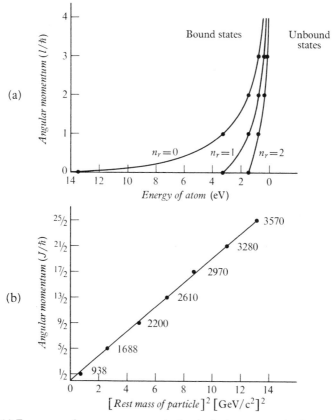

Fig. 21.19 (a) Energy-angular momentum graphs for the hydrogen atom for fixed values of the radial quantum number. Stable states are shown as dots. As the angular momentum increases the energies decrease and at $l = \infty$ the atom is unbound ($E = 0$) (from Ref. 21.7, Chew *et al.*).
(b) The Regge trajectory for non-strange baryon resonances of isobaric spin $T = \frac{1}{2}$ and ordinary spin $\frac{1}{2}$ (nucleon), 5/2...25/2. The trajectory becomes a straight line if mass squared is plotted as abscissa. Actual or predicted masses in MeV/c^2 are shown at the particle locations.

relationship between the mass of a particle and its intrinsic angular momentum. This is empirically plausible if one inspects the table of particles; mass increases frequently imply higher spin. Since spin is quantized the relationship with mass only has meaning at certain spin values, but a curve can be drawn between such points and is known as a *Regge trajectory*.

The connection between mass and angular momentum is shown in the theory of the hydrogen atom. The energy of a bound state of hydrogen is $E_n = -Rhc/n^2$ where $n = n_r + l + 1$ and n_r is a radial quantum number. If E_n is plotted against l (quantized) for given values of n_r, the trajectories shown in Fig. 21.19(b) are obtained, each of which indicates an approach of the system to an unbound state as $l \to \infty$. The permitted bound states are given by the intersection of the trajectories with the discrete angular momentum values. In the case of hadrons however, the trajectories also include unbound states, i.e. resonances which figure as rotational excitations of the ground state; Fig. 21.19(b) shows one of the lines for baryons and it is seen that for these particles the trajectory connects only states whose spins differ by two units. This is because exchange potentials have opposite signs in even and odd states, so that these lie on two separate lines. Other trajectories for baryons ($B = 1$) and for mesons ($B = 0$) exist and the Regge classification has been useful in suggesting spin values for new resonances once there is some indication of mass value.

21.4.6 *Effects of the electromagnetic and weak interaction*

The hadronic states exhibit both weak and electromagnetic decays when fast decay is not possible; the situation is exactly analogous to the decay of nuclear states in low energy nuclear physics. The characteristics of these two interactions may therefore be displayed to some extent in strong processes.

The main effects of the electromagnetic interaction have already been mentioned. It is involved in the electromagnetic structure of the nucleon and the pion, to which it is coupled through virtual photons. In both nuclear and subnuclear physics, the charge-dependent electromagnetic interaction is mainly responsible for energy differences between the states of an isobaric multiplet.

The current–current description of weak processes (Sect. 21.3.4) leads to the appearance of a term $\mathscr{J}\mathscr{J}$. This involves only strongly interacting particles and means (Ref. 21.10) that some of the properties of the weak interaction, e.g. parity non-conservation, should be found in strong processes though only to a very small extent. Formally, this is described by the introduction of a parity-violating term into the internucleon potential; reasonable assumptions indicate a strength of about 10^{-6} of the normal strong potential. The effect of such a potential is to create parity impurities in nuclear states, since these states basically arise from the strong interaction. States of impure parity will show parity-violating types of decay and a detailed search, for instance, has been made for α-decay between a $\mathscr{J} = 2^-$ state of ^{16}O and a

$\mathcal{J}=0^{+}$ state of ^{12}C. Alternative methods are based on the fact that a nuclear state with impure parity may decay radiatively through multipole transitions of the same $\mathcal{J}$ but opposite parity, e.g. $E1$ and $M1$ or $E2$ and $M2$. The basic relative probabilities of these two types of transition can be calculated from nuclear models and comparison with the ratio actually observed from a given nuclear state should then indicate the amplitude of parity admixture. The methods used for the radiative decay of bound states are observations of

(a) the asymmetry of decay of excited nuclei produced by the capture of polarized neutrons,
(b) the circular polarization of gamma radiation emitted by unpolarized nuclei. This arises because of the interference between electric and magnetic multipoles.

Small polarizations have been observed, notably in ^{181}Hf, but some disagreement between different groups of workers still exists (Ref. 21.10). There is no evidence that the theoretically predicted magnitude of the parity violating potential is seriously wrong.

21.5 The future of high energy physics

High energy physics advances our understanding of the *fundamental* structure of natural phenomena. From the description 'fundamental' must be excluded phenomena which are not fully understood merely because they are complex; solid state physics and the life sciences provide many examples of these. There is little doubt that such problems could in principle be solved by the application of the known methods of quantum mechanics. In high energy, or sub-nuclear, physics this is not the case, and the delineation of the boundaries of our uncertainty is the main object of current research in the field. It is perhaps fitting to remark that practical application of the results of high energy studies can be neither guaranteed nor excluded, and that the prime motivation for such studies is intellectual.

Progress in particle physics is a joint enterprise between experiment and theory. The successful classification of the interacting particles as deriving from representations of $SU(3)$, by which there is some unification of the roles of strange and non-strange particles, suggests that we may be on the brink of a deeper understanding, which could reveal the origin of the higher symmetries obeyed by the interactions. The mass splittings in the $SU(3)$ multiplets fit theoretical formulae so well that it has been proposed that these are really only 'fine structures' of extremely massive objects which are yet to be discovered. These may be the quarks of unitary symmetry; to make them, accelerators of a power beyond the present range will be necessary. Such accelerators may extend the range of observable particles, but equally we may approach an asymptotic limit and see an end to the particle spectrum. In either case it is to be hoped that a general picture of some simplicity will emerge in subnuclear physics as it has from time to time in other fields.

The interactions between particles are inextricably bound up with the properties of the particles themselves, but experimentally at least they may still be regarded as proper objects for investigation. Although the general idea of a quantized field, in which forces arise because of particle exchange, is likely to survive, it must be remembered that only in the case of quantum electrodynamics is there a complete and quantitative theory. The strong interaction cannot be treated by perturbation theory because of the size of the coupling constant and the exchange particle of the weak interaction remains to be discovered, again possibly as a result of experiments with a new high energy accelerator producing copious beams of energetic neutrinos.

The limits of validity of quantum electrodynamics obviously present an important challenge, since the theory is already checked to distances of the order of that at which strong interactions might be expected to have a perceptible effect. All interactions seem so far to be consistent with the *CPT* theorem, based on Lorentz invariance, and with dispersion relations, based on causality (i.e. that events take place after, not before, their initiating signals). It is, however, important to subject these general principles to the most exhaustive tests.

Within this framework dynamical problems at large momentum transfers need much more study, and the question of the limiting behaviour of the weak interactions as energy increases is an open one. Two important general methods which have been devised to overcome the difficulties of field theory for the strong interactions are (*a*) the *S*-matrix theory based on the analytic properties of the physically observable scattering amplitude (and underlying most of the theoretical discussion in this book) and (*b*) the algebra of current operators themselves. Each of these approaches, which are complementary, needs further development and testing.

The type of accelerator required to advance these studies should exceed those at present in operation by a factor of 20–30 in energy at least, because of the unfavourable return in centre of mass energy as bombarding energy increases.† The American 200–500 GeV accelerator at Batavia, Illinois and the European machine for a similar energy range are important steps, while the 70 GeV Russian machine at Serpukhov occupies a useful intermediate position. An accelerator of 1000 GeV has been given preliminary attention. In all these ventures, the basic technical principle remains that of the alternating gradient type of proton synchrotron, although many detailed technical advances including the use of superconducting magnets will doubtless appear in the future. The electron ring linear accelerator for protons is not yet at the stage of serious competition. Even if it were, it should be realized that much of the cost of these enormous undertakings arises in the experimental equipment such as large bubble chambers and computers with which data are gathered and analysed.

In conclusion Ref. 21.11, which surveys the status of high energy physics

† The colliding beam approach to this problem (Sect. 8.5) seems inevitably to be capable of only low event rates and will not furnish secondary beams.

in Europe in 1964, offers a compelling picture of the fascination, complexity and expense of the study of subnuclear physics.

References

1. SCHIFF, L. I. *Quantum Mechanics*, 3rd edn., McGraw-Hill, New York, 1968; SAKURAI, J. J. *Invariance principles and elementary particles*, Princeton Univ. Press, 1964; MARSHAK, R. E. and SUDARSHAN, E. C. G. *Introduction to elementary particle physics*, Interscience, New York, 1961; WILLIAMS, W. S. C. *An Introduction to Elementary Particles*, 2nd ed., Academic Press, New York, 1971; MUIRHEAD, H. *Physics of Elementary Particles*, Pergamon Press, Oxford, 1965; LANDAU, L. D. and LIFSHITZ, E. M. *Mechanics*, Ch. 2, Pergamon Press, Oxford, 1960.

2. WIGNER, E. P. 'Violations of Symmetry in Physics', *Scientific American*, Dec. 1965; HENLEY, E.M. 'Parity and Time Reversal Invariance in Nuclear Physics', *Ann. rev. nucl. Sci.*, **19**, 367, 1969.

3. KÄLLEN, G. *Elementary Particle Physics*, Addison Wesley, Reading, Mass., 1964.

4. EDEN, R. J. *High energy collisions of elementary particles*, Camb. Univ. Press, 1967; BURKHARDT, H. *Dispersion Relation Dynamics*, N. Holland Publishing Co., Amsterdam, 1969; LINDENBAUM, S. J. 'Forward Dispersion Relations' in *Pion–Nucleon Scattering*, ed. G. L. Shaw and C. T. Wong, Wiley-Interscience, New York, 1969; MOORHOUSE, R. G. 'Pion–Nucleon Interactions', *Ann. rev. nucl. Sci.*, **19**, 301, 1969.

5. SAKURAI, J. J. 'Weak Interactions', *Progr. nucl. Phys.*, **7**, 243, 1959; WU, C. S. 'The Universal Fermi Interaction and the Conserved Vector Current in Beta Decay', *Rev. mod. Phys.*, **36**, 618, 1964; RAMM, C. A. 'Neutrinos', *Science Journal*, April 1968; 'A Discussion on Neutrinos', *Proc. roy. Soc.*, A, **301**, 103, 1967.

6. BURHOP, E. H. S. (ed.), *High Energy Physics* (4 vols.), Academic Press, New York, 1967.

7. CHEW, G. F., GELL-MANN, M. and ROSENFELD, A. H. 'Strongly Interacting Particles', *Scientific American*, Feb. 1964; WILKINSON, D. H. 'The elementary particles', *Science Journal*, March 1966; DALITZ, R. H. *Strange particles and strong interactions*, Oxford University Press, 1962.

8. (a) MATTHEWS, P. T. and SALAM, A. 'Unitary Symmetry', *Proc. phys. Soc.*, **80**, 28, 1962; (b) MATTHEWS P. T. 'Particle Symmetry', *Endeavour*, **26**, 63, 1967; 'Unitary Symmetry' in Ref. 21.6, Vol. 1; (c) CHARAP, J. M., JONES, R. B. and WILLIAMS, P. G. 'Unitary Symmetry', *Rep. prog. Phys.*, **30**, 227, 1967.

9. MATTHEWS, P. T. 'Regge Poles and Diffraction Scattering', *Proc. phys. Soc.*, **80**, 1, 1962.

10. HAMILTON, W. D. 'Parity violation in electromagnetic and strong interaction processes', *Progr. nucl. Phys.*, **10**, 1, 1968.

11. 'A Discussion on Recent European Contributions to the Development of the Physics of Elementary Particles', *Proc. roy. Soc.*, A, **278**, 287, 1964.

Appendices

I Spherical harmonics

The functions $Y_l^m(\theta, \phi)$ are of frequent occurrence in atomic and nuclear physics. They provide a convenient means of describing any deformation of an initially spherical surface and can consequently be used to represent the angular 'shape' of wave functions. In particular, certain values of θ, ϕ will define nodal planes over which $Y_l^m(\theta, \phi)$ vanishes; the probability of finding a particle described by the wave function near these planes is small. A discussion of the properties of these functions is given in Ref. A1.1, p. 49, where it is shown that

$$Y_l^m(\theta, \phi) = (-1)^{(m+|m|)/2} \sqrt{\frac{2l+1}{4\pi} \frac{(l-|m|)!}{(l+|m|)!}} \times P_l^m(\cos \theta) \times e^{im\phi}$$

The function $P_l^m(\cos \theta)$ is an *associated Legendre function* which may be derived from the corresponding *Legendre function* (or polynomial) $P_l(\cos \theta)$ by a process involving repeated differentiation. The normalization factor provides that the squared modulus of the spherical harmonic, integrated over a sphere ($\theta \longrightarrow 0$ to π, $\phi \longrightarrow 0$ to 2π), shall be unity, i.e.

$$\int |Y|^2 \, d\Omega = 1 \quad \text{where } d\Omega = \sin \theta \, d\theta \, d\phi$$

The similar normalization for the Legendre functions is

$$\int P_l^2 \, d\Omega = \frac{4\pi}{2l+1}$$

while for the associated Legendre functions

$$\int (P_l^m)^2 \, d\Omega = \frac{4\pi}{2l+1} \frac{(l+|m|)!}{(l-|m|)!}$$

Expressions for the first few of these quantities are given in the following table:

Legendre function	Associated Legendre function	Spherical harmonic
$P_0(\cos\theta) = 1$	$P_0^0(\cos\theta) = 1$	$Y_0^0(\theta, \phi) = \sqrt{\dfrac{1}{4\pi}}$
$P_1(\cos\theta) = \cos\theta$	$P_1^0(\cos\theta) = \cos\theta$	$Y_1^0(\theta, \phi) = \sqrt{\dfrac{3}{4\pi}}\cos\theta$
	$P_1^{\pm1}(\cos\theta) = (1-\cos^2\theta)^{1/2}$	$Y_1^{\pm1}(\theta, \phi)$
		$= \mp\sqrt{\dfrac{3}{8\pi}}(1-\cos^2\theta)^{1/2}e^{\pm i\phi}$
$P_2(\cos\theta) = \tfrac{1}{2}(3\cos^2\theta-1)$	$P_2^0(\cos\theta) = \tfrac{1}{2}(3\cos^2\theta-1)$	$Y_2^0(\theta, \phi) = \sqrt{\dfrac{5}{16\pi}}(3\cos^2\theta-1)$
	$P_2^{\pm1}(\cos\theta)$	$Y_2^{\pm1}(\theta, \phi)$
	$= 3\cos\theta(1-\cos^2\theta)^{1/2}$	$= \mp\sqrt{\dfrac{15}{8\pi}}\cos\theta(1-\cos^2\theta)^{1/2}e^{\pm i\phi}$
	$P_2^{\pm2}(\cos\theta) = 3(1-\cos^2\theta)$	$Y_2^{\pm2}(\theta, \phi) = \sqrt{\dfrac{15}{32\pi}}(1-\cos^2\theta)e^{\pm 2i\phi}$

It will be noted that for a given l there are $2l+1$ spherical harmonics with different m-values.

The functions $Y_l^m(\theta, \phi)$ have nodal planes in the angular coordinate θ, obtained by setting $Y_l^m(\theta, \phi)=0$. There are also nodal planes in the coordinate ϕ, which are obtained by considering the zeros of the real forms

$$\tfrac{1}{2}(Y_l^m + Y_l^{-m}) \quad \text{and} \quad \frac{i}{2}(Y_l^m - Y_l^{-m})$$

which are equally good solutions of the second-order equation used to define Y. The total number of nodal planes in the angular coordinates is l. Thus for $l=1$ (p-state) we have wave functions with the angular dependence

$$\cos\theta$$

$$\sin\theta\cos\phi$$

$$\sin\theta\sin\phi$$

in each of which there is one nodal plane.

It follows from the definition of the spherical harmonics (Ref. A1.1) that if the operator which in wave mechanics represents the square of the total angular momentum of a particle is applied to a wave function $Y_l^m(\theta, \phi)$, the result is $l(l+1)\hbar^2 \times Y_l^m(\theta, \phi)$.

Reference

1. MANDL, F. *Quantum Mechanics*, Butterworth, 1957.

2 Multiple moments

A2.1 Classical moments of a charge distribution

Consider a static distribution of charges confined within a volume Ω (Fig. A2.1) surrounding the origin O. The electrostatic potential at the point P is given by

$$\phi_P = \sum_i \frac{e_i}{4\pi\varepsilon_0 |\mathbf{R} - \mathbf{r}_i|} = \sum_i \frac{e_i}{4\pi\varepsilon_0 \sqrt{R^2 + r_i^2 - 2Rr_i \cos\theta_i}} \tag{A2.1}$$

where $\mathbf{R}, \mathbf{r}_i$ are the position vectors, with absolute lengths R, r_i, of the points P and A (charge e_i) and θ_i is the angle between $\mathbf{R}$ and $\mathbf{r}$. For distances OP which are large compared with the linear dimension a of the charge distribution, i.e. for $R \gg r$ the denominator may be expanded in powers of r/R giving

$$\phi_P = \frac{\Sigma e_i}{4\pi\varepsilon_0 R} + \frac{\Sigma e_i r_i \cos\theta_i}{4\pi\varepsilon_0 R^2} + \frac{\Sigma e_i r_i^2 (3\cos^2\theta_i - 1)}{8\pi\varepsilon_0 R^3} + \cdots$$

$$= \phi_0 + \phi_1 + \cdots + \phi_L + \cdots \tag{A2.2}$$

in which the term ϕ_L is proportional to $1/(R^{L+1})$.

These potentials may be interpreted as follows:

ϕ_0 is the potential at P due to the resultant charge Σe_i,

$$\phi_1 = \frac{\Sigma e_i r_i \cos\theta_i}{4\pi\varepsilon_0 R^2} = \frac{1}{4\pi\varepsilon_0} \Sigma e_i \mathbf{r}_i \cdot \operatorname{grad} \frac{1}{R} \tag{A2.3}$$

is the potential at P due to the static *dipole moment* $\mathbf{D} = \Sigma e_i \mathbf{r}_i$ of the charge distribution,

ϕ_2 and subsequent terms represent contributions from quadrupole, octupole and higher moments of the system. Together, this set of moments located at the origin produces the same electrostatic field at a distant point as the charge distribution. As expressed in (A2.2) these moments are not generally useful because they involve θ_i, which is based on the arbitrary direction OP. It is possible, however, to define moments with respect to axes fixed in the charge distribution (perhaps aligned with a particular axis of symmetry) by transforming r_i and θ_i appropriately. In Cartesian coordinates the vector dipole

611

moment then has components

$$\mathbf{D} = \Sigma e_i x_i, \quad \Sigma e_i y_i \quad \text{and} \quad \Sigma e_i z_i$$

and the higher moments are symmetric tensors. The quadrupole moment for instance may be written

$$Q = \sum_i e_i \begin{vmatrix} xx, & xy, & xz \\ yx, & yy, & yz \\ zx, & zy, & zz \end{vmatrix} \tag{A2.4}$$

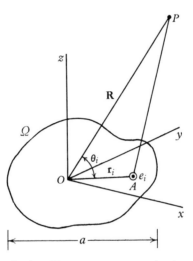

Fig. A 2.1 Static charge distribution. Charges e_i are supposed to be confined within a volume Ω of linear dimension $\approx a$.

where each term in the square array defines a partial moment, which must be included in the total quadrupole effect. The symmetries, and the relation $x_i^2 + y_i^2 + z_i^2 = r_i^2$ reduce the number of independent components from 9 to 5 (and for the Lth moment, to $2L + 1$). An equivalent expression for an element of the quadrupole tensor is

$$Q_{xy} = \sum_i e_i(3x_iy_i - r_i^2\delta_{x_iy_i}) \tag{A2.5}$$

where $\delta_{xy} = 0$ if $x \neq y$ and $= 1$ if $x = y$. Permutation of the coordinates gives the individual quadrupole moments. These are closely related to the five spherical functions necessary to describe the most general quadrupole deformation of a sphere (App. 1), as may be seen if the forms A2.5 are converted to polar coordinates using the relations.

$$x = r \sin \theta \cos \phi$$
$$y = r \sin \theta \sin \phi \tag{A2.6}$$
$$z = r \cos \theta.$$

The higher moments of the charge distribution become important when the resultant charge $\Sigma\ e_i$ is zero and the dipole term also vanishes. A quadrupole moment may be generated for instance by displacing and inverting a dipole so that both $\sum_i e_i$ and $\sum_i e_i\mathbf{r}_i = 0$ but $\Sigma\ e_ir_i^2$ is finite. Typical charge distributions illustrating these points are shown in Fig. 3.4. The 2^L-pole moment, appearing in the term ϕ_L in (A2.2) is a quantity of the order $\Sigma\ e_ir_i^L$.

If the charge distribution has spherical symmetry, all moments as defined by A2.2 vanish. If there is axial symmetry with respect to Oz the dipole moment vanishes but the quadrupole moment

$$Q_{zz} = \Sigma\ e_i(3z_i^2 - r_i^2) \tag{A2.7}$$

is finite. This leads immediately to the important case of a nucleus with axial symmetry; the intrinsic quadrupole moment with respect to this axis is

$$Q_0 = \int \rho(3z^2 - r^2)\,\mathrm{d}\tau \tag{A2.8}$$

where $\rho\,\mathrm{d}\tau$ is the charge within the volume element $\mathrm{d}\tau$. It is customary in nuclear physics to remove the dimensions of charge from this expression by use of the equation

$$Q_0 = \frac{1}{e}\int \rho(3z^2 - r^2)\,\mathrm{d}\tau \tag{A2.9}$$

The quadrupole moment is then measured in m^2 or barns. For an oblate nucleus Q is negative and for the prolate shape Q is positive.

A2.2 Radiation from time-varying moments

The classical calculation of the radiation from a charge distribution which varies sinusoidally with time is given in Refs. A 2.1–2.4. It is shown that the second term of (A 2.2) corresponds to the field of electric dipole $(E1)$ radiation while the third term leads not only to electric quadrupole $(E2)$, but also to magnetic dipole $(M1)$ fields. All higher moments (L) yield EL and $M(L-1)$ radiation and as far as (A 2.2) is concerned, these two types of radiation would be expected to arise in comparable intensity, which is much less than the intensity of radiation corresponding to moments of lower order unless these are abnormally small. The radiation from the magnetic multipole ML is weaker than that from the electric multipole EL by a factor $(v/c)^2$ where v is the average velocity of the oscillating charges.

A 2.3 Moments in quantum mechanics

In quantum mechanics an observable physical quantity such as an energy, or a momentum or a length, is represented by an *operator*, which indicates the performance of a certain mathematical transformation when applied to a function of coordinates. The average of a series of measurements on the

3 Oscillations in orbital accelerators

Consider a particle of charge e, mass m and velocity v moving in the median plane of a steady magnetic field of flux density B_z. The field is symmetrical about the z-axis and varies with radius according to the formula

$$B_z = B_0 \left(\frac{r_0}{r}\right)^n \tag{A 3.1}$$

where r_0 and B_0 are reference values. The field index n is then given by

$$n = -\frac{r}{B_z}\frac{\partial B_z}{\partial r} \tag{A 3.2}$$

The *radial equation of motion* is

$$\frac{d}{dt}(m\dot{r}) = \frac{mv^2}{r} - B_z ev \tag{A 3.3}$$

and the radius R of the equilibrium circular orbit (sometimes taken to be the reference value r_0) is therefore given by

$$\frac{mv^2}{R} = B_z ev \tag{A 3.4}$$

If the radius r of another orbit differs from R by a small amount x then

$$\begin{aligned}
m\ddot{x} &= \frac{mv^2}{R+x} - ev\left(B_z + x\frac{\partial B_z}{\partial r}\right)_{r=R} \\
&\approx \frac{mv^2}{R}\left(1 - \frac{x}{R}\right) - ev B_z\left(1 - \frac{nx}{R}\right) \\
&= -\frac{mv^2}{R^2}x(1-n) \tag{A 3.5}
\end{aligned}$$

since for the stable orbit $\ddot{R} = 0$ and $mv^2/R = B_z ev$.
It follows that

$$\ddot{x} + 4\pi^2 f^2(1-n)x = 0 \tag{A 3.6}$$

where $f = v/2\pi R$ is the revolution frequency in the stable orbit. This equation indicates a frequency of radial oscillation

$$f_r = f\sqrt{1-n} \tag{A 3.7}$$

The radial oscillation frequency is real if $n < 1$; if $n > 1$ (A 3.6) has an exponential solution and the orbit is not closed. This situation is realized in the peeler-regenerator method of particle extraction.

The *vertical equation of motion*, for a small axial displacement z out of the median plane may be written (Fig. 8.11)

$$\frac{d}{dt}(m\dot{z}) = B_r ev = ev_z \frac{\partial B_r}{\partial z} \tag{A 3.8}$$

where B_r is the radial component of the field at height z above the median plane on which, by symmetry $B_r = 0$. For a static field

$$\text{curl } \mathbf{B} = 0$$

so that for $z = 0$,

$$\frac{\partial B_r}{\partial z} = \frac{\partial B_z}{\partial r} = -\frac{nB_z}{R} \tag{A 3.9}$$

This gives

$$\frac{d}{dt}(m\dot{z}) = -nev\frac{z}{R}B_z = -nm\left(\frac{v}{R}\right)^2 z$$

from which

$$\ddot{z} + 4\pi^2 f^2 nz = 0 \tag{A 3.10}$$

and the frequency of vertical oscillation is

$$f_v = f\sqrt{n} \tag{A 3.11}$$

This frequency is real if $n > 0$; if $n < 0$ a circulating beam diverges vertically.

The radial and vertical oscillations were first studied in connection with the stability of betatron orbits, and are known as *betatron oscillations*. They are of period comparable with the period of revolution and must be distinguished from the slower *synchrotron* or *phase oscillations*.

In constant gradient accelerators n lies between 0 and 1, but certain resonance values for which f_r and f_v are small multiples of each other, or of the revolution frequency f, must be avoided. At these n-values energy will be exchanged between one type of motion and another and instability will usually result. These effects are especially important in synchrotrons because of the large number of turns necessary in the acceleration process. Resonances occur for $n = 0.25$ ($2f_v = f$), 0.75 ($2f_r = f$), 0.5 ($f_r = f_v$) and 0.2 ($f_r = 2f_v$) and the synchrotron n-value is usually chosen to lie between 0.5 and 0.75.

The connection between the motion of charged particles in an orbital accelerator and in a double focusing magnetic spectrometer (Sect. 7.1.3(*b*)) is now easily seen. For double focusing the resonance condition $f_r = f_v$ is exactly what is required since then the radial and vertical motions possess nodes (one of which may be at the source and one at the detector) at the same angular coordinate. The *n*-value necessary is thus 0·5 and the first node occurs at an angle $\pi(f/f_r) = \pi \sqrt{0\cdot5} = 254\cdot6°$ from the start of the oscillation. This is the angle required for double focusing in a spectrometer in which the source and detector are placed within the magnetic field.

4 Fission and the nuclear reactor

The discovery of the fission of uranium by slow neutrons and the main experimental phenomena are described in Sect. 14.1.5. A complete interpretation of the salient features of fission was given, within a few months of its discovery, by Bohr and Wheeler.[†] Their theory, which is based on the liquid drop model of the nucleus (Sects. **9.1**, **10.3**), is summarized in the following section.

A 4.1 The primary fission process

The fission reaction in ^{235}U

$$^{235}\text{U} + \text{n} \longrightarrow {}^{236}\text{U} \longrightarrow A + B + Q \qquad (A\ 4.1)$$

where A and B are unstable nuclei of mass number between say 70 and 170 and Q is an energy release of the order of 150 MeV appears superficially much different from more familiar types of slow neutron reaction proceeding through a compound nucleus, e.g.

$$^{14}\text{N} + \text{n} \longrightarrow {}^{15}\text{N} \longrightarrow {}^{14}\text{C} + {}^{1}\text{H} + Q' \qquad (A\ 4.2)$$

In fact the difference is only quantitative; the energy release Q, although much greater than Q', is calculable in the same way from the masses of the nuclei concerned and the reactions both proceed with observable yield only because there is sufficient energy available for the particles to overcome their mutual potential barrier.

The energy release Q in binary[‡] fission was calculated by Bohr and Wheeler by an application of the semi-empirical mass formula (Sect. **10.3**). This is still a useful approach because of its generality, although the masses of many fission products can now be estimated more reliably from knowledge of their decay chains. The results of these calculations may be qualitatively predicted from Fig. 10.1. The binding energy per nucleon is about 7·5 MeV for $A = 240$ and 8·5 MeV for $A = 120$; division of the former nucleus into two fragments which decay finally into stable species of mass 120 could therefore release about 240 MeV in all. A more detailed calculation gives the energy release for any specified pair of fragments. Since heavy nuclei contain

---

† N. Bohr and J. A. Wheeler, *Phys. Rev.*, **56**, 426, 1939.
‡ There are rare modes of fission in which three heavy particles (one usually an α-particle) are emitted.

excess neutrons in order to balance the Coulomb repulsion of the protons (Sect. **10.4**), the fission fragments initially formed contain too many neutrons for stability and undergo successive β^--decays. This process, and occasionally neutron emission from excited nuclei formed as a result of β-decay, accounts for part of the overall energy release (Sect. **A 4.2**).

Meitner, Hahn and Strassmann found that a 23-minute β^--activity was induced in uranium by neutrons of a few volts energy and pointed out that the observed cross-section for this process would greatly exceed $\pi\lambda^2$ for the incident neutron if ^{235}U were the isotope responsible. The activity was therefore attributed to a resonant radiative capture process in ^{238}U. The complex distribution of fission product activities found with thermal neutrons was not observed for neutrons of the energy required for resonant capture in ^{238}U; Bohr and Wheeler therefore suggested that fission by thermal neutrons took place in the rare isotope ^{235}U only, present in normal uranium to an extent of 1 part in 140. Fission in uranium was also observed with neutrons of energy above 1 MeV, with a cross-section too large to permit the effect to be ascribed to ^{235}U (because of its low abundance). This process, with a threshold of about 1 MeV, was therefore assigned to ^{238}U. These conclusions were confirmed by experiments with separated isotopes and were shown to be theoretically reasonable when calculations were made of the fission threshold in terms of a classical model.

We consider the compound nucleus (A, Z) formed by the absorption of a neutron by a heavy target nucleus as a spherical drop of incompressible fluid. The atomic mass $M(A, Z)$ is supposed to be given by the semi-empirical mass formula (10.10) and is greater than the sum of the masses of two constituent fragments by an energy equivalent Q; the question is, under what circumstances will fission into the two fragments take place? The stability of the initial nucleus against fission depends critically on the relative importance of the short-range nuclear forces and the long-range Coulomb forces, whose effects are represented in the semi-empirical mass formula by the terms with the coefficients a_v, a_s, a_c. In the case of a postulated very heavy nucleus $(A \approx 300$ say) the Coulomb force is the more important and the nucleus, once formed, can fly apart as two fragments almost instantaneously. This is illustrated in curve (a) Fig. A 4.1, which shows the potential energy of the two fragments as a function of their separation. For a lighter nucleus $(A = 236)$ the repulsion of the protons may be more than compensated by the short-range attractions providing that the nucleus (A, Z) has an economical shape, i.e. one in which it presents a minimum surface so that the maximum effect of the main binding term in (10.10) can be felt. For such a nucleus the potential energy graph as a function of fragment separation is as shown in curve (b) Fig. A 4.1; for small deformations from the spherical shape the initial nucleus is stable but for larger deformations the free surface is so much increased that the effect of short-range attractions is reduced and the Coulomb repulsion is no longer compensated. We thus arrive at the concept of a 'fission barrier' or 'critical energy for fission' E_f at which separation of fragments, already possible energetically with the energy release Q, becomes rapid. Before the deformation corresponding to E_f is reached, separation of fragments may still take place, but slowly, because of the necessity for penetration of the fission barrier. This process is analogous to the emission of α-particles from radioactive nuclei, and is described as *spontaneous fission*. For a still lighter nucleus ($A \approx 100$) the fission barrier is higher (Fig. A 4.1, curve (c)). In each case shown in Fig. A 4.1 the two fragments, once formed, repel each other, and at infinite separation have a kinetic energy determined by the mass change (apart from prompt emission of energy in the form of γ-rays or neutrons).

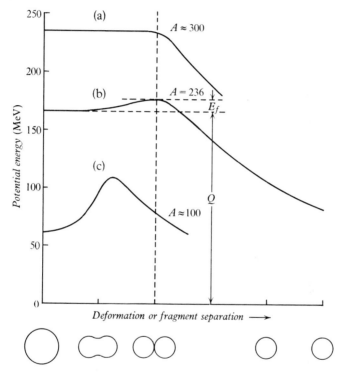

Fig. A 4.1 Critical energy and energy release in fission. The left-hand part of the curves shows the potential energy of a nucleus as a function of deformation; the right-hand part shows the potential energy as two fragments separate under their mutual Coulomb repulsion.

The critical energy for fission E_f, which must be supplied by the incident particle in the case of induced fission, depends on the quantity Z^2/A for the compound nucleus. This parameter enters because the surface correction to the main binding energy term is proportional to $A^{2/3}$, and the Coulomb potential energy is proportional to $Z^2/A^{1/3}$. Since the fission of ^{238}U requires a neutron of energy about 1 MeV the E_f value for ^{239}U must be at least 6 MeV, i.e. the neutron separation energy of 4·8 MeV together with the incident kinetic energy. This fact enabled Bohr and Wheeler to estimate that a nucleus would be instantaneously fissile if

$$\frac{Z^2}{A} \geqslant 47 \cdot 8 \qquad\qquad (A\ 4.3)$$

and to calculate the energy E_f, at which the potential energy reaches its maximum value, for nuclei with Z^2/A less than this limiting value. For

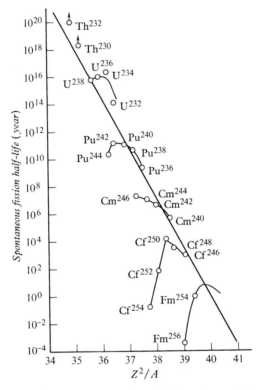

Fig. A 4.2 Spontaneous fission. The diagram shows the dependence of the half-value period of a nucleus, assumed to decay by fission only, on the parameter Z^2/A. (Reprinted by permission from *Nuclei and Particles* by E. Segrè, © 1964, W. A. Benjamin, Inc., Menlo Park, California.)

^{236}U (^{235}U + n) the critical energy appeared to be 5 MeV; since this is less than the neutron separation energy (6·4 MeV) it is made available by absorption of a neutron of zero energy, and ^{235}U, in contrast with ^{238}U, is fissionable by thermal neutrons. The increased neutron separation energy in the case of ^{236}U, as compared with ^{239}U, is due to the fact that the former is an even–even nucleus, which is more tightly bound in its ground state than an adjacent nucleus of even–odd type. It is a general rule that the thermally fissile nuclei form even–even compound nuclei by absorption of a slow neutron.

It is clear from these arguments and from Fig. A 4.1 that slow neutron induced fission will be confined to only a few nuclei. If A is too small, insufficient energy is made available by neutron capture to enable the nucleus to surmount the fission barrier; if A is too large, spontaneous fission already takes place. The occurrence of this mode of disintegration limits the number of transuranic nuclei that may be produced; Fig. A 4.2 shows how the lifetime for spontaneous fission decreases as the limiting value of Z^2/A is approached.

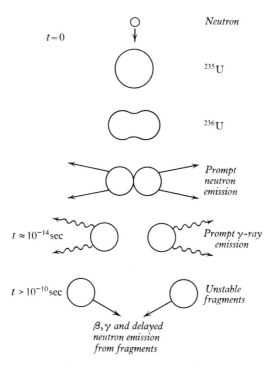

Fig. A 4.3 Schematic representation of the fission process in ^{235}U, showing emission of neutrons and γ-radiation and final decay of fission products. The time scale gives orders of magnitude only.

The fission process as envisaged by Bohr and Wheeler may now be represented schematically as shown in Fig. A 4.3. The two excited fragments of the compound nucleus, carrying many excess neutrons in comparison with stable nuclei of their particular charge, separate in a time short compared with their radiative lifetime of the compound state (i.e. $\ll 10^{-14}$ s). At about the instant of separation, a number of neutrons, known as prompt neutrons, are emitted, mainly from the moving fragments as the first stage of the process of decay towards stability. The discovery of these neutrons immediately suggested the possibility of a self-sustaining chain reaction in uranium. Further energy is lost by the emission of prompt γ-radiation from excited levels in typical γ-ray lifetimes of the order of 10^{-14} s. The fragments (Plate 14) are brought to rest in matter by the normal processes of energy loss for charged particles (Sect. **5.3**);† they are still neutron-rich and decay sequences involving β^--emission occur in the further process of reducing the neutron–proton ratio. A typical decay chain is that studied in part by Hahn and Strassmann

$$^{140}_{54}\text{Xe} \xrightarrow{16 \text{ sec}} {}^{140}_{55}\text{Cs} \xrightarrow{66 \text{ sec}} {}^{140}_{56}\text{Ba} \xrightarrow{40 \text{ hrs}} {}^{140}_{57}\text{La} \xrightarrow{218 \text{ sec}} {}^{140}_{58}\text{Ce} \text{ (stable)} \quad \text{(A 4.4)}$$

In some of these decay chains a β-process may leave a product nucleus so highly excited that neutron emission is a predominant alternative to β^--decay. An emission of delayed neutrons, with a half-life corresponding to that of the preceding β-emitter, then ensues. A well-known case is illustrated in Fig. A 4.4.

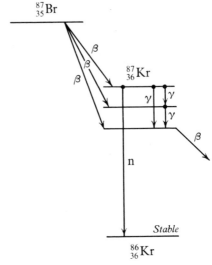

Fig. A 4.4 Delayed neutron emission from ^{87}Kr.

The importance of compound nucleus formation is implicit in the discussion just given of the fission mechanism, but it was not demonstrated objectively in the earliest experiments. It is now known that fission does indeed show the resonances (Fig. A 4.6) expected from a compound nucleus model, which might at first sight appear surprising in view of the collective motion which is obviously about to occur at the fission threshold. Present-day theories of fission, based on the collective model, are able to preserve both features.

A 4.2 Fission isomers

The potential energy curves of Fig. A 4.1 are based on the liquid drop model and must be improved by taking account of shell structure including pairing

† The fragments are actually incompletely ionized atoms. In slowing down they pick up electrons and their average charge decreases. Towards the end of their range the main mechanism of energy loss is a direct interaction between the nuclear charges of the fragment and of the absorber atoms (Ref. 5.4, Northcliffe).

effects. The necessity for this became apparent when it was found that certain excited americium isotopes exhibited spontaneous fission lifetimes of the order of milliseconds. The excitations concerned were not great enough to account for the reduced barrier factor; moreover the spins of the states were low so that gamma radiation to the ground state should have been the predominant decay mode.

These facts are now understood, following the calculations of Strutinsky (Ref. A 4.2), in terms of a double humped fission barrier produced by shell corrections to the nuclear binding energy. Instead of curve (*b*), Fig. A 4.1, potential energy curves of the type shown in Fig. A 4.5 occur in which may

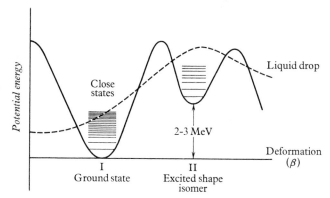

Fig. A 4.5 Fission isomers: The dotted line is the nuclear potential energy as a function of deformation, as given by the liquid drop model and shown in Fig. A 4.1. The full line shows the modification introduced by shell structure, which leads to two regions of 'stable' deformation I and II, each with its own set of levels.

be seen (*a*) a first minimum I corresponding to a stable deformation, as discussed for the collective model (ch. 12) and (*b*) a second minimum II representing a greater stable deformation. A heavy nucleus excited to 4 MeV by a nuclear reaction, e.g. by absorbing a neutron, would offer a much thinner barrier to a fission process (β increasing in Fig. A 4.5). On the other hand radiative decay to state I (β decreasing) would be inhibited by another barrier.

This picture explains the occurrence of fission, or shape, isomers. It also explains the so-called sub-threshold behaviour of fission excitation functions. These would be expected to show many close resonances corresponding to excitations in well I. In fact in the keV neutron range widely spaced groups of such levels are strongly enhanced;[†] this is due to overlap of the well-I states with the more widely separated states of well-II.

A 4.3 Characteristics of fissile materials[‡]

The characteristics of the important thermally fissile nuclei ^{233}U, ^{235}U and ^{239}Pu, which are required with high accuracy by the designers of nuclear

† E. Migneco and J. P. Theobald, *Nuclear Physics*, **A112**, 603, 1968.
‡ The material presented in this section has been kindly supplied by J. Walker.

reactors, have been obtained by an exhaustive series of experiments in many national laboratories. Continual intercomparison and checking of data takes place and the nuclear properties of these materials must be regarded as among the best established data of nuclear physics.

(a) *Production.* ^{233}U is obtained by slow neutron capture in thorium

$$^{232}\text{Th} + \text{n} \longrightarrow\ ^{233}\text{Th} + \gamma \xrightarrow{\beta^-}\ ^{233}\text{Pa} \xrightarrow{\beta^-}\ ^{233}\text{U} \qquad (\text{A } 4.5)$$

^{235}U is naturally available to the extent of 1 part in 140 of ordinary uranium.

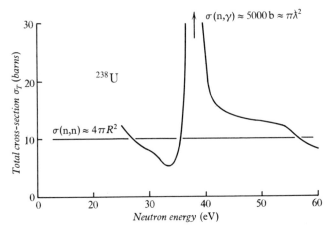

Fig. A 4.6 Total cross-section of ^{238}U for neutrons in the energy region 25–60 eV (Ref. A 4.3).

^{239}Pu is produced by capture of slow (resonance) neutrons in ^{238}U

$$^{238}\text{U} + \text{n} \longrightarrow\ ^{239}\text{U} + \gamma \xrightarrow{\beta^-}\ ^{239}\text{Np} \xrightarrow{\beta^-}\ ^{239}\text{Pu} \qquad (\text{A } 4.6)$$

The total cross-section for the interaction of slow neutrons with ^{238}U is shown over a certain energy range in Fig. A 4.6. At the resonance, only the neutron capture and neutron scattering are important, and the total cross-section is divided approximately equally between them.

(b) *Decay schemes*

Nucleus	Particle emitted	Energy (MeV)	Half-value period (yr)	Spontaneous fission rate (disintegrations $\text{g}^{-1}\,\text{s}^{-1}$)
^{233}U	α, no β	4·8	$1\cdot6 \times 10^5$	$< 2 \times 10^{-4}$
^{235}U	α, no β	4·4	7×10^8	3×10^{-4}
^{239}Pu	α, no β	5·1	$2\cdot4 \times 10^4$	$1\cdot0 \times 10^{-2}$

(*c*) *Energy released in fission of* ^{235}U (average values)

Kinetic energy of fission fragments	165 MeV
Prompt and delayed neutrons	5
Prompt γ-radiation	7
Delayed β-particles, associated antineutrinos and γ-radiation from radioactive decay	25
	202 MeV

If allowance is made for the fact that the antineutrinos do not deposit their energy in the fission source it follows that

$$3\cdot2 \times 10^{10} \text{ fissions s}^{-1} \text{ develop a power of 1 watt}$$

(*d*) *Mass and energy distribution of fission fragments.* A distribution of primary fragment masses, determined by extensive physical and chemical study of decay chains, is shown in Fig. A 4.7(a). Symmetric fission is rare in the slow

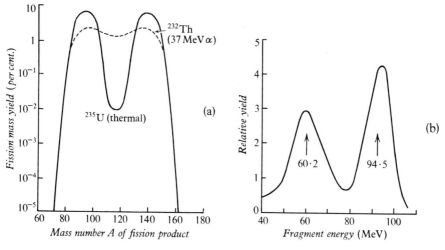

Fig. A 4.7 Fission fragments. (a) Mass distribution (Ref. A 4.4). (b) Energy distribution (Ref. A 4.1).

neutron induced phenomenon, but it is much more probable in high-energy ($E > 20$ MeV) fission. In general there are heavy and light fragments in each fission process; the energy distribution between these fragments is determined by the requirement of conservation of linear momentum and is shown in Fig. A 4.7(b).

(*e*) *Energy distribution of neutrons and gamma rays.* The fission neutron spectrum extending up to about 12 MeV is approximately that to be expected if evaporation from an excited nucleus takes place. It has been shown that the angular distribution of neutrons is consistent with evaporation from a moving fragment, as indicated in Fig. A 4.3. The fast neutron yields, ν, per thermal fission processes are:

^{233}U	2·51
^{235}U	2·47
^{239}Pu	2·91

More than 99 per cent of the neutrons are prompt.

The spectrum of γ-radiation is that to be expected as a result of transitions between the lower excited states of nuclei in the decay chains.

The delayed neutrons fall into six groups with periods in the range 0·2 to 55 s.

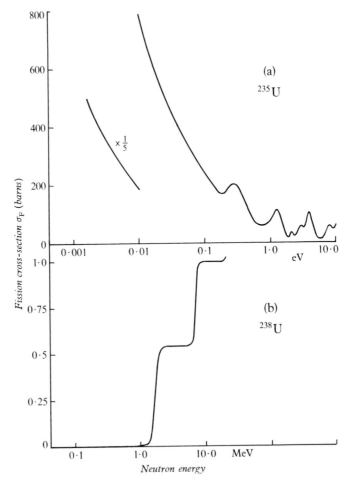

Fig. A 4.8 Fission cross-sections. (a) ^{235}U in the thermal and slow neutron energy range, showing $1/v$ law. (b) ^{238}U in the fast neutron energy range, showing threshold at about 1 MeV (Ref. A 4.3).

(*f*) *Cross-sections.* The total cross-section for the interaction of thermal neutrons with ^{235}U is 708 b. The processes which compete in building up this value are (for $v_n = 2200$ m s^{-1}):

(i) scattering (n, n) $\sigma(n, n) = 10$ b
(ii) capture (n, γ) reaction forming ^{236}U in ground state $\sigma(n, \gamma) = 108$ b
(iii) fission $\sigma_F = 590$ b

The variation of the fission cross-section with energy is shown in Fig. A 4.8(a), and the curve for ^{238}U above the fission threshold is given in Fig. A 4.8(b). The partial cross-sections for the competitive processes stand in the ratio of the corresponding widths at particular resonance levels and these are as follows for the level at 0·29 eV incident neutron energy:

$$\Gamma_n = 0.004 \text{ mV}, \quad \Gamma_\gamma = 31 \text{ mV}, \quad \Gamma_F = 110 \text{ mV}$$

The natural width of such levels is thus ≈ 100 mV and may be less than the instrumental width with which they are investigated or the Doppler width due to thermal motion of the target atoms.

(*g*) *Neutrons emitted per thermal neutron absorbed.* When a slow neutron is incident on fissile material absorption takes place by the (n, γ) reaction as well as by fission. Only the latter yields the additional fast neutrons which are essential to the generation of a chain reaction (Sect. **A 4.4**). The factor η is the ratio of fast neutrons produced to slow neutrons absorbed and may be written

$$\eta = v \frac{\sigma_F}{\sigma_F + \sigma(n, \gamma)} = \frac{v}{1 + \alpha} \tag{A 4.7}$$

where $\alpha = \sigma(n, \gamma)/\sigma_F$. Values of these quantities are as follows:

	$\sigma(n, \gamma) + \sigma_F$ (barns)	σ_F (barns)	η
^{233}U	593 ± 8	524 ± 8	$2 \cdot 29 \pm 0 \cdot 03$
^{235}U	698 ± 10	590 ± 15	$2 \cdot 08 \pm 0 \cdot 02$
(Natural Uranium)			$(1 \cdot 31 \pm 0 \cdot 01)$
^{239}Pu	1032 ± 15	729 ± 15	$2 \cdot 08 \pm 0 \cdot 03$

A 4.4 The chain reaction

The discovery of the emission of about 2·5 fast neutrons in each thermal fission of ^{235}U suggested the possibility that these neutrons could be used to produce further fissions and thus to generate a chain reaction. Such chain reactions are known in chemistry, and lead to an explosive development of chemical reactions. If such a process could be initiated between nuclear reactants the availability of enormous amounts of energy, greater than those known in chemical explosions by roughly the ratio of nuclear to chemical binding energies, could be envisaged. An uncontrolled release of such energy in a short time would constitute a nuclear explosion; controllable release of the same energy, over a longer time would provide a useful nuclear source of power. With these considerations in mind, physicists in 1939 commenced an intensive study of the conditions under which a chain reaction might become possible. The difficulty is essentially to prevent loss of the fission neutrons by processes other than further fission; such processes cannot be eliminated but their effect can be minimized by careful experimental design.

In 1939 the only fissile material available was natural uranium, in the form of uranium oxide. It is clear from Fig. A 4.8 that in this material fission in ^{235}U takes place with a high probability only with thermal neutrons, and it is therefore necessary to reduce the average energy of the fission neutrons from ≈ 1 MeV to $\approx \frac{1}{40}$ eV with a minimum loss of neutron flux. Since fast neutron cross-sections are of the order of nuclear areas, the mean free path for any type of energy-reducing collision in a solid is of the order of 0·5 m and a large system is therefore necessary. If this were just a large mass of natural uranium, slowing down would certainly take place, but the neutrons would in the end lose energy gradually because of the small transfer of energy per collision to the massive uranium nuclei. All would then at some stage reach an energy at which they could be absorbed resonantly by ^{238}U producing the unprofitable (n, γ) reaction (cf. Fig. A 4.6). In order to avoid these narrow 'resonance traps', the neutrons must be slowed down quickly by collisions in each of which a large amount of energy is on the average lost, and this can be achieved if the uranium is dispersed in a large volume of material containing light atoms to act as 'moderator'. The moderator used must be one in which neutron capture through the (n, γ) reaction is small. The earliest experiments of this type were made by Joliot, Halban and Kowarski in France. Using various geometrical arrangements of uranium oxide and water, surrounding a neutron source, they were able to show that the neutron density in the assembly was greater than that expected on the basis of the known number of neutrons released per fission. This could only be so if a convergent chain reaction were taking place, i.e. if each thermal neutron absorbed were giving rise to k (<1) new thermal neutrons which were available to produce further fissions. In such a situation the neutron density in the assembly would be expected to rise by a factor

$$1 + k + k^2 + \ \ldots \ = \frac{1}{1-k} \qquad \text{(A 4.8)}$$

Such a system thus *multiplies* the primary neutron flux, by a factor which may nowadays be very large, but so long as k, the reproduction factor, is less than 1 no continual build-up of flux or divergent reaction is possible.

Joliot, Halban and Kowarski realized that their initial experiments had in fact made it credible that a divergent chain reaction could be realized if neutron leakage and absorption could be further reduced. They therefore arranged to repeat their work with a heavy water moderator, since the (n, γ) reaction in deuterium has a much smaller cross-section than the same reaction in hydrogen. The story of the completion of this experiment is one of the epics of the Second World War. The 165 litres of heavy water necessary had been brought to France from Norway just before the invasion of that country, and it was sent on to England at the time of the fall of France in June 1940. Joliot remained in France but Halban and Kowarski completed their experiment in the Cavendish Laboratory, Cambridge, in December 1940 and produced in the words of a subsequent British Government statement, 'strong evidence that in a system composed of uranium oxide or uranium metal, with heavy water as the slowing down medium, a divergent slow-neutron chain reaction would be realized if the system were of sufficient size'. It is almost certain that, but for the war, such a system would have been achieved in France in 1941; in fact the first self-sustaining nuclear chain reaction in a reactor or 'pile' of natural uranium and graphite, was realized under the direction of Enrico Fermi at Chicago in 1942. The history of that project and its outcome in both military and economic fields may be read elsewhere (Ref. A 4.5–7).

A 4.5 Neutron economy in a reacting system: critical size (Ref. 4.8)

We now give a somewhat more quantitative account of these phenomena, dealing first with neutron multiplication in an infinite system containing natural uranium and a moderator which will be assumed to be pure graphite, and outlining a simple estimate of the critical size at which a finite system can sustain a chain reaction.

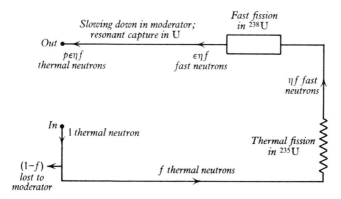

Fig. A 4.9 Sequence of processes following the birth of a neutron in a system containing moderator and uranium.

The essential condition for a divergent chain reaction in an infinite system is that the *reproduction factor* k_∞ shall be at least unity, so that the disappearance of a slow neutron is followed by processes which lead to the reappearance of a further slow neutron. The intermediate processes are represented diagrammatically in Fig. A 4.9, which takes as its starting point the appearance of a thermal neutron at some point within the system. The probability of the absorption of this neutron in uranium $(^{235}U + ^{238}U)$ is known as the *thermal utilization factor f*; the probability of the competing processes of absorption in moderator and in various constructional materials is $1 - f$. If v is the average number of fast neutrons produced by thermal fission of ^{235}U then the original thermal neutron will produce ηf fast neutrons where η is the quantity defined in Sect. **A4.3**(g).

The ηf fast neutrons are born in an environment of uranium, and slow down, both in uranium and in moderator by a variety of processes. Collisions of neutrons of average energy ≈ 1 MeV with abundant ^{238}U nuclei result in an inelastic scattering

$$^{238}U + n \longrightarrow ^{238}U^* + n' \qquad (A\ 4.9)$$

with a loss of energy of some hundreds of keV; there is also a small but not negligible chance of fast fission in ^{238}U, which is represented by the *fast fission factor* ε ($\approx 1 \cdot 02$). The fast fission factor is small because unless the uranium content in the assembly is very large, the main process of slowing down is due to elastic collisions between the neutrons and carbon atoms. About 100 collisions with carbon atoms are required to reduce the initial neutron energy to the thermal value; using known cross-sections or mean free paths it may be calculated that this process occupies about 10^{-3} s.

Although the neutrons lose perhaps 10 per cent of their energy in each collision this is insufficient to guarantee that the energy will not in some cases have a value corresponding to the resonance absorption peaks in the abundant ^{238}U (cf. Fig. A 4.6). We therefore introduce a *resonance escape probability* factor p, which gives the chance that a fast neutron reaches thermal energy as a result of the slowing down processes in the particular assembly. The total number of thermal neutrons produced for each thermal absorbed, i.e. the reproduction factor for the infinite system, is thus

$$k_\infty = \varepsilon \eta p f \qquad (A\ 4.10)$$

This is known as the four-factor formula. For the power-producing reactors of the U.K.A.E.A. at Calder Hall the lattice constants are

$$\eta = 1 \cdot 31, \quad \varepsilon = 1 \cdot 02, \quad p = 0 \cdot 89, \quad f = 0 \cdot 88$$

giving a value

$$k_\infty = 1 \cdot 05 \qquad (A\ 4.11)$$

In order to design an efficient reacting system p and f should be made large. The thermal utilization can be increased by increasing the amount of uranium in the system, but this reduces the amount of carbon and hence decreases p, since the slowing down process will then be more gradual and the resonant capture more serious. If the reacting system consists of a homogeneous mixture of carbon and uranium, the values of f and p are fixed by the relative concentration. It is, however, more favourable to use 'lattices' in which the uranium is concentrated into rods or lumps. In this case, owing to the high cross-section for resonant capture, neutrons of this energy are absorbed unprofitably in only the surface layers of the uranium, and the bulk of the fissile material does not contribute to resonance losses, although it is available for the desired thermal fission processes, which are less attenuated in the surface of the lump. All practical reactors using natural uranium and graphite are therefore based on a lattice; a typical arrangement is shown in Fig. A 4.10. The calculation of the factors f and p for a lattice is a matter of some complication and must be checked by direct experimental observation using non-reacting assemblies of similar construction but of small size.

It is not difficult to obtain a reproduction factor k_∞ greater than unity if pure materials are used. It becomes in fact very easy if uranium enriched in ^{235}U is employed. This, however, may not be economically feasible for large power-producing installations. For systems of finite size, the actual value of k_∞ is of extreme importance since for each thermal neutron absorbed only $k_\infty - 1$ neutrons can be permitted to escape from the system and the smaller the system the greater the likelihood of escape. A finite reactor with a given lattice therefore has a critical size below which the chain reaction will not diverge; this can be seen for a simple case as follows.

Consider a spherical system (Fig. A 4.11) and suppose that a distance M, known as the *migration length*, is taken to represent crudely the path length

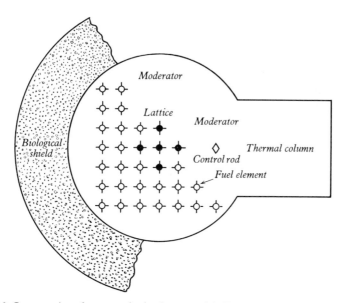

Fig. A 4.10 Cross-section of a reactor lattice (not to scale). Fuel elements (e.g. U) are embedded in a moderating medium (e.g. C) and the whole is surrounded by a massive concrete radiation shield. In most reactors a graphite neutron reflector (not shown) is installed between the lattice volume and the concrete shield in order to minimize the neutron losses. The thermal column (of moderator) is a region in which the ratio of thermal to fast neutrons increases with increasing distance from the lattice.

of a neutron between birth as a fast neutron and absorption as a thermal. Only neutrons which originate within the shell of thickness M will have an appreciable chance of escaping from the system, and to obtain orders of magnitude we assume that all such neutrons do escape. The ratio of this number to the number of neutrons which do not escape is approximately

$$\frac{\text{volume of shell}}{\text{volume of sphere}} \times \frac{\text{neutron density in shell}}{\text{neutron density in sphere}} \approx \frac{3M}{R} \times \frac{M}{R} = \frac{3M^2}{R^2}$$

since we may assume that the neutron density distribution is as shown in the lower part of Fig. A 4.11.

The fraction of neutrons lost by leakage is thus $1 - 3M^2/R^2$ and the effective reproduction constant is consequently

$$k_{\text{eff}} = k_{\infty}\left(1 - \frac{3M^2}{R^2}\right) \tag{A 4.12}$$

The critical radius R_c at which k_{eff} becomes equal to 1 is therefore given by

$$R_c^2 = \frac{3M^2 k_{\infty}}{(k_{\infty}-1)} \approx \frac{3M^2}{(k_{\infty}-1)} \tag{A 4.13}$$

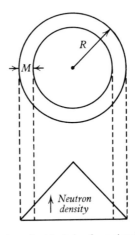

Fig. A 4.11 Calculation of critical size for a simple system (Ref. A 4.8).

For $k_{\infty} = 1 \cdot 05$ and $M = 0 \cdot 5$ m (for graphite) $R_c = 3 \cdot 85$ m. Obviously a reactor in which k_{∞} is larger will be correspondingly smaller. A reactor employing *fast* fission as the basic process will also be smaller because of the smaller migration length M.

A 4.6 Construction and operation of a reactor; reactor types

A critical assembly in which high levels of neutron density are to be reached must be provided with adequate biological screening. In most reactors this takes the form of a thick concrete box enclosing the reactor, including the arrangements for the insertion of samples for irradiation, instruments and the control rods. These are usually massive bars of cadmium, which has a high cross-section for the absorption of thermal neutrons by the (n, γ) reaction. In the control of a reactor the cadmium rods are withdrawn until the reproduction constant k rises above unity. The reactor neutron density ρ then increases in accordance with the equation

$$\frac{d\rho}{dt} = \frac{\rho(k-1)}{t_0} \qquad \text{(A 4.14)}$$

where t_0 is the neutron lifetime from birth to absorption, $\approx 1 \cdot 4 \times 10^{-3}$ s. The density thus rises exponentially with a time constant T, where

$$T = \frac{t_0}{(k-1)} \qquad \text{(A 4.15)}$$

so that even a very small (1 per cent) excess reactivity leads to a time constant of as little as $\frac{1}{10}$ s. Such a rapid rise of pile power would be difficult to manage but fortunately the existence of the delayed neutrons, with periods of up to 55 s, greatly increases the effective multiplication time, and permits mechanical adjustment to the control rods to be made. In bringing a reactor up to a prescribed level the control rods are thus first adjusted to give a small excess reactivity and then inserted to bring $k-1$ to zero when the required neutron density has been reached; the reactor then operates at constant power.

In an *uncontrolled* reacting system, or bomb, an enriched material (^{235}U or ^{289}Pu) is suddenly given a large excess reactivity by bringing together two masses to form a volume which is considerably greater than the critical size. The reaction proceeds by fast fission with a neutron mean free path of perhaps a few cm only and the generation time is only about 4×10^{-9} s. From eq. A 4.15 the neutron density increases with a time constant of about 4×10^{-8} s for a 10 per cent excess reactivity and the number of fission processes, each releasing about 200 MeV, will increase by a factor of about 10^{11} in 1 μs. This is the condition obtained in the explosion of a fission bomb.

It is not proposed here to give a list of the many types of production and research reactor which have been developed. We may, however, note the following versions, which have become important:

(i) The natural uranium–graphite lattice, used in the U.K.A.E.A. power production programme.

(ii) The natural uranium–heavy water lattice which can be smaller than (i) because of the lower absorption, and consequently yields a higher flux density. Such reactors are mainly found in national research establishments.

(iii) The enriched uranium–ordinary water lattice which leads to the extremely simple type of construction known as the swimming pool.

(iv) The enriched uranium–heavy water lattice, also mainly found in national research establishments, which is especially designed for the production of high neutron densities at low powers.

These reactors differ considerably in the measures adopted to remove the heat and if necessary to convert it into electrical power. The average power levels and central thermal neutron fluxes† for a number of reactors are shown in the following table:

TABLE A 4.1 Nuclear Reactor Performance

Reactor	Type	Central flux (neutrons $m^{-2} s^{-1}$)	Power (kW)
BEPO (Harwell)	Uranium–graphite	1×10^{16}	4000
Calder Hall	Uranium–graphite	$1 \cdot 6 \times 10^{17}$	200 000
NRX (Chalk River)	Uranium–heavy water	5×10^{17}	10 000
Brookhaven (1962)	Uranium–graphite	5×10^{16}	25 000
DIDO (Harwell)	Enriched uranium–heavy water	2×10^{18}	13 000

The nuclear reactor is of prime importance in nuclear physics as a source of thermal neutrons (Refs. 4.9, 4.10). The highest thermal fluxes exist at the centre of the pile, but are accompanied by neutrons of all other energies in the fission spectrum. This may not be a disadvantage if methods of monochromatization are used. If a pure thermal flux, of lower level, is required the 'thermal column', which is essentially a graphite reflector extension passing through the shield wall may be used (Fig. A 4.10).

† Defined for this purpose as ρv where ρ is the number of neutrons per unit volume and $v = 2200$ m s^{-1}.

References

1. WHITEHOUSE, W. J. 'Fission', *Progr. nucl. Phys.*, **2**, 120, 1952; FRASER, J. S. and MILTON, J. C. D. 'Nuclear Fission', *Ann. rev. nucl. Sci.*, **16**, 379, 1966.
2. STRUTINSKY, V. M. *Nucl. Phys.*, **A95**, 420, 1967; TSANG, C. F. and NILSSON, S. G. *Nucl. Phys.*, **A140**, 275, 1970; WILKINSON, D. H. *Comments on Nuclear and Particle Physics*, **2**, 146, 1968.
3. HUGHES, D. J. and SCHWARTZ, R. B. *Neutron Cross Sections*, Brookhaven National Laboratory Report BNL 325, 1958.
4. EVANS, R. D. *The Atomic Nucleus*, McGraw-Hill, 1955.
5. GOLDSCHMIDT, B. 'Hans Halban', *Nucl. Phys.*, **79**, 1, 1966.
6. SMYTH, H. D. 'Atomic Energy for Military Purposes', *Rev. mod. Phys.*, **17**, 351, 1945.
7. *Peaceful Uses of Atomic Energy* (Geneva International Conference Series, ed. J. G. Beckerly), 1955/6.
8. LITTLER, D. J. and RAFFLE, J. F. *An Introduction to Reactor Physics*, Pergamon Press, 1955.
9. MAIER-LEIBNITZ, H. and SPRINGER, T. 'Thermal Reactor Neutron Beams', *Ann. rev. nucl. Sci.*, **16**, 207, 1966; WALKER, J. 'Nuclear Reactors as Research Instruments'), *Rep. progr. Phys.*, **30**, 285, 1967.
10. VICK, F. A. 'Explorations in Atomic Energy', *Contemp. Phys.*, **8**, 41, 1967.

5 The coherent neutron scattering length for hydrogen

A 5.1 Definitions

A neutron and a proton can collide in singlet $(S=0)$ or triplet $(S=1)$ spin states, according as the intrinsic spins are opposed or in the same direction. Of the four possible arrangements of the individual spin components $\frac{1}{2}\hbar$ with respect to an axis, three belong to the triplet state and one to the singlet. The total cross-section for the interaction of slow neutrons with protons (by scattering) is therefore, as noted in Sect. 18.2.2,

$$\sigma_{el}^0 = 4\pi(\tfrac{3}{4}l_t^2 + \tfrac{1}{4}l_s^2) \qquad (A\ 5.1)$$

where l_s and l_t are the scattering lengths for the singlet and triplet states. This is the cross-section that would be measured by passing slow neutrons through a randomly oriented assembly of protons. For a single proton–neutron encounter, however, we may define a *coherent* scattering length†

$$l_{coh} = \tfrac{3}{4}l_t + \tfrac{1}{4}l_s \qquad (A\ 5.2)$$

which is necessary in the discussion of interference phenomena. This quantity is defined for a free proton, but if the proton is bound in a solid lattice the reduced mass in the neutron–proton collision is doubled and so also is the scattering length. This accounts for the rise in the neutron–proton scattering cross-section at low neutron energies when this is deduced from measurements made on hydrogen contained in heavy molecules; the neutron energy is then insufficient to break the chemical bonds and the cross-section increases (Fig. 18.5). For the same reason the coherent scattering length for the diffraction of slow neutrons from crystals containing hydrogen is

$$l_H = 2l_{coh} = \tfrac{1}{2}(3l_t + l_s) \qquad (A\ 5.3)$$

Determination of σ_{el}^0 and l_H will thus permit l_t and l_s to be obtained separately.

---

† We follow here the nomenclature of Chapter 18 in preference to the more conventional use of f or a for this quantity. Strictly the scattering *amplitude* f_0, as used in Chapter 14, is energy dependent and is complex unless δ_0 (18.17) is $0°$ or $180°$. This is a good approximation for slow neutron scattering, and for zero energy neutrons $(k \longrightarrow 0)$ the scattering amplitude becomes the scattering *length* $l\ (= -f)$, which is a constant for a given interaction.

A 5.2 Scattering of slow neutrons by ortho- and para-hydrogen

The possibility of using these two forms of hydrogen to establish the spin dependence of nuclear forces (which makes l_t different from l_s) was first discussed by Schwinger and Teller.† If the forces between a neutron and proton depend on the relative spin orientation, then the scattering from ortho- and from para-hydrogen should differ providing that there is some coherence

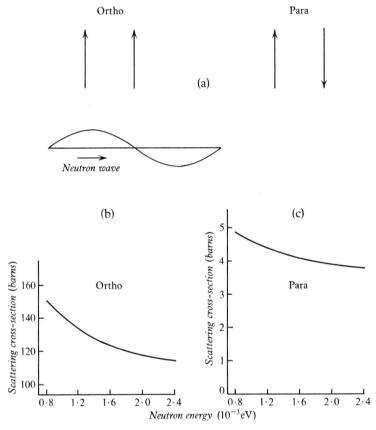

Fig. A 5.1 Scattering of slow neutrons by ortho- and para-hydrogen. (a) Spin orientations. (b), (c) Variation of scattering cross-section with neutron energy (Sutton *et al.*, *Phys. Rev.*, **72**, 1147, 1947).

between the waves scattered from the two protons. This will be so if the wavelength of the incident neutrons is long compared with the internuclear separation in hydrogen, i.e. $\lambda \gg 0.078$ nm. The analysis of the scattering cross-sections in terms of the singlet and triplet amplitudes is complicated by the possibility of inelastic scattering, which can always take place for ortho-hydrogen, with conversion to the para- form (and increase of neutron energy)

† J. Schwinger and E. Teller, *Phys. Rev.*, **52**, 286, 1937.

and can take place for parahydrogen with conversion to the ortho- form for neutrons of energy greater than 0·023 eV. Such neutrons can be obtained by velocity selector or filtration techniques (Ref. A 5.1).

The appearance of the ortho- and para- molecules to an incident neutron of arbitrary spin direction is represented in Fig. A 5.1(a). For parahydrogen the two protons appear uncorrelated and the coherent scattering length (neglecting the reduced mass effect) is double the length for a single proton, i.e.

$$l_{para} = 2l_{coh} = 2(\tfrac{3}{4}l_t + \tfrac{1}{4}l_s) \tag{A 5.4}$$

If the neutrons are of such a low energy ($E_n < 0·023$ eV, $T < 90$ K, $\lambda > 0·2$ nm) that there is no para $\longrightarrow$ ortho conversion, the total cross-section for parahydrogen scattering is proportional to $(l_{para})^2$. For orthohydrogen, an interference term depending on $(l_t - l_s)$ appears in addition, and a term also involving this difference occurs to represent the stimulated ortho $\longrightarrow$ para transition. Measurement of the orthohydrogen cross-section as well as that for parahydrogen enables both quantities $(3l_t + l_s)$ and $(l_t - l_s)$ to be found. Recent values of the scattering lengths, obtained in experiments with gaseous hydrogen, are given at the end of this section; the observed cross-sections (Fig. A 5.1(b), (c)) immediately show that the two types of molecule scatter differently so that the neutron–proton interaction is spin-dependent. It is also clear that the parahydrogen cross-section is very small, which indicates that the singlet and triplet scattering lengths are of opposite sign, as would be expected for a virtual singlet state (l_s negative).

The relative size of σ^0_{para} and σ^0_{ortho} renders accurate determinations by this method difficult, because of the possibility of contamination of the parahydrogen by the ortho- form. Alternative methods of finding l_{coh} have therefore been devised.

A 5.3 Crystal diffraction method

Crystal diffraction experiments with slow neutrons require strong sources and accurate measurements could not be made until nuclear reactors became available for experimental purposes. A typical apparatus for detecting Bragg reflections from a powdered crystalline sample using neutrons rendered monochromatic by a preceding crystal reflection is shown in Fig. A 5.2(a). This technique has been found superior to the use of a single crystal because imperfections in such crystals make the calculation of absolute intensities difficult. Figure A 5.2(b) shows the powder diffraction pattern obtained with a sample of sodium hydride (NaH) for neutrons of energy 0·0724 eV ($\lambda = 0·1$ nm).

The intensities of the diffraction spectra obtained with slow neutron beams depend on the coherent scattering lengths for the scattering nuclei. These add as amplitudes for each lattice point and create beams in just the directions for Bragg reflection; for a polycrystalline scatterer, these directions lie on the surface of cones. In principle an absolute measurement of intensity will give the coherent scattering length, using formulae well known from the theory of X-ray scattering, but in practice it is better to calibrate the crystal spectrometer by using an element such as carbon (in diamond) whose coherent length is known independently. For this purpose a monoisotopic substance of zero nuclear spin would be ideal since the coherent length could then be obtained directly from the elastic cross-section measured in a transmission experiment.

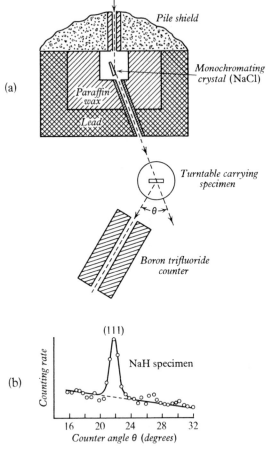

Fig. A 5.2 Slow neutron diffraction. (a) Experimental arrangement showing monochromator (Ref. A 5.4). (b) Typical results for diffraction from NaH, with neutrons of energy 0·0724 eV (Shull *et al.*, Ref. A 5.4).

In their analysis of the diffraction peaks from NaH, Wollan and Shull (Ref. A 5.4) had first to consider the probable structure of this crystal, and then the relative signs of the bound scattering lengths l_H and l_{Na} for the two atoms concerned. From the relative intensities of several peaks, including that indicated in Fig. A 5.2(b), they concluded that Na and H scatter neutrons in opposite phase, i.e. the intensity of the peak shown in the figure is proportional to $(l_H - l_{Na})^2$. By making use of a previously determined value for l_{Na} the hydrogen scattering length was found; the result is quoted at the end of this section.

A 5.4 Reflection of slow neutrons from liquid mirrors

The crystal diffraction method of determining the coherent length for hydrogen is open to the objection that correction must be made for the effect of thermal vibrations in destroying coherence. This can be avoided by a method based on another optical property of slow neutrons, that of total reflection from a mirror surface at small glancing angles.

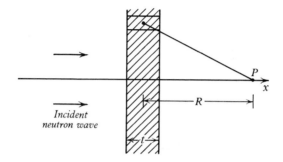

Fig. A 5.3 Calculation of refractive index for slow neutrons.

The existence of a critical angle for neutron reflection may be understood in exactly the same way as that for X-rays. For a material containing scattering centres in random positions the coherently scattered wave interferes with the incident wave in such a way that the resultant wave appears to have a different wavelength in the material. This endows the medium with a refractive index which may be calculated by the following method, using the Fresnel construction formula of optical diffraction: Consider the amplitude of the wave transmitted through a small thickness t of material at a point P (Fig. A 5.3) and imagine a series of Fresnel half-period zones drawn on the material for the point P. The amplitude for an infinite sheet is known to be half that due to the first zone, of area $\pi R \lambda$, and if the scattering amplitude, assumed independent of angle is f, then the amplitude at P for the total number of centres in the half-period zone is

$$\frac{i}{2} \frac{\pi R \lambda N t f}{R} \times \frac{2}{\pi} = i N t \lambda f \qquad (A\ 5.5)$$

where N is the number of centres per unit volume and we allow for the $1/R$ dependence of amplitude on distance. The $2/\pi$ factor averages the first zone amplitude at the point P and the quantity $i\ (=\sqrt{-1})$ represents the $\pi/2$

phase displacement. The scattered wave at P adds to the incident wave to create the transmitted wave and the transmitted amplitude is thus

$$A_\mathrm{T} = A_0(1 - iNt\lambda f) \approx A_0 e^{i\delta}$$

where A_0 is the incident amplitude. The phase-lag thus introduced is

$$\delta = -Nt\lambda f$$

In terms of the macroscopic refractive index n of the slab, we also have

$$\delta = \frac{2\pi}{\lambda}(n-1)t$$

and by comparing these two expressions we find

$$n = 1 - \frac{N\lambda^2 f}{2\pi}$$

which essentially contains the *optical theorem* (Sect. 14.2.2). For neutrons f is the bound coherent nuclear scattering length l, taken positive for $180°$ phase shift. For X-rays, f is the amplitude of a wave scattered by an electron multiplied by an atomic form factor for the forward direction. We thus see that if f is positive the refractive index is less than unity and total *external* reflection may be expected. This is the normal case for X-rays and for neutrons scattered by a large number of elements. For hydrogen, however, the coherent amplitude is negative† and total *internal* reflection would be expected. In either case the critical angle θ_c for a neutron wavelength λ is given by

$$\cos\theta_\mathrm{c} = n \quad \text{or} \quad 1 - \frac{\theta_\mathrm{c}^2}{2} = 1 - \frac{\lambda^2 N l_\mathrm{H}}{2\pi}$$

whence

$$\theta_\mathrm{c} = \lambda\sqrt{\frac{N l_\mathrm{H}}{\pi}} \approx 10' \text{ of arc}$$

and a determination of $\theta_\mathrm{c}/\lambda$ yields l_H.

The advantage of mirror reflection methods is that the coherent effect is determined by the average properties of the surface, i.e. it does not depend on the *structure* of the surface and liquids, solids or gases may be used. The effect is uncomplicated by thermal motion since the scattering is predominantly in the forward direction, and in the case of liquid the scattering from the surface can be shown to be elastic. In the actual application of the mirror method by Hughes, Burgy and Ringo (Ref. A 5.5 and Fig. A 5.4), liquid hydrogen was not used because of low intensity and because of the inconvenience of the total *internal* reflection. Instead, liquid hydrocarbon mirrors

---

† This fact, established by the ortho/parahydrogen measurements, is also shown conclusively by the mirror experiments.

were used; the addition of carbon, whose coherent amplitude is well known, converts the total reflection into an *external* phenomenon. By studying the intensity of the reflected beam at angles near critical as a function of the carbon to hydrogen ratio of the liquid hydrocarbon mirror, the coherent scattering length for hydrogen was obtained in terms of that for carbon; the result is given below.

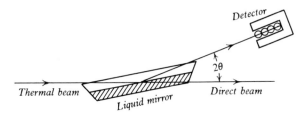

Fig. A 5.4 Scattering of thermal neutrons from a liquid hydrocarbon mirror (Ref. A 5.1).

A 5.5 Values of l_H

Values for l_H obtained by the methods just described are:

Method	Reference	l_H (10^{-15} m)
Ortho/parahydrogen scattering	Squires and Stewart (Ref. A 5.3)	-3.80 ± 0.05
Crystal diffraction	Shull *et al.* (Ref. A 5.4)	-3.96 ± 0.2
Liquid mirrors	Hughes, Burgy and Ringo (Ref. A 5.5)	-3.78 ± 0.02

References

1. HUGHES, D. J. *Neutron Optics*, Interscience, 1954; see also *Ann. rev. nucl. Sci.*, **3**, 93, 1953.
2. BACON, G. E. *Neutron Diffraction*, 2nd edn., Camb. Univ. Press, 1967.
3. SQUIRES, G. L. and STEWART, A. T. *Proc. roy. Soc.*, A, **230**, 19, 1955.
4. WOLLAN, E. O. and SHULL, C. G. *Phys. Rev.*, **73**, 830, 1948; SHULL, C. G., WOLLAN, E. O., MORTON, G. A. and DAVIDSON, W. L. *Phys. Rev.*, **73**, 842, 1948.
5. HUGHES, D. J., BURGY, M. T. and RINGO, G. R. *Phys. Rev.*, **84**, 1160, 1951.

6 The principle of detailed balancing

(Ref. A 6.1)

The scattering amplitude $f(\theta)$ for the reaction

$$X + a \longrightarrow Y + b \qquad (A\ 6.1)$$

is most generally expressed in terms of the elements of a *transition matrix, T*, which connects the initial state with all possible final states.† If this matrix is invariant under the operation of time reversal, then for spinless particles the forward and backward matrix elements are equal, i.e.

$$T_{ab} = T_{ba} \qquad (A\ 6.2)$$

For particles with spin, time reversal inverts the spin direction, and strict reciprocity for reactions between individual spin states is not necessarily to be expected.

In most experiments, spins are not directly detected, and an average must be taken over magnetic substates. Under these circumstances we shall assume that the probabilities of forward and backward reactions are equal. The observed cross-sections then take the form

$$d\sigma_{ab} \propto \frac{1}{(2s_a + 1)(2s_X + 1)} \sum |T_{ab}|^2 \, d\Omega \qquad (A\ 6.3)$$

$$d\sigma_{ba} \propto \frac{1}{(2s_b + 1)(2s_Y + 1)} \sum |T_{ba}|^2 \, d\Omega \qquad (A\ 6.4)$$

in which the spin factors give the averaging over the initial states and the summations run over all spin states of the initial and final system. Assuming that these sums of squares are equal for the forward and backward reactions, and introducing the kinematic proportionality factors, we obtain

$$p_a^2(2s_a + 1)(2s_X + 1)\frac{d\sigma_{ab}}{d\Omega} = p_b^2(2s_b + 1)(2s_Y + 1)\frac{d\sigma_{ba}}{d\Omega} \qquad (A\ 6.5)$$

where p_a and p_b are the c.m. momenta of particles a and b in the reaction. This result is usually described as the principle of detailed balancing or microscopic reversibility. It does not hold for separate spin states.

† The T-matrix is derived from the more general S-matrix by the exclusion of elastic scattering processes. It is the S-matrix to which the properties of analyticity, unitarity and crossing symmetry mentioned in Sect. 21.1.3, apply.

The result may easily be obtained for total cross-section from eq. (15.34) for a reaction passing through a well-defined state of a compound nucleus.

Reference

1. See DE BENEDETTI, S. *Nuclear Interactions*, Wiley, 1964, page 329; KÄLLEN, G. *Elementary Particle Physics*, Addison-Wesley, 1964, page 43; Ref. A 2.4, page 530.

7 The Dalitz plot (Refs. A 7.1, 7.2)

The *Dalitz plot* is a two-dimensional phase space diagram describing a three-body final state arising in either a decay process, e.g.

$$\omega \longrightarrow \pi^+ + \pi^- + \pi^0$$

$$K^+ \longrightarrow \pi^+ + \pi^+ + \pi^-$$

or an interaction, e.g.

$$\pi + p \longrightarrow \pi + \pi + p$$

$$K^- + p \longrightarrow \Lambda^0 + \pi^+ + \pi^-$$

$$^{11}B + p \longrightarrow \alpha + \alpha + \alpha$$

Examination of this plot gives information about the energy dependence of the probability for the process and may very clearly indicate the formation of resonant states among groups of the three final particles. If there are more than three particles in the final state, and if resonant states exist, it may still be possible to use the Dalitz plot. Thus for the reaction

$$K^+ + p \longrightarrow (K^+ \pi^-) + \pi^+ + p$$

a plot of the effective mass of $(K^+\pi^-)$ against that of $(\pi^+ p)$ may reveal the formation of $K^{*0}(890)$; a Dalitz plot may then be made for events including K^{*0}, π^+ and p.

Two forms of the Dalitz plot are frequently used. The first, sometimes known as a scatter plot, may be illustrated by considering the production of a final state containing three particles of *unequal mass* among which is distributed a total energy E in the c.m. system. If for each event observed, say in a bubble chamber, the c.m. kinetic energy T_1 of one particle is plotted against the kinetic energy T_2 of another, all points lie within a kinematically defined region (Fig. A 7.1(a)). The boundaries of this region correspond to collinear situations, as shown in the Figure. All other permitted momenta correspond to points within the boundaries. The value of the plot resides in the fact, not shown here, that if there is no tendency for particular energies $(E_2 + E_3)$ of the pair 2, 3, say to be favoured, then a sufficiently large number of observed points will assume a uniform distribution. If however 2, 3 form a resonance with effective mass M_{23} then particle 1 is more frequently found with the corresponding momentum, which is unique for a fixed c.m. energy because it corresponds to a two-body reaction. A band then appears on the Dalitz plot; the first hyperon resonance Y_1^* (1385) (Sect. 20.5.4, Fig. 20.8) was found in this way. The position of this band of course depends on the available energy and this is a disadvantage of this type of plot in that it does not directly indicate the resonance mass. This particular difficulty is avoided if

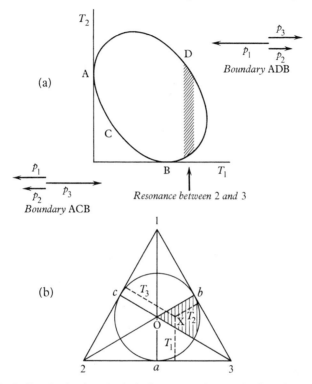

Fig. A 7.1 The Dalitz plot for three-body final states: (a) Scatter plot for a final state containing three particles of unequal mass. The centre of mass momenta are related by the equation

$$p_3^2 = p_1^2 + p_2^2 + 2p_1 p_2 \cos \theta_{12}$$

Observed kinetic energies of particle 1 and 2 (c.m.) lie within the contour given by $\cos \theta_{12} = \pm 1$. (b) Triangular plot for a final state containing three particles of equal mass m. The c.m. kinetic energies corresponding to a point X are represented by perpendicular distances to the sides of the triangle. The perpendiculars $1a$, $2b$, $3c$ represent the energy $E - 3m$ where E is the total c.m. energy in the final state. Possible energy distributions (non-relativistic) are represented by points within the inscribed circle.

a plot of the squares of effective masses is made, i.e. a plot of $M_{13}^2 = (E_1 + E_3)^2 - (\mathbf{p}_1 + \mathbf{p}_3)^2$ against $M_{23} = (E_2 + E_3) - (\mathbf{p}_2 + \mathbf{p}_3)^2$ instead of T_2 against T_1 (with a suitable shift of origin). A resonance band on the plot is then independent of total centre of mass energy, and only the boundaries of the plot change when the available energy is altered.

If the masses of the three particles in the final state are *equal*, as in the decay of the ω, it is customary to use a triangular coordinate system for the

Dalitz plot, Fig. A 7.1(b). This is appropriate because the physical situation for a constant decay energy E is that the three kinetic energies T in the c.m. system must satisfy the equation

$$T_1 + T_2 + T_3 = E - 3m = \text{constant}$$

If $T_1 T_2 T_3$ are represented as the perpendicular distances of a point X from the sides of an equilateral triangle, this condition is fulfilled since the sum is equal to the height of the triangle. Energy conservation is then satisfied for every point within the triangle, but momentum conservation determines the boundaries of the plot. In the non-relativistic approximation, this condition confines points to lie within the inscribed circle of the triangle, as shown in Fig. A 7.1(b); for relativistic energies the circle is distorted. Again, if the probability for the process is energy independent a uniform distribution of points, corresponding to different partitions of energy, is expected over the Dalitz plot. Because of the symmetry of the coordinate system the plot may in effect be folded on itself, so that all points lie within one sector such as the shaded area in Fig. A 7.1(b); an example is given in Fig. 20.10, Sect. 20.6.1.

Particular regions of the triangular Dalitz plot correspond to simple energy or momentum distributions. Thus at the centre O, $T_1 = T_2 = T_3$, along the lines Oa Ob, Oc two particles have equal energy, and at the boundary of the plot the particles are collinear (one being at rest in the c.m. system at points a, b, c). In an actual physical process, invariance under rotation or reflection may forbid the population of certain regions of the Dalitz plot and inspection of the plot may then lead to spin-parity assignments. In the case of the decay of the ω-meson,† the Dalitz plot shows a vanishing of density along the boundary. This can be shown to be characteristic of the three pion decay of a particle of spin-parity 1^-.

References

1. DALITZ, R. H. *Phil. Mag.*, **44**, 1068, 1953. The relativistic version of the plot is discussed by FABRI, E. *Nuovo Cim.*, **11**, 479, 1954.
2. HAGEDORN, R. *Relativistic Kinematics*, Benjamin, 1963.

† B. C. Maglic, L. W. Alvarez, A. H. Rosenfeld, and M. L. Stevenson, *Phys. Rev. Lett.*, **7**, 178, 1961.

8 Elementary particles

The information in these tables has been mainly taken from the article 'Review of Particle Properties', by the Particle Data Group, *Physics Letters*, **33B**, 1, 1970. Strangeness (S) is listed rather than hypercharge $(Y = S + B)$. Parity is not uniquely defined for fermions although the relative intrinsic parity of a fermion–antifermion pair is odd. Conventionally the proton, neutron and Λ-particle are assigned even parity. The particles in Table 2, which decay by the weak or electromagnetic interaction, are often described as *stable*. Table 1 contains only a selection of the many known meson and baryon resonances. Antiparticles are included in Table 2; their lifetime is assumed to be equal to that of the corresponding particle.

1. Strongly-decaying elementary particles

Group	Spin-parity $\mathscr{J}^\pi$	Isobaric spin T	Symbol	Mass MeV/c²	Strange-ness S	Main decay modes
Vector mesons $B = 0$	1^-	1	ρ (765)	765	0	$\pi\pi$
		0	ω (784)	783·7	0	$\pi\pi\pi$
		0	ϕ (1019)	1018·8	0	$K\bar{K}$
		$\frac{1}{2}$	K^* (892)	892·1	± 1	$K\pi, \bar{K}\pi$
Mesons $B = 0$	0^-	0	η' (958)	958	0	$\eta\pi\pi$
	2^+	1	$A2_L, A2_H$	1280, 1320	0	$\rho\pi, K\bar{K}$
		0	f_0 (1260)	1264	0	$\pi\pi$
Excited baryons $B = 1$	$\frac{1}{2}^-$	0	Λ (1405)	1405	-1	$\Sigma\pi$
	$\frac{3}{2}^+$	$\frac{3}{2}$	Δ (1236)	1236	0	$N\pi$
		1	Σ (1385)	1383	-1	$\Lambda\pi, \Sigma\pi$
		$\frac{1}{2}$	Ξ (1530)	1530	-2	$\Xi\pi$
	$\frac{3}{2}^-$	$\frac{1}{2}$	N (1520)	1510–1540	0	$N\pi, N\pi\pi$
		0	Λ (1520)	1518	-1	$\Sigma\pi, \bar{K}N, \Lambda\pi\pi$
		1	Σ (1670)	1670	-1	$\Sigma\pi, \Lambda\pi, \bar{K}N$
	$\frac{5}{2}^+$	$\frac{1}{2}$	N (1688)	1688	0	$N\pi, N\pi\pi$
		0	Λ (1815)	1820	-1	$KN, \Sigma\pi$

2. Elementary particles that do not decay strongly

Group	Name	Spin-parity J^π	Isobaric spin T, T_z	Symbol	Mass MeV/c^2	Strangeness S	Mean lifetime s	Main decay modes
	Photon	1^-	—	γ	0	—	stable	
Leptons $B=0$	Neutrinos	$\tfrac{1}{2}$	—	ν_e, ν_μ	0	—	stable	
				$\bar{\nu}_e, \bar{\nu}_\mu$	0	—	stable	
	Electrons		—	e^-	0·51	—	stable	
				e^+				
	Muons		—	μ^-	105·7	—	$2{\cdot}2\times10^{-6}$	$e^- \nu\bar{\nu}$
				μ^+				$e^+ \nu\bar{\nu}$
Mesons $B=0$	Pions	0^-	1, 0	π^0	135·0	0	$0{\cdot}76\times10^{-16}$	$\gamma\gamma$
			1	π^+	139·6	0	$2{\cdot}6\times10^{-8}$	$\mu^+\nu$
			-1	π^-				$\mu^-\bar{\nu}$
	Kaons		$\tfrac{1}{2}, \tfrac{1}{2}$	K^+	493·8	1	$1{\cdot}2\times10^{-8}$	$\mu^+\nu, \pi^+\pi^0, \pi\pi\pi$
			$-\tfrac{1}{2}$	K^-		-1		$\mu^-\bar{\nu}, \pi^-\pi^0, \pi\pi\pi$
			$\tfrac{1}{2}, -\tfrac{1}{2}$	K^0	497·8	1	$K_S^0\ 0{\cdot}86\times10^{-10}$	$\pi\pi$
			$\tfrac{1}{2}$	$\bar{K}^0$		-1	$K_L\ 5{\cdot}4\times10^{-8}$	$\pi\pi\pi, \eta\pi\nu, e\pi\nu$
	Eta	0, 0		η^0	549	0	$2{\cdot}5\times10^{-19}$	$\gamma\gamma, 3\pi^0, \pi\pi\pi$
Baryons $B=\pm1$	Nucleons	$\tfrac{1}{2}$	$\tfrac{1}{2}, \pm\tfrac{1}{2}$[a]	$p^+, \bar{p}^-$	938·3	0, 0	stable	
			$\tfrac{1}{2}, \mp\tfrac{1}{2}$	$n^0, \bar{n}^0$	939·6	0, 0	932	$pe\bar{\nu}$
	Lambda hyperon		0, 0	$\Lambda^0, \bar{\Lambda}^0$	1115·6	$-1, +1$	$2{\cdot}5\times10^{-10}$	$p\pi, n\pi$
	Sigma hyperons		1, ±1	$\Sigma^+, \bar{\Sigma}^-$	1189·4	$-1, +1$	$0{\cdot}8\times10^{-10}$	$p\pi, n\pi$
			1, 0	$\Sigma^0, \bar{\Sigma}^0$	1192·5	$-1, +1$	$<10^{-14}$	$\Lambda\gamma$
			1, ∓1	$\Sigma^-, \bar{\Sigma}^+$	1197·4	$-1, +1$	$1{\cdot}49\times10^{-10}$	$n\pi$
	Cascade hyperons		$\tfrac{1}{2}, \pm\tfrac{1}{2}$	$\Xi^0, \bar{\Xi}^0$	1314·7	$-2, +2$	3×10^{-10}	$\Lambda\pi$
			$\tfrac{1}{2}, \mp\tfrac{1}{2}$	$\Xi^-, \bar{\Xi}^+$	1321·3	$-2, +2$	$1{\cdot}7\times10^{-10}$	$\Lambda\pi$
	Omega hyperon	$\tfrac{3}{2}$	0, 0	$\Omega^-, (\bar{\Omega}^+)$	1672·5	$-3, (+3)$	$1{\cdot}3\times10^{-10}$	$\Lambda K, \Xi\pi$

[a] The two T_z values refer to the two particles in Column 5.

9 Fundamental constants

The constants and conversion factors are taken from the article on this subject by B. N. Taylor, W. H. Parker and D. N. Langenberg, *Rev. mod. Phys.*, **41**, 375, 1969. Errors in the constants, usually not more than a few parts per million, have been omitted. Symbols are generally in accord with the recommendations of the report 'Signs, Symbols and Abbreviations', *The Royal Society*, London, 1969.

The permeability of the vacuum, $\mu_0 (= 1/c^2\varepsilon_0)$ is taken to be $4\pi \times 10^{-7}$ J s^2 C^{-2} m^{-1} exactly and the permittivity of the vacuum ε_0 is then $(8\cdot85419 \pm 0\cdot00002) \times 10^{-12}$ J^{-1} C^2 m^{-1}.

The atomic mass scale is based on the mass of the isotope ^{12}C (Sect. **10.1**). A comprehensive table of atomic masses is given by J. H. E. Mattauch *et al.*, *Nucl. Phys.*, **67**, 1, 1965.

Quantity	Symbol	Value
General		
Velocity of light	c	$2\cdot99793 \times 10^8$ m s^{-1}
Avogadro's number	N_A	$6\cdot02217 \times 10^{23}$ mol^{-1}
Electron charge	e	$1\cdot60219 \times 10^{-19}$ C
Faraday constant	$F = N_A e$	$9\cdot64867 \times 10^4$ C mol^{-1}
Electron rest mass	m_e	$9\cdot10956 \times 10^{-31}$ kg
Electron charge to mass ratio	e/m_e	$1\cdot75880 \times 10^{11}$ C kg^{-1}
Classical electron radius	$r_0 = \dfrac{\mu_0 e^2}{4\pi m_e}$	$2\cdot81794 \times 10^{-15}$ m
Thomson cross-section	$\frac{8}{3}\pi r_0^2$	$6\cdot65245 \times 10^{-29}$ m^2
Gas constant	R	$8\cdot31434$ J mol^{-1} K^{-1}
Boltzmann constant	$k = R/N_A$	$1\cdot38062 \times 10^{-23}$ J K^{-1}
Planck constant	h	$6\cdot62620 \times 10^{-34}$ J s
	$\hbar$	$1\cdot05460 \times 10^{-34}$ J s
Compton wavelength of electron	$h/m_e c$	$2\cdot42631 \times 10^{-12}$ m
Compton wavelength of proton	$h/m_p c$	$1\cdot32144 \times 10^{-15}$ m
Atomic and nuclear masses		
Atomic mass unit	a.m.u.	$1\cdot66053 \times 10^{-27}$ kg
Atomic mass of electron	—	$5\cdot48593 \times 10^{-4}$ a.m.u.
Atomic mass of proton	M_p	$1\cdot007277$ a.m.u.
Atomic mass of hydrogen atom	M_H	$1\cdot007825$ a.m.u.
Ratio of proton mass to electron mass	—	$1836\cdot11$
Atomic mass of neutron	M_n	$1\cdot008665$ a.m.u.
Rest mass of proton	m_p	$1\cdot67261 \times 10^{-27}$ kg
Rest mass of neutron	m_n	$1\cdot67492 \times 10^{-27}$ kg
Neutron–H atom mass difference	$M_n - M_H$	$0\cdot782$ MeV

Nuclear physics

Quantity	Symbol	Value
Spectroscopic constants		
Rydberg constant	$R_\infty = \dfrac{m_e e^4}{8h^3 \varepsilon_0^2 c}$	$1 \cdot 09737 \times 10^7 \ \text{m}^{-1}$
Bohr radius	$a_0 = \dfrac{h^2 \varepsilon_0}{\pi m_e e^2}$	$5 \cdot 29177 \times 10^{-11} \ \text{m}$
Fine structure constant	$\alpha = \dfrac{\mu_0 e^2 c}{2h}$	$7 \cdot 29735 \times 10^{-3}$
Magnetic quantities		
Bohr magneton	$\mu_B = \dfrac{eh}{4\pi m_e}$	$9 \cdot 27410 \times 10^{-24} \ \text{A m}^2$
Nuclear magneton	$\mu_N = \dfrac{eh}{4\pi m_p}$	$5 \cdot 05095 \times 10^{-27} \ \text{A m}^2$
Electron magnetic moment	μ_e	$9 \cdot 28485 \times 10^{-24} \ \text{A m}^2$
Proton magnetic moment	μ_p	$1 \cdot 41062 \times 10^{-26} \ \text{A m}^2$
Gyromagnetic ratio of proton (corrected for diamagnetism of water)	γ_p	$2 \cdot 67520 \times 10^8 \ \text{rad s}^{-1} \ \text{T}^{-1}$
Conversion factors		
1 a.m.u.		$\begin{cases} 1 \cdot 66053 \times 10^{-27} \ \text{kg} \\ 931 \cdot 481 \ \text{MeV} \end{cases}$
1 electron mass		$0 \cdot 511004 \ \text{MeV}$
1 proton mass		$938 \cdot 259 \ \text{MeV}$
1 eV		$\begin{cases} 1 \cdot 60219 \times 10^{-19} \ \text{J} \\ 2 \cdot 41797 \times 10^{14} \ \text{Hz} \\ 8 \cdot 06547 \times 10^5 \ \text{m}^{-1} \\ 1 \cdot 16049 \times 10^4 \ \text{K} \end{cases}$
Energy–wavelength conversion		$1 \cdot 23985 \times 10^{-6} \ \text{eV} \times \text{m}$

Problems

Chapter 1

1.1 A neutron beam is diffracted by a lattice containing planes separated by 3×10^{-10} m in such a way that the incident and diffracted beams are both inclined at 30° to these planes. Calculate the energy of the neutrons in electron volts. (University of Cambridge 1946)

1.2 Calculate the de Broglie wavelength of:
(a) an electron of energy 1 MeV, [$8\cdot8 \times 10^{-13}$ m]
(b) a proton of energy 1000 MeV, [$7\cdot3 \times 10^{-16}$ m]
(c) a nitrogen nucleus of energy 140 MeV. [$6\cdot5 \times 10^{-16}$ m]

1.3 Express the gravitational force on an electron in terms of (a) an electric field, (b) a transverse magnetic field, for an electron velocity of 100 m s^{-1}. [$5\cdot5 \times 10^{-11}$ volts m^{-1}; $5\cdot5 \times 10^{-13}$ T]

1.4 'The approximate strength of the nuclear force in the attractive region is comparable to the attraction between two opposite charges of magnitude 3 e' (Weisskopf, Ref. 1.16). Using a dimension of $2 \times 10 >^{15}$ m calculate the corresponding nuclear potential energy (6·5 MeV).

Chapter 2

2.1 If two radioactive nuclei A and B, produced in the process of nuclear fission, are characterized by the disintegration constants λ_1 and λ_2, and if the probability that a time less than T elapses between the subsequent disintegrations of A and B is represented by $W(T)$, show that

$$W(T) = 1 - \frac{1}{\lambda_1 + \lambda_2} (\lambda_1 e^{-\lambda_2 T} + \lambda_2 e^{-\lambda_1 T})$$

(University of Cambridge 1942)

2.2 A sample of uranium oxide (U_3O_8), freshly prepared from an old uranium mineral, was found to emit 20·5 α-particles per milligram per second. Comment on this result. (U = 238, O = 16, $\lambda^{238}U = 4\cdot8 \times 10^{-18}$ s^{-1}). (University of Cambridge, 1944)

2.3 What proportion of ^{235}U was present in a rock formed 3000×10^6 years ago, given that the present proportion of ^{235}U to ^{238}U is 1/140? (University of Liverpool, 1954)

2.4 An atom of mass 226·10309 a.m.u. is α-active, decaying to one of mass 222·09397 a.m.u. Estimate the velocity of the emitted α-particle given that the mass of the helium atom is 4·00260 a.m.u. and $c = 3 \times 10^8$ m s^{-1}. (University of Hull, 1957)

2.5 The maximum permissible concentration of ^{24}Na (half-life 15 h) in water is 8 μCi l^{-1}. It is necessary to dispose of an effluent containing 200 μCi l^{-1} produced continuously at the rate of 30 l h^{-1} by feeding it continuously through a holding tank. Assuming perfect mixing, what would be the minimum capacity of the tank if only a further sixfold dilution of the overflow may be safely assumed? (University of Birmingham, 1959)

2.6 Calculate the relative proportions of the nuclides ^{238}U, ^{226}Ra and Rn in an ore in which equilibrium has been reached. [4.3×10^{11}; 1.6×10^5; 1]

2.7 Given that the lifetimes of the uranium isotopes ^{238}U and ^{234}U are 4.5×10^9 years and 2.6×10^5 years calculate the relative abundance of these two nuclides in nature, assuming that one is produced by decay from the other. [0·006%]

2.8 If one gramme of natural chlorine was bombarded by 10^7 thermal neutrons s^{-1} m^{-2} for 10^6 years, calculate the resultant activity of each of the radioactive isotopes of chlorine from the data given below:

(Atomic weight of chlorine = 35·5; natural chlorine contains 75·4 per cent of the isotope of mass number 35 and 24 per cent of the isotope of mass number 37; the cross-sections for thermal neutron capture are 44×10^{-28} m^2 for ^{35}Cl and 0.56×10^{-28} m^2 for ^{37}Cl. The half-life of ^{36}Cl is 4.4×10^5 yr; the half-life of ^{35}Cl is 38·5 min. Avogadro's number is 6.0×10^{23} per mole.) (University of Keele, 1963)

2.9 Three bodies A, B, C of a radioactive series have decay constants λ_a, λ_b, λ_c and the period of A is very long compared with that of B and of C. Show that in time t after the separation of B, short compared with the lifetime of B and C, the quantity of C formed from A is given by

$$C = \tfrac{1}{2}\lambda_b \lambda_c C_0 t^2$$

where C_0 is the ultimate equilibrium amount of C.

2.10 Show that if two consecutive radiations from a long-lived parent have decay constants λ_1 and λ_2, the apparent mean life of the second radiation is $(1/\lambda_1) + (1/\lambda_2)$.

2.11 Show that if the delayed coincidence method (p. 337) is applied to a radioactive sequence A $\longrightarrow$ B $\longrightarrow$ C where the decay constants of the disintegrations of B and C are each λ, then the delayed signals from C, with respect to the formation of B, are distributed according to the formula $\lambda^2 t e^{-\lambda t}$

Chapter 3

3.1 Calculate the Zeeman displacement per tesla for the *s*-state of hydrogen. [46·7 m^{-1} T^{-1}]

3.2 Derive an expression for the orbital frequency of an electron in a Bohr orbit with principal quantum number n. Show that the frequency of radiation emitted according to Bohr's postulate in a transition between states n_1 and n_2 approaches the orbital frequency when n_1 and n_2 are both large.

Chapter 4

4.1 One of the lines of the hydrogen spectrum of wavelength 486·2 nm is accompanied by a weak line due to deuterium at a wave length which is shorter by 0·1313 nm. Show that this observation is consistent with the known mass of the deuterium atom.

4.2 The frequency of the hyperfine transition $(F=0 \longrightarrow F=1)$ in the hydrogen atom is 1426×10^6 Hz. Estimate the average distance between the electron and proton in the atom, assuming only the values of h, the Bohr magneton and the magnetic moment of the proton. [0·038 nm]

4.3 Estimate the frequency at which nuclear magnetic resonance might take place in atomic hydrogen in free space, given the following information:

Magnetic moment of the proton $= 1·4 \times 10^{-26}$ A m^2
Magnetic moment of the electron $= 9·27 \times 10^{-24}$ A m^2
Radius of the first Bohr orbit in hydrogen atom $= 0·53 \times 10^{-10}$ m. [530 MHz]

4.4 Calculate the Stern–Gerlach (angular) deflection in a field of gradient 500 T m^{-1} and length 0·10 m for:
(*a*) a beam of sodium atoms,
(*b*) a beam of neutrons, each of thermal velocity $(kT=0·025$ eV). [2·2°, 4×10^{-5} radians]

4.5 How many magnetic substates are expected for an orthohydrogen molecule in its ground state, assuming that spin and orbital motion are decoupled? [9]

Calculate the maximum deflection experienced by such a molecule when it passes through an inhomogeneous field of length 0·05 m and gradient 500 T m^{-1} at a velocity of 500 m s^{-1}. [$2·48 \times 10^{-5}$ m]

4.6 Show that if the proton may be represented as a uniform object with angular momentum quantum number $s = \frac{1}{2}$, the absolute magnetic moment is

$$\tfrac{1}{2}\sqrt{3}\, \mu_N.$$

Chapter 5

5.1 An α-particle $(M_1=4)$ passes through the gas of an expansion chamber and is scattered through an angle of 56° by collision with a heavier nucleus, which is observed to move off at an angle of 54° with the original direction of the α-particle. Find the probable mass number of the struck nucleus $(M_2 \approx 12)$.

5.2 Show that in the collision of an α-particle $(M_1=4)$ with a proton $(M_2=1)$ the maximum angle of scattering is 14° 30'.

5.3 Using symbols as in Fig. 5.1 draw a momentum diagram for the laboratory system and show that

$$v_2 = v_1 \frac{M_1}{M_2} \frac{\sin \theta_L}{\sin (\theta_L + \phi_L)}$$

5.4 A target of lithium ($M_2 = 7$) is bombarded by protons ($M_1 = 1$) of energy 5 MeV. Calculate:
(*a*) the energy of protons scattered elastically through an angle of $90°$, [3·75 MeV]
(*b*) the energy of protons observed at $90°$ which have excited the 0·48 MeV state of ^{7}Li. [3·33 MeV]

5.5 A particle M_1 of energy E_1 bombards a nucleus M_2 producing particles M_3 and M_4, with an energy release Q. Show that if particles M_3 are observed at angle θ, their energy E_3 is given by the equation:

$$(M_3 + M_4)E_3 - 2\sqrt{M_1 M_3 E_1 E_3} \cos \theta = (M_4 - M_1)E_1 + M_4 Q$$

5.6 Calculate approximately the number of primary pairs of ions per millimetre path produced by a proton of energy 1 MeV in helium at N.T.P. (assume $I_0 = 24·5$ eV). [26]

Suggest how formula (5.43) might be modified to predict the ionization produced by a 1 MeV electron and estimate a value for this, also in helium at N.T.P. [0.055]

5.7 Show that, except for small ranges, the straggling of a beam of ^{3}He particles is greater than that of a beam of ^{4}He particles of equal range. (University of Cambridge, 1940)

5.8 Calculate the closest distance of approach of an α-particle of energy 6 MeV to a gold nucleus ($Z = 79$). [$3·8 \times 10^{-14}$ m]

For what angle of scattering would an α-particle of energy 9 MeV approach the nucleus to the same distance? [$60°$]

5.9 A finely collimated beam of α-particles of energy 10 MeV bombards a metal foil (thickness 2 g m^{-2}) of a material of atomic weight 107·8. A detector of area 10^{-4} m^2 is placed at a distance of 0·1 m from the foil so that scattered α-particles strike it normally on the average at an angle of $60°$ with the direction of incidence. When the incident α-particles represent a current of 1 μA the counting rate in the detector (which is 100 per cent efficient) is 6×10^4 particles s^{-1}. Calculate the atomic number of the element forming the scattering foil. [46]

5.10 An α-particle is found to have a range of 300 μm in a photographic emulsion. What range would you expect for (*a*) a ^{3}He nucleus, (*b*) a ^{3}He nucleus each of the same initial velocity as the α-particle? [225 μm; 900 μm]

5.11 The α-particles from ThC′ have an initial energy of 8·8 MeV and a range in standard air of 8·6 cm. Find their energy loss per cm in standard air at a point 4 cm distant from a thin source. (University of Liverpool, 1954)

5.12 A triton (^{3}He) of energy 5000 MeV passes through a transparent medium of refractive index $n = 1 \cdot 5$. Calculate the angle of emission of Cherenkov light. [44°] Calculate also the number of photons emitted per m path between wavelengths 400 nm and 600 nm using, e.g. 5·51. [19,500]

5.13 A proton enters a rectangular block of glass of refractive index 1·47 along a line perpendicular to one pair of faces. For what energy will the Cherenkov light be totally reflected at the opposite face? [1600 MeV]

5.14 Discuss the significance of the relation of the mass absorption coefficient for soft X-rays in light elements ($0 \cdot 02^{-1}$ m^2 kg^{-1}) to the Thomson scattering cross-section $(8\pi/3)r_0^2$.

5.15 The range of protons in C2 emulsion is given in the following table (range in µm, energy in MeV).

Range	0	50	100	150	200	250	300	350	400	450	500
Energy	0	2·32	3·59	4·61	5·48	6·27	7·01	7·69	8·32	8·91	9·47

Draw graphs of the range–energy relations for deuterons and helium 3 particles. (University of Birmingham, 1961)

5.16 Show that the differential cross-section $\sigma_L(\theta_L)$ for scattering of protons by protons in the laboratory system is related to the corresponding quantity in the centre of mass system by the equation: $\sigma_L(\theta_L) = 4 \cos \theta_L \sigma(\theta)$. (University of Birmingham, 1962)

5.17 Calculate the number of photons produced per m of air ($n = 1 \cdot 000293$) by the Cherenkov effect from the path of a relativistic electron in the wavelength range $\lambda = 350$ to 550 nm. [27]

Calculate also the energy below which Cherenkov radiation is not observed. [20·3 MeV]

5.18 A beam of X-rays is attenuated by a factor of 0·64 in passing through a block of graphite ($A = 12$) of thickness 0·01 m. Estimate the atomic number of carbon from this information.

5.19 The cross-section for the reaction:

$$^{10}B + n \longrightarrow {}^7Li + {}^4He$$

is 4×10^3 barn. Calculate the fraction of a ^{10}B layer which disappears in a year in a flux of 10^{16} neutrons m^{-2} s^{-1}. [0·12]

5.20 Calculate for the Compton scattering of X-rays of wavelength 0·01 nm:
(*a*) the wavelength of scattered radiation at 45°. [0·0107 nm]
(*b*) the velocity of the corresponding recoil electron. [$5 \cdot 3 \times 10^7$ m s^{-1}]

5.21 The cross-section for two-quantum annihilation of non-relativistic positrons may be written

$$\sigma = \pi r_0^2 \frac{c}{v}$$

where v is the positron velocity and r_0 is the classical electron radius. Calculate the annihilation rate in carbon ($Z = 6$, $A = 12$, density $= 2 \cdot 22 \times 10^3$ kg m^{-3}) and the mean life in this material. [$R = 4 \cdot 9 \times 10^9$ s^{-1}, $2 \cdot 04 \times 10^{-10}$ s]

5.22 Using eq. (5.37) for a positron of velocity 10^8 m s^{-1}, and the result of problem 5.21, compare the rate of energy loss by ionization in carbon with the rate of loss by annihilation. [Annihilation/ionization $\approx 1/1600$]

5.23 Show that in the annihilation of positrons at rest by negative electrons of momentum p the angle between the two annihilation quanta is $\pi - p_{\perp r}/mc$ where $p_{\perp r}$ is the momentum of the electron at right angles to the line of flight of one of the quanta.

Show further that the energy of the two quanta is $mc^2 \pm \frac{1}{2} p_{\parallel} c$ approximately where $p_{\parallel}$ is the momentum of the electron along the line of flight of one of the quanta. Calculate the angle and the energy difference between the quanta for annihilation in a solid in which the average electron energy is 10 eV. [$\pi - 0 \cdot 36°$, 3200 eV]

5.24 Show that the threshold quantum energy for the production of an electron–positron pair in the field of a free electron is $4mc^2$. (Calculate the invariant mass $E^2 - p^2 c^2$ for the initial system using relativistic formulae and set it equal to $3\,mc^2$.)

5.25 In the elastic collision of a particle of mass M_1 with a particle of mass $M_2\,(<M_1)$, the heavier particle cannot be scattered through an angle greater than θ_L^m. Show that

$$\sin \theta_L^m = \frac{M_2}{M_1}$$

and that the corresponding angle in the centre-of-mass system is given by

$$\cos \theta^m = -\frac{M_2}{M_1}$$

Show also that the corresponding laboratory angle of projection of M_2 is given by

$$\sin \phi_L^m = \sqrt{\frac{M_1 - M_2}{2M_1}}$$

5.26 For the conditions of the previous problem, when the particles M_1 are scattered through the greatest possible angle, show that their energy is

$$E_1 \frac{M_1 - M_2}{M_1 + M_2}$$

where E_1 is the energy before scattering.

5.27 Show that the velocity v_p of protons projected forward by α-particles incident with velocity v_α is given by $v_p = 1 \cdot 6 v_\alpha$.

5.28 In the annihilation of a fast positron of kinetic energy T by a free electron at rest, two quanta are usually produced. Show that when $T \gg mc^2$ the energy of the annihilation radiation observed in the forward direction is

$$E_\gamma = T + \tfrac{3}{2} mc^2$$

5.29 In the slow neutron reaction

$$^{10}B + n \longrightarrow {}^7Li + {}^4He + Q$$

the energy release Q is 2·79 MeV. Calculate the initial velocity of the ^{4}He nucleus. [$9·4 \times 10^6$ m s^{-1}]

Assuming that the reaction takes place in nitrogen at N.T.P., find the initial primary ionization due to the α-particle, taking I_0 (eq. (5.43)) to be 15. [$2·65 \times 10^6$ ion pairs m^{-1}]

Chapter 6

6.1 A cylindrical ionization chamber, filled with BF_3 gas, captures 100 thermal neutrons per second ($^{10}B + n \longrightarrow {}^7Li + {}^4He + 2·8$ MeV). Assuming that all the reaction products are stopped in the gas calculate the ionization current produced. If the anode voltage is increased to give a gas amplification of 11, could the neutron be detected by an amplifier of 50 pF input capacitance and 3 mV sensitivity? (University of Birmingham, 1961)

6.2 The γ-rays from a radioactive source are observed using a scintillation spectrometer and the pulse height spectrum shows peaks at 46·3 V, 39·9 V and 33·5 V, together with a continuum. Assuming that the source emits only one γ-ray, calculate the energy of this γ-ray and the upper limit of the continuum. (Rest energy of the electron $= 0·511$ MeV.) (University of Birmingham, 1957)

6.3 Calculate the electric field at the surface of the wire of a proportional counter with a wire radius $\frac{1}{10}$ mm and a cylinder radius 0·01 m when 1500 V is applied between the two. [3250 kV m^{-1}]

Assuming that ionization by collision begins at a field of 2250 kV m^{-1} and that the mean free path of an electron in the counter is 5×10^{-6} m, calculate the multiplication factor. [512]

6.4 A cylindrical Geiger counter with a wire of diameter 0·2 mm and a cylinder of diameter 0·02 m is operated with a potential difference of 1500 volts. If the mean free path of an electron in the gas filling is 10^{-5} m and if the ionization potential is 15·7 volts, express in mean free paths the radial thickness of the region over which ionization by collision is possible. [$10·7\lambda$]

6.5 Verify the statement on p. 162 that the drop radius corresponding to the maximum of the curve in Fig. 6.14 is 6×10^{-10} m. Use $\gamma = 76 \times 10^{-3}$ N m^{-1} and assume that the drops carry one electronic charge. Show further that the radius of such a drop from which no evaporation to surroundings in equilibrium with a plane liquid surface takes place is about 4×10^{-10} m.

Chapter 7

7.1 Neutrons from the d + d reaction, of energy 4·0 MeV, are detected by an instrument based on recoil protons. If the instrument has a threshold corresponding to protons of 100 keV what fraction of recoil events will not be recorded? (University of Birmingham, 1962)

7.2 Calculate the flight times for 1 keV and 1 eV neutrons for a flight path of 10 m. [22·9, 723 μs]

7.3 Find the value of Bρ (T m) for the following:

(*a*) an α-particle of energy 8 MeV (He^{++}), [0·405]

(*b*) an electron of energy 5 MeV, [0·018]

(*c*) a proton of energy 100 MeV, [1·5]

(*d*) an oxygen ion of energy 160 MeV (^{16}O^{4+}). [1·8]

7.4 By using arguments similar to those on pp. 188–189 show generally that in the case of the scattering of a neutron by a nucleus, the energy distribution of the recoil nuclei in the laboratory system has the same functional form as the angular distribution of the scattered neutron in the centre-of-mass system.

7.5 In the counting of charged particles discrimination between particles of different mass may be obtained if the signals from a thin counter (giving dT/dx) and a stopping counter (giving T) are multiplied. Show that the value of $T(dT/dx)$ is to a first approximation independent of velocity, and find the ratio of this quantity for deuterons and ^{3}He particles. [1/6]

7.6 A time-of-flight spectrometer has a pulse width Δt and a flight path l. Show that the energy spread ΔE at mean energy E, corresponding to this pulse width, is proportional to $E^{3/2} \Delta t$. Evaluate the energy resolution of such a spectrometer for neutrons of energy 100 eV if $\Delta t/l = 0·1$μs m^{-1}. [2·8 per cent]

Chapter 8

8.1 The peak potential difference between the dees of a cyclotron is 25 000 V and the magnetic field is 1·6 T. If the maximum radius is 0·3 m, find the energy acquired by a proton in electron volts and the number of revolutions in its path to the extreme radius. [11·0 MeV, 220]

8.2 Calculate the velocity of a proton of energy 10 MeV as a fraction of the velocity of light. [0·15] How long would a proton take to move from the ion source to the target in a uniform 10 MeV accelerating tube 3 metres long? [1·4 × 10^{-7} s]

8.3 A pulse of 10^{10} particles of single charge is injected into a cyclic accelerator and is kept circulating in a stable orbit by the application of a radiofrequency field. What is the mean current when the radiofrequency is 7 MHz? [11.2 mA]

8.4 A magnetic field of 1·2 T is being explored by the 'floating wire' method. If the current in the wire is 2 amperes and the tension is set at 0·5 kg weight, what is the energy of the proton whose trajectory is followed by the wire? [252 MeV]

8.5 Assuming that the output voltage drop in a cascade generator of n stages is given by:

$$\Delta V = \frac{2}{3} n^3 \frac{i}{fC}$$

derive an expression for the *n*-value for which the terminal voltage is a maximum.

Calculate:

(*a*) the number of stages,

(*b*) the output voltage,

(*c*) the ripple voltage $\frac{1}{2}n(n+1)\ i/fC$

for a cascade generator with $C=0.02$ μF, $f=200$ Hz, transformer peak voltage 110 kV, $i=4$ mA. [11, 1533 kV, 66 kV]

8.6 A belt system of total width 0·3 m charges the electrode of an electrostatic generator at a speed of 20 m s^{-1}. If the breakdown strength of the gas surrounding the belts is 3 MV m^{-1} calculate:

(*a*) the maximum charging current, [1·6 mA]

(*b*) the maximum rate of rise of electrode potential, assuming a capacitance of 111 pF and no load current. [14·4 MV s^{-1}]

8.7 A cascade generator operating from a 500 Hz mains supply delivers a current of 1 mA at an output voltage of 1 MV. What is the percentage fluctuation when the load is taken from a *RC* circuit consisting effectively of a 5MΩ resistance and a condenser stack of 0·001 μF capacity? [0·064%]

8.8 Calculate the cyclotron frequency for non-relativistic deuterons in a field of 1 T. [7·65 MHz]

Calculate also the frequency at which nuclear magnetic resonance with $\Delta m=2$ would be observed for deuterons at rest in the same field. [13 MHz]

8.9 Protons of energy 100 MeV with their spins aligned in the direction of motion enter a transverse magnetic field. Calculate the ratio between the angle of deviation of the proton and the angle through which the spin is turned because of the magnetic moment of the particle. [0·3]

8.10 Show that the orbit radius of a singly-charged particle of kinetic energy T and rest mass M_0 in a magnetic field B is:

$$R = \frac{T}{eBc}\left(1+\frac{2M_0c^2}{T}\right)^{1/2}$$

Calculate this radius for a proton of kinetic energy equal to its rest mass and a field of 1·25 T. [4·32 m]

8.11 Consider a synchrocyclotron which accelerates protons to an energy of 420 MeV in a field of 1·9 T. Calculate for the energies of 100 and 420 MeV,

(*a*) the velocity of the protons, [$\beta=0.43, 0.72$]

(*b*) the radius of the orbit, [0·78 m, 1·71 m]

(*c*) the mass of the proton, in units of the rest mass. [1·11, 1·45]

8.12 For the synchrocyclotron specified in problem 8.11 calculate:

(*a*) the synchronous frequency for energies 0, 100 and 420 MeV, [29, 26·1, 20 MHz]

(*b*) the time taken for a particle to reach the final energy for an increment of 0·01 MeV per turn, [1770 μs]

(*c*) the maximum repetition rate at which the machine can be operated. [565 Hz]

8.13 Deuterons of energy 15 MeV are extracted from a cyclotron at a radius of 0·51 m by applying an electric field of 6 MV m^{-1} over an orbit arc of 90°. Calculate the equivalent reduction of magnetic field and the resulting increase in orbit radius Δr. [0·16 T, 0·053 m]

8.14 In the previous problem verify that if the electric field is applied for an angle θ, the *maximum* orbit separation is

$$2 \Delta r \sin \theta/2$$

and that this occurs at an angle $\pi/2 + \theta/2$ beyond the entrance to the deflector.

8.15 From the formula

$$\frac{\Delta E}{mc^2} = \frac{4\pi}{3} \frac{r_0}{R} \left(\frac{E}{mc^2}\right)^4$$

where $r_0 = \mu_0 e^2/4\pi m_e$, calculate the energy loss per turn for an electron in a synchrotron in which the electrons are moving with an energy of 5 GeV on a radius of 26 metres. [2·1 MeV]

Calculate also the power rating of the oscillator necessary to supply this loss for a bunch of 10^{12} electrons. [625 kVA]

8.16 In high energy physics, it is customary to measure momenta in the unit MeV/c. Using the relativistic formula

$$E^2 = (T + M_0 c^2)^2 = p^2 c^2 + M_0^2 c^4$$

find the momentum (in MeV/c) of
(*a*) a 1000 MeV proton, [1696]
(*b*) a 1 MeV proton, [43·4]
(*c*) a 10 MeV photon, [10]
(*d*) a 100 MeV electron. [101]

8.18 Calculate the value of β ($= v/c$) for protons of the following energies: 100 MeV, 500 MeV, 1000 MeV. [0·43, 0·76, 0·88]

Chapter 9

9.1 Starting from Schrödinger's equation find the number of bound *s*-states for a particle of mass 2200 electron masses in a square well potential of depth 70 MeV and radius $1·42 \times 10^{-15}$ m ($\hbar/mc = 3·85 \times 10^{-13}$ m, $mc^2 = 0·51$ MeV). (University of Glasgow, 1959)

9.2 Using the relativistic relation between momentum and energy find the minimum kinetic energy of (*a*) an electron, (*b*) a proton confined within a dimension of 7×10^{-15} m (assume $\Delta p \times \Delta r = \hbar$). [28 MeV, 0·42 MeV]

9.3 Show that the kinetic energy of a relativistic electron confined within a box of linear dimension R may be written

$$T = \frac{\hbar^2}{m_0 R^2 (1 + \gamma)}$$

where $\gamma = E/m_0 c^2$ and E is the total energy.

9.4 Show that in a Fermi gas of neutrons, the number of degenerate states with energy less than $\hbar^2 k^2 / 2M$ is $(2/9\pi)(kR)^3$, where R is the linear dimension of the containing volume.

9.5 Write down the expected configuration of the following nuclei, omitting closed shells or subshells:

$$^{27}_{13}\text{Al}; \; ^{29}_{14}\text{Si}; \; ^{40}_{19}\text{K}; \; ^{93}_{41}\text{Nb}; \; ^{157}_{64}\text{Gd}. \quad [d^5_{5/2}; \; s_{1/2}; \; d_{3/2} + f_{7/2}; \; g_{9/2}; \; h^3_{9/2}]$$

9.6 No decay process linking the nuclei $^{150}_{60}\text{Nd}$ and $^{150}_{61}\text{Pm}$ is known. The (atomic) mass difference between the nuclei $^{150}_{60}\text{Nd}$ and $^{150}_{62}\text{Sm}$ has, however, been measured as 3633 ± 4 µu on the carbon scale (McLatchie *et al.*, *Phys. Lett.*, **10**, 330, 1964) and the accepted decay energy for the beta process $^{150}_{61}\text{Pm} \longrightarrow \; ^{150}_{62}\text{Sm}$ is $3\cdot46 \pm 0\cdot03$ MeV. Determine the relative stability of ^{150}Nd and ^{150}Pm and the decay energy. Suggest a possible mode of decay. [^{150}Pm decays to ^{150}Nd with an energy release of 76 ± 30 keV. Probably K-electron capture]

Chapter 10

10.1 In a mass spectrograph comparison of H^+ with He^+ the interval of mass observed corresponded to $73\cdot73$, expressed in the usual units, while a comparison of H_2D^+ with He^+ gave the corresponding interval $269\cdot1$, helium being on the lighter side in each instance. Given that the packing fraction of helium is $5\cdot4$, calculate the packing fractions of H and D. (University of Cambridge, 1943)

10.2 Given that the mass defect curve falls from $+0\cdot14$ mass units for uranium to $-0\cdot06$ mass units in the middle of the periodic table, estimate in kWh the energy which could theoretically be obtained from 1 kg of ^{235}U. [$2\cdot8 \times 10^7$]

10.3 How much energy is necessary to split up an α-particle ($M_\alpha = 4\cdot00260$ a.m.u.) into its constituent nucleons ($M_n = 1\cdot008665$ a.m.u., $M_H = 1\cdot007825$ a.m.u.). [$28\cdot2$ MeV]

10.4 Assuming that the binding energy of an even-A nucleus may be written

$$-B(A, Z) = f(A) + \frac{0\cdot083}{A}\left(\frac{A}{2} - Z\right)^2 + 0\cdot000627\frac{Z^2}{A^{1/3}} \pm 0\cdot036A^{-3/4}$$

($+$ for Z odd, $-$ for Z even) determine the number of stable nuclides of mass $A = 36$. Take $A^{1/3} = 3\cdot3$, $A^{1/4} = 2\cdot45$ and confine your attention to the range $13 \leqslant Z \leqslant 20$. (University of Birmingham, 1961)

10.5 In the carbon atom the K edge is at 300 eV and the L_1, L_{II} edges at 60 eV. The atomic mass is $12\cdot0000$. Calculate:

(a) the mass of the carbon nucleus in a.m.u. neglecting electron binding, [$11\cdot99671$]

(b) the percentage correction introduced when electron binding is allowed for. [10^{-5}]

Chapter 11

11.1 If, in the usual notation, $R = 1.45 \times 10^{-15} A^{1/3}$ m, calculate a value for the maximum kinetic energy of the positrons emitted in the decay

$$^{15}_{8}O \longrightarrow {}^{15}_{7}N + \beta^{+} + \nu$$

indicating any assumption made. (University of Hull, 1958)

11.2 Estimate the energy of the μK X-ray for a phosphorus nucleus $(Z = 15)$. (Ionization potential for hydrogen $= 13$ eV, $m_\mu = 207\ m_e$.) (University of Birmingham, 1962)

11.3 Calculate the radius of the π-mesic atom s-orbit for the nucleus calcium $(Z = 20)$. Assuming a nuclear radius parameter $r_0 = 1.2 \times 10^{-15}$ m, show that the orbit lies outside the nucleus. Calculate the density of nucleons available for interaction and the mean free path corresponding to a cross-section $\pi(\hbar/\mu c)^2$ where $\mu = 273 m_e$. [1.6×10^{-14} m]

11.4 For any given muonic atom, calculate the n-value of the muonic orbit which is just inside the electronic K-shell. [14]

11.5 Write down the equations which describe the scattering of a high energy electron (energy E) by a nucleus of mass M. Show that if $E \gg mc^2$ the recoil energy of the nucleus (treated non-relativistically) is

$$\frac{E^2(1 - \cos\theta)}{Mc^2 + E(1 - \cos\theta)}$$

where θ is the angle through which the electron is scattered.

11.6 Show that the electrostatic potential $U(r)$ at distance r from the centre of a sphere containing a uniform density of positive charge is

$$U(r) = \frac{q}{4\pi\varepsilon_0 R}\left[\frac{3}{2} - \frac{1}{2}\left(\frac{r}{R}\right)^2\right] \quad \text{for } r \leqslant R$$

if q is the total charge in a sphere whose radius is R.

Evaluate the electric field strength for all values of r and show that it is continuous at the boundary $r = R$. (Evans, *The Atomic Nucleus*, p. 45)

Chapter 12

12.1 Suppose that it were conceivable that a photon of energy 100 keV should be emitted from a point on the surface of a nucleus (in fact the photon wavelength makes such a concept unreasonable). Find the Doppler spread of energy in the photon spectrum due to classical rotation of the nucleus with one quantum of angular momentum. Assume a nuclear radius of 7×10^{-15} m and $A = 125$. [0.1 keV]

12.2 The masses of the atoms ^{10}Be, ^{10}B and ^{10}C are 10.0135, 10.0129 and 10.0168 in a.m.u. Calculate in a.m.u. the masses of the bare nuclei of these atoms, as shown relatively in Fig. 12.1(*a*) [10.0115, 10.0104, 10.0138] Adjust the masses of ^{10}Be and ^{10}B with respect to ^{10}B by correcting for the mass difference between the neutron and proton [10·0101, 10·0104, 10·0152], and express the difference between the ground state energies in MeV. [^{10}Be $- ^{10}$B $= -0·3$ MeV, ^{10}C $- ^{10}$B $= 4·47$ MeV]

Now use the expression for Coulomb energy change given on p. 286 (with $\frac{3}{5}(e^2/4\pi\varepsilon_0 r_0) = 0·58$ MeV), to calculate the true excitation of the ground states of ^{10}Be and ^{10}C with respect to ^{10}B. This should correspond to the position of the first $T = 1$ level of ^{10}B as indicated in Fig. 12.1(*b*). [2·1 MeV, 1·5 MeV]

Chapter 13

13.1 A nucleus X emits a beta-particle forming a residual nucleus Y in an excited state of energy 250 keV. What is the minimum energy of the beta transition necessary to permit the gamma radiation from Y to be absorbed resonantly in an external nucleus Y at rest? Neglect thermal motions and natural line widths and assume that the gamma radiation is emitted before the nucleus Y loses any energy by collision. [0·56 MeV]

13.2 In an experiment to determine the lifetime of an excited nucleus by the recoil-distance method, the counting rate was found to decrease by a factor of 2 for a source displacement of 0·07mm. If the mean lifetime of the decaying state is 7×10^{-11} s, find the velocity of recoil. [1·4 × 10^6 m s^{-1}]

13.3 The 14·4 keV γ-ray transition in ^{57}Fe has been extensively investigated using the Mössbauer effect (see, for example, Hanna *et al.*, *Phys. Rev. Lett.*, **4**, 177, 1960) and it is found that the ($I = \frac{1}{2}$) ground state is split with an energy difference corresponding to a source (Doppler shift) velocity of 3·96 mm s^{-1}. Calculate the internal magnetic field at the ^{57}Fe nucleus, assuming that the ground state magnetic moment is 0·0955 nuclear magnetons. [31·5 T]

Chapter 14

14.1 Neutrons are produced by bombarding a heavy hydrogen compound with 0·9 MeV deuterons. Calculate the energy of the neutrons emitted from the target at an angle of 115° to the bombarding beam and show that the energy spectrum at this angle is practically independent of target thickness. (Energy released in the reaction is 2·2 MeV.) (University of Cambridge, 1946)

14.2 An aluminium foil was bombarded with the 7·3 MeV α-particle beam from a cyclotron. Observation of protons emitted in the resulting reaction at 90° to the incident beam revealed the presence of groups with energies 9·24, 6·98, 5·55 and approximately 4·4 MeV. No γ-rays were observed in coincidence with the highest energy group. State the reaction equation and from the available data derive the Q-values. (University of Hull, 1960)

14.3 In the photodisintegration of the deuteron, the neutron and proton are in general projected with unequal energies. Deduce an expression showing how the kinetic energy of the proton depends on its angle of projection (with respect to the direction of incidence of the quantum) and calculate the angle for the special case in which proton and neutron have the same energy. Assume equal masses for the heavy particles. (University of Cambridge, 1941)

14.4 A thin hydrogenous target is bombarded with 5 MeV neutrons, and a detector is arranged to collect those protons emitted in the same direction as the neutron beam. The neutron beam is replaced by a beam of γ-rays; calculate the photon energy needed to produce protons of the same energy as with the neutron beam. (University of Birmingham, 1959)

14.5 Calculate the mass of the neutron from the following data: Threshold for the reaction

$$^2H(\gamma, n)^1H = 2\cdot225 \pm 0\cdot002 \text{ MeV.}$$

Mass spectrometer doublet $2^1H_1^+ - {}^2H^+ = 1\cdot5480 \pm 0\cdot0021 \times 10^{-3}$ a.m.u. $M_H = 1\cdot007825 \pm 0\cdot000003$ a.m.u. $[M_n = 1\cdot008666]$

14.6 The reaction $^{34}S(p, n)^{34}Cl$ has a threshold at a proton energy of $6\cdot45$ MeV. Calculate (non-relativistically) the threshold for the production of ^{34}Cl by bombarding a hydrogenous target with ^{34}S ions. [219·3 MeV]

14.7 Neutrons are produced by bombarding a target nucleus of mass number A with protons. The threshold energy for the reaction in the laboratory system is E_T. Show that when the incident energy increases above E_T by a small amount δ, the neutrons are emitted into a forward cone of semi-angle $A\sqrt{\delta/E_T}$. Calculate this angle for $A = 26$, $E_T = 5$ MeV and $\delta = 1$ keV. [21°] Show also that the neutron energy at threshold is $E_T/(A+1)^2$ and evaluate this for the case quoted. [6·9 keV] Show further that neutrons are just observed at 180° when the incident energy reaches $E_T(A^2/A^2 - 1)$ and evaluate this for the case quoted. [5007·4 keV]

14.8 Calculate the quantum energy which would be necessary, according to the hypothesis of a Compton effect, to produce (*a*) recoil protons of energy $5\cdot7 \times 10^6$ eV, [55 MeV], (*b*) recoil nitrogen nuclei of velocity 4×10^6 m s^{-1}. [90 MeV]

14.9 Calculate the perpendicular distance of the line of flight of a 10 MeV proton from the centre of a nucleus when it has an orbital angular momentum, with respect to this point, of $3\hbar$. Calculate also the distance for a photon of energy 5 MeV and angular momentum $\hbar$. [$4\cdot3 \times 10^{-15}$ m, $3\cdot9 \times 10^{-14}$ m]

14.10 The nuclear disintegration $^{12}C(n, n') 3^4He$ is observed in a nuclear emulsion, and in one particular event the three α-particles (energies E_1, E_2, E_3) are coplanar. If the energy release in the reaction is Q show that the kinetic energy of the incident neutron is given by

$$\frac{1}{8Mp^2} [2M(E_1 + E_2 + E_3 - Q) + p^2 + q^2]^2$$

where p and q are the total momenta of the α-particles parallel and perpendicular to the direction of the incident particle.

Chapter 15

15.1 A photon cannot be absorbed completely by a free electron, since this particle cannot exist in states of excitation. A complex particle of mass M may, however, absorb the photon. Show that the energy of excitation of M is

$$Mc^2 \left\{ \sqrt{1 + \frac{2h\nu}{Mc^2}} - 1 \right\}$$

where $h\nu$ is the photon energy.

15.2 Show that in the scattering of a particle M_1 by target nucleus M_2 the linear momentum transfer to the nucleus M_2 is the same in both the laboratory and centre-of-mass systems of coordinates.

Calculate this momentum transfer for the scattering of 65 MeV α-particles by ^{16}O nuclei in the case that the angle at which the recoil nucleus is projected is 60°. Express the result (*a*) in kg m s^{-1} and (*b*) as an inverse length, by dividing by $\hbar$. [2.95×10^{-19} kg m s^{-1}; 2.81×10^{15} m^{-1}]

15.3 If a target nucleus has mass number 24 and a level at 1·37 MeV excitation, what is the minimum proton energy required to observe scattering from this level? (University of Birmingham, 1957)

15.4 A lithium target is bombarded with homogeneous protons of controllable energy. If a sharp rise in the yield of radiation from the process

$$^{7}\text{Li} + {^{1}\text{H}} \longrightarrow {^{8}\text{Be}} + \gamma$$

is observed at a proton energy of 441 keV calculate the excitation energy of the corresponding resonance level in ^{8}Be. (^{7}Li $= 7.0160$, ^{1}H $= 1.0078$, ^{8}Be $= 8.0053$, 1 a.m.u. $= 931$ MeV.) (University of Birmingham, 1956)

Chapter 16

16.1 The maximum energy E_{max} of the electrons emitted in the decay of the isotope ^{14}C is 0·156 MeV. If the number of electrons with energy between E and $E + dE$ is assumed to have the approximate (non-relativistic) form:

$$n(E)\, dE \propto E^{1/2}(E_{max} - E)^2\, dE,$$

find the rate of evolution of heat by a source of ^{14}C emitting 3.7×10^7 electrons per second. (University of Cambridge, 1953)

16.2 Calculate the maximum energy of the positron spectrum associated with the decay of $^{13}_{7}$N to $^{13}_{6}$C. Assume that the atomic masses of $^{13}_{7}$N and $^{13}_{7}$C are 13·00574 and 13·00335 a.m.u. respectively, that the mass of an electron is 5.5×10^{-4} a.m.u. and that 1 a.m.u. is equivalent to 931 MeV. (University of Hull, 1958)

16.3 ^{37}A decays by electron capture with a Q value of 0·82 MeV; the recoil energy of ^{37}Cl produced by this decay is 9.7 ± 0.8 eV. Show that these data are consistent with a zero rest mass for the neutrino.

16.4 Calculate the maximum electron and proton energies in the decay of the neutron. What is the energy of the proton when the electron has half its maximum energy? (^{1}n $= 1.008665$, ^{1}H $= 1.00783$, e $= 5.486 \times 10^{-4}$ a.m.u., 1 a.m.u. $= 931$ MeV.) (University of Birmingham, 1956)

16.5 A radioactive isotope of copper has two stable neighbouring isobars. The one is nickel, with atomic number 28 and atomic mass 63.9297 a.m.u. The other is of zinc with atomic number 30 and the atomic mass 63.9332 a.m.u. The mass of an electron is 0.00055 a.m.u. [0.512 MeV]

The following radiations are observed from a sample of this isotope of copper:

(*a*) negative beta particles of 0.57 MeV,
(*b*) positive beta particles of 0.66 MeV,
(*c*) gamma rays of 1.35 MeV,
(*d*) X-rays that are characteristic of nickel.

Coincidence counting experiments (with a resolving time of 10^{-6} s) show coincidences only between the X-ray and the gamma rays.

Deduce what you can about the decay scheme. (University of Keele, 1963)

16.6 The reaction ^{34}S(p, n)^{34}Cl has a threshold at a laboratory proton energy of 6.45 MeV. Calculate (non-relativistically) the upper limit of the positron spectrum of ^{34}Cl, assuming $mc^2 = 0.51$ MeV, n $-^1$H $= 0.78$ MeV. [4.45 MeV]

16.7 The nucleus ^{11}C disintegrates mainly by the emission of a positron, but electron capture is also possible, with an energy release of 2 MeV. Write down the equation for the electron capture process, and calculate the initial energy of the recoil nucleus, assuming that the energy of the K-edge in ^{11}B is 187 eV. [^{11}C(e$^-$, ν)^{11}B, 381 eV]

16.8 At what distance from the centre of a ^{238}U nucleus is the α-particle of its radioactive decay released with zero kinetic energy? The disintegration energy is 4.27 MeV. [6.1×10^{-14} m]

16.9 A nucleus of mass M captures an electron and the resulting energy release is Q. Show that the nucleus recoils with a kinetic energy of approximately

$$\frac{Q^2 - m_\nu^2\, c^4}{2(Mc^2 + Q)}$$

where m_ν is the mass of the neutrino (which has been shown to be ≈ 0 by a measurement of the recoil energy in the case of ^{7}Li).

16.10 On the assumption that the energy distribution in a low-energy allowed β-spectrum may be approximated by the formula given in problem (16.1) show that the mean kinetic energy of the spectrum is $\frac{1}{3}$ of the maximum energy.

Show also that for a high-energy allowed spectrum, if the electron rest mass may be disregarded, the mean energy of the spectrum is $\frac{1}{2}$ of the maximum.

16.11 If two α-emitting nuclei, with the same mass number, one with $Z=84$ and the other with $Z=82$, had the same decay constant and if the first emitted α-particles of energy 5·3 MeV, estimate the energy of the α-particles emitted by the second. (University of Birmingham, 1964)

16.12 In the so-called ξ-approximation, sometimes used in beta-particle theory, it is assumed that the Coulomb energy of an electron at the nuclear radius is very much greater than the end-point energy of the beta spectrum. Evaluate the ratio of this Coulomb energy to end-point energy for the cases of RaC [β-end point 3·18 MeV] and $^8_3$Li (β end-point 12 MeV) assuming that the nuclear radius is given by $1·2 \times 10^{-13} A^{1/3}$. [5.2, 0·15]

16.13 The nucleus of ThC ($^{212}_{83}$Bi) emits an α-particle of 6·06 MeV energy, leaving the residual nucleus in a 40 keV excited state. The resulting electromagnetic transition leads to internal conversion and Auger electrons. Calculate the difference between the Auger electron energy observed at an angle of 0° and 180° with the direction of emission of the α-particle, assuming that the energy of the unshifted line is 7·6 keV. [2·5%]

Chapter 17

17.1 Assuming the expression given on p. 465 for the angular correlation between the two photons in the transition $0 \longrightarrow 1 \longrightarrow 0$, show that the most probable angle between these two photons is 55°.

17.2 A beam of protons of intensity I and polarization P is to be used in a study of the left–right asymmetry produced by elastic scattering. Show that it is desirable to choose conditions such that $P^2 I$ is a maximum.

Chapter 18

18.1 A particle of mass M and energy E is scattered by a square well potential of depth V_0 and radius a. Show that the s-wave phase shift is given by

$$\tan \delta = \frac{k}{k'} \tan k'a - \tan ka$$

where

$$k^2 = \frac{2ME}{\hbar^2} \qquad k'^2 = \frac{2M}{\hbar^2}(E+V_0)$$

Derive an expression for the cross-section when E is small. (University of Glasgow, 1958)

18.2 Using the tabulated values of the magnetic moment of the neutron and proton, calculate the force between these two particles in a triplet state at a separation of 3×10^{-15} m and the work required, on account of this force, to bring the neutron from infinity to this distance from the proton. Assume that the spins always point along the line joining the particles. [1 N, 6250 eV]

Chapter 19

19.1 A gamma ray interacts with a stationary proton and produces a neutral pion. Show that the threshold energy of the gamma ray is E_γ where

$$E_\gamma = m_\pi c^2 \left(1 + \frac{m_\pi}{2m_p}\right)$$

(University of Birmingham, 1969)

19.2 Pairs of protons and antiprotons can be produced in the bombardment of stationary protons by proton beams (*a*) directly by the reaction $p+p \longrightarrow p+p+p+\bar{p}$ and (*b*) by first producing π-mesons and then a secondary collision of the meson with another stationary proton, $\pi + p \longrightarrow p+p+\bar{p}$. Taking the mass of the π-meson to be one-seventh of the mass of the proton (or antiproton) find the least energy of the incident proton for which either reaction becomes possible. (University of Birmingham, 1960)

19.3 The pion of a π-mesic atom when absorbed by a proton at rest may decay into a neutron and a neutral pion. Find the momenta of the decay products.

The neutral pion itself subsequently decays into two photons. Find the maximum and minimum possible momenta of such photons in the rest system of the original atom. (University of Birmingham, 1957)

19.4 A particle of mass m_i strikes a particle of mass m_s at rest in the laboratory frame. Several particles emerge from the collision with total mass M_f. Show that the threshold energy for the reaction in the laboratory frame is

$$T = \frac{M_f^2 - (m_i + m_s)^2}{2m_s} c^2$$

(University of Birmingham, 1957)

19.5 In their centre-of-mass system two protons, each of total energy $\gamma_c mc^2$ collide elastically and each suffers deflection through angle θ. Show that in the laboratory system, where one proton was initially at rest, the incident proton has total energy

$$E = \gamma_L mc^2 = (2\gamma_c^2 - 1)mc^2$$

and that its angle of deflection ϕ_1 is given by

$$\tan \phi_1 = (\tan \tfrac{1}{2}\theta)/\gamma_c$$

If ϕ_2 is the angle of motion of the recoil proton show that

$$\tan \phi_1 \tan \phi_2 = \frac{2}{1 + \gamma_2}$$

(University of Birmingham, 1956)

19.6 From the isobaric spin analysis (Sect. 19.2.3(*d*)) show that if $f_{3/2} \approx f_{1/2}$ then the total cross-sections for the $\pi^+ p$ and $\pi^- p$ interaction are equal.

19.7 Write down the isobaric spins T (and spin components T_z) for the particles in the reactions

$$p + d \longrightarrow {}^3H + \pi^+$$
$$\longrightarrow {}^3He + \pi^0$$

Using the Clebsch–Gordan coefficients required, show that the ratio of the π^+ to π^0 differential cross-section at a given angle is $2/1$.

$$\langle 1 \tfrac{1}{2} 1 \ -\tfrac{1}{2} | \tfrac{1}{2} \tfrac{1}{2} \rangle = \sqrt{\tfrac{2}{3}}, \ \langle 1 \tfrac{1}{2} 0 \tfrac{1}{2} | \tfrac{1}{2} \tfrac{1}{2} \rangle = -\sqrt{\tfrac{1}{3}}.$$

19.8 Show that the g-factor for muonium (Sect. **19.3**(d)) in the hyperfine state $F = I + \tfrac{1}{2}$ is

$$g = g_I \frac{2I}{2I+1} + g_s \frac{1}{2I+1}$$

where I is the muon spin, $s \ (=\tfrac{1}{2})$ is the electron spin, and g_I, g_s are the corresponding muon and electron g-factors.

19.9 A particle of rest mass M, momentum p, and total energy E collides with a stationary particle of mass m. Find the velocity of the c.m. system, in which the total momentum vanishes. If in this system the momentum of each particle is reversed by the collision, show that the final momentum of the first particle in the original system is

$$\frac{M^2 - m^2}{M^2 + m^2 + (2Em/c^2)} \cdot p$$

(University of Birmingham, 1959)

19.10 A π^0 meson of rest mass m_0 has velocity v relative to the laboratory frame of reference. If θ is the angle between the direction of one photon and the direction of v, show that the probability distribution for θ is

$$\frac{c^2 - v^2}{(c - v \cos \theta)^2} \frac{1}{2} \sin \theta \ d\theta$$

What is the energy of one of the photons when $\theta = 0$? (University of Cambridge, 1967)

19.11 The following reactions are observed for identical incident momenta of the pions:

$$\pi^+ + p \longrightarrow \rho^+ + p$$
$$\pi^- + p \longrightarrow \rho^- + p$$

Assuming the isotopic spin of the ρ-meson is 1, derive expressions for the two total cross-sections in terms of the amplitudes of the total isotopic spin eigenfunctions involved. If the reaction occurs only in the state for which total isotopic spin is $3/2$, what does your result give for the ratio of the two total cross-sections? (University of Durham, 1969)

19.12 Use eq. (5.59) to verify the statement (Sect. 19.2.2) that if two particles have ranges in emulsion which stand in the same ratio as their momenta, then this is also the ratio of rest masses.

19.13 Neutral mesons are produced by the reaction pp $\longrightarrow$ ppπ^0. Show that the threshold proton energy in the laboratory system is

$$\left(\frac{m_\pi^2}{2m_p} + 2m_\pi\right) c^2$$

19.14 By using the Clebsch–Gordan coefficients

$$\langle \tfrac{1}{2}\tfrac{1}{2}\tfrac{1}{2} -\tfrac{1}{2}\,|\,1\;0\rangle = \frac{1}{\sqrt{2}} = \langle \tfrac{1}{2}\tfrac{1}{2}\tfrac{1}{2} -\tfrac{1}{2}\,|\,0\;0\rangle$$

or otherwise verify that in nucleon–nucleon scattering the total cross-section in the state of $T=0$ is related to the (np) and (pp) total cross-sections by the equation

$$\sigma(T=0) = 2\sigma(\text{np}) - \sigma(\text{pp}) \quad \text{[see Ref. 19.8, page 184].}$$

Chapter 20

20.1 Prove that after a beam of momentum p has passed through an electrostatic separator of length L and field strength E, particles of masses m_1 and m_2 are separated by a distance

$$\delta s = \frac{L^2 eE}{4(cp)^3} \left[(m_1 c^2)^2 - (m_2 c^2)^2\right]$$

What length of separator is required to give $\delta s = 1$ cm for 5 GeV/c^2 pions and kaons if a field strength of 50 kV cm^{-1} is used? (University of Birmingham, 1969)

20.2 A K$^-$ beam is produced in an internal target of a proton synchrotron. The K$^-$ mesons have a momentum $3m_K c$, and the K$^-$ meson at rest has a half life for decay of 10^{-10} s. Find at what distance from the target the K$^-$ beam intensity is reduced by a factor 4, assuming loss by decay only. (University of Birmingham, 1962)

20.3 The deuteron beam of kinetic energy T from a synchrotron is used to bombard a deuterium gas target. The particles produced in the reaction d + d $\longrightarrow$ $\alpha + \pi^+ + \pi^-$ are observed. Find an equation for the maximum momentum q of the α-particle in the c.m. system of the final particles. Find also an expression for the maximum and minimum momentum of the α-particle in the laboratory system in terms of T and q. (University of Birmingham, 1962)

20.4 A beam of K$^+$ mesons travelling with velocity $v = \sqrt{3/2}\,c$ in the laboratory emits μ^+ mesons of velocity $u = \tfrac{1}{2}c$ at an angle of 30° to the beam direction. Deduce, starting from the Lorentz transformation, the velocity and angle at which these μ^+ mesons were emitted in the rest frame of the K$^+$ mesons. (University of Birmingham, 1963)

20.5 A lambda hyperon decays at rest into a proton and a negative π-meson. Use relativistic formulae to calculate the kinetic energies of the decay products. If the hyperon were to decay in flight, with a velocity $\beta = 0.4$, what would be the laboratory angle between the decay products if the c.m. angle of the proton was $90°$?

Explain, for this decay in flight, why the proton has a small maximum angle of emission in the laboratory system, whereas the pi meson can come out backwards, relative to the line of flight of the hyperon. (University of East Anglia, 1969)

20.6 Σ^0-hyperons are produced in the capture of stopped K^- mesons by protons ($K^- p \longrightarrow \Sigma^0 \pi^0$). In the subsequent decay of the Σ^0 ($\longrightarrow \Lambda^0 \gamma$) the Λ^0 particles are emitted isotropically in the rest system of the Σ^0. Show that the energy spectrum of the Λ-particles in the laboratory system is uniform.

Chapter 21

21.1 In the scattering of K^- mesons by protons, there is a pole in the scattering amplitude corresponding to the lambda particle ($K^- + p \longrightarrow \Lambda^0$). Show that at this pole the total energy of the K^- meson in the c.m. system is

$$(m_\Lambda^2 - m_p^2 + m_K^2)/2m_\Lambda$$

Show also that the corresponding (unphysical) laboratory energy is

$$(m_\Lambda^2 - m_K^2 - m_p^2)/2m_p$$

21.2 In the scattering of pions by nucleons, the nucleon pole in the s-channel (Sect. 21.1.3) occurs at a pion c.m. energy of $\mu^2/2m$. Find the (unphysical) total energy of the pion in the laboratory system $[-\mu^2/2m]$.

21.3 Show that in the scattering of high energy electrons (E) by protons (M), the momentum transfer for an angular deflection θ in the laboratory system is

$$|t| = \frac{4E^2 \sin^2 \frac{1}{2}\theta}{1 + (2E/M)\sin^2 \frac{1}{2}\theta}$$

Appendices

1 The cross-section of the isotope of uranium of mass number 235 for the capture of thermal neutrons is about 500×10^{-28} m^2, which is about 500 times greater than the total cross-section of the nucleus. How is the capture cross-section determined, how can the existence of so large a cross-section be understood, and what is the importance of this large cross-section for the design of uranium reactors? (University of Cambridge, 1955)

2 Calculate the binding energy of the ground state of a positronium atom ($e^+ + e^-$), assuming that the ionization potential of hydrogen is 13.5 eV. [6.8 eV] Calculate also the orbit radius. [1.06×10^{-10} m]

3 An intimate mixture of uranium 235 and graphite is required for certain experiments. The graphite is known to be contaminated with one part per million by weight of boron 10. What is the maximum fraction by weight of uranium 235 in the mixture if the infinite multiplication constant is not to exceed unity?

$$\sigma_a = 4 \times 10^{-3} \text{ barn for carbon}$$
$$= 3 \cdot 8 \times 10^3 \text{ barn for boron}$$
$$= 7 \cdot 0 \times 10^2 \text{ barn for uranium}$$
$$\sigma_f = 5 \cdot 8 \times 10^2 \text{ barn for uranium}$$

Assume 2·5 neutrons per fission and ignore the effects of fast neutrons. (University of Birmingham, 1961)

4 In the Fermi mass formula the surface and Coulomb energy terms may be written as $a_s A^{2/3}$ and $a_c(Z^2/A^{1/3})$ respectively where $a_s = 13 \cdot 0$ MeV and $a_c = \frac{3}{5}(e^2/4\pi\varepsilon_0 r_0) = 0 \cdot 58$ MeV. The nuclear radius R may be assumed equal to $r_0 A^{1/3}$. Calculate the value Z^2/A at which the division of a nucleus into two equal parts (symmetric spontaneous fission) becomes possible, assuming that this happens when the energy release is equal to the mutual potential energy of the fragments at the instant of formation. (University of Birmingham, 1962)

5 An aqueous solution of a plutonium salt is to be stored in a tank having a square base and rectangular sides. What is the maximum permissible area of the base if the arrangement must never become critical? (σ_a for Pu = 1030 b, $\sigma_f = 730$ b, $\nu = 2 \cdot 9$, $M^2 = 0 \cdot 032$ m^2 (migration area)). (University of Birmingham, 1963)

6 For the neutron-induced fission of ^{235}U leading to ^{148}La and ^{88}Br, the masses of the nuclides in a.m.u. are:

$$^{235}\text{U } 235 \cdot 112 \qquad ^{148}_{57}\text{La } 147 \cdot 989$$
$$\text{n } 1 \cdot 009 \qquad ^{88}_{35}\text{Br } 87 \cdot 961$$

Calculate the release of energy per fission of this type. [159 MeV]

7 With the following data for ^{137}Cs:

mass of uranium irradiated $= 1 \cdot 00$ g
irradiation time $= 1$ week
thermal neutron fission cross-section $= 4 \cdot 2$ b
fission yield of ^{137}Cs $= 6\%$
counting rate $= 100$ counts min^{-1}
chemical yield 50%
counter efficiency $= 15\%$
half life of ^{137}Cs $= 33$ years,

calculate the effective slow neutron flux. (University of Birmingham, 1958)

8 A source consisting of 1 μg of ^{242}Pu is spread thinly over one plate of an ionization chamber. Alpha-particle pulses are observed at the rate of 80 s^{-1}

and spontaneous fission pulses at the rate of 3 per hour. Calculate the half life of ^{242}Pu and the partial decay constants for the two modes of decay. (University of Birmingham, 1958)

9 A 20 megaton hydrogen bomb produces 100 kg of neutrons. Assuming all of these to be utilized in the reaction ^{14}N(n, p)^{14}C and the ^{14}C to be converted to carbon dioxide and dispersed throughout the atmosphere, estimate the effect on the background of a 6-litre proportional counter filled to 3 atmospheres pressure with atmospheric carbon dioxide. (Half-life for ^{14}C = 5570 years.) (University of Birmingham, 1958)

10 The binding energy of a nucleus of atomic number Z and mass number A is often written in the form

$$B(Z, A) = a_v A - a_s A^{2/3} - a_c \frac{Z^2}{A^{1/3}} - a_a \frac{(A - 2Z)^2}{A} \pm \delta$$

where a_v, a_s, a_c, a_a and δ are empirically determined constants. Given that $a_s = 13$ MeV and $a_c = 0.6$ MeV calculate the energy release in the fission of the nucleus $^{238}_{92}$U into two identical nuclei. (University of Birmingham, 1959)

11 An empirical expression for the fall in γ-ray activity after a very short pulse of fission processes is:

$$\text{Number of photons} = 1.9 \times 10^{-6} t^{-1.2} \text{ s}^{-1} \text{ fission}^{-1}$$

where t is the time in days after the pulse. Calculate the number of curies of γ-activity in a rod of uranium 10 days after being taken out of a reactor where it has been producing a constant power due to fission of 10 kW for 100 days. (University of Birmingham, 1957)

12 Cadmium has a resonance for neutrons of energy 0.178 eV and the peak value of the total cross-section is about 7000 b. Estimate the contribution of scattering to this resonance. (University of Birmingham, 1958)

13 For a thermal reactor a typical neutron lifetime is 1.4×10^{-3} s. Calculate the change in power level in a thermal reactor in 1 s for an excess reactivity of 1 per cent, neglecting the effect of delayed neutrons. [× 1250]

14 Calculate the glancing angle for Bragg reflection of neutrons of 101 eV energy from the (111) planes of LiF ($d = 0.232$ nm). [0.35°]

15 Fast neutrons are slowed down in a moderator of an element with mass number A. Show that the maximum fractional energy loss per collision is $4A/(A+1)^2$.

16 Calculate the electrostatic field at a distance of one Bohr radius from a proton. [515 GV m^{-1}]

Index

Index

FALL 77

INVENTORY 1983